Biometrical Genetics
Analysis of Quantitative Variation

Biometrical Genetics
Analysis of Quantitative Variation

Darbeshwar Roy

Alpha Science International Ltd.
Oxford, U.K.

Biometrical Genetics
Analysis of Quantitative Variation
410 pgs. | 50 figs. | 107 tbls.

Darbeshwar Roy
Dean, Agriculture
Bihar Agricultural University
Sabour, Bhagalpur

Copyright © 2012

ALPHA SCIENCE INTERNATIONAL LTD.
7200 The Quorum, Oxford Business Park North
Garsington Road, Oxford OX4 2JZ, U.K.

www.alphasci.com

All rights reserved. No part of this publication may be reproduced, stored in a retrieval system, or transmitted in any form or by any means, electronic, mechanical, photocopying, recording or otherwise, without prior written permission of the publisher.

ISBN 978-1-84265-710-2

Printed in India

To the Memory of
my teacher
(Late) Professor J.L. Jinks

Preface

Plant breeder manipulates various morphological, physiological, anatomical, biochemical, and chemical characters during breeding programme. He manipulates these traits in diploid, polyploid or haploid species with different mating systems such as self fertilizing, cross fertilizing or asexual propagation. Further, he either works with a few genotypes or populations of genotypes. Lastly, he either works with natural variation or creates variation by various means. The various types of traits mentioned above can be grouped into qualitative, quasi-quantitative and quantitative trait. Analysis of qualitative traits is done through the use of Mendelian tools, the classical ratios. Quantitative traits can be analysed following biometrical tools using basic generations and multiple mating designs. Consequences of inadequacy of testers in Triple Test Cross and selection of a particular mating design for estimation of components of variance have been described. Comparison of triallel with quadrallel mating designs in terms of estimable components of variation is added. Application of AMMI model, GGE biplot techniques for prediction of the performance and stability of a variety from multilocational trial data and selection of environment(s) for conducting selection using graphs has been described. Methods of estimation of inbreeding coefficient have been described. Also added are some very pertinent figures related to drift and rate and effect of inbreeding. Response to selection in tetraploid has been described. Relation between physical and genetic distance has been elaborated.

Variation can not be studied only at the phenotypic level but also at gene's immediate product (enzymes/isozymes) or at DNA sequence level. Variation at DNA sequence level can be studied with the help of molecular-markers. The analysis of variation is essential for its successful exploitation. The information on genetical architecture of the trait which is helpful in designing the breeding methodology includes number of genes determining a trait types of gene action and interaction, linkage, reciprocal differences, heterosis and inbreeding depression, genotype × environment interaction, etc. Analysis of qualitative traits requires simple mathematical skill whereas the analysis of quantitative traits requires knowledge of Mendelian genetics and basic statistics. This book describes the analysis of quantitative variation using Birmingham and Edinburgh notation. The mathematics applied and the statistical derivations worked out are explained in a very simple and stepwise manner so that the reader can solve problems in plant genetics. The latest developments in the field of study of quantitative variation have been the use of molecular markers in gene mapping, QTL analysis and marker added selection.

A chapter on "Problem in Biometrical Genetics" has been included in order to make the teaching of biometric genetics more effective and thus having realization of the importance of biometrical genetics in practical plant breeding.

I would like to thank Dr. M.J. Kearsey, School of Biological Science, University of Birmingham, England for suggesting to incorporate more on molecular marker technology, QTL analyses and marker assisted selection.

Excellent teaching by (late) Professor J.L. Jinks, Dr. M.J. Kearsy, Dr. M.J. Lawrence, Dr. J.S. Gale, Dr. B.W. Barnes, Dr. H.S. Pooni, Dr. C.E. Caten, Dr. G.H. Jones, Dr. J.H. Croft, Dr. C.M. Thomas and Dr. G.G. Henshew of the Department of Genetics at the University of Birmingham, that the author received while doing Ph.D. during 1980-83, has been a source of inspiration.

I express my indebtness to Shri N.K. Mehra, M/s Narosa Publishing House, for his interest and attention to my proposal.

Finally, I would like to thank my wife Veena, daughter Cynthia and son Shishir, for bearing with me while I used to work in the evenings.

Darbeshwar Roy

Contents

Preface		*vii*
1. Genes, Chromosomes and Characters		**1.1**
1.1	The Gene	1.1
1.2	Biochemistry of the Gene	1.2
1.3	Chromosomes	1.5
1.4	Characters	1.6
1.5	Protein Synthesis	1.6
	References	*1.8*
2. Analysis of Qualitative Traits		**2.1**
2.1	Gene Action and Interaction	2.1
	2.1.1 Dominance	2.1
	2.1.2 Overdominance	2.2
	2.1.3 Co-dominance	2.3
	2.1.4 Multiple alleles	2.3
	2.1.5 Epistasis or non-allelic interaction	2.3
2.2	Application of Laws of Probability	2.5
	2.2.1 Application to cases of two or more genes	2.6
	2.2.2 Types of probability	2.6
	2.2.3 Binomial probability	2.7
	2.2.4 Multinomial probability	2.7
2.3	Test of Goodness of Fit	2.8
	2.3.1 Two factor segregation	2.8
	2.3.2 Test of heterogeneity of families	2.9
	2.3.3 Testing of independence	2.9
2.4	Experimental Size	2.10
	2.4.1 Distinction between two hypotheses	2.10
2.5	Linkage and Pleiotrophy	2.12
	2.5.1 Detection of linkage	2.12
	2.5.2 Estimation of recombination frequency	2.13

	2.5.3	Efficiency of different types of families	2.14
	2.5.4	Factors affecting the recombination frequency	2.15
	2.5.5	Pleiotropy	2.15
2.6		Extra Chromosomal Inheritance	2.16
	2.6.1	Criterion for extra-chromosomal inheritance	2.16
	2.6.2	Cell organelleles	2.17
2.7		Sex-Determination	2.19
2.8		Expression of Qualitative Gene(s)	2.20
	2.8.1	Penetrance and expressivity	2.21
2.9		Estimation of Gene Frequency	2.21
	2.9.1	Incomplete dominance	2.21
	2.9.2	Multiple alleles	2.21
2.10		Estimation of Genetic Diversity	2.22
		References	*2.22*

3. Analysis of Quantitative Traits — 3.1

3.1		Theories of Quantitative Variability	3.3
3.2		Description of Normal Distribution Curve	3.4
	3.2.1	Mean	3.5
	3.2.2	Mode and median	3.6
	3.2.3	Variance and covariance	3.6
	3.2.4	Range	3.6
	3.2.5	Skewness and kurtosis	3.6
3.3		Threshold Variability	3.8
3.4		Meristic Variability	3.9
3.5		Poisson Distribution	3.9
3.6		Modifications of Binomial and Poisson Distributions	3.10
	3.6.1	Gamma distribution	3.10
	3.6.2	Hypergeometric distribution	3.10
3.7		Biomodal and Multimodal Distributions	3.11
		References	*3.11*

4. Statistical Estimations, Tests, Models and Designs — 4.1

4.1		Estimation and Inference	4.1
	4.1.1	Methods of estimation	4.1
4.2		Test of Significance	4.2
	4.2.1	Chi-square test	4.3
	4.2.2	*F*-test	4.3

	4.2.3	The *t*-test	4.4
	4.2.4	Z-distribution	4.5
	4.2.5	Relationship between different distributions	4.5
	4.2.6	Non-central chi-square	4.5
	4.2.7	Non-central F	4.5
4.3	Statistical Models and Expectation of Mean Squares		4.6
	4.3.1	Fixed model	4.6
	4.3.2	Random model	4.6
	4.3.3	Mixed model	4.7
4.4	Experimental Designs and Calculation of Mean Squares		4.7
	4.4.1	Randomized Complete Block Design (RBD)	4.7
	4.4.2	Nested design	4.8
	4.4.3	Cross-classified design	4.9
	4.4.4	Factorial experiment	4.10
	4.4.5	Analysis of co-variance	4.11
4.5	Correlation and Regression Analysis		4.12
	4.5.1	Correlation coefficient	4.12
	4.5.2	Regression analysis	4.13
	4.5.3	Joint regression analysis	4.14
	4.5.4	Multiple regression	4.15
	4.5.5	Partial correlation and regression	4.15
	4.5.6	Multiple correlation	4.15
4.6	Orthogonal Polynomial		4.15
4.7	Non-Parametric Statistics		4.16
4.8	Multivariate Analysis		4.17
	4.8.1	Principal component analysis	4.17
	4.8.2	Factor analysis	4.18
	4.8.3	Canonical analysis	4.18
	4.8.4	Discrimination and classification	4.19
	4.8.5	Principal co-ordinate analysis	4.19
4.9	Growth Curves		4.20
	4.9.1	Competition	4.22
	References		*4.23*

5. Analysis of Means — 5.1

5.1	Scaling Tests	5.2
5.2	Estimation of Parameters	5.3
5.3	Epistasis or Non-allelic Interaction	5.4
5.4	Estimation of Coefficient of Dispersion	5.6

5.5	Concept of Average Effect of Gene	5.7
5.6	Other Models of Estimation of Genetic Effects	5.9
5.7	Detection of Epistasis by other Methods	5.10
	References	*5.12*

6. Analysis of Variance and Covariance — 6.1

6.1	Partitioning of Variance	6.1
6.2	Estimation of Variances in Generations Derived from Cross of two Pure Breeding Lines	6.1
6.3	Covariance/Correlation	6.3
6.4	Estimation of Variances in Random Mating Population	6.4
6.5	Development of Progenies	6.4
6.6	Estimation of Components of Variation	6.5
6.7	Expectations of σ_w^2, σ_b^2, σ_m^2 (or σ_f^2),	6.7
	6.7.1 Covariance between relatives in random mating population	6.9
6.6	NCM III and TTC	6.10
	6.6.1 NCM III	6.10
	6.8.2 TTC (Triple Test Cross)	6.10
	6.8.3 Consequences of inadequacy of testers in triple test cross	6.12
6.9	Variants of TTC	6.13
	6.9.1 TTC with population of inbred lines	6.13
	6.9.2 Triple test cross (selfed families)	6.13
6.10	Model Fitting to Mean Squares	6.14
6.11	Comparison of Efficiency of the Mating Designs	6.14
	6.11.1 Selection of a particular mating design	6.20
6.12	Uses of Estimates of Components of Variation	6.20
6.13	Negative Estimates of Variance Components	6.20
6.14	Comparison of Variability	6.20
6.15	Partitioning of Covariance/Correlation	6.21
	6.15.1 Partitioning of correlation coefficient	6.21
	6.15.2 Calculation of additive genetic correlation	6.22
	6.15.3 Standard error of correlation coefficient	6.22
6.16	Intra-Class Correlation	6.23
6.17	Repeatability	6.23
	References	*6.24*

7. Diallel Analyses — 7.1

7.1	Full Diallel Analysis	7.1

7.2	Half Diallel Analysis	7.3
7.3	Genetical Structure of Quantitative Traits	7.3
	7.3.1 Wr – Vr analysis	7.4
7.4	Estimation of Genetical Components of Variation	7.6
	7.4.1 F_2 Diallel	7.8
7.5	Diallel Analysis with Heterozygous Parents	7.9
	7.5.1 Identification of heterozygous parent(s)	7.10
7.6	Combining Ability Analysis	7.10
	7.6.1 Estimation of GCA and SCA effects	7.11
	7.6.2 Estimation of GCA and SCA variances	7.12
7.7	Partial Diallel	7.14
	7.7.1 Number of crosses per parent in partial diallel	7.15
7.8	Number of Parents, Family Size and Number of Replications in Diallel	7.16
7.9	Reference Population in Diallel	7.17
7.10	Comparison with other Mating Designs	7.18
7.11	Assumptions under Diallel Analysis	7.18
7.12	Varietal Diallel	7.19
7.13	Triallel Analysis	7.20
	7.13.1 Order of lines in three-way hybrids	7.22
	References	7.22

8. Genotype × Environment Interaction — 8.1

8.1	Definition	8.1
8.2	ANOVA for Genotype × Environment Interaction	8.2
8.3	Types of Environment	8.4
8.4	Approaches to Analysis of $g \times e$ Interaction	8.5
	8.4.1 Regression analysis	8.5
	8.4.2 Analysis of non-linear genotype × environment interaction	8.15
	8.4.3 Limitations with regression models	8.17
8.5	Multivariate Analysis of Genotype × Environment Interaction	8.18
	8.5.1 Principal component analysis	8.18
	8.5.2 Biplot method	8.18
	8.5.3 GGE bilot	8.20
8.6	Non-Parametric Approach	8.22
	8.6.1 Cluster analysis	8.22
	8.6.2 Ranks of genotypes	8.24
	8.6.3 Study of environmental factors	8.25

	8.7	Genetical Analysis of Genotype × Environment Interactions	8.26
		8.7.1 Generations derived from cross of two pure breeding lines	8.27
		8.7.2 Multiple mating design	8.30
		8.7.3 Interaction between epistatic gene action and environment	8.32
	8.8	Genetic Control of Environmental Sensitivity	8.33
		8.8.1 Theories of environmental sensitivity	8.33
		8.8.2 General or specific nature of genotypes involvement in environmental sensitivity	8.35
		8.8.3 Choice of selection environment and the effect of the chosen environment on the properties of the selected lines	8.35
		References	*8.36*
9.	**Analysis of Reciprocal Differences**		**9.1**
	9.1	Maternal Effects	9.1
	9.2	Estimation of Maternal Effects	9.2
		9.2.1 Generations derived from the cross of two pure breeding lines	9.2
		9.2.2 Multiple mating designs	9.4
	9.3	Detection of Epistasis and Linkage of Epistatic Genes in the Presence of Reciprocal Differences	9.5
	9.4	Reciprocal Effects Model	9.6
	9.5	Covariances between Relatives	9.11
	9.6	Sex Linkage	9.11
	9.7	Sex Limitation	9.13
		References	*9.16*
10.	**Linkage and Epistasis**		**10.1**
	10.1	Linkage	10.1
	10.2	Effect of Random Mating on Linkage Disequilibrium	10.2
	10.3	Effect of Linkage on Variances of Advancing Generations Derived by Selfing Cross between Pairs of Purebreeding Lines	10.2
	10.4	Detection of Linkage Disequilibrium	10.4
	10.5	Generation that Yields an Approximate Estimate of DVF_∞	10.5
	10.6	Effect of Linkage on Covariances	10.6
	10.7	Linkage vs. Pleiotropy	10.7
	10.8	Epistasis	10.7
	10.9	Effect of Non-allelic Interaction on Estimation of Variance Components	10.9
		10.9.1 Biases in D and H components obtained from T.T.C.	10.9
	10.10	Comparison of Tests of Epistasis	10.10

	10.11	Epistasis in Random Mating Population	10.10
	10.12	Detection of Linked Epistasis	10.13
		References	*10.14*

11. Heterosis and Inbreeding Depression   11.1

     11.1   Heterosis   11.1
     11.2   Theories of Heterosis   11.1
     11.3   Expressing Heterosis in Terms of Biometrical Parameters   11.3
            11.3.1   Generation means   11.3
            11.3.2   Variances   11.4
     11.4   Effect of Linkage Disequilibrium on Dominance Ratio   11.4
     11.5   Effect of Interaction on Dominance Ratio   11.5
     11.6   Environmental Heterosis   11.6
     11.7   Dominance Ratio   11.7
     11.8   Maternal Effects in Heterosis   11.7
     11.9   Heterosis in Population Cross   11.9
     11.10   Misinterpretation of Heterosis   11.10
     11.11   Genetic Distance/Divergence   11.10
            11.11.1   Estimation of genetic distance   11.11
            11.11.2   Cluster formation   11.12
            11.11.3   Intra- and inter-cluster distances   11.12
            11.11.4   Use of genetic parameters in the estimation of genetic distance   11.13
     11.12   Inbreeding Depression   11.13
     11.13   Inbreeding in Random Mating Population   11.14
     11.14   Estimation of Inbreeding Coefficient   11.16
     11.15   Other System of Inbreeding   11.19
            References   *11.21*

12. Polyploids and Haploids   12.1

     12.1   Definition and Classification   12.1
     12.2   Biometrical Genetics of Autotetraploids   12.2
     12.3   Scaling Tests   12.4
     12.4   Variances and Covariances   12.4
     12.5   Random Mating Population   12.5
     12.6   Triploids   12.6
     12.7   Haploids   12.8
            12.7.1   Production of haploids   12.8
            12.7.2   Identification of haploids   12.9

		12.7.3	Advantages and disadvantages	12.9
		12.7.4	Biometrical genetics of haploids	12.11
	12.8	Doubled Haploids		12.12
		12.8.1	Analysis of means	12.12
		12.8.2	Analysis of variances	12.13
		12.8.3	Estimation of additive and additive × additive interaction variances	12.14
		12.8.4	Estimation of dominance variance and the degree of dominance	12.15
		12.8.5	Linkage disequilibrium	12.15
		12.8.6	Estimation of recombination value	12.16
		12.8.7	Skewness and kurtosis	12.16
		12.8.8	Estimation of number of effective factors	12.17
		12.8.9	Covariances/correlation	12.17
		12.8.10	Generation of deriving *DH* lines	12.19
		12.8.11	Efficiency of *DH* method	12.19
		References		*12.20*

13. Competition   13.1

	13.1	Definition and Types	13.1
	13.2	Yield-Density Relationship	13.2
	13.3	Mixture Diallels	13.3
	13.4	Intergenotypic Competition	13.4
		13.4.1 Estimation of a, b_m and b_d	13.4
		13.4.2 Partitioning of C-values	13.6
	13.5	Nature of the Underlying Genetical Control of Competition and Competitive Ability	13.7
	13.6	Competition in Natural Population	13.8
	13.7	Estimation of Outcrossing Rate	13.8
	13.8	Detection of Competition	13.11
		References	*13.11*

14. Environmental Variation   14.1

	14.1	Definition and Types	14.1
	14.2	Estimation of Environmental Variation	14.1
	14.3	Individual Plant Randomization	14.2
	14.4	Sampling Variation	14.3
	14.5	Other Experimental Designs	14.3
	14.6	Estimation of Experimental Error	14.4
	14.7	Augmented Design	14.4

14.8	Environmental Covariance	14.5
	References	*14.5*

15. Heritability — 15.1

15.1	Definition and Types	15.1
15.2	Heritability a Characteristic of Population	15.2
15.3	Estimates of Heritability	15.2
15.4	Effect of $G \times E$ Interaction	15.3
15.5	Uses of Estimates of Heritability	15.3
15.6	Changing Estimates of Heritability	15.4
15.7	Problems in use of Estimate of Heritability	15.4
	References	*15.4*

16. Estimation of the Number of Effective Factors — 16.1

16.1	Approaches to Estimation of Number of Genes	16.1
16.2	Chromosome Assay	16.2
16.3	Statistical Properties of Distribution	16.2
16.4	Estimate of k using Dominance Genetic Effect and Dominance Genetic Variance	16.3
16.5	Use of F_2 Derived Generation Variances	16.3
16.6	Estimation of Number of Effective Factors in Random Mating Population	16.3
16.7	Genotype Assay	16.5
	16.7.1 Theory	16.5
16.8	Use of Dihaploids in the Estimation of Number of genes	16.7
	16.8.1 Use of range and genetical variance of an F_1, F_2 derived *DH* populations	16.8
	16.8.2 Use of variances of F_2-derived population	*16.8*
	16.8.3 Genotype assay	16.9
16.9	Estimation of the Number of Effective Factors in Haploids	16.9
16.10	Method for Locating Genes	16.9
	References	*16.9*

17. Analysis of Skewness and Kurtosis — 17.1

17.1	Genetical Causes of Skewness and Kurtosis	17.1
17.2	Information on Gene Action and Interaction	17.1
17.3	Estimation of Coefficients of Skewness and Kurtosis in Population of Pure Breeding Lines	17.2

17.4	Effects of Random Environmental Variation and Genotype × Environment Interaction	17.3
17.5	Estimation of Skewness and Kurtosis in Dihaploid Population	17.4
17.6	Effect of Skewness on Selection	17.4
	References	*17.5*

18. Transformation of Scale   18.1

18.1	Tests of Non-normality	18.1
18.2	Relation of Variability to Mean	18.2
18.3	Scale Effect	18.3
18.4	Variance Stabilizing Transformation	18.3
18.5	Poisson Distribution	18.3
18.6	Binomial Distribution	18.3
	References	*18.4*

19. Genetic Structure of Population   19.1

19.1	The Hardy-Weinberg Equilibrium		19.1
	19.1.1	Testing goodness of fit	19.1
	19.1.2	Extension of H. W. equilibrium to cases of multiple alleles, sex-linked genes and polygenic inheritance	19.2
	19.1.3	Sex-linked loci	19.3
	19.1.4	Polygenic traits	19.3
19.2	Changes in Gene Frequency		19.4
	19.2.1	Effects of non-random mating	19.4
	19.2.2	Self-fertilization	19.4
	19.2.3	Sib mating	19.4
	19.2.4	Assortative mating	19.4
	19.2.5	Negative assortative mating	19.5
19.3	Mutation		19.5
	19.3.1	Balance between mutation and selection	19.6
19.4	Selection		19.8
	19.4.1	Complete elimination of recessive homozygotes	19.9
	19.4.2	Balance between selection and inbreeding	19.10
19.5	Competitive Selection		19.10
19.6	Migration		19.11
	19.6.1	Balance between selection and migration	19.11
	19.6.2	Models for studying the population structure	19.11

19.7	Drift		19.12
	19.7.1	Effective population size	19.14
	19.7.2	Random genetic drift in natural populations	19.15
19.8	The Founder Principle		19.16
19.9	Gametic Selection		19.16
19.10	Meiotic Drive		19.17
19.11	Genetic Load		19.17
	19.11.1	Segregational load	19.17
	19.11.2	Mutational load	19.19
	References		*19.19*

20. Selection Theory — 20.1

20.1	Response to Selection		20.1
	20.1.1	Selection differential	20.2
	20.1.2	Response due to selection in tetraploid	20.3
20.2	Correlated Response		20.3
20.3	Path Analysis		20.5
	20.3.1	Determination of residual variability	20.6
	20.3.2	Path coefficient vs correlation coefficient	20.6
	20.3.3	Application of path analysis	20.7
20.4	Yield and Yield Components		20.7
20.5	Types of Selection		20.8
	20.5.1	Efficiency of different methods	20.8
	20.5.2	Combined selection	20.10
20.6	Selection Criterion		20.10
	20.6.1	Single trait selection	20.10
	20.6.2	Multitrait selection	20.10
20.7	Comparison of Efficiency of Different Methods		20.13
20.8	Selection Limit		20.13
20.9	Natural Selection		20.14
	20.9.1	Stabilizing selection	20.15
	20.9.2	Directional selection	20.16
	20.9.3	Disruptive selection	20.16
	References		*20.18*

21. QTL Analysis — 21.1

21.1	QTL Analysis		21.1
	21.1.1	Linkage between QTL and molecular marker	21.2
	21.1.2	Mapping methods	21.2

	21.2	Models for Estimating Genetic Effects of QTL	21.10
	21.3	Mapping Populations	21.19
	21.4	Population Size	21.20
	21.5	Experimental Design	21.20
	21.6	Mapping in Polyploids	21.22
		21.6.1 Methods of linkage analysis	21.24
	21.7	Genetic Mapping	21.25
	21.8	Information from QTL Mapping	21.28
	21.9	Comparative Mapping and Orthologous Poly Genes	21.32
	21.10	Identification of Links between Genotype and Phenotype	21.33
		References	*21.35*

22. Matrix — 22.1

	22.1	Definition and Types	22.1
	22.2	Matrix Operations	22.1
	22.3	Dispersion Matrix	22.5
	22.4	Orthogonal Matrix	22.6
	22.5	Diagonalization of Matrix	22.6
		References	*22.6*

23. Problems in Biometrical Genetics — 23.1

Index *I.1-I.5*

1

Genes, Chromosomes and Characters

Although Meischer's work in the 1870's showed that the chromosome is largely composed of protein and nucleic acid but nucleic acid seemed ruled out at the site of specificity by its supposedly monotonous tetraploid structure (Levene) and it was assumed upto the middle of 1940's that the specificity resided in the pattern of successive amino acids (i.e. protein). But with the discovery of the transformation principle (Avery, McLeod and McCarty, 1944) and later experiments of Hershy and Chase (1952) in *Diplococcus pneumoniae* showed that DNA is the hereditary material. Further experiments of Hershy and Chase demonstrated that DNA was the material transmitted between parent and offspring. By early 1950's X-ray studies on DNA by Wilkins, Franklins and others indicated a well organized multiple-stranded fibre about 22 Å in diameter that was also characterized by the presence of groups spaced 3.4 Å apart along the fibre and a repeating unit every 34 Å. Then came the demonstration of consistency of DNA per chromosome set (Boivin, Vendrely and Vendrely 1948; Mirsky and Ris 1949 and Swift, 1950). This followed Chargaff's demonstration of the equality or near equality in the numbers of guanine and cytosine nucleotides and of adenine and thymine nucleotides in the DNA extracted from a variety of organisms.

1.1 The Gene

Mendel in his experiments on pea used the term 'Mermal' to denote dominant and recessive alternatives. Bateson translated this into 'character' and according to Bateson and Punnet's hypothesis the dominant alternative was seen as due to the presence of unit character where as the recessive as absence of this. But soon it was established that the hereditary differences in characters tend to split up into many independent pairs of alternatives and that the effect of these was often difficult to identify as characters. Through the work of Johannsen (1903, 1909) and multiple factor hypothesis of Emerson and East (1913), East and Hayes (1911), Hayes (1913) and East (1916) it came to be recognised that unit (genetic) factor is a better term than unit character and a little later Johannsen's term 'gene' was accepted as a distinct one for the basic unit of heredity.

Each gene occupies a specific position (called the locus, plural loci) on a chromosome sutton (1903). Further, a gene can be said to be particular representative of a locus considered by itself and allele, in the sense of Bateson's (1909) definition of 'allelomorph' is used for a particular representative in relation to others. Alleles may be complementary or non-complementary and in extreme cases may seem to have no more in common in their differences from type than random non-alleles although one can usually trace a thread of physical similarity of some sort through an allelic series. Studies of antigen and self-incompatibility mechanism show that there may be hundred of alleles at a locus (multiple alleles).

The demonstration by cytological study of the persistence of individualized chromosomes and their pairing at synapsis (Wenrich, 1916) and Belling's demonstrations (1928) of as many as 2000 such particles in favourable plant material and further demonstration of some 5000 identifiable bands in the salivary gland chromosomes of *Drosophila* and

1.2 Biometrical Genetics–Analysis of Quantitative Variation

the location of particular loci in or near such bands by Painter (1934) and Bridges (1922) led to general acceptance of the particulate nature of loci and in other words locus was considered as a physical entity.

1.2 Biochemistry of the Gene

The gene is made up of nucleic acid, the genetic material. There are two main groups of nucleic acid- DNA (Deoxyribonucleic acid) and RNA (Ribonucleic acid). In all organisms except plant viruses, polio, where the genetic material is RNA, the genetic material is DNA. Nucleic acid consists of one or a sequence of nucleotides. Thus nucleotides are the basic unit of nucleic acid. Each nucleotide is made up of sugar, phosphate and nitrogenous base. Sugar and the nitrogenous base complex is known as nucleosides. There are four different kinds of bases, namely, adenine, cytosine, thymine and guanine. Adenine and guanine are purine bases whereas thymine and cytosine are pyrimidine bases. DNA differs from RNA in that the sugar is ribose in place of deoxyribose and uracil replaces thymine in RNA. The C_1 of the carbon is linked to the nitrogenous

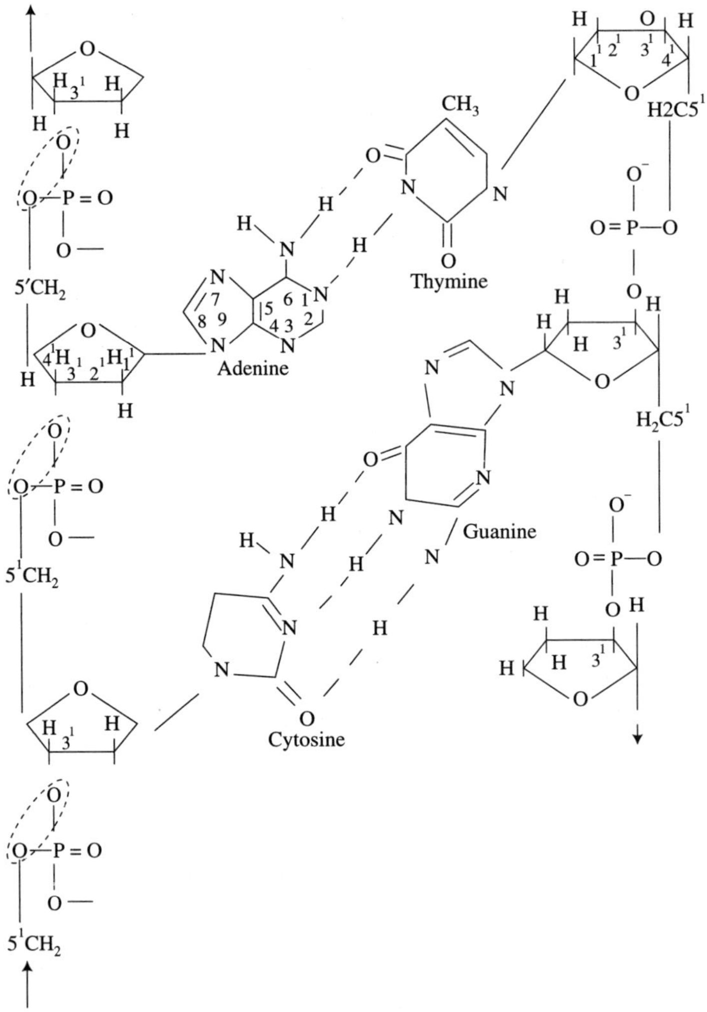

Fig. 1.1 Phosphodiester bonds and the linkage between the two strands of DNA molecule.

base at the 9 position for the purines or 1 position for pyrimidine for a nucleoside. Although phosphoric acid can be linked to hydroxyl groups at 2′, 3′ and 5′ of ribose and 3′ and 5′ of deoxyribose sugar, respectively, nucleotides are linked by phosphodiester bonds between the 3′ and 5′ positions of the sugar as shown in Fig. 1.1. Watson and Crick (1953) proposed the model of the DNA structure based on X-ray diffraction studies as shown in Fig. 1.2. DNA usually consists of two complementary and oppositely directed but interwined helices, each with a sugar-phosphate backbone carrying at each level a purine or pyrimidine group with hydrogen bonding to its complements on the other strand. The diameter of the DNA molecules is 20 Å. The coiling of the double helix is right handed and a complete turn occurs at every 34 Å. The two strands under certain cell conditions separate and each of the two strands which serves as a template for the formation of a new strand, synthesizes its complementary strand and thus provides the mechanism through which any specific pattern would tend to be maintained indefinitely and one in which any change in this pattern, accidental or induced, would be followed by an equally persistent multiplication of the new pattern. The DNA replication thus is semiconservative. Most DNA molecules are of the double helical type but DNA in the bacteriophage $\phi \times 174$, is circular and single stranded.

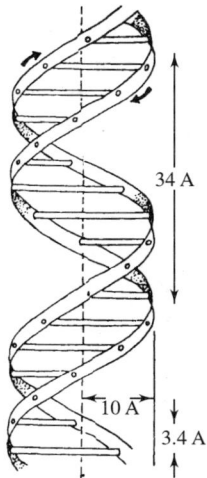

Fig. 1.2 Double stranded helix configuration of DNA.

Advances in the analysis of fine structure of gene started during this period where many supposedly single loci in the same species were successfully split. In some cases subdivision of a system of supposed allele or complex loci with multiple effects seemed to yield components with single effect which led to the hypothesis that gene is made up after all of ultimate particles that behaves as units both structurally and physiologically. But thorough analysis of the effects in cases in which supposed loci have actually been divided by crossing over, have not indicated any clear-cut separation into physiological simple units. Map based on complementarity (*cis-trans* test) does not fully agree with those based on crossing over (Giles 1965; Green 1965). Further analysis in the direction of complete abolution of the classical structural gene suggested that disruption by crossing over or by other means may occur between any two nucleotide pairs of the DNA molecules. Ponte corvo and Roper (1956) while working with the white super locus of *Drosophila melanogaster* suggested the possibility of unlimited divisibility. Demerec (1957) and his associates while working with *Salmonella* found succession of physiological units corresponding to several successive steps in the synthesis of such substances such as histidine, cytosine, methionine and tryptophan. This contrasts with the usual wide scattering in the chromosomes of the loci controlling such metabolic steps in *Neurospora* and *Aspergillus*, although sporadic cases of clustering are now known in these and other organisms. Each of the physiological unit is capable of breakage (and inactivation) at many points and thus seems to be composed of many structural genes. Thus one gene-one enzyme hypothesis may hold for some enzymes but not for all enzymes. Benzer (1957), Stresinger and Franklin (1957) and Edger and Epstein (1965) studied the subdivision of blocks concerned with a single physiological process in phage. Benzer introduced the term 'cistron' for a physiological unit (interpreted to be such by the criterion of a difference between *cis* and *trans*-recombinants of internal mutations) and 'recon' for a block within which crossing over has not been detected (i.e., a classical structural gene). He also proposed the term 'muton' for a unit of mutation. We have seen that a classical gene was assumed to contain a vast number of descrete sites at which mutation might occur. Benzer's analysis of a very large number of mutations of T_4 phage of *Colon bacillus* where all mutants were characterized by inability to multiply in one or more host strains, indicated that crossing over may occur between any two nucleotides (probably about 10^{-4}). The structural gene (recon) thus seems to be reduced to a single nucleotide although a physiological unit (cistron) may have more than 400 nucleotides. As we have seen that maps based on complementation do not always correspond to ones based on crossing over but then there may be apparent complementation in *trans*-recombinants between mutant sites that are located between ones that show more or less failure of complementation. Recombinally separable forms of a gene within a cistron are referred to as heteroalleles. Also, a muton may be delimited to a single nucleotide or some part of a nucleotide. Different forms of a mutationally defined gene are called homoalleles. This ambiguity in the definition of the cistron, coupled with the reduction of the classical structural gene (recon) and the mutational unit (muton) to a single polypeptide make it seem best to follow Demerec using the term 'gene locus' for a unit of DNA that determines a single polypeptide molecule even though divisible by rare crossing over. Thus one gene-one enzyme hypothesis (Beadle and Tatum, 1941) has now been replaced by one gene-one polypeptide chain hypothesis.

Jacob and Monod (1961) found in bacteria a succession of co-operating loci (operons, an operator plus structural genes under its control is called an operon) that are called into action or suppressed in unison by a terminal member (the operator) which itself is controlled by a repressor, the regulatory substance produced by a regulatory gene. Thus the whole block behaves to some extent as a physiological unit at a higher level than the gene locus. The regulatory gene controls structural genes relative to the tuning, level and frequency of the expression in various tissues. McClintock discovered the Ac-Ds system in corn. She called those units as controlling elements. These are regulatory genes that regulate the activity of other structural genes. The controlling element Ds acts on its immediate genetic neighbour whereas Ac acts at a distance. These elements have the property of transposibility i.e., these elements could physically shift their position within the chromosome or between chromosomes. These transposable elements are called transposons and they contain the defined nucleotide sequence which promote their transposition. They also carry one or more genes alongwith them.

We have seen that the genetic map of nucleotide changes should correspond with a mutational map of amino acid changes although in many eukaryotes the distances between nucleotide mutations need not correspond to distances between mutant amino acids. Further, there is a possibility that physiologically significant patterns in the DNA may be separated by long inert-stretches if the chromosome is assumed to consist of a succession of physiological units. Patterson (1934) and his associates' results from *Drosophila* can be accounted for if there is twice as much inert as active material. It has been found that each of the β like globin genes is made up of DNA sequence of more than 1500 base pairs long which yields the complementary mRNA sequence but the mature mRNA is about 800 base pairs long which is obtained after a series of steps of processing. In most eukaryotic cells the mRNA during transfer from nucleus to cytoplasm goes through RNA processing i.e. the precursor mRNA which is originally longer is shortened after processing. It was found that the precursor mRNA contains non-coding intervening sequences called introns that are inter-spersed between the protein/polypeptide coding sequences called exons. During the processing there is excision of introns but addition of a methylated guanosine cap to the 5' end and the frequent addition of poly (A) tail to the 3' end.

In many, if not most, taxonomic groups, differences in genome size are due to difference in the numbers of a relatively few specific nucleotide sequence. Those nucleotide sequences that make disproportionate contributions to the total nuclear DNA content belong in the part of genome now characterized as repetitive DNA. There is presence of substantial numbers of repeated DNA sequence in eukaryotic genome.

In most higher plants less than half of the DNA content in the nucleus represents the unique sequence type that comprise individual structural genes that code for a particular enzyme or protein. Even here only a small fraction of the unique sequence has been found to code for protein. More than half of the nuclear DNA is repetitive in nature. In the repetitive class, certain sequences repeat between 10 and 1,000 times, other sequences repeat between 1000 and 10,000 times and some even repeat from 10,000 to 1,000,000 times. Repetitive sequence is called satellite DNA whereas inverted repetitors are called palindromes. Highly repetitive DNA is thought to have no function and that is junk DNA that has been accumulated through evolutionary time or that it is parasitic or selfish DNA which having gotten into the genome, perpetuate itself while not greatly reducing the fitness of the host (Orgel and Crick, 1980: Doolittle and Sapienza, 1980). Intermediately repetitive DNA is more heterogenous than highly repetitive DNA. Some of the smaller multigene families are highly clustered whereas the larger families tend to be dispersed throughout the genome. Some duplicate loci through recurrent mutations may evolve entirely new functions while others may evolve into gene families in which the products of duplicate loci may retain related functions. There are thus two classes of duplicate genes: those which can be detected through the analysis of their specific, usually, protein products and second, those detected only as repeated nucleotide sequence in the genome. The genes themselves exist in only one or a few copies whereas in case of repeated nucleotide sequences their numbers range from several scores to several millions. Thus gene which can be defined as a stretch of DNA, is a functionally complex array which includes non-transcribed flanking sequences, transcribed but not translated leaders, transcribed but non-translated introns, exons that actually specify amino acid sequence and within these at every third position a redundant base whose identity can usually change without alterning the codon. The functional heterogeneity thus lies within a stretch of genome corresponding to about 0.01 centimorgan.

Most of the DNA manipulated by plant breeders probably is of the single copy type. Most likely, plant breeders are manipulating regulatory DNA. Multiple copies of a particular gene have a potential for unequal crossing over. With unequal crossing over one can obtain from a few copies of DNA sequence to many or from few to fewer. Thus multiple copy genes are the store house of variability and the amount of variability in any breeding scheme depends on the number of copies of genes. With multiple copies the fixation of a genotype may neither be realized. This may have important implications in long term selection experiments where the variability is never exhausted (Phillips, 1981).

1.3 Chromosomes

The nucleic acid is present as nucleic acid-protein complex known as nucleo protein on the chromosomes. Proteins and nucleic acid are called macromolecules. The two major classes of proteins are the histones and the non-histones. The proteins are basic in nature whereas the nucleic acid is acidic. The eukaryotic chromosomes after staining look like a dark network called chromatin (nucleic acid plus protein) under the light microscope. Some parts of chromosome (or, whole chromosome as in case of supernumerary or B chromosome) which appear heavily stained are called heterochromatin whereas the lightly stained parts are euchromatin. This dark staining heterochromatic material is usually found interspersed between lightly stained euchromatic portions along the length of most chromosomes. Heitz (1928) showed that the differences in the staining result at least partly from the degree of coiling of chromosomes during cell cycle. Heterochromatin usually remains condensed whereas euchromatin undergoes cycle of condensation and unravelling during prophase which is recognisably different from heterochromatin. There are two distinct classes of heterochromatin. Even in the heterochromatic region some heterochromatic regions remain condensed throughout entire life cycle whereas the other heterochromatic regions change their staining characteristics either in different cell stages or in different cells. The former is called constitutive heterochromatin whereas the later is called facultative heterochromatin. The constitutive heterochromatin and the highly repetitive DNA are synonymous. Heterochromatin may be responsible for increasing crossing over, changing the position of crossing over, reducing pairing between homeologous chromosomes and increasing pairing between homologous chromosomes. Heterochromatin (B chromosomes, telomeres and knobs) may modify the chromosome behaviour.

Certain species of both plants and animal can carry two types of chromosomes, the basic complement and the so-called B chromosomes. The B chromosomes vary in number from one individual to another, some carrying no B chromosomes at all. They are not essential for the successful growth and development of organism and their variation in number does not have the drastic effects on phenotype as we find in case of basic complement. The B chromosomes appear to be genetically inert or subinert and show the properties of being heterochromatic. The B chromosomes in maize are small, heterochromatic and with no known function. They tend to increase in numbers upto the point at which they start to affect the plant viability adversely. B chromosomes show non-disjunction and preferential fertilization of the egg which constitute the B chromosomes accumulation mechanism. They have shown to affect nuclear processes in sorghum and morphology and fertility in rye (Muntzing, 1943). They have also been shown to influence the recombinations in certain regions of the genome. The B chromosome increases chiasma frequency in many plant species including maize (Bell, 1982) and has been shown to enhance recombination repair which is based on the observation in maize that B chromosomes increase resistance to γ-irradiation induced DNA damage (Stanb, 1984).

1.4 Characters

The phenotype that we observe is a result of interaction between the genotype and the environment. The genes which consititute the genotype produce structural and enzymatic proteins which in turn influence all subsequent cell metabolisms and development. The development includes all those activities which result in cell, tissue, organ and system differentiation and integration as shown in Fig. 1.3. All these processes interact with environmental factors such as light, temperature, nutrients, etc.

The character that we see is a result of complicated chain and network of interacting processes to which numerous gene products contribute. Genes determine the metabolism and development and thus determine the morphological and physiological characters of the organism and also the life cycle.

The characters can be studied at phenotypic as well as genotypic level. Morphological, physiological or biochemical and chemical characters can be studied at phenotypic level in the plants or in the seeds. Plant's morphological characters include root, shoot/branches, leaves, inflorescence characteristics whereas the seed characteristics can be morphological, chemical or biochemical. At genotype level either we can study traits such as enzymes (or isozymes)/ proteins which are the immediate products of gene or we can have the nucleotide sequence itself as a trait in the DNA.

1.5 Protein Synthesis

For the integrated system and development to function, it is essential to regulate DNA replication, the process of protein synthesis and all subsequent cell metabolism.

The process of transferring genetic information from DNA to protein includes 1. transcription (in which mRNA is copied or transcribed from a DNA strand through RNA polymerase) 2. translation (in which a particular nucleotide sequence in the mRNA is translated into a particular amino acid sequence (polypeptide chain) with the help of ribosome) and the consequences to the organism of change in the types and amount of different protein produced (regulation of this process).

Proteins are polypeptide chains of amino acids. The information specifying the amino acids sequence of protein resides in the sequence of nucleotides in the DNA molecule. The nucleotide sequence of DNA is transcribed as messenger RNA which has a nucleotide sequence complementary to the DNA nucleotide sequence. The mRNA thus carries the genetic message specifying amino acid sequence and thus mRNA serves as a template for protein synthesis. The codon specifying a amino acid is triplet (three mRNA nucleotides) and there is a complementary anticodon in the tRNA molecule. Several different

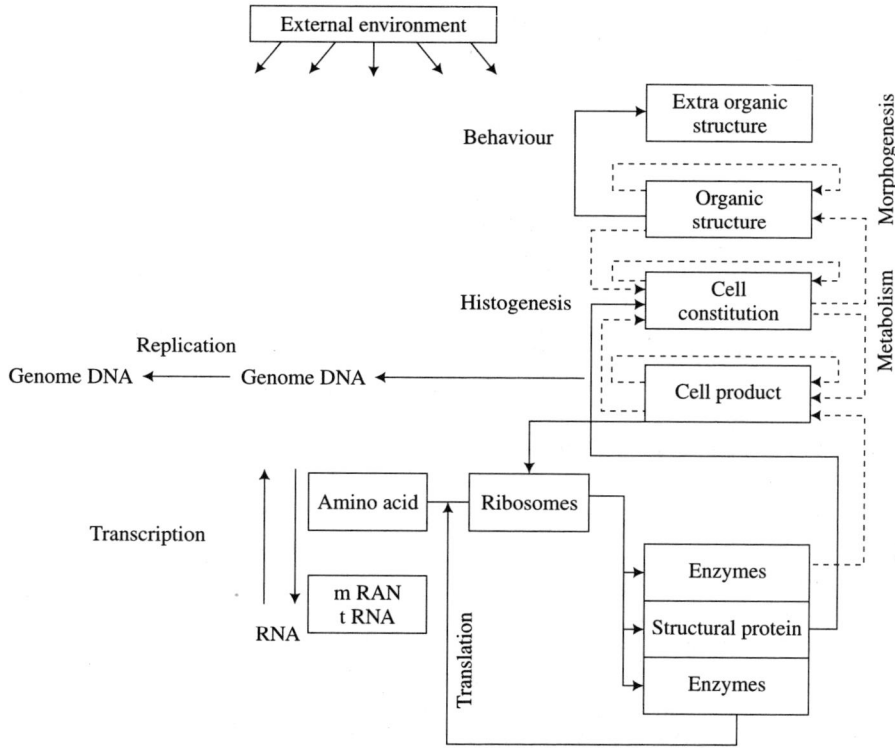

Fig. 1.3 Interaction between genome and environment at successive levels of organization (Wright, 1968).

groups of three nucleotides can code the same amino acid and this property of code is called the degeneracy of the code. Also, codon shows the property of universality in that the same code applies to organism as different as bacteria, tobacco and man and hence probably to all organisms. The transcription of mRNA as well as other forms of RNAs such as transfer RNA (tRNA) and ribosomal RNA (rRNA) takes place in the nucleus. The RNAs then move into the cytoplasm where the protein synthesis takes place. The mRNA in the cytoplasm becomes associated with ribosomes which are composed of protein and rRNA and are sites of proteins synthesis. The amino acids in the cytoplasm are activated by ATP and then esterified to their corresponding tRNAs and thus activated amino acid tRNA complexes (amino acyl tRNAs) are formed. The activated amino acid tRNA complex is catalysed by enzyme specific for each amino acid. The mRNA and the initiator tRNA, Met-tRNA are bound to the ribosome and starts the initiation of protein synthesis. The formyl methionyl tRNA initiates the protein synthesis in bacteria and a number of cell organelleles (mitochondria and chloroplast) whereas in mammals and plant cells similar function is served by non-formylated Met-RNA. An initiating nucleotide sequence, the initiating codon (AUG or GUG) on mRNA directs the binding of an initiating amino acyl rRNA on the ribosome. A second amino acyl tRNA is then attached to a decoding ribosome site at the direction of codon at 3′ side of the initiating mRNA sequence. Following the attachment of amino acyl tRNAs a peptide bond is formed under the influence of a peptide forming enzyme in which the carboxyl groups of the initiating amino acyl tRNA is attached to the amino acid group of the first amino acyl tRNA. The polypeptide chain is lengthened by sequential addition of a new amino acyl residue enzymatically transferred from activated amino acyl tRNA complexes each of which is bound to the ribosome in response to specific codon in the mRNA. Once a peptide bond has been formed the ribosome moves along the mRNA which brings the next codon in the position for alignment of the next amino acyl tRNA. The tRNA with specific amino acid becomes oriented to the mRNA on the ribosome according to the complementary nucleotide sequences. The translation of the codons of mRNA takes place in 5′ to 3′ direction. The polypeptide chain is completed when the termination codons on mRNA are reached. The terminating codons in mRNA are UAA, UAG and UGA. Thereafter the newly synthesized protein is released from the ribosome. The secondary, tertiary and quarternary structure of the protein develops as a result of the amino acid sequence and the environmental conditions.

References

Avery, O.T., Me, Leod, C.M. and McCarty. 1944. Studies on the chemical nature of the substance inducing transformation of pneumococcal types. Induction of transformation by deoxyribonucleic acid fraction isolated from Pneumococcus type 3. 1. Exp. Med., **76:** 137–158.

Bateson, W. 1909. Mendel's principles of Heredity. Cambridge, At the University Press, 1930.

Bateson, W. and Punett, R.C. 1905. Experimental studies on the physiology of heredity. Rept. to the Evolution Com. of Roy Soc. 2.

Belling, J. 1928. The ultimate chromosomes of lilium and Aloe with regard to the number of genes. Univ. Calif. Publ. in Botany, **14:** 307–318.

Benzer, S. 1957. The elementary units of heredity. In: Symposium on the chemical basis of heredity, eds. W.D. McElroy and B. Glass, pp. 70–93, Baltimore, The Johns Hopkins Press.

Boivin, A., Vandrely, R., and Vandrely, C. 1948. L' aiic L' acid desoxyribo nucleique in noyan cellulaire depositaire des caracters her-editaires: arguments dordre analylique, C.R. Acad. Sci. **226:** 1061–1063.

Demerec, M. 1957. A comparative study of certain gene loci in Salmonella. Cold Spring Harbour Symp. Quant. Biol., **21:** 113–121.

East, E.M. 1916. Studies on size inheritance in Nicotiana. Genetics, **1:** 164–176.

East, E.M. and Hayes, H.K. 1911. Inheritance in maize. Conn. Agr. Exp. Sta. Bull 167, pp. 142.

Edgar, R.S. and Epstein, R.H. 1965. Conditional lethal mutations in bacteriophage T 4. Proc. XI lnt. Congr. Genetics (1963), 2, 2–16, Oxford: Pergamon.

Emerson, R.A. and East, E.M. 1913. The inheritance of quantitative character in maize. Bull. Agr. Exp. Sta. Nebraska No.2, 118 pp.

Flavell, R. 1980. The molecular characterization and organization of plant chromosomal DNA sequences. Ann. Rev. Pl. Physiol., **31:** 569–596.

Gardner, E.J., Simmons, M.J. and Snustad, pp. 1991. Principles of Genetics, John Wiley, New York.

Giles, N.H. 1965. Genetic fine structure in relation to function in Neurospora. Proc. XI Congr. Genet. (1963), **2:** 16–30, Oxford: Pergamon Press.

Green, M.M. 1965. Genetic fine structure in Drosphila. Proc. XI Congr. Genet. (1963), **2:** 37–49. Oxford, Pergamon Press.

Heitz. E. 1928. Das Heterochromatin der Moose. Jahrb. Wiss. Bot., **69:** 762–818.

Hershy, A.D. and Chase, M. 1952. Independent functions of viral protein and nucleic acid in growth of bacterio phage. J. Gen. Physiol., **36:** 39–56.

Jacob, F. and Monod, 1. 1961. Genetic regulatory mechanisms in the synthesis of protein. J. Mol. Biol., **3:** 318–356.

Jones, R.N. and Ress, H. 1982. B. Chromosomes, Academic Press, New York.

Lapitan, N.L. V. 1992. Organization and evaluation of higher plant nuclear genomes. Genome, **35:**171–181.

Lehninger, A.L. 1984. Principles of Biochemistry. New Delhi, CBS Publications.

Levine, L. 1969. The Biology of the Gene. Mosby and Toppan.

McClintock, Barbara. 1952. Chromosome organization and genetic expression. Cold Spring Harbour Symp. Quant. Biol., **16:** 15–47.

McClintock, Barbara 1956. Controlling elements and the gene. Cold Spring Harbour Symp. Quant. Biol., **21:** 197–216.

Meischer, F. 1871. On the chemical supposition of pus cells. Hoppe Seyler's Med. Chem. Untersuch., **4:** 441–460.

Mendel, G. 1865. Versuche uber Pflanzen-Hybriden. Verh Natur Forsch Ver Brunn IV, 3–47.

Mertz, E.T., Bates, L.S. and Nelson, O.E. 1964. Mutant gene that changes protein composition and increase lysine content of maize endosperm. Science, **145:** 279–280.

Mirsky, A.L. and Ris, H. 1949. Variable and constant components of chromosomes. Nature, **178**: 83–84.

Painter, T.S. 1934. A new method for the study of chromosome aberr at lions and the plotting of chromosome maps in Droshila melanogaster. Genetics, **19**: 175–185.

Strickberger, M. W. 1996. Genetics, 3rd edn. Prentice Hall of India. New Delhi.

Swift, H.R. 1950a. The constancy of desoxyribose nucleic acid in plant nuclei. Proc. Nat. Acad, Sci., **36**: 643–654.

Watson, J.D. and Crick, F.H.C. 1963. Genetic implications of the structure of desoxyribonucleic acid. Nature, **17**: 964–969.

Wenrich, D.H. 1916. The spermatogenesis of Phrynotettix magnus with special reference to synapsis and the individuality of the chromosome. Bull. Museum Compar. Zool., Harvard, **60**: 57–133.

Wilkins, M.H.E 1963. The molecular configuration of nucleic acids. Science, **140**: 941–950.

Analysis of Qualitative Traits

A qualitative trait shows discontinuous variation and is under control of one or few genes and where the effect of each gene is discernible. One uses Mendelian tools to analyse discrete variation. One studies counts and ratios in the different advancing generations (F_2, F_3, etc. and backcross) following a cross between two pure breeding lines different in qualitative character(s) and the following informations are obtained.

(i) How many genes are governing particular qualitative character?
(ii) How many alleles are there at each locus?
(iii) What type of gene action is involved?
(iv) If there is epistasis then what type of epistasis is there?
(v) Whether the genes show linkage or pleiotropy. If there is linkage then whether the linkage is in coupling phase or repulsion phase. What is the value of recombination? How the genes are arranged with respect to one another? What are the relative distances between different genes? Finally, which chromosome carries gene(s) for a particular qualitative trait?
(vi) Whether the qualitative character shows maternal inheritance or sex linked inheritance.
(vii) Whether the Mendelian genes show the genetic background effect on its expression and also how it interacts with the physical environment.
(viii) Determination of gene frequency in a population.

2.1 Gene Action and Interaction

2.1.1 Dominance

Considering one locus with two alleles *(A, a)* system when F_1 of the two parents *(AA and aa)* resemble either of the two parents, it is a case of complete dominance. When the phenotype of F_1 falls in between the two parents it is a case of partial dominance but when it falls outside the range of the two parents gene action can be said to be of over dominance type. In case of complete dominance the F_2 population will have dominant and recessive phenotypes in the ratio 3:1. The test cross population will have individuals in 1:1 proportion whereas in BC_1 population, all individuals will be of dominant phenotype and in BC_2 the ratio of dominant to recessive phenotype will be 1:1 as shown in Fig. 2.1. The terms recessive and dominant merely indicate which homozygote is more easily distinguishable from the heterozygote.

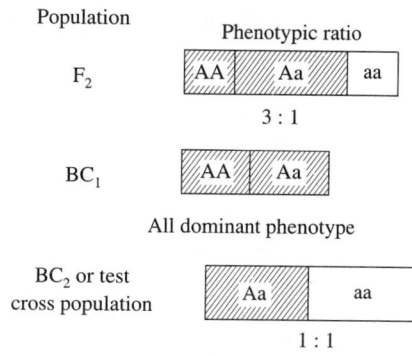

Fig. 2.1 Shaded area indicates dominant phenotype and unshaded the recessive phenotype.

2.1.1.1 Molecular Basis of Dominance

There is a chain of reactions between the primary gene product (assumed to be an enzyme) and the observed character. At the molecular level the primary gene products of the most wild type alleles showing dominance in diploid are mostly enzymes or other proteins necessary for the proper metabolic functioning of the organism whereas mutant gene showing recessive character produces products that are often non-functional or only partially functional. Very active primary gene product is more like than one with slight activity to exhaust the available amount of substrate and leads to complete dominance. Conversely, a very slight primary effect on metabolic processes is more likely to be represented at all subsequent steps by agents that do not deplete their substrate much and thus lead to incomplete dominance of ultimate effect. In other words, if one representative of the gene uses up the substrate as rapidly as supplied with the gene will be completely dominant. Under the various theories (bottleneck hypothesis, partitioning or multiplicative hypothesis and the intermediate between the two) Wright (1968) showed that there cannot be less than exact semi-dominance of a character due to an active gene product over partial or complete inactivation but there may be any degree up to complete dominance (complete inactivation of the gene does not lead to complete absence of the character). In multiple allelic series it is possible under the bottleneck hypothesis that all may exhibit semi-dominance with each other but it is more likely that the higher ones will pass through near dominance to complete dominance of the highest. Under other hypotheses the degree of dominance rises systematically with increasing activity but theoretically never quite reaches completeness with the highest.

Interaction between the chains of processes initiated by genes at the same and other loci are, however, the rule with most characters and give rise to the phenomenon of dominance. Degree of the dominance and alteration of it are necessarily of importance in population genetics.

2.1.2 Overdominance

Overdominance is shown by characters concerned with biological fitness such as size, productivity and viability. The superiority of heterozygote relative to homozygote has been called the overdominance hypothesis (Shull and East, 1908). There are many examples in which the superiority of heterozygotes lies in producing either active or effective or optimum or excess quantity of gene product in comparision to homozygous parents which produce either inactive or in effective or too small quantity of required product. In Drosophila the white eyed gene(w) in heterozygous condition ($W + Iw$) causes a marked increase in the amount of certain fluorescent pigments over both white and wild type homozygotes. The two different alleles in heterozygous condition may produce a more effective product. This can be explained as if the two alleles separately in homozygous state produce defective (multi-chained) protein while in the heterozygous condition may complement each other within a single multi-chained protein and thus resulting in an active product as has been found in case of alkaline phosphatase mutation in *E. coli*. Schwartz and Laughner (1969) in maize demonstrated that heterozygote possesses an alcohol dehydrogenase enzyme that is both active and stable in comparison to inactive or unstable homozygous parental enzymes. Emerson (1948) in *Neurospora* found that the heterokaryon (Pab+/pab) produces an optimum amount of *p*-aminobenzoic acid for growth in comparison to either homokaryon. If homozygotes produce two different gene products, the heterozygote may produce the two different products in the same individual. This is a possibility in sickel cell heterozygote which may be producing two haemoglobins, sickel haemoglobin to protect the individual from malaria infection and normal haemoglobin which prevents homolytic anaemia.

Heterozygotes have been thought to be developmentally stable meaning there by that their development is normally unaffected by environmental stress. In other words they are buffered or canalized which specifies protection against variability in comparison to the homozygous inbred parents. But then in extreme environmental conditions such as drought or very poor nutritional environment hybrids may not show such characteristics. Similarly, inbred lines (homozygotes) have been found showing developmental stability comparable to hybrids. Since the homozygous parents involved in cross may differ at a number of loci, it is difficult to determine the exact dominance relation of particular individual genes and thus it is essential to distinguish between the spurious and genuine overdominance.

2.1.3 Co-dominance

The phenomenon of dominance does not exist where the observed characters are the immediate product of the two alleles formed independently, without significant competition. This situation called co-dominance is found with allelic haemoglobins and other proteins. Co-dominance is also the rule for the apparent factors of the complex specificities of antigens that somehow reflect specificities of the genes. But there is an exception in that the self incompatibility alleles of many plants, each allele, in the style, function independently of the other in diploids but there are interactions between the alleles of the diploid pollen grains in tetraploids. Apparent co-dominance has also been found in some allelic series, determining characters that are presumably more remote from primary gene action than those above.

2.1.3.1 Antibody-antigen reaction

In MM, NN and MN and AO, BO, OO and AB blood groups where co-dominance occurs the difference between alleles, interaction of gene products can be recognized by studying the antibody-antigen relationship. The different alleles M, N produce different antigens which react with the antibodies produced in the host in response to the antigen and thereby reduce the harmful effects. These different antibodies are specific for a particular antigen and each different antigen causes the development of a different antibody (also called immunoglobin) which can be isolated from the sera and antisera. Immunoglobins (Ig) are composed of four polypeptide chains held together by disulphide bonds. The two identical polypeptide chains are heavy and are about 450 amino acids long whereas the other two identical chains are light and are about 200 amino acid long. Each pclypeptide chain possesses a variable (v) domain about 100 amino acids long that provides antibody specificity towards particular antigens and different immunoglobin cell lines produce different variable regions. In case of ABO system an allele produces antigen, an enzyme, called transferase. It is a sugar protein compound. The A allele produces an N-acetyl galactosaminyle transferase whereas the B allele produces galactosyl transferase. In case of O allele no terminal transferase enzyme appears to be produced. The difference between these three alleles thus lies in the presence of different kinds of terminal galactose sugar or the absence of the sugar molecule in the sugar-protein complex. The O allele is recessive to A or B because of the result of a defective or absence of enzyme whose phenotypic effect is masked in heterozygous condition (AO, or BO). AB individual shows dominance as both antigens are present in AB heterozygote.

2.1.4 Multiple alleles

A gene is made up of a number of nucleotides and that the mutational unit is a single nucleotide i.e. the variation can arise at each of the nucleotide position along its length. Thus we can have more than only two possible kinds of allele in a gene. These multiple alleles then differ in the sequence of nucleotides and will then differ in the sequence of amino acids in the polypeptide chain i.e. in term of gene product. In other words these multiple alleles will differ in enzyme or protein differences. These differences can easily be detected through starch or polyacrylamidegel electrophoresis that measures the mobility of an enzyme or protein in an electric field. Depending upon their molecular size and electrical charge (+ or –) enzymes/proteins will separate and move towards either the negative or positive pole. The positions occupied by these molecules on the electrical gradient are identified as bands when the gel is treated with agents (dyes) that can assay molecular or enzymatic activity or is exposed to U.V. light. Thus each genotype produces an identifiable electrophoretic pattern of allozymes. Thus one can estimate genetic variability in the population.

2.1.5 Epistasis or non-allelic interaction

Interaction effects are least likely to occur where the observed character is closely related to primary

2.4 Biometrical Genetics—Analysis of Quantitative Variation

gene action as it is in the case of allelic differences in protein/enzyme composition although serological responses of analysis are not so close but are also largely independent of the rest of the genome. Interaction effect does occur with respect to the ultimate products of chain of metabolic processes in which each step is controlled by a different locus. Thus it is implied that interaction effects are universal in the more complex character like yield that traces to such processes. The product at one of the steps in such a chain of processes may be subject to destruction upto certain level, by substance tracing to other genes. A gene that would otherwise exhibit incomplete dominance over an inactivated allele may actually be recessive to the later because the heterozygote falls below this threshold (Wright, 1927, 1929).

With two loci A, B and each with two alleles (A, a and B, b) determining a trait, the phenotypic ratios corresponding to different types of interaction in the F_2 will be as given in Table 2.1.

Table 2.1 Phenotypic ratios in two factors segregation with no epistasis and different types of epistasis

Genotypes	A-B-	A-bb	aaB-	aabb
Classical ratio (No epistasis)	9	3	3	1
Dominant epistasis	12		3	1
Recessive epistasis	9	3	4	
Duplicate genes with cumulative effect	9		6	1
Duplicate		15		1
Complementary	9		7	
Dominant and recessive interaction	13		3	

When gene at one locus masks the expression (or effect) of another it is called the phenomenon of epistasis. With no epistasis the classical dihybrid phenotypic ratio in the F_2 will be 9:3:3:1 and thus there are four phenotypic classes. But with epistasis the phenotypic classes with dihybrid parents will be less than four in the F_2 population. In case of complementary epistasis (9:7) the actions of these two genes are complementary i.e. together both dominant alleles complement each other and produce a different phenotype whereas recessive homozygous of a gene is epistatic to the other and produces the same phenotype as recessive homozygous at both loci. Under this hypothesis each gene controls a step essential for the production of product. In case of duplicate epistasis (15:1), dominant allele at either of the two loci produces the same phenotype as dominant alleles at both loci whereas the recessive homozygous at both loci produces a different phenotype. Recessive epistasis (9:3:4) occurs when the recessive homozygous at A locus is epistatic to the B locus and thus produces the same phenotype as produced by recessive homozygous at both loci. A-B- and A-bb produce two different phenotypes. In case of dominant epistasis dominant allele at A locus is epistatic to the B locus and produces one phenotype whereas aaB- and $aabb$ produce two different phenotypes. The ratio 9:6:1 results when dominant allele at either loci produces the same phenotype whereas dominant alleles at both loci and recessive homozygous at both loci produce two different phenotypes. The ratio 13:3 is obtained when the dominant allele at one locus (A) and the recessive homozygous at the other locus (bb) produce the same phenotype whereas the dominant allele at B locus (aaB-) produces a different phenotype.

2.1.5.1 Test for allelism

The *cis-trans* test for allelism can be used to see whether the two mutants having similar effects represent the two different genes or they are different alleles of the same gene. The two mutations located on the same homologous chromosomes are in *cis*-position ($AB/++$) whereas when located on opposite homologous chromosomes are in *trans*-position ($A+/+B$). This functional test for allelism is based on the principle that complementation occurs between the mutations in different cistron A, B (inter-cistronic complementation) whereas within a cistron (A or B) there is no complementation. The two mutants in *trans*-position are usually allelic if they produce a mutant (recessive) phenotype but non-allelic if they form a wild type (dominant). The partial or complete restoration of wild type in the later case is called the phenomenon of complementation.

2.2 Application of Laws of Probability

Considering one locus with two alleles the F_1 of the two parents (AA, aa) will be heterozygous (Aa) which upon selfing produces 3/4 of plants having the dominant and 1/4 the recessive character. This Mendelian proportion of 3/4 is a statistical probability. Let us see what happens when the heterozygote (Aa) is allowed to self-fertilize. The Aa plant will produce pollen and egg cells. A pollen cell will contain either A or a gene and we will have equal quantities of pollen containing A and a gene. In other words half the pollen produced will have A gene and the other will have a gene. In terms of probability, the probability of occurrence of pollen with A gene or a gene is 1/2. Likewise, the probability of an egg cell containing A or a gene is 1/2. The statistical probability is defined as the limiting value of the relative frequency with which some event occurs and thus $p(A)$ or $p(a)$ is 1/2. In the above example there are only two events or outcomes possible at each experiment but in principle a number of events or outcomes is possible. Now if the experiment is repeated n times and supposing that the event A has occurred $n(A)$ times, the proportion of time on which it has occurred is $n(A) = p(A)$ which will be the probability of occurrence of event A. If n is increased to infinity, the $p(A)$ will become closer to its expected limiting value. This hypothetical value is called the statistical probability of A and is denoted by $p(A)$ or $p(a)$.

As the pollen and egg cells unite at random that is to say independently of the gene which they contain, the joint probability of the two genes occurring together will follow the rules of multiplication which states that 'if A and B are the two statistically independent (i.e. the occurrence of A does not alter the probability that B will occur) events then the probability that A and B will occur is equal to the probability that A will occur multiplied by the probability that B will occur' and

$$p(A \text{ and } B) = p(A) \times p(B)$$

Thus the following matings will occur with the same probability of $\frac{1}{2} \times \frac{1}{2} = \frac{1}{4}$

Pollen	$P(A)$ or $p(a)$	Egg cell	$p(A)$ or $p(a)$	Offspring	Probability
A	$\frac{1}{2}$	A	$\frac{1}{2}$	AA	$\frac{1}{2} \times \frac{1}{2} = \frac{1}{4}$
A	$\frac{1}{2}$	a	$\frac{1}{2}$	Aa	$\frac{1}{2} \times \frac{1}{2} = \frac{1}{4}$
a	$\frac{1}{2}$	A	$\frac{1}{2}$	Aa	$\frac{1}{2} \times \frac{1}{2} = \frac{1}{4}$
a	$\frac{1}{2}$	a	$\frac{1}{2}$	aa	$\frac{1}{2} \times \frac{1}{2} = \frac{1}{4}$

Now the probability that AA or Aa or aa will occur or in other words the probability of offspring will be equal to the sum of their separate probabilities. This follows from the law of addition of probabilities which states that 'if A and B are mutually exclusive events, that is, if one occurs, the other cannot occur'; then the probability that either A or B will occur is equal to their separate probability' and so

$$p(A \text{ or } B) = p(A) + p(B)$$

Thus the probability that the above mating will produce an offspring (either AA, Aa or aa) equals 1.0 $\left(=\frac{1}{4} + \frac{1}{4} + \frac{1}{4} + \frac{1}{4}\right)$. The probability of 1.0 indicates that the event is a certainty and a probability of 0.0 indicates that the event is impossible. Thus the probability has a range from 0 to 1.0. The probability that the offspring would be a dominant genotype (either AA or Aa) will be equal to $\frac{3}{4}$ $\left(=\frac{1}{4} + \frac{1}{4} + \frac{1}{4} = p(AA) + p(Aa) + p(Aa)\right)$ and that of a recessive genotype would be 1/4. Similarly, the probability that the offspring would be a dominant heterozygous genotype (Aa) will be $\frac{1}{2}\left(=\frac{1}{4} + \frac{1}{4}\right)$.

In case of test cross (Aa × aa) the male and female gametes and offspring will be produced with the probabilities given below:

Female gametes	Probability of occurrence	Male gametes	Probability of occurrence	Offspring genotype	Probability
A	$\frac{1}{2}$	a	1	Aa	$\frac{1}{2} \times 1 = \frac{1}{2}$
a	$\frac{1}{2}$	a	1	aa	$\frac{1}{2} \times 1 = \frac{1}{2}$

Thus the probability of occurrence of *Aa* genotype in the progeny will be $\frac{1}{2}$ and also the probability of occurrence of *aa* type individual will be $\frac{1}{2}$.

2.2.1 Application to cases of two or more genes

The above approach can be extended to cases of two or more genes segregating independently. In case of one gene with two alleles and complete dominance, the dominant and recessive phenotypes appear in F_2 with 3/4 and 1/4 probabilities, respectively. Considering the two loci together with independent segregation four phenotypes will appear in the F_2 with $\frac{9}{16}, \frac{3}{16}, \frac{3}{16}$ and $\frac{1}{16}$ probabilities, respectively as shown below:

A-locus		B-locus		Phenotypes		Joint probability
Phenotype	Probability	Phenotype	Probability	A	B	
Dominant	3/4	Dominant	3/4	Dom.	Dom.	$\frac{3}{4} \times \frac{3}{4} = \frac{9}{16}$
Recessive	1/4	Recessive	1/4	Dom.	Rec.	$\frac{3}{4} \times \frac{1}{4} = \frac{3}{16}$
				Rec.	Dom.	$\frac{3}{4} \times \frac{1}{4} = \frac{3}{16}$
				Rec.	Rec.	$\frac{1}{4} \times \frac{1}{4} = \frac{1}{16}$

The phenotype showing dominance at both loci will occur with a probability of 9/16, the phenotype showing dominance at one locus but recessive at the other locus and vice versa will have a probability of 3/16 each and the phenotype showing recessive phenotype at both loci will appear with a probability of 1/16. Also as we know from the consideration of one locus case that dominant homozygous genotype *AA* appears with a' probability of 1/4 and therefore the probability of occurrence of a dominant homozygous at both loci will be $\frac{1}{16} \left(\frac{1}{4} \times \frac{1}{4} \right)$. Similarly, it can be shown that with three genes segregating independently, eight phenotypes will occur in the F_2 with probabilities as $\frac{27}{64}, \frac{9}{64}, \frac{9}{64}, \frac{9}{64}$ and $\frac{1}{64}$, respectively.

In the two factor test cross (*Aa Bb* × *aabb*), the four phenotypes will appear with probability 1/4 each as shown below.

Similarly it can be shown that with three genes segregating independently eight phenotypes will occur in the test cross population with a probability of 1/8 each.

2.2.2 Types of Probability

There are two types of probability. *Priori* probability and *posterior* probability. The *priori* probability is the probability that can be specified before hand from the nature of the event. For example, the probability of obtaining a head or tail in the tossing of a coin is 1/2. The probability based on genetic theory of obtaining dominant and recessive phenotypes in the F_2 of one factor cross is 3/4 and 1/4, respectively. Thus a *priori* probability estimates the chance of obtaining a specific result in a particular mating. This type of probability is involved in Mendelian inheritance. The *priori* probability can be used in prediction only if the events under consideration are independent of each other. The *posterior* or empiric probability is the probability which is calculated from certain evidence (i.e. by counting the number of times a given event occurs in a certain number of cases). This type of probability is involved in quantitative inheritance. Bayes theorem can be expressed in the formula, posterior probability α priori probability × likelihood where likelihood (i.e. the probability of occurrence) can be defined as the statistical probability of obtaining observed results given that the hypothesis is correct. Thus this theorem enables us to evaluate the probability that the hypothesis is true. If the observation comes from a continuous distribution rather than from a discontinuous distribution as in the above case their likelihood is defined as their joint probability density.

A-locus		B-locus		Joint phenotype		Joint probability
Phenotype	Probability	Phenotype	Probability	A	B	
Dominant	1/2	Dominant	1/2	Dom.	Dom.	$\frac{1}{2} \times \frac{1}{2} = \frac{1}{4}$
Recessive	1/2	Recessive	1/2	Dom.	Rec.	$\frac{1}{2} \times \frac{1}{2} = \frac{1}{4}$
				Rec.	Dom.	$\frac{1}{2} \times \frac{1}{2} = \frac{1}{4}$
				Rec.	Dom.	$\frac{1}{2} \times \frac{1}{2} = \frac{1}{4}$

2.2.3 Binomial Probability

If there is an event with just two possible outcomes A and a as we have seen in the above case with probabilities p and q where $p + q = 1.0$, the probability of an event A occurring r times in n trials is the binomial probability.

$$p(A = r) = \frac{n!}{n!(n-r)!} p^r q^{n-r}$$

and r is said to follow binomial distribution.

This $p(r)$ is the likelihood which will usually depend on one or more unknown parameters and here p is unknown.

$n! = n(n-1)(n-2) \ldots 4, 3, 2, 1$ and $r = 0, 1 \ldots n$ and $0!$ is one. The mean and variance of binomial distributions are

$$\text{Mean} = pn$$
$$\text{Variance} = pqn$$

These above formulae are appropriate in case of analysis of observed numbers of individuals in the two classes. In case of analysis of the proportions of individuals observed to fall into two classes, the following formulae will be used.

$$\text{Mean } (\mu) = p \text{ and}$$
$$\text{Variance } (\sigma^2) = pq/n$$

Now the probability of obtaining 5 dominant phenotype and no recessive phenotype in a family of 5 individuals in the test cross ($Aa \times aa$) will be

$$P(5) = \frac{5!}{5!(5-5)!} \left(\frac{1}{2}\right)^5 \left(\frac{1}{2}\right)^0$$
$$= 3.1\%$$

This result is equal to that obtained from assuming that the occurrence of one dominant phenotype is independent of the occurrence of another dominant phenotype and thus that all the 5 individuals will be of dominant phenotype. The probability will be equal to $\left(\frac{1}{2}\right)^5 = 3.1\%$. Similarly the probability of obtaining 3 dominant and one recessive phenotype in a family of 4 individuals of a test cross ($Aa \times Aa$) will be equal to

$$P(3) = \frac{4!}{3!(4-3)!} \left(\frac{3}{4}\right)^3 \left(\frac{1}{4}\right)^1$$
$$= 0.66\%$$

where the probability of obtaining a dominant phenotype in F_2 is 3/4 whereas the probability of obtaining a recessive phenotype is 1/4.

There are two ways in which an event say A will occur r times. First event A will occur r times in succession and then a will occur $n - r$ times in a row. This will have a probability of $p^r q^{n-r}$. Secondly, there can be another sequence in which event A will appear r times and the number of such sequences will be $\frac{n!}{r!(n-r)!}$. Therefore, the overall probability of event A occurring r times $p(A = r)$ without regard to their order of appearance is $P(A = r) = \frac{n!}{r!(n-r)!} p^r q^{n-r}$ The term $\frac{n!}{r!(n-r)!}$ is called the binomial coefficient the values of which may be found by substitution of successive value of r ($r = 0, 1, 2 \ldots$). This formula is in fact the rth term in the expansion of $(p + q)^n$.

$$(p + q)^n = \sum_{0}^{n} \frac{n!}{r! \, n-r!} p^r q^{n-r}$$
$$= P(n) + P(n-1) + \ldots + P(1) + P(0)$$

This corresponds to the different possible combinations of dominant and recessive progeny that can be obtained. Considering a family size of 2, we can have the dominant and recessive phenotypes appearing in the F_2 from the cross ($Aa \times Aa$) with the probabilities given below:

Dominant	Recessive	Probability	Terms of binomial expression
2	0	0.56	p^2
1	1	0.374	$2pq$
0	2	0.062	q^2

Since
$(p + q)^2 = (3/4 + 1/4)^2 = 1 = 0.56 + 0.374 + 0.062$

2.2.4 Multinomial probability

An event can have three or more possible outcomes as we can see in case of dihybrid or trihybrid cross where the F_1 is $Aa\ Ba$ and $Aa\ Bb\ Cc$, respectively which upon selfing produces four and eight phenotypes in the F_2 in the ratio of 9:3:3:1 and

27:9:9:9:3:3:3 and in the test crosses (Aa Bb × aabb), (AaBb Cc × aabbcc) in the ratio of 1:1:1:1 and 1:1:1:1:1:1:1:1, respectively. Now if p_i is the probability that a member belongs to the ith class and if n members are chosen randomly, the probability that x_1 belongs to the first class, x_2 belongs to the second class and so on is

$$P(x_1, x_2, \ldots x_r) = \frac{n!}{x_1! \, x_2! \ldots x_r!} p_1^{x_1} p_2^{x_2} \ldots p_r^{x_r}$$

Here $x_1, x_2 \ldots x_r$ are said to follow multinomial distribution which is an extension of the binomial distribution. The probabilities are the terms of the expansion of $(p_1 + p_2 + \ldots P_r)^n$ where $p_1, p_2 \ldots p_r$ are the probabilities of occurrence of individuals of 1, 2, … rth classes. In case of F_2 of family size n produced by selfing of $Aa\,Bb$ the probability that W belongs to the first phenotypic class, X belongs to the second class, Y belongs to third class and Z belongs to fourth phenotypic class is

$$P(W, X, Y, Z) = \frac{n!}{W! \, X! \, Y! \, Z!} p^W q^X r^Y s^Z$$

where p, q, r and s are the probabilities of occurrences of individuals of four phenotypes, respectively.

2.3 Test of Goodness of Fit

Whether F_2 data show one, two or more factors segregation is tested using χ^2 which tests the goodness of fit. Assuming one factor segregation with complete dominance the expected frequencies of the two classes (dominant and recessive) of individuals will be in the ratio of 3:1 in the F_2 population and 1:1 in the test cross population. Now supposing n_1 and n_2 being the number of individuals in the two classes and let $n_1 + n_2 = n$, the expected number of individuals in the two classes will be $\frac{3n}{4}$ and $\frac{n}{4}$ in F_2 and $\frac{n}{2}$ and $\frac{n}{2}$ in test cross population as given below. The class I represents the individual belonging to dominant phenotype whereas class II represents the individual belonging to recessive phenotype.

	Phenotypic class		
	I	II	Total
Observed number	n_1	n_2	$n_1 + n_2 = n$
Expected number (F_2 population)	$\frac{3n}{4}$	$\frac{n}{4}$	$\frac{3n}{4} + \frac{n}{4} = n$
Expected number (Test cross population)	$\frac{n}{2}$	$\frac{n}{2}$	$\frac{n}{2} + \frac{n}{2} = n$

The chi-square (χ^2) is calculated as

$$\chi^2 = \Sigma \frac{(O - E)^2}{E}$$

where O and E refer to the observed and expected number of individual in a class and Σ stands for summation over all classes. The number of class in the present case is two. The χ^2 has one degree of freedom. The degree of freedom can be defined as the number of classes which can be fitted arbitrarily which is equal to the number of classes minus one which is 1 in the present case. It can also be defined as the number of independent comparisons that can be made in the data. Comparisons between observations whose values are uninfluenced by changes in each other are called independent comparisons (Mather, 1943). The χ^2 calculated is compared with the table value of χ^2 and if the calculated value of χ^2 is less than the table value, the hypothesis that the segregation has taken place according to the theoretical expectation is accepted. In other words the data show a good fit to the expectations.

2.3.1 Two Factor Segregation

In case of two loci segregation there will be four phenotypic classes with frequencies in the ratio of 9:3:3:1 in the F_2 and 1:1:1:1 in the test cross population. If n_1, n_2, n_3 and n_4 are the observed values of the different phenotypic classes with $n_1 + n_2 + n_3 + n_4 = n$ then the expected number of individuals in the four phenotypic classes will be $\frac{9n}{16}, \frac{3n}{16}, \frac{3n}{16}$ and $\frac{n}{16}$ in case F_2 population and $\frac{n}{4}, \frac{n}{4}, \frac{n}{4}$ and $\frac{n}{4}$ in testcross population. The χ^2 is calculated and compared with the table value in the way as described above. The degree of freedom here is 3 (= 4 – 1).

The general formula for testing a two class segregation whose expected frequencies ratio is 1:1 and the observed frequency ratio is $a_1 : a_2$ is

$$\chi^2 = \frac{(a_1 - la_2)^2}{ln}$$

where $n = a_1 + a_2$. This formula takes the following forms for the commonly occurring ratios of 1:1, 3:1, 9:7 and 15:1, respectively.

Ratio	Formula
1 : 1	$x^2 = (a_1 - a_2)^2/n$
3 : 1	$x^2 = (a_1 - 3a_2)^2/3n$
15 : 1	$x^2 = (a_1 - 15 a_2)^2/15n$
9 : 7	$x^2 = (a_1 - \frac{9}{7} a_2)^2/n$

2.3.2 Test of heterogeneity of Families

As in practice we are making a number of crosses, we have a number of either F_2 families or testcross families. Then we would like to test
1. whether or not the segregation in a family conforms to the expectation.
2. whether or not the families are homogeneous i.e. is there agreement between observed and the expected frequency considering all families together.

Suppose there are m ($i = 1, m$) families. For each of the m families we calculate χ^2 with one degree of freedom. Comparison of calculated χ^2 value with the table value will test whether or not segregation in a family conforms to the expectation. Thereafter we calculate the observed number of individuals in the phenotypic classes over all families. Now χ^2 with one degree of freedom using this overall families data as shown below considering one factor segregation is obtained.

Family	Phenotypic classes		χ^2 value
	I	II	
$i = 1$	–	–	χ_1^2
.			
.			
.			
m	–	–	χ_m^2
Total	T_1^*	T_2^*	χ^2

T_1^* and T_2^* are the total number of observed individuals in the two classes.

From these χ^2 values we test the following items.

Item	D.F.	χ^2
Overall	1	$\chi_{(1)}^2$
Heterogeneity	$m - 1$	$\chi_{(m)}^2 - \chi_{(1)}^2$

$$\chi_{(m)}^2 = \sum_{i=1}^{m} \chi_i^2$$

If heterogeneity is significant it indicates faulty classification. Experimentation which includes poor classification or partial selection of stronger genotype due to competition resulting from high density planting or limited nutrients supply, shows significant heterogeneity item.

2.3.3 Testing of Independence

In case of two factors segregation we would like to test whether or not one factor segregates independently of the other factor. χ^2 test provides a test of independence. Supposing two factors A and B the data can be arranged in a two way table usually called contingency table. In the simplest case we can have 2 × 2 contingency table in which the data can be classified into 2 classes (A_1, A_2) considering A factor and B_1 and B_2 considering B factor as shown in Table 2.2.

Table 2.2 A 2 × 2 contingency table

		B factor		
		B_1	B_2	Total
A factor	A_1	a	b	$a + b$
	A_2	c	d	$c + d$
	Total	$a + c$	$b + c$	$a + b + c + d$

Let a, b, c and d be the number of observed individuals in these quadrates. The expected number of individual in each of these quadrates will be as follows:

For a the expected number will be $\dfrac{(a + b)(a + c)}{(a + b + c + d)}$

For b the expected number will be $\dfrac{(a + b)(b + d)}{(a + b + c + d)}$

For c the expected number will be $\dfrac{(c + d)(a + c)}{(a + b + c + d)}$

For d the expected number will be $\dfrac{(c + d)(b + d)}{(a + b + c + d)}$

The χ^2 which is here called a contingency χ^2 is calculated as

$$\frac{(a+b+c+d)(ad-bc)^2}{(a+b)(c+d)(a+c)(b+d)}$$

The degree of freedom is $v = (r-1)(c-1)$ where r is the number of classes considering A factor and c is the number of classes considering B factor. This is the number of quadrates whose frequencies can be arbitrarily assigned and equals the total number of quadrates $(r \times c)$ minus $(r + c - 1)$, the number of quadrates from which the frequencies can be determined from the quadrates totals. In the present case of 2×2 contingency table $v = 1 \times 1 = 1$ as $r = 2$ and $c = 2$.

2.3.3.1 Yates' correction for continuity

The distribution of χ^2 is a continuous distribution whereas the distribution of the frequencies is discontinuous. As the sample size increases the discontinuous frequency distribution takes the form of continuous χ^2 distribution. That is why the expected quadrate frequencies should not be less than 4 in order to avoid irregularities due to this discontinuity. In case of 2×2 contingency table the observed numbers of the most numerous classes are reduced by half and those of least numerous classes are added by half. For example, if $ad - bc$ is positive, 1/2 is subtracted from a and d and if $ad - bc$ is negative, 1/2 is added to b and c. It does not change the margin totals. The above formula can then be redefined in the following way:

$$\chi^2_{(1)} = \frac{\left(|ad-bc| - \frac{1}{2}n\right)^2}{(a+b)(c+d)(a+c)(b+d)}$$

where $|ad - bc|$ refers to the absolute value of this quantity. Thus if $(ad - bc)$ is positive $\frac{1}{2}n$ is subtracted from it and if $(ad - bc)$ is negative, $\frac{1}{2}n$ is added to it. Here $n = a + b + c + d$.

2.4 Experimental Size

If our objective is to isolate at least one recessive *(aa)* in the progeny of selfed heterozygote *(Aa)* or one *AABB* genotype in the progeny of cross *Aa Bb ×* *Aa Bb* or to distinguish between one and two factors interactions such as between 1:1 and 3:1 or 9:7 and 3:1 then we would require to raise a minimum number of progeny so as to ensure the isolation of such individuals or to reduce the probability of misclassification. When the aim is to isolate at least one recessive *(aa)* individual in the progeny of selfed heterozygote *(Aa)*, we know that the probability of obtaining at least one recessive in the progeny of inbreeding heterozygote is 1/4, then the probability that all the individuals in a family of n size will be recessive is $\left(\frac{1}{4}\right)^n$. Considering that n is large enough to contain at least one recessive *(aa)* in 99 cases out of 100 (a probability of 0.990) the n will be calculated as follows:

$$\left(\frac{1}{4}\right)^n = \frac{99}{100}$$

Taking logarithm this becomes

$$\log\left(\frac{1}{4}\right)^n = \log\frac{99}{100}$$

i.e.

$$n \log\left(\frac{1}{4}\right)^n = \log 99 \times 10^{-2}$$

which upon simplification yields the value of n as 16.0. Table 2.3 shows the number of individuals which should be raised in a progeny in order that a certain type, expected to form a known fraction of the progeny may be expected to occur with a chosen level of probability, at least once.

2.4.1 Distinction between two hypotheses

There are two methods for calculating the required family size (n) when we want to distinguish between two hypotheses i.e. either between 1:1 and 3:1 or between 3:1 and 9:7 ratios.

(i) Method of χ^2
(ii) Standard error test of significance

2.4.1.1 Method of χ^2

To be able to distinguish between the two ratios n, the family size, has to be so large that on either hypothesis χ^2 should have a low probability. If we choose 5% as the level of significance the value for χ^2 at 1. d.f. is 3.841 and thus the above calculated χ^2

Analysis of Qualitative Traits

Table 2.3 Family size required for obtaining at least one individual with an expected level of probability

Fraction expected	Family size Level of probability						
	0.900	0.950	0.980	0.990	0.995	0.998	0.999
$\frac{1}{2}$	3.3	4.3	5.6	6.6	7.6	9.9	10.0
$\frac{1}{4}$	8.0	10.4	13.6	16.0	18.4	21.6	24.0
$\frac{1}{8}$	17.2	22.4	29.3	34.5	39.7	46.6	51.7
$\frac{1}{16}$	35.7	46.4	60.6	71.4	82.1	96.3	107.0
$\frac{1}{32}$	72.5	94.4	123.2	145.1	166.9	195.7	217.6
$\frac{1}{64}$	146.2	190.2	248.4	292.4	336.4	394.6	430.6
$\frac{1}{3}$	5.7	7.5	9.6	11.4			
$\frac{1}{9}$	19.5	25.4	33.2	39.1			
$\frac{1}{27}$	61.0	79.4	103.6	122.0			

* Adapted from "The Measurement of Linkage in Heredity" by K. Mather.

value should be equal to 3.841. Thus considering 1:1 and 3:1 ratio

$$\frac{(a_1 - a_2)^2}{n} = 3.841$$

and

$$\frac{(a_1 - 3a_2)^2}{3n} = 3.841$$

Now as $a_1 + a_2 = n$, so $a_2 = n - a_1$. Putting the value of a_2 as $n - a_1$ the above equations upon simplifying will yield the value of n, the required family size, as

$$n = 3.841(2 + \sqrt{3})^2 = 53.5$$

Thus, at least 54 plants in F_2 should be raised in the order to distinguish between the above two hypotheses. The general formula for determining the minimum experimental size for distinguishing a I_1:1 segregation from that of I_2:1 segregation at 5% level of significance is

$$n = \frac{3.841(1 + \sqrt{l_1 l_2})^2}{(\sqrt{l_1} - \sqrt{l_2})^2}$$

2.4. 1.2 Method of standard error test

In this method the underlying principle is that the value of ratio of difference between observed and expected ratio to the standard error of difference should be at least 1.96 so as to be significant at 5% level of significance. The standard error of the number of recessive would be $\sqrt{n/4}$, $\sqrt{3n/16}$ and $\sqrt{\frac{63n}{256}}$ in case of 1:1, 3:1 and 9:7 ratios, respectively. Thus in case of distinguishing between 1:1 and 3:1 ratios we have the following two equations considering a_1 and a_2 as the number of individuals in the two classes.

$$\frac{a_1 - n/2}{\sqrt{n/4}} = 1.96$$

and

$$\frac{\frac{3n}{4} - a_1}{\sqrt{\frac{3n}{16}}} = 1.96$$

which upon solving yields the value of n as

$$n = [(1.96(\sqrt{2} + \sqrt{3})]^2 = 53.4$$

In case of distinction between 3:1 and 9:7 ratios the equations will take the form as

$$\frac{a_2 - 1/4n}{\sqrt{\frac{3n}{16}}} = 1.96$$

and
$$\frac{\frac{7}{16}n - a_2}{\sqrt{\frac{63n}{265}}} = 1.96$$

These equations upon solving will yield the value of n as 94.32. Thus 94 plants will be required in order to distinguish between 3:1 and 9:7 ratios.

2.5 Linkage and Pleiotrophy

Genes on a chromosome are in linear order and form a linkage group. The number of linkage groups in a species corresponds to its number of chromosomes in haploids and all the genes on a chromosome in the heterogametic sex show sex linkage. Linkage was discovered by Bateson and Punnett in 1906. The intensity of linkage varies between from 1.0 (tight linkage) to partial linkage and to 0.5 (no linkage). The detection and estimation of linkage(more exactly, the recombination frequency, p) is done through various methods (Mather, 1963), namely (i) linear model, (ii) maximum likelihood (Fisher), (iii) product formula (Fisher and Balmukund, 1928; Immer, 1930) and (iv) Emerson's method.

2.5.1 Detection of linkage

Considering two loci A, B with A dominant over a and B dominant over b and with parents $AABB$ and $aabb$ there will be four phenotypic classes in F_2 or in the test cross generation (generation derived through backcrossing of F_1 to its recessive parent, $aabb$). Assuming no linkage, the ratio of the different phenotypic classes will be in the proportion 9:3:3:1 and $\frac{1}{4}:\frac{1}{4}:\frac{1}{4}:\frac{1}{4}$, respectively. The deviation from 9:3:3:1 or $\frac{1}{4}:\frac{1}{4}:\frac{1}{4}:\frac{1}{4}$ will indicate the presence of linkage but then we know that this deviation can be due to deviation in segregation at either locus A or B or both, so we must be sure that segregation at both loci when considered independently is normal i.e. 3:1 ratio in case of F_2 population and 1:1 in case of testcross population, respectively. The general linear model for detecting the deviation from the expected segregation frequency takes the form as

$$C = k_1 a_1 + k_2 a_2 + k_3 a_3 + k_4 a_4$$

where k_1, k_2, k_3 and k_4 are the coefficients which may vary according to whether the objective is to test the segregation at either locus or to detect the linkage and a_1, a_2, a_3 and a_4 are the observed fequency of the four phenotypic classes. And $\frac{\chi^2}{V(C)}$ is distributed as χ^2 with one degree of freedom. The expected value of C becomes

$$C = m_1 a_1 + m_2 a_2 + m_3 a_3 + m_4 a_4 = 0$$

with the variance as $V(C) = n \Sigma mk^2$ where m_1, m_2, m_3 and m_4 are the expected frequencies of the four phenotypic classes and the total frequency, $n = a_1 + a_2 + a_3 + a_4$. The values of k_1, k_2, k_3 and k_4 in various linear models take the form as given in Table 2.4.

The model for testing the deviation from 3:1 segregation at A takes the form as

$$C_1 = a_1 + a_2 - 3a_3 - 3a_4$$

Table 2.4 Values of k_1, k_2, k_3 and k_4 in different linear models

Population	Phenotypic classes			
	A-B-	A-bb	aaB-	aabb
F_2 Population				
Expected frequency	$m_1(9)$	$m_2(3)$	$m_3(3)$	$m_4(1)$
Observed frequency	a_1	a_2	a_3	a_4
Models				
C_1	1	1	−3	−3
C_2	1	−3	1	−3
C_3	1	−3	−3	9
Test cross population				
Expected frequency	$m_1\left(\frac{1}{4}\right)$	$m_2\left(\frac{1}{4}\right)$	$m_3\left(\frac{1}{4}\right)$	$m_4\left(\frac{1}{4}\right)$
Models				
C_1	1	1	−1	−1
C_2	1	−1	+1	−1
C_3	1	−1	−1	+1

where $k_1 = k_2 = 1$ and $k_3 = k_4 = 3$ with $V(C_1) = n(9 + 3 + 27 + 9)/16 = 3n$.

Thus χ_A^2 which tests the deviation of segregation from 3:1 ratio at A locus becomes

$$\chi_{A(1df)}^2 = \frac{C_1^2}{V(C_1)} = \frac{(a_1 + a_2 - 3a_3 - 3a_4)^2}{3n}$$

The model for testing the deviation from 3:1 segregation at B locus takes the form as

$$C_2 = a_1 - 3a_2 + a_3 - 3a_4$$

where $k_1 = k_3 = 1$ and $k_2 = k_4 = -3$ with variance as

$$V(C_2) = 3n$$

Thus χ_B^2 which tests the deviation of segregation from 3:1 ratio at B locus becomes

$$\chi_{B(1df)}^2 = \frac{C_2^2}{V(C_2)} = \frac{(a_1 - 3a_2 + a_3 - 3a_4)^2}{3n}$$

The model for testing the deviation of joint segregation from its expectation of no linkage (9:3:3:1) takes the form as

$$C_3 = a_1 - 3a_2 - 3a_3 + 9a_4$$

where $k_1 = 1$, $k_2 = k_3 = -3$ and $k_4 = 9$ with variance as

$$V(C_3) = 9n$$

Thus χ^2 linkage which tests the joint segregation from its expectation of no linkage becomes

$$\chi_{L(1df)}^2 = \frac{C_3^2}{V(C_3)} = \frac{(a_1 - 3a_2 - 3a_3 + 9a_4)^2}{9n}$$

Thus we see that the χ^2 with 3 degrees of freedom which tests for the joint deviation for all of the observed frequencies from the expected frequencies is calculated as

$$\chi_{3(df)}^2 = \frac{(a_1 - m_1)^2}{m_1} + \frac{(a_2 - m_2)^2}{m_2} + \frac{(a_3 - m_2)^2}{m_3} + \frac{(a_4 - m_4)^2}{m_4}$$

and is partitioned orthogonally into three parts, Thus

$$\chi_{3(df)}^2 = \chi_{A(1df)}^2 + \chi_{B(1df)}^2 + \chi_{L(1df)}^2$$

Similarly, in the testcross population

$$\chi_A^2 = (a_1 + a_2 - a_3 - a_4)^2 / n$$

$$\chi_B^2 = (a_1 - a_2 + a_3 - a_4)^2 / n$$

$$\chi_{Linkage}^2 = (a_1 - a_2 - a_3 + a_1)^2 / n$$

The general formula for detecting linkage assuming one factor segregating into ratio $l_1:1$, and another factor in the ratio $l_2:1$, respectively will be

$$\chi^2 = \frac{(a_1 - l_2 a_2 - l_1 a_3 + l_1 a_4)^2}{l_1 l_2 n}$$

where a_1, a_2, a_3 and a_4 are the observed proportion of individuals in the F_2 or any other generation.

2.5.2 Estimation of recombination frequency

Considering the above example of two parents, AA BB and aabb and assuming p_1 and p_2 being the recombination frequency in the male and female parents (AaBb) the gametes produced by these parents would have the frequencies as

Gametes	AB	Ab	aB	ab
male	$\frac{1}{2}(1-p_1)$	$\frac{1}{2}p_1$	$\frac{1}{2}p_1$	$\frac{1}{2}(1-p_1)$
female	$\frac{1}{2}(1-p_2)$	$\frac{1}{2}p_2$	$\frac{1}{2}p_2$	$\frac{1}{2}(1-p_2)$

and the frequencies of the four phenotypic classes would be as

A-B-
$$\frac{1}{4}[2 + (1-p_1)(1-p_2)]$$

A-bb
$$\frac{1}{4}[1 - (1-p_1)(1-p_2)]$$

aaB-
$$\frac{1}{4}[1 - (1-p_1)(1-p_2)]$$

aabb
$$\frac{1}{4}[1 - (1-p_1)(1-p_2)]$$

Thus the expected frequencies of all the four phenotypic classes depend on the quantity $(1-p_1)(1-p_2)$ and assuming $p(1-p_1)(1-p_2) = \theta$, the frequencies in the four phenotypic classes would become

A-B-	A-bb	aaB-	aabb
$\frac{1}{4}(2+\theta)$	$\frac{1}{4}(1-\theta)$	$\frac{1}{4}(1-\theta)$	$\frac{1}{4}\theta$

In case of F_2 families the linkage function for the calculation of χ^2 is

$$C_3 = a_1 - 3a_2 - 3a_3 + 9a_4$$

The expected value of this function in terms of θ is

$$C_3 = \frac{n}{4}(16\theta - 4)$$

Thus θ can be estimated from the equation

$$n(4\theta - 1) = a_1 - 3a_2 - 3a_3 + 9a_4$$

or $\theta = \frac{1}{4}(2a_1 - 2a_2 - 2a_3 + 10a_4)$

The variance of θ is estimated as

$$V\theta = \frac{1 + 6\theta - 4\theta^2}{4n}$$

2.5.2.1 Method of Maximum Likelihood

In this method the value of θ is obtained by maximising the logarithm of the likelihood expression, L with respect to θ

$$L = C + a_1 \log m_1 + a_2 \log m_2 + a_3 \log m_3 + a_4 \log m_4$$

where C is a constant. By differentiating this expression with respect to θ and equating it to zero the estimate of θ is obtained as

$$\frac{dL}{d\theta} = 0 + a_1 \frac{d \log m_1}{d\theta} + a_2 \frac{d \log m_2}{d\theta} + a_3 \frac{d \log m_3}{d\theta} \ldots + a_i \frac{d \log m_i}{d\theta}$$

Thus in case of F_2, maximization leads to the equation

$$\frac{dL}{d\theta} = \frac{a_1}{2+\theta} - \frac{a_2}{2-\theta} - \frac{a_3}{1-\theta} + \frac{a_4}{\theta} = 0$$

which upon simplification leads to the quadratic equation

$$2a^4 + [a_1 - 2(a_2 + a_3) - a_4]\theta - n\theta^2 = 0$$

from which the θ is calculated. From the general quadratic equation such as $a + bx + cx^2 = 0$, the x is estimated as

$$X = \frac{-b \pm \sqrt{b^2 - 4ac}}{2c}$$

Now as $\theta = (1 - p_1)(1 - p_2)$, assuming $p_1 = p_2$, the recombination frequency, P is estimated as

$$p = \sqrt{1 - \theta} \text{ where } \theta = (1 - p)^2$$

The variance of θ, $V\theta$ is obtained as

$$V\theta = \frac{2\theta(1 - \theta)(2 + \theta)}{n(1 + 2\theta)}$$

from which the variance of recombination frequency, Vp is estimated as

$$Vp = \frac{V\theta}{4\theta}$$

In case of testcross families, maximization leads to

$$\frac{dL}{d\theta} = \frac{a_1}{1-\theta} + \frac{a_2}{\theta} + \frac{a_3}{\theta} - \frac{a_4}{1-\theta} = 0$$

This method of estimation of p leads to the formula which is in universal use for the backcross data of this type.

$$p = \frac{a_2 + a_3}{n} \text{ in case of coupling phage linkage and}$$

$$p = \frac{a_1 + a_4}{n} \text{ is case of repulsion phase linkage with}$$

standard error of p, $SE(p) = \sqrt{\frac{p(1-p)}{n}}$

2.5.2.2 Product ratio method

In this method the ratio of the product of extreme classes (a_1, a_4) to the product of middle classes (a_2, a_3) $\frac{a_1 a_4}{a_2 a_3}$ is equated with its expected value $\frac{m_1 m_4}{m_2 m_3}$ from which θ is estimated. In case of F_2 families the equation for estimation of θ takes the form as

$$Q = \frac{a_1 a_4}{a_2 a_3} = \frac{m_1 m_4}{m_2 m_3} = \frac{(2+\theta)\theta}{1 - 2\theta + \theta^2}$$

from which θ is estimated as

$$\theta = \frac{1 + Q - \sqrt{1 + 3Q}}{Q - 1}$$

The formula for the estimation of variance of θ, $V(\theta)$ is the same as that in case of method of maximum likelihood.

2.5.2.3 Emerson's method

In this method the θ in case of F_2 families is estimated as

$$\theta = \frac{(a_1 + a_4) - (a_2 - a_3)}{n}$$

and $V(\theta) = \frac{1 - \theta^2}{n}$

where $n = a_1 + a_2 + a_3 + a_4$. The variance of $p \cdot vp$ is the same as that of method of maximum likelihood.

2.5.3 Efficiency of Different Types of Families

Mather (1963) compared the relative efficiency of F_2 and back cross families in providing information about the estimate of linkage. He showed that F_2 gave twice more information about p, the recombination value as a backcross. At $p = 0.5$, backcross generation was as good as F_2 with incomplete dominance but when p is small F_2 gives more precise estimate of p than backcross in either phase (coupling or repulsion). When dominance is complete the F_2 was as good as backcross when p was very small in coupling phage (tight linkage) but provided less information in light repulsion phase.

2.5.4 Factors affecting the recombination frequency

The crossing over takes place in 4-strand phase. The recombination frequency between genes has been shown to be affected by various genetic and non-genetic factors like sex, maternal age, temperature, nutritional factor, chemical and radiation effect, cytoplasmic effect, background genotype effect, chromosome structure effect and centromere effect. Crossing over is usually completely suppressed in male in Drosophila. It also tends to vary with the age of the mother in an irregular way (Bridgs, 1927). Plough (1917) found high rate of crossing over with the increase of the temperature. Thoday and Bram (1956) and Lawrence (1958) showed factors determining the rate of crossing over residing in the cytoplasm. Increased concentration of metallic ions has shown to decrease the rate of crossing over. Antibiotics like mitomycine, actinomycin and others have shown an increase in crossing over (Suzuki, 1965a). X-ray irradiation increases crossing over in Drosophila (Whittinghill, 1937, Levin, 1955). The rate of crossing over between the specified loci varies with the change in the background genotype as has been shown in Drosophila (Levin and Levin, 1955). The change in chromosome structure occurs through loss, gain or rearrangement of a particular section of chromosome. Inversion brings about a complete absence of detectable single crossovers within the inverted regions in Drosophila and this acts like crossover suppressor (Startevant and Beadle, 1936). However, the inhibition in crossing over in one chromosome increases the frequency of crossing over in other normal non-homologous chromosomes. There is inhibition in the neighbourhood of the ends of rearrangement. The crossing over between loci when placed close to centromere has shown reduced frequency.

When more than two loci are involved in the same linkage group the loci are arranged in a unique order by putting those with the lowest amount of recombination next to each other. In case of three loci for which all recombination percentages are small the largest percentage tends to be the exact sum of other two but with the larger ones it falls off from this because of double crossover. A crossover in one region, however, interferes with crossing over in adjacent regions so that the percentage of double crossover tends to be less (at short distances) than expected from the product of total probability for the two components regions. This interference is measured as one minus the coefficient of coincidence which is the ratio of the observed double crossover to the expected double crossover. At fairly long distances the ratio approaches unity. The value of coincidence falls and the value of interference rises when the distance between loci decreases.

2.5.5 Pleiotropy

The manifold effects of a gene is called pleiotropy (Gruneberg, 1938). Pleioptropy can be genuine or spurious. Pleiotropy will be the genuine one if the two effects trace to two different primary action of the genes and spurious if they trace to the same primary mode of action or when one effect traces to the other. Considering the one gene-one polypeptide hypothesis the genuine pleiotropy does not exist unless the polypeptide is assumed and consists of a number of components each giving different effect. Thus in a physiological sense there can be no genuine pleiotropy and the apparent pleiotropy for which we have many examples is spurious. But then we can have situation where the same gene mutation showing susceptibility to one disease and resistance to another disease as in case of mutation for sickel-cell anaemia, which shows resistance to malaria and thus contributing differently to the selective value for an individual. This can be considered as a genuine pleiotropic effect from the view point of population genetics.

2.6 Extra Chromosomal Inheritance

The total hereditary material can be grouped into two: hereditary material of chromosome complement (genome) and hereditary material of extra chromosomal complement (plasmon). The plasmon includes plastogenes of plastids, kinetogenes of kinetosomes, chondriogenes of mitochondria and centrogenes of centrioles and together they are called plasmagenes. The genome comprises chromosomes which are the seats of the nuclear genes.

As the cell size and content are usually extremely different in the uniting gametes of higher organism except for the nuclei the consistently different progenies of reciprocal crosses have been due to some asymmetry in the gene content of gametes or with a delay of a generation in phenotypic expression. We have example of sex-determination mechanism (sex linkage) where the reciprocal crosses may differ because of differences in the number of representatives of one or more of the chromosomes in the uniting gametes. For example, if one sex is XX and the other is XY or XO, the offspring of the homogametic sex (XX) should be alike in reciprocal crosses but those of heterogametic sex (XY or XO) may be different. In many cases of reciprocal differences in F_1 or later generations, it has been found that all of the offspring reflect the maternal genotype instead of their own. This might be due to (i) maternal influence either on the egg or on later stages as in the seeds of flowering plants and the embryos and fetuses of mammals and (ii) if the gene products act in the oocyte with effects that carry through the reduced egg and to the zygote (effect of maternal genotype) and as a consequence of this maternal phenotype need not agree with maternal genotype as we can see in the case of snail, *Limnaea peregra* where dextral coiling is dominant over sinistral according to the maternal genotype but the direction of coiling cannot be a consequence of maternal phenotype. Also, the occasional occurrence of mixed offspring in which the sinistral gene can be shown to have mutated indicates that the character is due to gene action in the oocyte (Diver and Anderson-Kotto, 1938).

When two strains say A and B are crossed, the F_1 $(A \times B)$ is not similar to RF_1 $(B \times A)$ in phenotype and this reciprocal difference can be due to cytoplasmic influence or physiological relationship between a saprophyte and embryos or seeds it nourishes (maternal effect). The reciprocal differences observed above could be due to maternal effect or cytoplasmic factor or both. Maternal effect is transmitted through the egg cytoplasm for only one generation, since in the following generation new egg cytoplasm is formed according to the pattern of new maternal genotype. In contrast to the maternal effect, the reciprocal differences caused by extra chromosomal factors do not usually disappear after one generation but may persist as long as the extrachromosomal factors can prepetuate itself and thus the effects are assumed to be transmitted unchanged. The reciprocal differences thus can be transient or persistent. Transient reciprocal differences which can be attributed to maternal nutrition (environments) is prevalent in animal kingdom. In plants the maternal influence on early growth characters is usually traceable to the maternal effect in seed, the F_1 kernel and the difference between selfed and the F_1 lines will show the paternal effect. Maternal influence may be ephemeral or may persist throughout the lifespan of the individual. The persistent reciprocal difference can be due to: (i) some degree of maternal inheritance (through unequal contribution of cytoplasmic determinants from the female and male gemetes to zygote) and (ii) the appearance of a novel phenotype when a cross is made in one direction but not in other. Here the reciprocal differences involve characteristics not present in either parent. Reciprocal difference can appear for the first time in F_2 or backcross generation, for example male sterility in flax. The reciprocal difference is here due to interaction between chromosomal genes and maternal extra-chromosomal constituents.

2.6.1 Criterion for extra-chromosomal inheritance

Extra-chromosomal factors can be said to be involved if there is absence of segregation at meiosis or when the segregation fails to follow Mendelian laws. Cytoplasmic factor is involved if a trait in a progeny after repeated backcrossing to one of the parents shows the maternal character. Extra-chromosomal elements are supposed to be involved if they show specific induction and high rate of recovery of mutants when treated by mutagens. Homokaryons isolated from heterokaryons showing trait (s) changing according to its cytoplasmic types rather than according to the nuclear genotype points to the involvement of cytoplasmic factors. Finally, extrachromal factor is involved if it is shown to be associated physically with the trait. But then there are characters such as the configuration of the surface of a cell and the pattern of cortical granules in *Paramecium aurella* (Sonneborn, 1970) where the parental pattern serves as a template for that of offspring and is independent of nucleus or extrachromosomal DNA control. Kappa particles conditioning killer trait in paramecium, sigma, an RNA virus found in Drosophila making it sensitive to CO_2 and the sex ratio endosymbionts, a micoplasma, the presence of which in Drosophila causes the production of excess female offspring are examples of non-Mendelian heredity which have been turned out to be infectious and interpretable as due to virus and other parasites (Wright, 1968). The association between algae and fungi in lichens is one example of beneficial symbiosis. Symbionts may exhibit an infection like transmission with a hereditary continuity of their own. Characteristics controlled by genes in the cytoplasm include male sterility, chlorophyll formation and vigor.

Also, we have examples of delays in the expression of gene as described by Jenning (1940) in paramecium. Crosses between the two strains differing in size resulted in exconjugants similar to the parent from which they came and thus showed maternal effect but after 22 to 30 fission without autogamy the lines resulted were of the same size irrespective of from which parent they came and which was determined according to the common zygote nucleus. This cytoplasmic lag implies some autonomy of the cytoplasm before reconstitution under genic influence. Late morphological characters such as polychaeta in *Drosophila funetris* (Timofeeff Ressovsky, 1935), number of lumber vertebrae in mice (Russell and Green 1943) and fused tail in mice (Reed, 1937) are examples of delay of a generation in the expression of Mendelian genes and suggest oocyte effect rather than maternal effect.

There are a considerable number of cases of male sterility that behave as if transmitted by the cytoplasm where the persistent male sterility implies transmission down the straight female line. Also, there are a considerable number of examples in which persistent matroclinous heredity has been observed with respect to characters other than chlorophyll defect or male sterility. Crosses between species and sub-species have shown not only maternal influence but also in some cases superficial paternal inheritance. The reciprocal differences have been due to the fact that the cytoplasm of a species may aquire highly persistent properties that make it incompatible with the genome of a relative species (nucleo cytoplasmic incompatibility).

2.6.2 Cell organelleles

The cell organelleles include mitochondria, plastids, centrioles, kinetosomes, microsomes, lysosomes, golgibodies, etc. Non-Mendelian segregatin can alsobe due to gene conversion, somatic mutation, aneuploidy and paramutation.

2.6.2.1 Mitochondria

Mitochondria are the seats of respiratory activities. Mitochondria from all eukaryotes function similarly in ATP production and certain aspects of intermediary metabolism. Cytochromes which are involved in electron transport chain are located in mitochondria and chloroplast. Mitochondria together with plastids and glyoxysomes are involved in photosynthesis, amino acid metabolism and the fat synthesis. Mitochondria are also involved in responses to environmental and pathological stress and are themselves responsive to harmonal signal (Hanson and Day, 1980). The increase of free proline found in certain water-stressed plants involve the mitochondrially located enzymes such as proline oxidase, glutamate hrydrogenase and Δ' pyrroline-5-carboxylase dehydrogenase (Hanson and Hitz, 1982). Likewise mitochondrial malate dehydrogenase enzyme may affect plant viability. The cytoplasmic DNA differs from nuclear DNA in their different guanine and cytosine contents bands at different points in the centrifuge tube (Cscl density gradient configuration) and appears mostly as circular double stranded DNA similar to that found in bacterial prokaryotes. The mitochondrial DNA codes for only a few peptides. The amount of DNA in mitochondria is sufficient for coding perhaps 20 average sized protein in addition to mitochondrial ribosomes and transfer RNA. The cytoplasmic DNA is not fully independent of nuclear DNA for its replication, transcription, translation and enzyme function. Mitochondrial respiratory enzyme, cytochrome C oxidase, is composed of seven subunits, four coded by nuclear genes and three coded by mitochondrial genes. Mitochondrial genes are known to code specifically for at least six protein components of respiratory enzyme complexes. At least 90% of the mitochondrially located polypeptides are coded by nuclear genes. The mitochondria of most higher plants are maternally inherited. Mitochondria which is small in size in comparison to proplastid can be transmitted via ovule or pollen. Biparental inheritance of mitochondrial genome has been reported in *Brassica* (Erickson and Kemble, 1990). The trait cytoplasmic male sterility is encoded by mitochondrial genome. Recombination between mitochondrial genomes have been found in a majority of species combination (for example, in *Brassica, Nicotiana*) which opens the possibility of exchange of mitochondrial genes between species. Thus only a small part of a foreign genome can be introgressed in the cultivated species through mitochondrial recombination as in case of trait like cytoplasmic male sterility.

2.6.2.2 *Chloroplast*

Chloroplasts which contain chlorophyll are the sites of photosynthesis. Chloroplasts contain DNA, RNA and ribosomes. The chloroplast genome possesses sufficient DNA to code for 200 or more medium sized protein and probably could contain the genes for many of the plastid isozymes. Besides ribulose diphosphate carboxylase or nitrite reductase and multi protein structures such as ribosomes or ATPase complex, enzymes of EMP and oxidative pentose phosphate pathways many enzymes of TCA and amino acid metabolism and DNA synthesis are found in plastid as well. The carbon fixation enzyme, RUBP case, is polymeric enzyme composed of large and small subunits. The larger subunits are encoded by chloroplast DNA whereas the smaller subunits are encoded by the chromosomes. Although mitochondria and chloroplasts contain their own DNA, they depend on structural genes located in the nucleus for the synthesis of chlorophylls and carotenoids and some of the enzymes involved in the photosynthesis and respiration. The nuclear genes such as *iojap* affects the formation of chloroplast ribosonmes and thus prevents the translation of chloroplast mRNA into protein, causing colourless plants. Plastid inheritance in higher plants is predominantly or exclusively maternal but in genera such as *Medicago* and *Pennesetum* there is occasional biparental plastid inheritance and thus there is potential for selection for biparental inheritance (Smith, 1989). U.V. radiation can increase the biparental inheritance. Nuclear gene mutations have been found to produce the same effect. Nuclear genes in some of the higher plants (*Oenothera, Pelargonium*) also appear to be involved in controlling the extent of biparental inheritance. The most frequent behaviour of plastid populations in a fusion product is a random segregation of parental type in successive cell generation in accordance with Michaelis (1967) theory. Recently there is report of occurrence of plastid genome recombination in higher plants after somatic hybridization between representatives of *Nicotiana* and *Solanum*. Isolation of plastid recombinants and determination of the frequency of recombination is a problem because of the lack of plastid mutant and also the selective pressure cannot be created at the protoplast level. Although sexual crossing is the ideal way of transferring the nucleus of one species into cytoplasm of another, alloplasmic plants are produced by a series of backcrossing and thus is time consuming. The same can be achieved in one step in nature by a process called spontaneous androgenesis which is a form of apomixis in which the male gamete when put into the embryo sac develops directly into embryo without fertilization with the egg. However, the frequency of occurrence of spontaneous androgenesis is very low and further needs suitable marker for screening of large population of plants. All organelles such as mitochondria, chloroplasts, controles, kinetosomes, kinetoplasts differ in their genetic constitution and thus they provide the organism with great versality to survive drastic environmental changes. It has been found that both mitochondria and chloroplast DNA of corn share a sequence of 12,000 nucleotides which points to the possibility of transfer of section of DNA between organellele and nucleus and between different organelleles and this is possible by means of transposable element. The genetic elements that can move into or out of the main chromosome of a cell are called episomes.

Thus the interrelationship among nucleus, plasmon and all cell inclusions including viruses must be established before these special extra-nuclear effects can be considered profitable in plant breeding.

2.7 Sex-Determination

There are varieties of genetical mechanisms of sex determination in plants.

(i) The same cell has the ability to produce male and female gametes as in case of vast majority of monoecious flowering plants. But in case of corn, *Zea mays,* a monoecious crop plant a recessive gene called tassel seed (*ts*) in homozygous condition (*tsts*) produces pistillate structure with no production of pollen whereas there is another recessive gene called silkless (*sk*) which in homozygous condition produces staminate structure (the ear is without pistil). The gene '*ts*' is epistatic to gene '*sk*'. Thus a plant with *ts/ts, sk/sk* genotype is pistillate while one with ts^+/ts^+, *sk/sk* or ts^+/ts, *sk/sk* is staminate and the monoecious plants is of $ts^+/-$, $sk^+/-$-genotype.

(ii) There is single gene difference determining the sex as in case of dioecism. In asparagus, *Asparagus offlcinales,* the sex is under control of one locus with two alleles, *M* and *m*. *mm*, the homozygotic recessive is female whereas *Mm* is male and *MM* is super male. The sex in fruit papaya, *Carica papaya,* is under control of one locus with three alleles (M_1, M_2 and *m*) Storey (1941). *mm* is female, M_2m is hermaphrodite and M_1m is male. M_1M_1, M_1M_2 and M_2M_2 genotypes are presumably lethal as they do not appear in the progeny. From breeding view point the crosses $M_2m \times mm$ and $M_2m \times M_2m$ are useful as they produce maximum number of fruits bearing trees. There is occurrence of sex reversal which (in male and hermaphrodite) can be attributed to interaction between certain genetic factors (presence of modifiers) and environmental factors. The sex expression in cucumber is under control of three loci: *m*, *F* and *a,* respectively. The *m* locus controls the specificity of the stimulus to develop primordial staminate and pistillate flowers whereas *m+/-* genotypes are strictly declinous. The *F* locus determines the degree of female tendency. The *F* allele is partially dominant over F^+ and intensifies femaleness. This locus is strongly influenced by environmental factors (photoperiod, temperature, etc.) and background genetic changes. The *aa* at a locus intensifies male tendency and the male intensification is dependent on genotype at *F* locus. Combinations of genotypes at these three loci yield different basic sex types as shown in the following table.

Genotypes $m^+/-$, F^+/F^+, *a/a* and *m/m*, F^+/F^+, *a/a* are typically androecious. As the expression of sex is under influence of environmental factors,

Phenotype	Genotype		Locus
	m	*F*	*a*
Androecious	–/–	F^+/F^+	a/a
Monoecious	m^+/m^+	F^+/F^+	–/–
Hermaphrodite	m/m	F/F	–/–
Gynoecious	m^+/m^+	F/F	–/–

the sex type of cultivar will differ in different environments. In other words, there is instability of sex type. Further, as the genes at other than three major loci influence sex expression the genotype identical for three major genes will show marked variability in different genetic background and special care should be taken while breeding a particular type. Sex expression in many species is often strongly inclueced by photoperiod but no simple relationship exists. Either maleness or femaleness can be promoted by either short days or long days depending upon species. This is even true for different cultivars of cucumber (*Cucumis sativus*). The sex expression of cucumber line can be altered through application of chemicals such as silver nitrate

(AgNO$_3$) or ethephon (2-chloroethyl phosphoric acid). Spraying of gynoecious plants with AgNO$_3$ will induce male flower which then can serve as male parents in gynoeciousx gynoecious hybrids. Ethephon will promote female flower in both monoecious and andromonoecious cultivars. Gibberellic acids, the hormones particularly GA_4 and GA_7 have been found effective than GA_3 in inducing male flower in an otherwise strictly female flowering lines.

(iii) The chromosomal differences determine the sex as in case of *Melandrium album* where the pistillate plants are XX and staminate plants are XY. The chromosomal determination of sex is here similar to that observed in Drosophila and man. This model of sex determinatin is also found in *Actidia* (Kiwi fruit) where diploids and hexaploids produce male and female in the ratio of 1:1.

There is yet another type of sex determination in plants *Lychnis dioeca* (Mather, 1948) where the background genotype acts as a switching mechanism besides XY difference. The switching mechanism determines which of the two paths development will follow in a given individual.

In some organism such as *Bonellia,* the sex is determined by environment where the larvae develop into males in isolation but into females if exposed to secretions from females. In insects, Hymenoptera, the fertilized eggs from diploid individuals develop into female while unfertilized eggs from haploid individuals develop into males. Here the sex determination depends on the occurrence or other wise of fertilization in the other half of the sexual cycle. The various factors like environments, chromosomes segregation at meiosis or the occurrence of fertilization act as a switch in sex determination. In some organism the sex is dependent on the stage of development as in certain pulmonates in which each individual develops first into male and then transforms into a female or in some cases the reverse (Wright, 1968). Besides some monoecious species have developed genetic or cytoplasmic mechanism like self incompatibility (Lewis, 1954; Mather, 1973) which is as efficient in enforcing cross fertilization as would be exhibited by dioecism. This system as we shall see later is self perpetuating like X-Y system of sex-determination.

2.8 Expression of Qualitative Gene(s)

The phenotype of an individual is the result of interaction between genotype and environment. The genes constitute the genotype and genes could be qualitative, quasiquantitative or quantitative in action. The expression of particular gene depends on the internal environment and the external (physical) environment. The internal environment includes the effect of background genotype i.e. the gene effects at other loci, sex and the age of the individual and the cytoplasmic environment which constitutes the chemical environment. We have known examples where the expression of a gene at a particular locus depends on the gene effect at other loci, the gene products of other loci (a case of phenylketonuria disease in man). In case of sex limited trait such as milk yield the genes for which are although present in both sexes, but its expression is limited to one sex only. The same is with sex controlled traits like baldness and gout which are more common in one sex. The external environment includes not only temperature, light, nutrition but also pre-natal and post-natal environment provided by the mother. The example of role of maternal environment in determining the phenotype of an individual is more in animal kingdom. One can see interaction between the foetus and the maternal environment. The survival of a particular genotype as a result of blood group incompatibility between mother and offspring is one example. In primrose the flower is red at room temperature but white flower develops when the temperature is raised to above 86°F. Likewise there are examples of temperature dependent phenotypic expression in Drosophila in case of number of facetes in the Bareye mutant and fur color in mammals. There are examples of gene showing resistance or susceptibility reaction at different temperature (see chapter 33). In the extreme case the environment may modify the development of individual to an extent that its phenotype resembles the effect of a particular gene although the environment effect is not inherited. Such an individual is called phenocopy. The reverse is also true. Diabetes individuals are the phenocopies of normal individuals when treated with insulin which constitutes the environment. The genes thus rely on the environment for their own activity and effect. Thus speaking of gene activity in isolation i.e. without specifying or at least implying a particular environment has little relevance.

2.8.1 Penetrance and expressivity

The terms penetrance and expressivity have been used to characterize the expression of a particular gene. The term penetrance can be defined as the proportion of individuals having the expected phenotype determined by particular gene(s). Even a major gene can escape detection if there is incomplete penetrance which means that the affected phenotype is not always expressed or if the affected phenotype can result from other genes or from environmental causes. The penetrance can take values from 0 to 100%. The term 'expressivity' is referred to as the degree of expression of a particular gene in an individual. In general, the average degree of expression of a trait increases with the percentage of occurrence. Penetrance and expressivity are terms related to threshold characters (Stern, 1960). Waddington (1955) showed-multifactorial differences determining these characters. The continuous variability beyond the threshold for abnormality of a character can be explained using the terminology the developmental canalization of normal condition.

2.9 Estimation of Gene Frequency

When we are interested in estimating gene frequency we take a random sample of plants from the population and classify into dominant and recessive phenotypes. Now, if u is the frequency of A allele and v is the frequency of a allele the expected frequencies of the three genotypes, AA, Aa and aa in the population will be u^2, $2uv$ and v^2 at equilibrium but from the sample of n plants we have the observed number of dominant phenotypes say a_1 (the expected value being $u^2 + uv$) and a_2, the recessive phenotypes (the expected value being v^2). Therefore, the gene frequency of allele, a becomes

$$v = \sqrt{\frac{a_2}{n}} = \sqrt{\frac{nv^2}{n}}$$

and the frequency of A allele is calculated as $u = 1 - v$ as $u + v = 1.0$. The standard error for these estimates of gene frequency will be

$$SE(v) = SE(u) = \sqrt{\frac{1 - u^2}{4n}}$$

2.9.1 Incomplete dominance

In the above case we have assumed that A is dominant over a and thus Aa and AA are indistinguishable but then we can have situation when A and a show co-dominance and thus we have three genotypes showing three different phenotypes. The best example is the MN blood system in man where MM, MN and NN genotypes can be recognised. Suppose n individuals are sampled from a population. Considering diploid inheritance there will be $2n$ genes in the sample. Let the observed number of individuals in each group be a_1 a_2 and a_3, respectively. The estimate of the relative frequency of M allele would be

$$u = \frac{2a_1 + a_2}{2n}$$

and that of v allele would be

$$v = 1 - u = \frac{a_2 + 2a_3}{2n}$$

The standard error of the gene frequency estimate would be

$$SE(u) = SE(v) = \sqrt{\frac{uv}{2n}}$$

2.9.2 Multiple alleles

In case of ABO system of blood groups the four phenotypes OO, AA and AO, BB and BO and AB can be recognised although we can't distinguish between AA and AO individuals, and between BB and BO individuals. OO individuals can be recognised only in the absence of both A and B whereas AB individuals can be detected even in the presence of one another as A and B show codominance. The observed number of individuals in a sample size of n along with the expected frequency of different phenotypes are given below:

Phenotypes	Genotype	Expected frequency	Observed number
0	00	ω^2	a_1
A	AO, AA	$u^2 + 2u\omega$	a_2
B	BO, BB	$v^2 + 2v\omega$	a_3
AB	AB	$2uv$	a_4

where $n = a_1 + a_2 + a_3 + a_4$, u is the frequency of A allele, v is the frequency of B allele and ω is the frequency of 0 allele ($\omega = 1 - u + v$) and $u + v + w = 1.0$. The estimate of gene frequency is obtained through maximum likelihood method. The log likelihood expression is

$$L = a_1 \log \omega^2 + a_2 \log (u_2 + 2u\omega) + a_3 \log (V_2 + 2v\omega) + a_4 \log 2uv$$

Partial differentiation with aspect to be u and v will lead to equation for estimation of u and v.

$$\frac{\delta L}{\delta u} = -\frac{2a_1}{\omega} + \frac{a_2 + a_4}{u} - \frac{a_2}{u + 2\omega} - \frac{2a_3}{v + 2\omega} = 0$$

and

$$\frac{\delta L}{\delta v} = \frac{2a_1}{\omega} - \frac{2a_2}{u + 2\omega} - \frac{a_3 + a_4}{v} - \frac{a_4}{v + 2\omega} = 0$$

As these equations can not easily be solved the approximate trial value of u and v can be substituted in the above equation and if $\frac{\delta L}{\delta u}$ and $\frac{\delta L}{\delta v}$ come to zero then these trial values are the maximum likelihood estimate but if they are not zero efficient correction can be made in the trial values of gene frequencies.

From the observed number of phenotypes in different classes the trial gene frequency can be calculated as

$$\omega t = \sqrt{\frac{a_1}{n}}$$

$$ut = \sqrt{\frac{a_1 + a_2}{n}}$$

and $\quad vt = I - ut - \omega t$

The maximum likelihood estimates of u and v are

$$u = ut + \delta u$$
$$v = vt + \delta v$$

To estimate δu and δv we need to have the amount of information about u (Iuu) which is the rate of change of $\frac{\delta L}{\delta u}$ on u when u is near to its trial value. Likewise we calculate Ivv and Iuv.

$$Iuu = n \Sigma \frac{1}{m} \left(\frac{\delta m}{\delta u} \right)^2$$

$$Ivv = n \Sigma \frac{1}{m} \left(\frac{\delta m}{\delta v} \right)^2$$

$$Iuv = n \Sigma \frac{1}{m} \left(\frac{\delta m}{\delta u} \right) \left(\frac{dm}{dv} \right)$$

where m is expected frequencies of different classes. The equations for calculation of u and v are:

$$Iuu\, \delta u + Iuv\, \delta v = ut$$
$$Iuv\, \delta u + Ivv\, \delta v = t$$

From these above equations u and v are calculated and then finally gene frequencies are calculated. We can use the estimates of u and v as ut and vt and by iteration we can obtain new estimates of δu and δv and thus new estimates of u and v till maximum likelihood estimates are obtained.

2.10 Estimation of Genetic Diversity

Estimation of genetic diversity using quantitative traits is discussed in chapter 11.

References

Allard, R.W 1956. Formulas and tables to facilitate the calculation of recombination values in heredity. Hilgardia. **24**: 235–278.

Bridges, C.B. 1927. The relation of age of the female to crossing over in the third chromosome of Drosophila melanogaster. J. Gen. Physiol., **8**: 689–700.

Burt, A. and Bell, G. 1987. Mammalian chiasma frequencies as a test of two theories of recombination. Nature, **326**: 803–805.

Emerson, S. 1948. A physiological basis for some suppressor mutations and possibly for one gene heterosis. Proc. Nat. Acad. Sci., **34**: 72–74.

Gardner, EJ., Simmons, MJ. and Snustad, P.P. 1991. Principles of Genetics, John Wiley, New York.

Gruneberg, H. 1938. Some new data on the grey lethal mouse. J. Genet., **36**: 153–170.

Jennings, H.S. 1940. The cell and cytoplasm in protozoa. The cell and Protoplasm, AAAs Publ. No. **14**: 44–55.

Levine, L. 1969. The biology of the Gene. Mosby and Toppan.

Lewis, D. 1954. Comparative incompatibility in Angiosperms and Fungi, Adv. Genet., **6**: 235–285.

Mather, K. 1938. The measurement of Linkage in Heredity. Methuen, London.

Plough, H.H. 1917. The effects of temperature on crossing over and non-disjunction in Drosophila. J. Exp. Zool., **24**: 147–208.

Reed, S.C. 1937. The inheritance and expression of fused, a new mutation in the mouse. Genetics, **22**: 1–13.

Russell. WL. and Green, E.L. 1943. A skeletal difference between reciprocal F hybrids of a cross between two inbred strains of mice. Genetics, **28**: 87.

Schwartz. D. and Laughner, W.J. 1969. A molecular basis for heterosis. Science, **166**: 626–627.

Sinnoott, E.W., Donn, L.c. and Dobzhansky, T. 1958. Principles of Genetics. McGraw-Hill and Kogakusha.

Sonneborn, T.M. 1970. Gene action in development. Proc. Royl. Soc. London (B), 347–366.

Stansfield, WD. 1969. Schaum's outline of Theory and Problems of Genetics. McGraw-Hili, New York.

Stern, C. 1960. Yogt and the terms Penetrance and Expressivity. Amer. J. Human Genet., **12**: 141.

Strickberger, M. W. 1996. Genetics, 3rd edn. Prentice-Hall of India, New Delhi.

Sturtevant, A.H. and Beadle, G.W 1936. The relations of inversions in the X chromosome of Drosophila melanogastcr to crossing over and non-disjunction. Genetics, **21**: 554–664.

Timofeeff-Ressovasky, N.W. 1934. The experimental production of mutations. BioI. Rev., **9**: 411–456.

Waddington, C.H. 1955. On a case of quantitative variation on either side of the wild type. Zeit. Abst. Vererb., **87**: 208–228.

Wright, S. 1927. The effects in combination of the major color factors of the guinea pig. Genetics, **12**: 530–569.

Wright, S. 1929. Fisher's theory of dominance. Amer. Nat., **63**: 274–279.

Analysis of Quantitative Traits

A quantitative trait is one which is under many genes control, all the genes having small and similar effects, acting in additive fashion and is more influenced by the environment. Quantitative variation can be more easily understood by the following example. With one locus two alleles system and complete dominance F_2 population will consist of 3 genotypes and 2 phenotypes, respectively. But with additive effects of genes there will be 3 phenotypes with the frequency distribution as 1:2:1 as shown in Figure 3.1a. With the increase of number of loci determining a particular character there will be an increase in the number of genotypes and phenotypes in the F_2 in the manner given in Table 3.1 and the phenotypic distributions would look like as shown in Figure 3.1 (a.d). There would be 5,7 and 9 phenotypic classes with 2,3 and 4 genes, respectively and with k genes there would be $2^k + 1$ classes. The frequencies of different genotypes will be in proportion to the coefficients in the expansion of $(p + q)^{2k}$ which takes the form as

Table 3.1 Number of genotypes and phenotypes in F_2 corresponding to the number of loci segregating assuming additive effects and with complete dominance

Number of loci	Number of genotypes	Number of phenpotpes (additive effect)	Number of phenotypes (complete dominance)
1	3	3	2
2	9	5	4
3	27	9	8
4	81	17	16
5	243	33	32
10	59,049	1025	1024
k	3^k	$2^k + 1$	2^k

$$p^{2k} + 2kp^{2k-1}q + ... + 2kpq^{2k-1} + q^{2k}$$

which in essence describes a biomial distribution. k is the number of loci and p and q are the frequencies of allele A and a, respectively with $p + q = 1.0$. Thus considering one locus with 2 alleles $(p + q)^2$ will take the form as

$$p^2 + 2pq + q^2$$

where p^2, $2pq$ and q^2 being the frequencies of genotypes AA, Aa and aa, respectively. With two loci, $(p + q)^4$ will become

$$p^4 + 4p^3q + 6p^2q^2 + 4pq^3 + q^4$$

It also indicates that the genotypes having either all increasing or decreasing alleles will have their probability of occurrence the least with the increase of number of loci. The value of individual genotype will be equal to $\sum_{i=1}^{k} d_i$, where d_i is the additive effects of the ith loci (it may include dominance as well as epistatic effects). So long as the number of genes controlling a character is few and each gene having large effects as in case of Mendelian genes in classical genetics the phenotypes and in turn the genotypes can be recognised. The phenotypic distinction reveals the underlying genotypic differences. But as the number of genes governing a character increases, assuming each gene having small, equal, cumulative effect (without considering intra-allelic and inter allelic interaction and with no interaction with environment when these are there) there would be more classes with smaller difference between them which would be difficult to distinguish. With 10 or more loci segregating and with dominance, the distribution of phenotypes is difficult to distinguish from that of genes showing additive effects. Considering error of measurement similar to class difference it will be difficult to recognise the discontinuity. Moreover, as the quantitative trait is subject to environmental variation, its effects will be the bluring of edges of genetic discontinuity (Figure 3.1d) and thus variation would look like continuous and takes the shape of a normal frequency curve (Figure 3.2). The curve is high in the centre which represents the high frequencies of the most common combination of genes and which tapers off equally at both extremes for the rarer combinations and this is symmetrical. The normal distribution curve for quantitative trait can be explained by central limit theorem of probability. It states that the mean of sample of k drawn from any population (continuous or discrete) with mean, μ, and finite variance, σ^2, will have a distribution that approaches as k becomes infinite, the normal distribution with mean, μ, and variance, σ^2/k. In other words if a character is a result of the sum of large number of independent effects, all of which are of the same general order of magnitude then this character will be normally distributed. As these independent effects could be due to genetic or environmental factors, so the mere fact that a trait is normally distributed does not say anything whether it is due to genetic or environmental influence. A continuous distribution does not necessarily mean that segregation of a large number of polygenes is involved. Major effect genes may be modified by the action of minor genes to produce a continuous distribution of resistance levels and as few as 3 or 4 controlling genes may be present (Thompson, 1975).

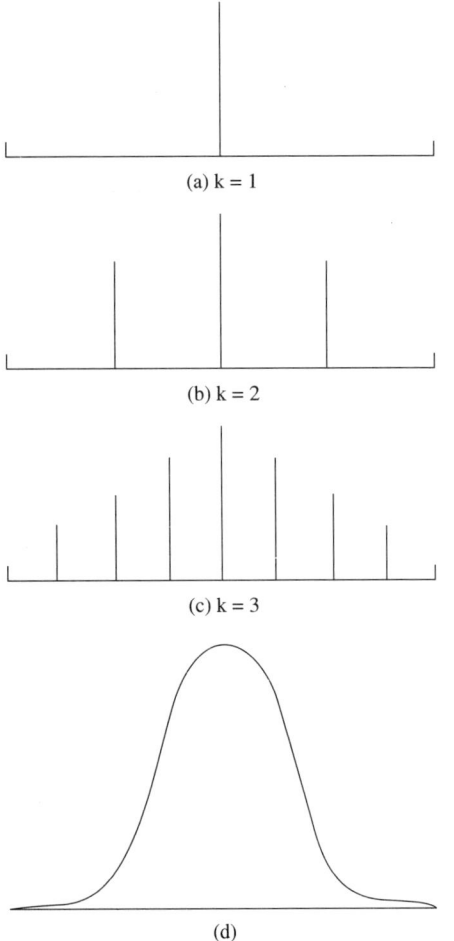

Fig. 3.1 (a-d) Shape of the binomial distribution of differing k, the number of loci.

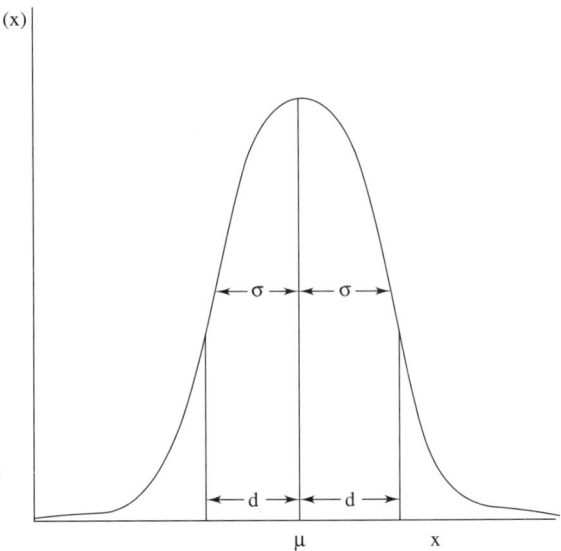

Fig. 3.2 Normal distribution curve.

In statistical terms character is a variable and a numerical variable which takes different values with different probabilities is called a random variable. Variables are of two types: discrete variables (all qualitative/quasi-quantitative traits which only take the integral values 0, 1, 2) and continuous variables which result from metrical measurement of individuals and can therefore, take any value within a certain range.

3.1 Theories of Quantitative Variability

Although Mendel in 1865 proposed the laws of heredity, the statistical nature of continuous variations became clear after the rediscovery of Mendel's law by Correns, de Vries and Tschermak in 1900. During these years the study of continuous variation and of evolutionary processes was being developed first by Darwin and later by Galton, Pearson and others. Darwin in his theory of evolution (1859) had emphasized the importance of small cumulative changes in evolutionary process. Galton in 1889 distinguished sharply between the inheritance of continuous and step wise variability. Galton and Pearson's work with human physical data showed continuous variation to be partly heritable. Soon after the rediscovery of Mendel's law, controversy arose between the biometricians led by Pearson and the Mandelians led by Bateson, as the former regarded continuous variation as implying discontinuous genetic variation. The discordance arose because of failure to discriminate sharply between the contributions of heredity and environment. The phenotype of a quantitative character is a consequence of both genetic and environmental sources. But the controversy ended when Yule in 1906 observed that the continuously varying characters could be inherited in a Mendelian way provided there were a large number of genes involved and provided each had a small effect. Thus Mendelian theory could apply to study continuous variation. Johannsen (1903, 1909) while working for his pure line theory demonstrated that the character seed weight of a number of lines in beans (*Phaseolus vulgaris*), each maintained by self fertilization, displayed quantitative variation. The variation was due to both heritable and non-heritable factors. The variation among individuals plants within a line will be all environmental whereas the variation between individuals from different lines could be due to heredity and environment. The variation from the two causes could be distinguished by progeny test and there is more than one factor involved in the determination of this character. Thus his experiments demonstrated that although the phenotypic variation is continuous, the underlying genetic variation is discontinuous and the continuous phenotypic variation is realized through the action of environment.

At the same time Nilsson-Ehle (1909) working with wheat and oat, explained the variation in grain colour in F_2 and F_3 generation from red-glumed versus white-glumed cross with the help of three independent hereditary factors, each having similar effect and acting in additive fashion. Thus it was a complete Mendelian analysis of quantitatively varying characters. East (1910) demonstrated similarly that the variation in grain colour (ranging from yellow to white) was determined by two independent factors.

On the basis of their results Nilsson-Ehle and East independently put forward the multifactorial theory (or Mather's polygenic concept) for explaining quantitative variation. East (1916) listed eight points based on the results from the study of F_1 and F_2 and other selfed generations derived from the cross between two varieties of *Nicotiana longiflora* that differed in corolla tube length.

1. F_1 from the two pure breeding homozygous/lines (P_1 P_2) should be comparable to the parental strains in uniformity.
2. Like parental strains all the variation within F_1 population should be environmental.
3. The variability of F_2 population should be much greater than the F_1 population.
4. Parental types P_1, P_2 and F_1's should be recovered from large F_2 population.
5. In certain cases individuals should show a more extreme deviation (transgressive variation) than is found in the frequency distribution of either parent.
6. Individuals from different points in frequency distribution differing markedly in their means and modes.

7. Individuals from the same or different points in the F_2 frequency distribution curve should give F_3 populations differing in variability.
8. In F_3 and other succeeding selfing generations the variability of any family can never be greater than the population from which it came.

Thus, like the Mendelian genes, the genes for quantitative traits should show the segregation, dominance, epistasis and linkage. In general it is almost impossible to identify individual genotypes among the offspring of a cross. Our inability of easily and neatly detecting and following the individual effects in the phenotype gets complicated because of the variation arising from segregation and recombination at other loci (background genetic variation) and the environmental variation but then these problems can be overcome by growing large number of a particular genotype and use of marker gene. One reliable method of detecting the presence of a gene for quantitative trait on a particular chromosome would be to determine a linkage relationship between that trait and a distinguishable Mendelian locus (so called marker locus). Sax (1923) reported linkage between seed size showing continuous variation and seed colour under control of a single major gene in *Phaseolus vulgaris*. Later Rasmusson (1935) reported linkage between flower color a single gene determined trait and flowering time under control of polygenes. The work of Penrose (1938, 1946) is considered as being pioneering in this field. He attempted to detect linkage between a marker locus determining a qualitative character and another locus which affects meristic (graded) character. Then work started on determining, locating and measuring the specific components of the polygenic system and their numbers (effective factors) in *Drosophila melanogaster* by chromosome assay using marker gene (Mather, 1942; Mather and Harrison, 1949; Breese and Mather, 1957, 1962; Thoday 1961, 1979 and Spickett and Thoday 1966). Wehrhahn and Allard (1963) used inbred backcross line method to detect and measure the effects of the individual genes involved in the inheritance of ear-emergence in wheat. Law (1966, 1967) used chromosome substitution, recombination and inbreeding method to locate genes controlling yield and other quantitative traits in wheat. Now molecular marker technology is being used to determine, locate and estimate the effect of individual quantitative trait loci, QTLs.

3.2 Description of Normal Distribution Curve

The quantitative trait follows a bell-shaped normal distribution curve (Figure 3.2), which is continuous and symmetric and where the frequency function of a normally distributed variable, x with mean μ and variance σ^2 is

$$f(x) = \frac{1}{\sigma\sqrt{2\pi}} \exp\left(\frac{-(x-\mu)^2}{2\sigma^2}\right)$$

Here the factor $\frac{1}{\sigma\sqrt{2\pi}}$ ensures that the area under normal curve is 1 and $f(x)$ the probability density function, depends on the departure of x from its mean (μ) value expressed as proportion of standard deviation, σ, and x has a value of $-\infty < x < \infty$. The area under the density function between any two points say X_1 and X_2 that is to say the integral of the function between them represents the probability that the random variable, X, will lie between these two values.

$$\text{Prob } (X_1 < X \leq X_2) = \int_{x_1}^{x_2} f(X)\, dx$$

If dx is very small measurement in x, the density function will be practically constant between $x \pm dx$ then the probability that X will lie in this small interval is very nearly $f(x)\, dx$ which is the area of a rectangle with height $f(x)$ and width dx. Thus $f(x)$ may be thought of as representing the probability density at x. There is a relationship between the frequency of certain values (expressed as deviation from the mean, SD) and the area occupied by these values in the normal curve. If the curve is broken into sections (Figure 3.3) and the area measured, it can be shown that 68% of the area of the normal curve falls within $\pm 1\sigma$ of the mean whereas 95% and 99 per cent of the area within $\pm 2\sigma$ and $\pm 3\sigma$ of the mean, respectively.

The characteristics of normal curve which are also called the descriptive measures and which in turn will be the properties of a quantitative trait are as follows:

1. Measures of central tendency (mean, mode and median),
2. Measures of dispersion or variability (range, variance, standard deviation, standard error),
3. Measures of symmetry (skewness) and peakedness (kurtosis).

3.2.1 Mean

The arithmetic mean or simply the mean is the first degree statistic. The sample estimate of mean ($\hat{\mu}$) is

$$\overline{X} = \sum_{i=1}^{N} X_i/N$$

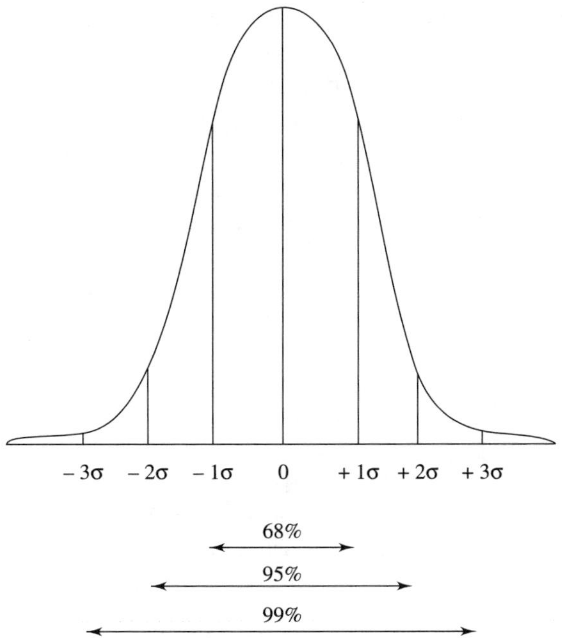

Fig. 3.3 Relation between the standard deviation and the area under the normal curve.

where X_i is the measurement of ith individual ($i = 1, 2, \ldots N$) and N is the sample size. In case of frequency data $\overline{X} = \Sigma f\, Xi/N$, where f is the frequency of the ith individual. We can have different means with same variance as shown in Figure 3.4.

The harmonic mean of a set of positive variables $X_1\, X_2, \ldots X_N$ is defined as to be the number X_n whose reciprocal is $\dfrac{1}{X_n}$ is the arithmetic mean of $\dfrac{1}{X_1}, \dfrac{1}{X_2}$, $\ldots \dfrac{1}{X_N}$ the reciprocals of the variables.

$$\frac{1}{X_n} = \frac{1}{N}\sum_{i=1}^{N}\frac{1}{X_i}$$

It is a useful measurement when the quantity of importance exerts its influence in the denominator rather than numerator. For example, as we will see later, the random genetic drift is proportional to the reciprocal of the population numbers and the effective population size which is the harmonic mean and is calculated as the reciprocal of the average of the reciprocals of a series of population numbers and if the population size fluctuates its effective population size will be dominated by a few bottlenecks.

The geometric mean X_g of it set of N positive variables $X_1, X_2, \ldots, X_N$ is defined as the positive Nth root of their product.

$$X_g = (X_1\, X_2\, X_3\, \ldots\, X_N)^{1/N}$$

to the reciprocal of the population number. It can also be calculated as the antilog of the average of the logarithms of the variables. The distribution of variable can be asymmetrical but the distribution of log variables is often nearly symmetrical and in this condition the use of geometric mean is more appropriate than arithmetic mean.

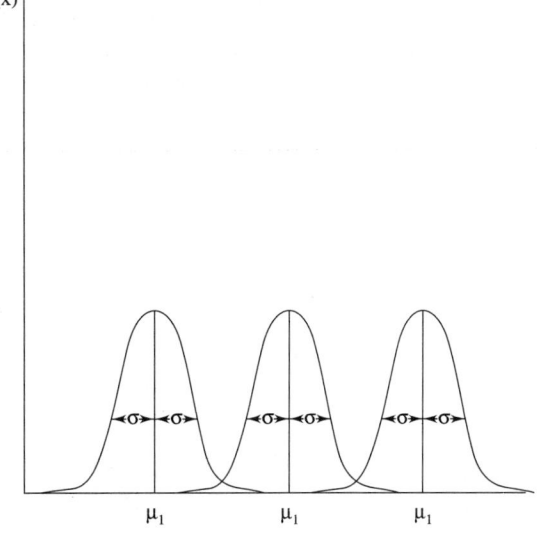

Fig. 3.4 Normal distributions with different means but same variance.

For the given variables $X_1, X_2, \ldots X_N$, arithmetic mean > geometric mean > harmonic mean. Thus the arithmetic mean will be dominated by the large numbers whereas geometric mean will be dominated by the small numbers.

3.2.2 Mode and Median

Mode is the value with highest frequency in the distribution whereas median refers to the halfway point in the distribution when the measurements of individuals are arranged in either ascending or descending order. The distribution with two or more modes (peaks) are called bimodal and multimodal distribution, respectively.

These statistics have not statistical or genetical significance. If the distribution is highly asymmetrical or has more than one peak, median is more appropriate than mean.

3.2.3 Variance and Covariance

These are called second degree statistics. The sample estimate of variance ($\hat{\sigma}^2$) is

$$S^2 = \Sigma (X_i - \overline{X})^2/N - 1$$

or $$S^2 = \Sigma f (X_i - \overline{X})^2/N - 1$$

in case of frequency data. The estimate of variance can not be negative.

The shape of the curve is determined by the amount of variance. We can have same mean but different variances as shown is Figure 3.5.

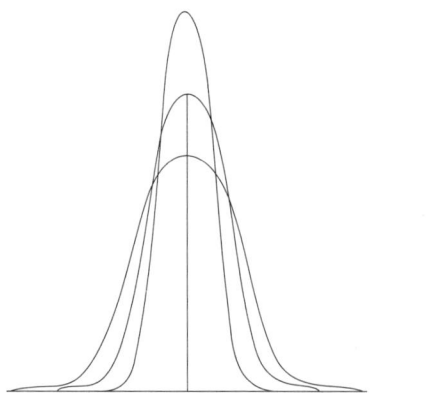

Fig. 3.5 Normal distributions with different variances but same mean.

The covariance between two variables X and Y is estimated as

$$\text{Cov}(X, Y) = \Sigma (X_i - \overline{X})(Y_i - \overline{Y})/N - 1$$

The covariance provides a measure of association between two variables.

3.2.4 Range

It is the difference between largest and the smallest measurements of the individuals in the sample. It is a poor measure of the dispersion and thus has no statistical or genetical usefulness but for small samples from the normal population the range can be shown to be as efficient for estimating standard deviation as is the sample standard deviation

3.2.5 Skewness and Kurtosis

The non-normality of any distribution is measured by its skewness and kurtosis and these are called 3rd and 4th degree statistic, respectively.

Skewness takes sign, positive or negative (Figure 3.6), according to whether there is an excess of individuals of low values close to mean with fewer higher values spread far above it or the reverse or in other words whether the curve has a longer tail on the right side or the curve is more elongated to the left side (Figure 3.6). Figure 3.7 shows frequency distributions differing in skewness. There is an empirical relationship between mean, mode and the median. In case of symmetrical distribution

$$\text{mean} = \text{mode} = \text{median}$$

In asymmetrical distribution

$$\text{mean} > \text{median} > \text{mode}$$

when skewness is positive, and

$$\text{mean} < \text{median} < \text{mode}$$

when skewness is negative.

For a distribution of moderate skewness

$$\text{mean} - \text{mode} = 3 \, (\text{mean} - \text{median})$$

But a better test of symmetry is made by another statistics g_1 called Karl Pearson's coefficient of skewness which measures the direction and the extent to which the distribution is distorted from symmetrical distribution. The sample estimate of skewness is

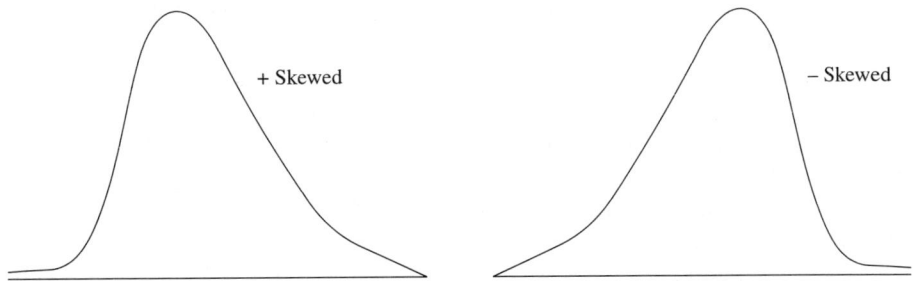

Fig. 3.6 Positively and negatively skewed distributions.

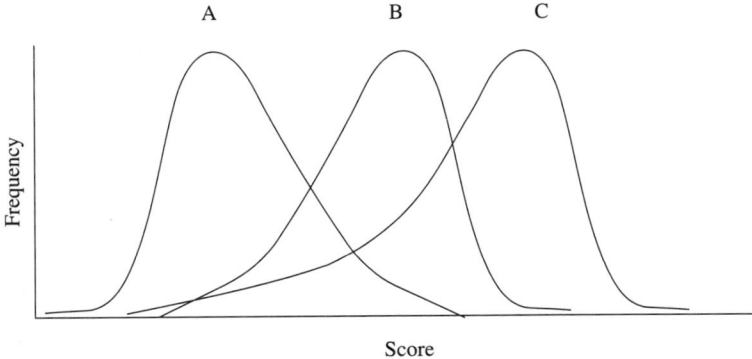

Fig. 3.7 Three frequency distributions differing in skewness.

$$g_1 = (\Sigma(X_i - \overline{X})^3/N)/(\Sigma(X_i - \overline{X})^2 N)$$
$$\times \sqrt{\Sigma(X_i - \overline{X})^2_N}$$

For a symmetrical distribution $g_1 = 0$. There are a number of factors for a quantitative trait not being symmetrical about its mean and these are those that prevent the random distribution of the segregating genes or the genotypes and are discussed in the Chapter 19.

The fourth degree statistics, kurtosis, is a statistical measurement of the relative peakedness (or flatness) of a distribution in terms of the projected frequencies in the central peak in relation to the density of the observations falling in its shoulders and tails. The sample estimate of kurtosis is

$$g_2 = \left(\frac{(X_i - \overline{X})^4 /N}{(\Sigma(X_i - \overline{X})^2 /N)} \right)$$

For a normal distribution $g_2 = 3$, so in order to give values of positive or negative kurtosis depending upon whether too many individuals are close to mean and at tail ends or the reverse, this normal distribution value of 3 is subtracted from the g_2. In other words, $g_2 - 3$ will be a measure of the kurtosis. When the curve is more peaked than normal distribution (Fig. 3.8—solid curve), g_2 will be greater than 3 and $g_2 - 3$ will be + ve and the curve is called leptokurtic. When the curve is flatter than the bell shaped normal curve as shown in Figure 3.8 (dotted line) its g_2 will be less than 3. i.e. $g_2 - 3$ will be negative and the curve is called platykurtic. The normal distribution is spoken of as mesokurtic which means that it falls between leptokurtic and platykurtic distributions. Figure 3.9 shows frequency distribution differing in kurtosis.

The sample estimates of the means, variances, skewness and kurtosis are called statistics whereas the corresponding quantities for the population are called parameters. Each of these statistics can be described which we will see in the later chapters as sum of heritable and non-heritable component and interaction variation. Thus heritable component

3.8 Biometrical Genetics–Analysis of Quantitative Variation

in turn can be defined in terms of gene action and interactions and the non-heritable components appropriate to the experimental design.

There are traits which are polygenic in nature but show discontinuous or quasi continuous distribution (Gruneberg, 1952). There are three types of quantitative traits: metric, meristic and threshold traits. Threshold traits are determined by multiple genetic and environmental factors.

3.3 Threshold Variability

In this type of variability the distribution consists of two almost or quite descrete parts, a weak normal distribution and a class of complete or incomplete failures of the character. The array of the genes determining a character not only may have a threshold below which it wholly fails to express but also a ceiling at which it is so fully developed that further increase in the factors produces no additional effects. The threshold arises from mathematical necessity (as in case of distribution of percentage occurrence in which there is an important class at 0% or at 100%) but often for physiological reasons, the threshold being that point at which a self regulatory (or homeostasis) process breaks down(Wright, 1968). The character such as resistance to diseases and insects in plants

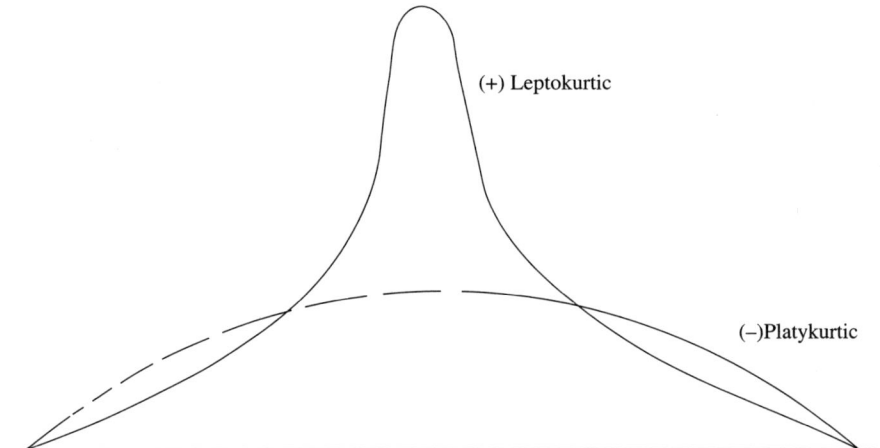

Fig. 3.8 Lepokurtic and platykurtic distributions.

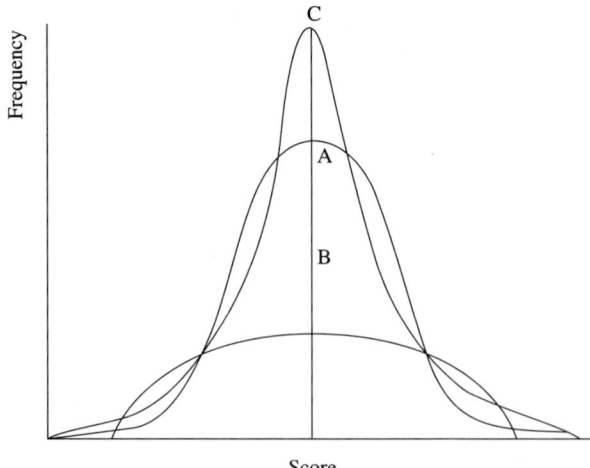

Fig. 3.9 Three frequency distributions differing in kurtosis.

shows this type of variability. Thus the threshold traits are either present a absent in any one individual. There are, characters like viability and fecundity (fitness) which constitute the selective value of an individual show this above type of variability. Infact, the selective value, a complex character, is a function of a number of characters besides involving viability, fecundity, mating, life span and size. Natural selection is usually directed towards an optimum type where phenotypic performance is close to the average. In such case genes with constant effect on a measurable character increases the total fitness of an individual in combination below the average (optimum) but decreases fitness above the average. Selective value as a character usually superimposes interaction effects of the most extreme sort upon whatever interaction effects there may be among genes with respect to the underlying characters. Selective value of characters with respect to natural selection usually falls off from an intermediate optimum which implies the existence of numerous selective peaks. If the selective value falls off directly with deviation from the mean in case of a variable with a normal distribution, the distribution of selective value is half normal distribution which is asymmetrical but as it is more likely, it falls off approximately as the square of the deviation from mean it is highly asymmetrical (Wright 1935).

3.4 Meristic Variability

Meristic trait is one for which the values are distinctly separated and only a limited number of values constitute the range of measurement i.e. the range is small. In such trait the increments in measurements are comparable in size to the difference between individuals. A meristic (or graded) character has a succession of thresholds and ceiling for each gene. The interval between thresholds, corresponds to physiologically equivalent steps. Thus there is a stepwise distribution between phenotypes. Such a character often shows normal or near normal distribution but not infrequently a particular number of genes may be present in unduly high frequencies. Meristic variability differs from continuous variability in the presence of natural scale. The character, number of kernel rows in maize, chaeta/ bristles number in Drosophila and number of neck vertebrae within a species of fish and litter size in mouse show this type of variability. The meristic variability has been shown to be due to both genetic and environmental causes. The character showing this type of variability is studied by taking counts. Characters showing quasi-quantitative variation can be analysed by studying means, variances, etc.

3.5 Poisson Distribution

The following can be said to follow Poisson distribution:
 (i) Distribution of random events where the number of possible events is much larger than mean.
 (ii) Number of bacterial colonies on plates or microbial spores in field plot where the requisite of independence is met.
 (iii) Variation in density of large organisms that are moving about in a uniform environment is wholly independent of each other. For example, insects count in a field plot.

Thus organisms or events which are distributed at random in space or time follow Poisson distribution.

With n large and $p = q = 0.5$ the genotypic frequency of AA, Aa and aa will be 0.25, 0.5 and 0.25, respectively but if $p \neq q = 0.5$ i.e. if p is less than 0.5 then the product of the expansion appears skewed and the degree of skewness is a function of the difference in the frequencies of A and a alleles. With p very small (p roughly comparable to $1/n$) and n large, the different genotypes will appear in the form of an extremely skewed distribution known as Poisson distribution which takes the form as

$$e^{-\mu} \frac{\mu^r}{r!}$$

where μ is the mean value of the distribution, e is the base of natural logarithm (2.718) and is a constant and $r = 1, 2, 3, 4$ events. This expression upon expansion takes the form as

$$P(X = r) = \frac{1}{e^{\mu}} \left[\frac{\mu^0}{0!}, \frac{\mu^1}{1!}, \frac{\mu^2}{2!}, \frac{\mu^3}{3!} \ldots \right]$$

i.e. the probability of no event $(r = 0)$ is $\frac{1}{e^u}$, the probability of one event $(r = 1) = \frac{u^1}{e^u}$; the probability of two events $(r = 2)$ is $\frac{\mu^2}{2!e^u}$. This distribution differs from binomial distribution in that r can take any value whereas in binomial r cannot be greater than $n + 1$. The sum of all the terms in the Poisson series from $r = 0$ to ∞ is unity. The characteristic of a variable, Y, following Poisson distribution is that the mean of the distribution (μ) equals to its variance (σ^2), i.e. $\mu = \sigma^2$. For testing the null hypothesis that the data conform to the Poisson distribution, χ^2 test of goodness of fit is used with $t - 2$ degrees of freedom. An additional degree of freedom is deducted from t, the number of classes as the mean frequency is an estimate of the average rather than a true expected value.

3.6 Modifications of Binomial and Poisson Distributions

When there is correlation between the occurrence of the events there will be found clusters consisting largely or wholly of one of the alternatives. Such distribution where a continuous random variable takes values between 0 and 1 is said to follow the Beta distribution with parameters p and q with the probability density

$$F(x) = \frac{x^{p-1}(1-x)^{q-1}}{\beta(p, q)} \quad 0 \le x \le 1$$

where $\beta(p, q)$ the integral of the numerator between 0 and 1 is the complete β function

$$\beta(p, q) = \frac{\Gamma(p)\Gamma(q)}{\Gamma(p+q)}$$

This distribution can be observed when the length of an insect can be considered as the sum of head, thorax and abdomen which are certain to be highly correlated because of common effect of general size. As the size is a polygenic trait, the total length may be expected to vary normally but if the common factor does not vary normally this will give rise to distribution deviating from the normal. If Y_1 and Y_2 are independent χ^2 variates with f_1 and f_2 degrees of freedom, respectively then $Y_1/Y_1 + Y_2$ follows the beta distribution with $p = \frac{1}{2}f_1$ and $q = \frac{1}{2}f_2$ with

$$\text{mean} = \frac{p}{pq}$$

and

$$\text{variance} = \frac{pq}{(p+q)^2(p+q+1)}$$

3.6.1 Gamma distribution

Poisson distribution showing a tendency towards clustering of events as a result of correlation implies that the mean expectation varies from sample to sample. These means are distributed from 0 to $+\infty$ and follow a gamma distribution and has the probability function.

$$P(x) = \binom{x+n-1}{x} P^n Q^x$$

$$= \binom{n}{x} P^n (-Q)^x$$

where

$$\binom{n}{x} = \frac{n(n-1)\ldots(n-x+1)}{x!}$$

This is the appropriate term in the binomial expansion of $P^n(1-Q)^{-n}$. The distribution for this reason is called the negative binomial distribution. The distribution of insects population on two dates in crop field serves an example. One the first date the insect population will show Poisson distribution whereas on the second date the insect counts show deviation from Poisson distribution owing to clumping effects. Most deviations from Poisson distribution increase Y, the number of the extreme classes which will thus result in an increase of the variance. The testing of null hypothesis is done using χ^2 with $t - 3$ degrees of freedom. The mean and variance of the negative binomial distribution are as follows:

$$\text{Mean} = n\,Q/P$$

and

$$\text{variance} = n\,Q/P^2$$

If n is an integer, the negative binomial distribution can be interpreted as the sum of the n independent geometric variables.

3.6.2 Hypergeometric distribution

Considering 2 alleles with $p(A) = q(a) = 0.5$ in the gene pool of N, where N is a small number, if n alleles are taken out at random the number of alleles in the sample will follow a binomial distribution if the alleles are replaced after they are taken out from the genepool and the probability will be $p = A/N$, but if the alleles are not replaced the proportion of A allele in the gene pool will change from one trial to another and will follow a hypergeometric distribution. This is a case of uniform negative correlation between components.

$$P(x) = \frac{\binom{A}{x}\binom{N-A}{n-x}}{\binom{N}{x}}$$

$x = 0, 1, 2,$ smaller of R or n

where $\binom{R}{x} = \frac{R!}{x! \, R-x!}$

The variance of this distribution (variance = $npq\left(\frac{N-n}{N-1}\right)$ is smaller relative to the means (np) than that of binomial distribution of the same size.

The segregation in polyploids is a case of this type of distribution. The gametic array in a hexaploid of constitution A^4a^2 will be as follows:

$$A_{aa}\binom{4}{1}\binom{2}{2}/\binom{6}{3} = 0.20$$

$$AAa\binom{4}{2}\binom{2}{1}/\binom{6}{3} = 0.60$$

$$AAA\binom{4}{3}\binom{2}{0}/\binom{6}{3} = 0.20$$

This distribution can be applied to cases in which there are more than two alternatives as in case of multiple alleles in polyploids.

$$fx_1 \, x_2 \, x_n = \binom{n \, q_1}{x_1}\binom{nq_2}{x_2}\binom{nq_3}{x_3}\binom{nq_n}{x_n}/\binom{n}{k}$$

3.7 Biomodal and Multimodal Distributions

The normal distribution is a unimodal (single mode or peak or hump) distribution whereas multimodal distributions have two or more than two peaks. If the several peaks are not due to sampling distribution i.e. if they are real, then the different peaks reflect the corresponding probability distribution and thus it can be said that the distribution is a composite one made up of several unimodal distributions. There are many examples of bi-modal distributions shown by quasi-quantitative traits (quantitative characters determined by a single pair of genes). The bimodal distribution often indicates a mixture of two populations but it can arise as in case of population showing dimorphism. A human population can show this type of distribution for height because of sexual dimorphism. If two populations with unimodal distributions are mixed together the resulting distribution may have two peaks if the two distributions are well separated (i.e. if the two species are widely different) but if the two distributions overlap (i.e. if the two populations are closely similar except in some respect) the resulting distribution may have rather only one broad peak. It is often difficult to decide whether the population is homogeneous or whether it is a mixture of two or more distinct sub-populations(heterogeneous). The variability in a homogeneous population is due to the operation of multiple factors. A separation according to high or low values of anyone of these would give two distributions with different mode (or means). If it is only when the differences due to one factor play an overwhelming role that heterogeneity is recognizable.

References

Allard, R.W. 1960. Principles of Plant Breeding. Wiley and Toppan.

Bateson, W. 1902. Mendel's Principles of Heredity London.

Breese, E.L. and Mather, K. 1957. The organization of polygenic activity within a chromosome in Drosophila. 1. Hair characters. Heredity, **11**: 373–395.

Bulmer, M.G. 1980. The Mathematical theory of Quantitative Genetics. Oxford University Press. Oxford.

Correns, C. 1900. Mendels Regel uber das verhalten der Nachkommenschaft Rassen Bastarde. Ber. Deutsch. bot., Ges., **18**: 158–168.

Darwin, C. 1859. The Origin of Species. John Murray, London.

East, E.M. 1910. The Mendelian interpretation of variation that is apparently continuous. Amer. Nat., **44**: 65–82.

East, E. M. 1915. Studies on size inheritance in Nicotiana. Genetics, 1:164-176. Bateson, W. 1902. Mendel's Principles of heredity. London.

Facloner, D.S. 1989. Introduction to Quantitative Genetics. 3rd edn. Longman, Burnt Mill.

Galton, F. 1889. Natural inheritance. Macmillan, London.

Henderson, C.R. 1988. Progress in statistical method applied to quantitative genetics since 1976. In: Proceedings of the 2nd International Congo on Quantitative Genetics, 1987, May 31 June, Releigh, NC. Sunderland, M.A. Sinauer Associates.

Johannsen, W.L, 1903. Uber Erblichkeit in Populationen und in reinen. Leninen, Gustav Fischer, Jena.

Kearsey, M. J. and Pooni, H. S. 1996. The Genetical Analysis of Quantitative traits. Chapman Hall, London.

Kempthorne, O. 1969. An Introduction to Genetic Statistics. Iowa state Univ. Press, Ames, Iowa.

Law, C.N. 1967. The location of genetic factors controlling a number of quantitative characters in wheat. Genetics, **56**: 445–461.

Levine, L. 1969. The biology of the Gena. Mosby and Toppan.

Mather, K. 1942. The balance of polygenic combinations. J. Genet., **43**: 309–336.

Mather, K. and Harrison, B.J. 1949. The manifold effect of selection. Heredity, **3**: 1–32 and 131–162.

Mather, K. and Jinks, J.L. 1982. Biometrical Genetics. 3rd edn. Chapman and Hall, London.

Mendel, G. 1866. Versuch uber Pflanzen–Hybriden.

Nilsson-Ehle, H. 1909. Kreuzungunter suchungen an Hafer and Weizen Lund.

Person, F. 1904. On a generalized theory of alternative inheritance with special reference to mendel's laws. Phil. Trans. A., 203: 53-86.

Rasmusson, J.M. 1935. Studies on the inheritance of quantitative characters in Pisum. I. Preliminary note on the genetics of flowering. Hereditas, **20**: 161– 180.

Sax, K. 1923. The association of size differences with seed coat pattern and pigmentation in Phaseolus vulgaris. Genetics, **8**: 552–560.

Spickett, S.G. and Thoday, J.M. 1966. Regular response to selection. 3. Interaction between located polygenes. Genetic Res., **7**: 96–121.

Thoday, J.M. 1961. Location of polygenes. Nature, **191**: 368–370.

Thompson. J. N. 1975. Quantitative variation and gene number. Nature, 258: 665-668. Levine, L. 1969. Biology of the Gene. The C.V. Mosby Company, St Louis and Toppan Company, Ltd, Tokyo.

Wright, S. 1968. Evolution and Genetics of Populations, Vol. I. Genetic and Biometric foundations. The University of Chicago Press. Chicago and London.

Yule, G.Y. 1906. On the theory of inheritance of quantitative compound characters on the basis of Mendel's Laws– A preliminary note. Rept. Third Intl. Conf. Genetics, 140–142.

4

Statistical Estimations, Tests, Models and Designs

4.1 Estimation and Inference

The different formulas for estimating a particular statistic are called estimators. A good estimator should meet the four criteria laid down by R.A. Fisher. The four criteria are: (i) consistency, (ii) unbiasedness (iii) minimum variance (or efficient) and (iv) sufficient. The estimation of parameters together with measures of reliability of these parameters consitute the solution of the problem of inference.

I. Consistency

When the sample size (n) is increased to infinity the estimate ($\hat{\theta}$) is very near to the true value of the parameter, θ. The large sample size can only ensure consistent estimate.

II. Unbiasedness

An estimator, $\hat{\theta}$, is said to be an unbiased estimator of θ if $E\hat{\theta} = \theta$ i.e. mean of all possible estimates of $\hat{\theta}$ equals the estimate of parameter, θ. So the bias $= \hat{\theta} - \theta$. The sources of bias are (i) wrong formula for an estimator (ii) faulty measurement of variable (iii) small size sampling and (iv) under representative sample, i.e. manner of sampling.

III. Minimum Variance

An estimator $\hat{\theta}$ is said to be a minimum variance estimator if $V(\hat{\theta})$ is smallest. As $V(\theta) = E(\hat{\theta} - E\hat{\theta})^2$, if $V(\hat{\theta})$ is small then the difference between $\hat{\theta}$ and $E\hat{\theta}$ will be small or in other words, $\hat{\theta}$ will be lying close to $E\hat{\theta}$. So among all possible estimators, the efficient estimator will be the one which is having minimum variance.

IV. Sufficient

An estimator is said to be sufficient estimator if it, even when the sample size is small, extracts the whole of the relevant information which the observations contain. Arithmetic means of samples from the normal distribution or from the poisson distribution series are examples of sufficient statistics.

4.1.1 Methods of estimation

There are different methods of estimation such as maximum likelihood, least square, minimum max, minimum chi-square and invariance estimation.

4.1.1.1 Maximum likelihood method

In a discrete distributions such as binomial or multinomial the probability of r success in n trials, $p(r)$, is

$$\frac{n!}{r!(n-r)!} p^r q^{n-r}$$

or

$$\frac{n!}{n!\, s!\, t!} p^r q^s r^t$$

These expressions then become the likelihood, l, function of the parameter p which is to be estimated. The objective now is to find the value of p which maximizes l. To maximize l we take the logarithm of l, the likelihood and if we maximize $\log l$, we are maximizing the likelihood, l itself.

Thus

$$L = \log l = \log \frac{n!}{r!(n-r)!} + r \log p + (n-r) \log(1-p)$$

In case of binomial distribution the term $\frac{n!}{r!(n-r)!}$ is constant. To find the maximum (or minimum) of a function with respect to a parameter is to differentiate it with respect to the parameter (p in this case) and set the derivates equal to zero, i.e.

put $\frac{dL}{dp} = 0$ and solve for p

Now $\frac{dL}{dp} = 0 + \frac{r}{p} - \frac{n-r}{q}$

$\therefore \quad \frac{r}{p} = \frac{n-r}{q}$

Thus $\frac{p}{q} = \frac{r}{n} - r$

and $\hat{p} = \frac{r}{n-r+r} = \frac{r}{n}$

In case of continuous distribution the likelihood is the probability density of the sample i.e.

$$l = \frac{1}{\sigma\sqrt{2\pi}} \exp\left(-\frac{1}{2}\frac{(x-\mu)^2}{\sigma^2}\right) dx_i$$

where $i = 1, 2, ... n$.

The maximum likelihood method maximizes the probability of obtaining the observed results (frequencies or sequences in case of nucleotide or amino acid sequences). MLH is quite tolerant of violation of its assumption.

4.1.1.2 Least squares

When we are studying continuously varying character in a population we are recording a number of observations say $Y_1, Y_2 ... Y_n$. Each observation consists of a mean and a random sampling error and thus the observation Y_1 takes the form

$$Y_i = \mu + \varepsilon_i$$

where $i = 1 ... n$ and which states that the ith observation is made up of mean and a sampling error. This is the simplest linear additive model. The mean may involve the single parameter μ or be the sum of a number of components associated with several effects or sources of variation. This model can be applied to the problem of estimating or making inferences about population means and variances.

The sample mean $\bar{Y}$ becomes

$$\bar{Y} = \sum_{i=1}^{N} Y_i = \frac{\sum_i (\mu + \varepsilon_i)}{n}$$

$$= \mu + \frac{\Sigma \varepsilon_i}{n}$$

The error is randomly distributed with mean zero. With large sample size $\bar{Y}$. will be close to μ as $\frac{\Sigma \varepsilon_i}{n} = 0$. The ε's can be estimated as

$$Y_i - \bar{Y} = \varepsilon_i$$

Thus $\sum_i \varepsilon_i^2 = \sum_i (Y_i - \bar{Y})^2$

$\Sigma(Y_i - \bar{Y})^2$ is minimum when $\mu = \bar{Y} = \frac{\Sigma Y_i}{n}$ Thus $\bar{Y}$ can be said to be a minimum sum of squares or least square estimate of μ.

Fisher's (1930) maximum likelihood method is particularly useful for gene frequency estimation and the least square methods are useful for estimating various sources of quantitative variation. Least square method also provides good estimate even if the observations do not follow a normal distribution. The two methods provide the same answer when the observations are approximately normally distributed.

4.2 Test of Significance

We have empirical observations such as frequency distribution, ratios or means, variances and other statistics of different families. We can also have their expected values based on the theoretical observation. Our objective would be test whether or not the observed descriptions conform to the null hypothesis. In other words, whether the proposed hypothesis for explaining the observed data should be accepted or rejected. This is the principle of hypothesis testing underlying all tests of significance. At this point then comes the question of level of significance i.e. what should be the level of departure from the hypothesis before we can reject our hypothesis. Although the decision on the particular level of significance depends on the significance of the result of the test the two generally accepted level of significance, α, are the 5 per cent and 1 per cent level of significance. The former indicates that above the level of probability ($p = 0.05$) of obtaining a result (which is expected if the experiment is repeated 100 times only on 5 occasions, the observed result will deviate from the expected) the hypothesis will be rejected whereas in case of later the hypothesis is rejected if the probability value of a result is above 0.01. When $p = 0.01$ the observed result deviates from expected one in 100 times. The level of significance thus signifies the level of admissible error. Statistical power – A testing procedure can be said to be robust if small or moderate departure from the normaility and constant variance do not seriously affect performance.

4.2.1 Chi-square test

It tests not only the deviation of a single frequency from an expected value but also gives a single overall test of a system (binomial or multinomial) of frequencies. Given the observed frequency (0) and the expected frequency (E), the deviation in standard form can be represented as

$$\chi_i = (0 - E)/\sigma = (0 - E)/\sqrt{E}$$

Such deviations have zero mean and unit variance. For a set of deviations, χ^2 takes the form as

$$\chi^2 = \frac{(0 - E)^2}{E}$$

χ^2 is thus the sum of squares of normally and independently distributed variables with zero mean and unit variance. For any normally distributed variable, Y, with mean μ and variance, σ^2,

$$\chi^2_{(1df)} = (Y_i - \mu)^2/\sigma^2$$

If $Y_1, Y_2, \ldots Y_n$ are independently and normally distributed variables with mean μ and variance σ^2 then

$$\chi^2_{(1\,df)} = (Y_i - \mu)^2/\sigma^2$$

$$\chi^2_{(n)} = \sum_{i=1}^{n} \frac{(Y_i - \mu)^2}{\sigma^2}$$

and

$$\chi^2_{(n-1)} = \sum_{i=1}^{n} \frac{(Y_i - \overline{Y})^2}{\sigma^2}$$

where $\overline{Y} = \frac{1}{n}\sum_{i=1}^{n} Y_i$ and $\sum_{i=1}^{n} (Y_i - \overline{Y})^2 = \sigma^2 \chi^2_{(n-1)}$. Thus the sum of squares in the ANOVA can be tested as χ^2

with $(K - 1)$ df and the probability value is doubled. Further, if X is distributed as $\chi^2_{(n)}$ and Y is distributed as $\chi^2_{(m)}$ then $(X + Y)$ is distributed as $\chi^2_{(n + m)}$.

χ^2 has a different distribution for each value of the degrees of freedom. χ^2 distribution is skewed with a range of value from $\chi^2 = 0$ to $\chi^2 = \infty$ With $n = 1$, the curve for χ^2 becomes the right half of a normal curve with a doubled ordinates but as the mean value approaches $\sqrt{n}$ its variance approaches $\sqrt{\frac{1}{2}}$ and the χ^2 distribution approaches normal distribution with value of n greater than 30. For values of $n > 30$ Fisher uses $Z = \sqrt{2\chi^2} - \sqrt{2n - 1}$ as the quantity, that has a unit standard deviation, in calculating the probability of values of χ^2. The χ^2 test can be used as a test of goodness of fit. It is also used as a test of independence besides testing the homogeniety of several independent estimates of population variance.

4.2.2 F-test

If X is distributed as $\chi^2_{(n-1)}$ and Y is distributed $\left(\chi^2_{(m-1)}\right)$ and if X and Y are constant then the ratio of the two χ^2's $= \left(\dfrac{\chi^2_{(n-1)}}{\chi^2_{(m-1)}}\right)$ is distributed as F (Snedecors' F) distribution where $n - 1$ and $m - 1$ are degrees of freedom.

$$F = \frac{\chi^2_{(n-1)}}{\chi^2_{(m-1)}}$$

As $\chi^2_{(n-1)} = \dfrac{nS_1^2}{\sigma_1^2}$ and $\chi^2_{(m-1)} = \dfrac{mS_2^2}{\sigma_2^2}$

$$F = \left(\frac{nS_1^2}{n - 1}\right)\bigg/\left(\frac{mS_2^2}{m - 1}\right)$$

$$= \hat{\sigma}_1^2/\hat{\sigma}_2^2$$

where $\hat{\sigma}^2$ is the unbiased estimate of σ^2 and σ_1^2 is always the larger of the two sample variances. Thus F can be defined as the variance ratio. The shape of the distribution of variance ratio varies according to the numerator (n) and denominator (m) degrees of freedom. Tests on variances such as F tests are very sensitive to departures from normality. The F test can be used for testing the homogenity of variances. In case of testing of homogeneity of variances through variance ratio test the probability of this F value is multiplied by 2 before drawing any conclusion. In case of Bartlett's test the χ^2 is obtained as

$$\chi^2_{n-1} = \left[\sum_{i=1}^{n} K_i \log e \, \bar{S} - \sum_{i=1}^{n} K_i \log e \, S_i^2 \right]$$

where $\bar{S}^2 = \dfrac{1}{\sum\limits_{i=1}^{n} K_i} = \left(\sum\limits_{i=1}^{n} K_i S_i^2 \right).$

n being the number of genotypes/generations involved in the experiment, K_i is the degree of freedom associated with the ith genotype and S_i is the corresponding mean squares, $\bar{S}^2$ the pooled variance is estimated as

$$\bar{S}^2 = \dfrac{\sum\limits_{i=1}^{n} K_i S_i}{\sum\limits_{i=1}^{n} K_i}$$

If χ^2 is not significant it shows that the variances are homogeneous.

4.2.3 The t-test

The t (Student's t) statistic is calculated as

$$t = \dfrac{X - \mu}{S\bar{X}} = \dfrac{\bar{X} - \mu}{\sqrt{SX^2/n}} = \dfrac{X/\sqrt{n}}{SX^2}$$

where $\bar{X} = \dfrac{1}{n} \sum\limits_{i=1}^{n} X_i$ is the sample mean and n is the number of observation which is less than 30. SX^2 is the sampling variance, μ the population mean and σ^2 the population variance.

With n being less than 30 there is danger of misinterpreting of significance as S is subject to sampling error but with n greater than 30 the error becomes sufficiently small to be safely ignored. t has a different distribution for each value of the degree of freedom. The t distribution is more peaked in the centre and has higher tails than the normal distribution.

As the sample size increases the t-distribution approach the normal distribution (or C-distribution). And t^2 equals

$$t^2 = \dfrac{(\bar{X} - \mu)^2}{\dfrac{SX^2}{n}} = \dfrac{n(\bar{X} - \mu)^2}{SX^2}$$

The t^2 thus can be said to be the ratio of the two independent χ^2 with 1 and n degrees of freedom.

The t test has the following usage:

1. When two sample means are to be compared the following test can be constructed.

(a) When the two independent sample means have equal variance.

$$t = \dfrac{\bar{X}_1 - \bar{X}_2}{S\bar{X}_1 - \bar{X}_2}$$

$$S\bar{X}_1 - \bar{X}_2 = \sqrt{\dfrac{S^2}{n_1} + \dfrac{S^2}{n_2}}$$

when $SX_1^2 = SX_2^2 = S^2$ and $n_1 \neq n_2$.

In case $n_1 = n_2$

$$S\bar{X}_1 - \bar{X}_2 = \sqrt{\dfrac{2S^2}{n}}$$

(b) When the two samples are from the two populations differing in variances ($\sigma_1^2 \neq \sigma_2^2$)

In this case $S\bar{X}_1 - \bar{X}_2 = \sqrt{\dfrac{SX_1^2}{n_1} + \dfrac{SX_2^2}{n_2}}$

2. Testing of the significance of an observed correlation coefficient (r).

If $X_1Y_1, X_2Y_2 \ldots X_nY_n$ are the random samples from a bivariate normal population the hypothesis that the population correlation coefficient is zero ($p = 0$) is tested using t statistic.

$$t_{(n-2)} = \dfrac{r\sqrt{n-2}}{\sqrt{1-r^2}}$$

where n is the number of pairs of observations.

3. Testing of the significance of an observed regression coefficient (b).

Given the above random sample the hypothesis that the regression coefficient of Y on X in the population ($\beta = 0$) is tested using t statistic.

$$t_{(n-2)} = (b - \beta) \sqrt{(n-2) \Sigma (X_i - \overline{X})^2 / \Sigma (Y_i - \overline{Y})^2}$$

4. Testing the significance of the difference between two independent correlation coefficients (r_1 and r_2)

This is done by using the statistic

$$Z' = \frac{Z_1 - Z_2}{\sqrt{\frac{1}{n_1 - 3} + \frac{1}{n_2 - 3}}}$$

where

$$Z_1 = \frac{1}{2} \log e \frac{1 + r_1}{1 - r_1}$$

$$Z_2 = \frac{1}{2} \log e \frac{1 + r_2}{1 - r_2}$$

which are approximately normally distributed with

$$\text{S.E.} = \frac{1}{\sqrt{n - 3}}.$$

4.2.4 Z-distribution

Fisher used the natural logarithm of the ratio of the two root mean squares and termed it as Z and its distribution is called Z distribution. The Z like t is calculated from estimated standard deviation and variances and like each value of C and t, each value of Z corresponds to some definite probability. The tabulation of Z requires a three dimensional table, the probability, the degrees of freedom for numerator and denominator.

4.2.5 Relationship between different distributions

The calculations of C and χ^2 require that the variance, σ^2 is known whereas in case of t and F, the estimated variance $\hat{\sigma}^2$ is used. The C is a special case of both t and χ^2 whereas Z is a special case of t, C and χ^2. χ^2 with large degrees of freedom is equivalent to $C (= \sqrt{2\chi^2 - 2N - 1})$. t with large N, the sample size is equivalent to C. The C has a numerator of 1 degree of freedom and a denominator fixed by the hypothesis. The t^2 is a variance ratio, $F(F = t^2)$ when $n = 1$ and m is finite. χ^2 is equivalent to variance ratio when m tends to infinity.

So far we have seen the distribution of variables. Now we will see the distribution of quadratic forms, the variances when the variables are normal.

4.2.6 Non-central chi-square

If $Y_1, Y_2, ... Y_n$ are independent normal variables with mean $\mu_1, \mu_2 ... \mu_n$, respectively and with variance equal to unity then $\sum_{i=1}^{n} Y_i^2 = w$ is distributed as the non-central χ^2 with n degrees of freedom and parameter $\lambda = \frac{1}{2} \Sigma \mu_i^2$. The frequency function of this distribution is

$$f(w) = \sigma^{-\lambda} \sum_{i=0}^{\infty} \frac{\lambda' w^{1/2}(n+2i) - 1 e^{-w/2}}{i! 2^{\frac{1}{2}(n+2i)} \Gamma \left(P + \frac{2i}{2} \right)} \quad 0 \leq w < \infty$$

$$= \sum_{i=0}^{\infty} e^{-\lambda} \frac{\lambda'}{i!} gn + 2i(w)$$

where $gn + 2i(w)$ is the central chi-square frequency with $n + 2i$ degrees of freedom. Since $e^{-\lambda} \lambda'/i_1$ is the ith term of poisson distribution, the non-central χ^2 can be seen as the sum of an indefinite number of chi-square frequencies each weighted by a term from the poisson distribution.

4.2.7 Non-central F

If a random variable w is distributed as $\chi'_{(n\lambda)}$ that is as the non-central chi-square with n degrees of freedom and parameter λ and if another variable Z is distributed as $\chi'^2_{(m,0)}$ that is the central χ^2 with m degrees of freedom and if w and Z are independent then the quantity $u = \frac{w}{n} \frac{m}{Z}$ is distributed as the non-central F distribution (F^1) with n, m degrees of freedom and with parameter λ. The frequency function of u

$$f(u) = \sum_{i=0}^{\infty} \frac{\Gamma \frac{(2i + n + m)}{2} \left(\frac{n}{m} \right)^{1/2(2i+n)} \lambda^1 e^{-\lambda}}{\Gamma(m_2) \Gamma \frac{(2i + n)i!}{2}}$$

$$\frac{u^{1/2}(2i+n-2)}{\left(1+\frac{nu}{m}\right)^{1/2(2i+n+m)}} \quad 0 \le u < \infty$$

This non-central F plays an important role in the theory of the power of tests in regression and analysis of variance.

4.3 Statistical Models and Expectation of Mean Squares

The analysis of variance technique developed by Fisher supplies information about the different sources of variation and depending upon the underlying model provides (i) estimate of the fixed effects, (ii) estimate of the variance components and (iii) test of hypothesis concerning the sources of variation in the model. Thus the ANOVA is a means of partitioning of a sum of squares into a set of quadratic forms. A statistical model describes the makeup of an observation. The observation Y_{ij}, the yield in ith block of the jth treatment in randomized complete block design takes the form as

$$Y_{ij} = \mu + b_i + t_j + e_{ij}$$

where μ is the mean, b_i the ith block effect ($i = 1 \ldots r$), t_j is the jth treatment effect ($j = 1 \ldots s$). It is assumed that all effects in the model are independently and normally distributed. It is essential to define these effects. e_{ij}s are the error which are independently and normally distributed with zero mean and variance, σ^2. The parameters in the model will depend on the type of material under testing and also the parameters will change depending upon the experimental design (completely randomized design, latin square, etc.). The model can be fixed, random or mixed.

4.3.1 Fixed model

When both treatments and blocks are fixed in an experiment the model follows a fixed model. By treatment and block fixed is meant that if we repeat the experiment we are using the same set of treatments and blocks. In other words, t_j's and b_i's are fixed variables and $\Sigma t_j = 0$, $\Sigma b_i = 0$ and $\varepsilon e_{ij} = 0$ The μ is no longer the population mean but the mean of all possible replications of experiment. Thus inferences can be drawn about this particular set of treatments.

Here the set of treatments (may comprise clones, varieties, families, etc.) is chosen by the experimentor because of certain merits possessed by each. The breeder may be interested in estimate of variability among a set of lines or among crosses of certain particular lines. With genotypes (s) and sites (t) fixed the expected mean squares takes the form as shown in Table 4.1

Table 4.1 Expected mean squares in case of fixed model

Item	d.f.	M.S.	E.M.S.
Genotypes	$s-1$	MSg	$\sigma_e^2 + rt\,K_g^2$
Sites	$t-1$	MSs	$\sigma_e^2 + rS K_S^2$
Genotypes × sites (F^*)	$(s-1)(t-1)$	MS$_{g \times s}$	$\sigma_e^2 + r K_{s \times g}^2$
Error	$st(r-1)$	MSe	σ_e^2
Total	$tsr-1$		

r = Number of replications.
*F = Fixed.
* Although the fixed effects have no variance but one can compute the variances of the fixed effects (K^2).

The different items in the ANOVA are tested as follows:
F (interaction) = MS$_{g \times s}$/MSe
Genotypes = MSg/MSe
Sites = MSs/MSe

4.3.2 Random model

When both blocks and treatments are random samples from a population of blocks and from a population of treatments, respectively the random model is followed. Here every time we repeat the experiment we choose a random sample of blocks and treatments. In other words we are not using the same set of blocks and treatments. The breeder/geneticist is interested in the variability estimate of the population as such but because he cannot accommodate the whole population into the experiment he takes a random sample representing that population. Thus b_i's and t_j's are assumed drawn from an infinite population with zero mean and variance σ^2 with the expectations $\varepsilon b_i = 0$ and $\varepsilon t_j = 0$ and $\varepsilon b_i^2 = \sigma_b^2$ and $\varepsilon t_j^2 = \sigma_t^2$. In case of interaction the interaction effect is also randomly distributed with zero mean and variance σ_{bt}^2. With genotypes and sites random the expectation of the mean squares takes the form as shown in Table 4.2.

The different items in ANOVA are tested as follows:

$$\text{Interaction} = MS_{g \times s}/MS_e$$

If this item is not significant then pooled error is calculated as $MS_{g \times s} + MS_e$ and both genotypes and sites items are tested against this pooled error but if the interaction item is significant then both genotypes and sites items are tested against the interaction.

Table 4.2 Expected means squares in case of random model

Item	d.f.	M.S.	E.M.S.
Genotypes	$(s-1)$	MSg	$\sigma_e^2 + r\sigma_{g \times s}^2 + rt\sigma_g^2$
Sites	$(t-1)$	MSs	$\sigma_e^2 + r\sigma_{g \times s}^2 + rS\sigma_s^2$
Genotypes × sites	$(s-1)(t-1)$	$MS_{g \times s}$	$\sigma_e^2 + r\sigma_{g \times s}^2$
Error	$ts(r-1)$	MSe	σ_e^2
Total	$tsr-1$		

4.3.3 Mixed model

This is the model which is actually followed in quantitative genetics/plant breeding. We have a fixed set of treatments (genotypes) but the blocks are a random sample from a population of blocks. In mixed model the interest lies in drawing inferences about the fixed treatments. Here the interaction effect is randomly distributed with zero mean and variance σ_{bt}^2. With genotypes fixed and sites random the expected mean squares takes the form as shown in Table 4.3.

Table 4.3 Expected mean squares in mixed model

Item	d.f.	M.S.	E.M.S.
Genotypes	$s-1$	MSg	$\sigma_e^2 + rs\sigma_s^2 + rt K_s^2$
Sites	$t-1$	MSs	$\sigma_e^2 + rs\sigma_s^2$
Genotype × site	$(s-1)(t-1)$	$MS_{g \times s}$	$\sigma_e^2 + r\sigma_{s \times g}^2$
Error	$ts(r-1)$	MSe	σ_e^2
Total	$tsr-1$		

The variance ratio test (F) is conducted as follows:
Interaction = $MS_{g \times s}/MSe$, Genotypes = $MSg/MS_{g \times s}$, Sites = MSs/MSe

Thus we see that the computation will be the same regardless of the model though the choice of error term and the type of inference will vary in different models.

Given treatments A, B, C, D at levels a, b, c and d with r replications, the expected value of mean squares for U, any treatment or interaction effect, can be worked out following the general rules (Snedecor and Cochran 1967).

1. The expected value of mean squares for an item say U will contain a term in σ^2 (an error) and a term in $\sigma^2 U$. It will also contain a variance term for any interaction in which (i) all the letters in U appear and (ii) all the letters in the interaction represent random effect.

2. The coefficient of the term in σ^2 is one. The coefficient of any other variance is r times the product of all letters a, b, c ... that do not appear in the set of capital letters A, B, C, ... specifying the variance.

4.4 Experimental Designs and Calculation of Mean Squares

In the experimental design the treatment combinations are applied to experimental units and the effect of treatment combination is measured (also called the response). The objectives in experimental designs are (i) to compare the effect of different treatment combinations and (ii) to quantify the effect of a particular treatment combination together with a standard error. In order to achieve these objectives it is essential that the selected experimental units should be as similar as possible but then the two similar experimental units may not reflect the normal variation encountered in practice. Thus extreneous sources of variation from the experimental units should be excluded but then the selected units must be representative of those encountered in practice.

4.4.1 Randomized complete block design (RBD)

Here we have a set of treatments (pure breeding lines, families or open pollinated populations) say n ($j = 1, 2, ... n$) raised in a randomized complete block design with a number of replications say r ($j = 1, 2, ... r$). The data y_{ij} obtained from this design can be presented as given on next page.

Treatment	Blocks		
	$b_i = 1$	2...	r
$t_j = 1$	Y_{11}	Y_{12}	Y_{1r} $Y_{1.} = \Sigma Y_{ij}$
.			
.			
.			
n	Y_{n1}	Y_{n2}	Y_{nr} $Y_{n.}$
	$Y_{.1} = \sum\limits_{j} Y_{ij}$		$Y_{.r}$ $Y_{...} = \sum\limits_{i}\sum\limits_{j} Y_{ij}$

The ANOVA takes the form as given in Table 4.4.

Table 4.4 ANOVA in RBD experiment

Item	d.f.	S.S.	M.S.	F
Blocks/replications	$r - 1$	SSr	MSr	MSr/MSe
Treatments	$n - 1$	SSt	MSt	MSt/MSe
Error	$(r - 1)(n - 1)$	SSe	MSe	
Total	$nr - 1$			

The different items in the ANOVA are calculated as follows:

Correction factor $(CF) = (\sum\limits_{i,j} Y_{ij})^2 / n \times r$

Total sum of squares $= \sum\limits_{i,j} Y_{ij}^2 - CF$

Treatment sum of squares $= \sum\limits_{j}(\sum\limits_{i} Y_{ij})^2 / r - CF$

Replication sum of squares $= \sum\limits_{i}(\sum\limits_{j} Y_{ij})^2 / n - CF$

Error sum of squares = Total sum of squares − (Treatment sum of squares + Replication sum of squares)

The critical difference (CD) is calculated as follows:

CD at 5% $= \sqrt{2} \times \sqrt{MS_e/r} \times t$ at 5%

CD at 1% $= \sqrt{2} \times \sqrt{MS_e/r} \times t$ at 1%

The coefficient of variation (CV) of the experiment is calculated as $(\sqrt{MS_e/r}/\sum\limits_{i,j} Y_{ij}/nr) \times 100$ and standard error of the treatment means (SE) will be $SE = \sqrt{MSe/r}$

4.4.2 Nested design

We can have situations where the treatments have hierarchial or nested structure. For example, in case of pedigree or single seed decent method of handling the segregating population each plant (or line) in the F_n generation can be traced back to the single F_2 plant. In the simplest case we can have a number of F_4 families ($j = 1 \ldots m$) from each of the n ($i = 1 \ldots n$) F_2 plants raised in field with r ($k = 1 \ldots r$) number of replications. The data can be arranged in the form given below:

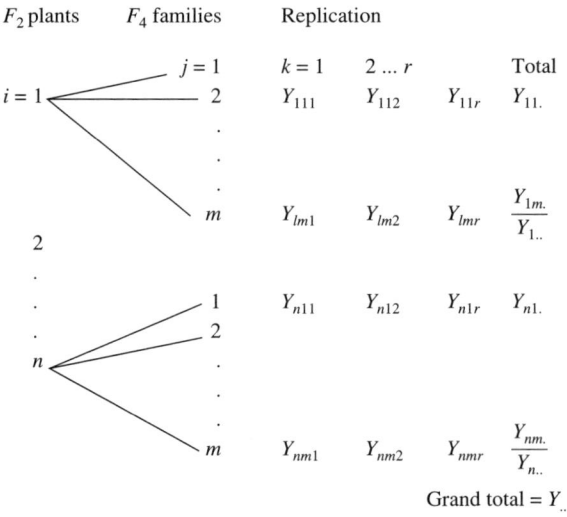

Grand total = $Y_{...}$

The ANOVA will take the form given in Table 4.5.

The sum of squares due to different items are calculated as follows:

Table 4.5 ANOVA in nested experiment

Item	d.f.	S.S.	M.S.	F
Between F_2 plants	$n - 1$	SSF_2	MSF_2	MSF_2 MSF_2/MS
Between F_4 families within F_2 plants	$(m - 1)n$	SSF_4/F_2	MSF_4/F_2	MSF_4/MSe
Within F_4 families	$nm(r - 1)$	SSe	MSe	
Total	$nmr - 1$			

Correction factor = $\left(\sum_{i,j,k} Y_{ijk}\right)^2 / nmr$

Total sum of squares = $\sum_{i,j,k} Y_{ijk}^2 - CF$

Between F_4 families within F_2 plants

Sum of squares = $\sum_{i,j} \left(\sum_k Y_{ijk}\right)^2 / r - CF$

Between F_2 plants

Sum of squares = $\sum_i \left(\sum_{i,j,k} Y_{ijk}\right)^2 / mr - CF$

4.4.3 Cross-classified design

It is similar to the factorial experiments where the set of treatments consists of all possible combinations of the levels of two or more factors. In the simplest case of two factors experiment we can have a set of lines or population crossed with another set of lines or populations. With two factors say A and B the model takes the form as

$$Y_{ijk} = \mu + A_i + B_j + (AB)_{ij} + e_{ijk}$$

The two factor interaction is called the first order interaction whereas the three factors interaction is called the second order interaction. If we have a set of varieties ($i = 1 \ldots s$) grown in locations ($j = 1 \ldots t$) then the data can be arranged in the manner given below:

	Location		
	$j = 1 \ldots$	t	Total
Treatment $t = 1$	Y_{11}	Y_{1t}	$Y_{1.}$
.			
.			
.			
s	Y_{s1}	Y_{st}	$Y_{s.}$
			$Y_{..}$

In the above two way table $Y_{ij} = Y_{ijk}/r$ where r is the number of replications. The ANOVA takes the form as given in Table 4.6.

Table 4.6 ANOVA in cross-classified experiment

Item	d.f.	S.S.	M.S.	F
Genotypes	$s - 1$	SSg	MSg	$MSg/MS_{g \times s}$
Locations	$t - 1$	SSe	MSe	MSe/MSe
Genotype × location	$(s-1)(t-1)$	$SSgl$	$MSgl$	$MSgl/MSe$
Error	$st(r-1)$	SSe	MSe	

The various sum of squares are calculated as follows:

Correction factor = $\left(\sum_{i,j,k} Y_{ijk}\right)^2 / str$

Total sum of squares = $\sum_{i,j,k} Y_{ijk}^2 - CF$

These above items are calculated from the replicated data. From the mean data i.e. from the two way table the different items are calculated as follows:

Correction factor = $\left(\sum_{j,k} Y_{ijk}\right)^2 / st$

Genotype sum of squares = $\sum_i \left(\sum_j Y_{ij}\right)^2 / t - CF$

Location sum of squares = $\sum_j \left(\sum_j Y_{ij}\right)^2 / s - CF$

Genotype × Location sum of squares
$$= \sum_{i,j} Y_{ij}^2 - \text{SS due to genotype}$$
$$- \text{SS due to location}$$

4.4.3.1 Partitioning of treatment sum of squares

Between treatment sum of squares, we might test $t_1, t_2 \ldots t_n = 0.0$ but then we might be interested in partitioning the between treatment sum of squares into a series of sum of squares which examine the treatments in more detail. For example, in case of randomized complete block with three treatments say T_1 T_2 and T_3 we would be interested in making the following comparisons.

(i) Comparison (C_1) which measures the difference in the effects of two rival treatments (T_1 vs T_2, T_2 vs T_3 or T_1 vs T_3).

(ii) Comparison (C_2) which measures the difference in effects of the two rival treatments vs no treatment i.e. $(T_1 + T_2)/2$ vs T_3, T_1 vs $(T_2 + T_3)/2$ or $(T_1 + T_3)/2$ vs T_2. Here the average of the two treatment effect is compared with the third treatment effect.

(iii) Comparison (C_3), the mean of overall result (the treatment effect) i.e. $(T_1 + T_2 + T_3)/3$.

The coefficients (K_1, K_2 and K_3) of the treatment totals in case of the above comparisons are as follows.

	Treatment totals		
	T_1	T_2	T_3
Comparisons			
$C_1(T_1$ vs $T_2)$	+1	−1	0
$C_2(T_1 + T_2)/2$ vs T_3	+1	+1	−2
$C_3((T_1 + T_2 + T_3)/3)$	+1	+1	+1

C_1, C_2 and C_3 are called contrasts or comparisons. When the contrasts are mutually independent or orthogonal they are called orthogonal contrasts. Comparisons must obey certain rules. For tests to be independent ΣK must be equal to zero. For comparisons to be independent the sum of the products of K's in pair must be zero. The number of comparisons cannot exceed the degrees of freedom for treatments. The sum of squares due to a comparison will be

$$C^2 / \left(r \sum_i k_i^2 \right)$$

which will always have one degree of freedom.

$C = K_1 T_1 + K_2 T_2 + K_3 T_3$ in the above case. Thus C is a linear function of the treatment totals. r is the number of replications on which the treatment total is based. The three sum of squares will thus be calculated as follows:

$C_1 SS = [(+1) \times T_1 + (-1) \times T_2)]^2 / 2r$

$C_2 SS = [(+1) \times T_1 + (+1) \times T_2 + (-2) \times T_3]^2 / 6r$

$C_3 SS = [(+1) \times T_1 + (+1) \times T_2 + (+1) \times T_3]^2 / 3r$

If treatment means are used instead of treatment totals then r will not appear in the denominator. The ANOVA in this analysis will take the form given in Table 4.7.

Table 4.7 Partitioning of treatment sum of squares

Item	d.f.
Treatments	2
C_1	1
C_2	1
C_3	1
Error	1

The between treatments sum of squares will be

$$C_1 SS + C_2 SS$$

4.4.4 Factorial experiment

In case of single level of two treatments say A and B there will be four treatments such as A, B, interaction AB and 0, the above treatment totals the following information can be obtained.

(i) The effect of treatment

$$A = (A + AB) - (B + 0)$$

(ii) The effect of treatment

$$B = (AB + AB) - (A + 0)$$

In other words, the effect of a treatment, say X, equals average of observations using that treatment X minus the average of observations not using X.

(iii) The interaction,

$$AB = (AB + 0) - (A + B)$$

In case of two levels of three treatments (2^3 factorials experiment and here $n = 3$) say nitrogen N (O, N), phosphorus P (O, P) and potash K (O, K), information on effects of N, P, K, two factor interactions such as NP, NK and PK and three factor interaction NPK can be calculated using various comparisons (C's) with the following coefficients:

With number of replications $r = 4$, the ANOVA takes the form as given in Table 4.8.

Table 4.8 ANOVA in factorial design

Item	d.f.
Replications	$r - 1 = 3$
Treatments	7
N	1
P	1
K	1

	N	O				N		P
	P		O		P	O		
	K	O	K	O	K	O	K	O
Total	OOO	OOOK	OOOP	OPK	OON	ONK	ONP	NPK
Comparision								
$C_1(N)$	+1	+1	+1	−1	−1	−1	−1	−1
$C_2(P)$	+1	+1	−1	−1	+1	+1	−1	−1
$C_3(K)$	+1	−1	+1	−1	+1	−1	+1	−1
$C_4(NP)$	+1	+1	−1	−1	−1	−1	+1	+1
$C_5(NK)$	+1	−1	+1	−1	−1	+1	−1	+1
$C_6(PK)$	+1	−1	−1	+1	+1	−1	−1	+1
$C_7(NPK)$	+1	−1	−1	+1	−1	+1	+1	−1

NP	1
NK	1
KP	1
NPK	1
Error	8 × 3 = 24

Thus, the treatment sum of squares which has got 7 degrees of freedom is partitioned into seven sum of squares each having one degree of freedom.

If two and three factor interactions are not significant then they are included in the error.

4.4.5 Analysis of co-variance

Analysis of covariance is used to control error and increase experimental precision and to adjust variety means on the basis of plant stand. The assumptions underlying analysis of covariance are the same as that of analysis of variance. The X's are fixed and measured without error and the regression of Y and X after removal of treatment and block effects is linear and is independent of treatment and block effects (Steel and Torrie, 1980). With X being plant stand and Y the yield, the analysis of covariance will take the following form for n varieties with r number of replications.

Character	X		Character	Y	Character X and Y
Item	d.f.	S.S.		S.S.	S.S.
Replications	$r − 1$	SSr		SSr	SSr
Varieties	$n − 1$	SS_x		SS_y	SS_{xy}
Error	$(n − 1) \times (r − 1)$	SS_{Ex}		SS_{Ey}	SS_{Exy}

The regression coefficient $b_{y/x}$ is calculated as

$$b_{y/x} = SS_{xy}/SS_x$$

The variety mean is adjusted in the following way:

$$\hat{Y}_i = \overline{Y}_{i.} - b_{y/x}(\overline{X}_{i.} - \overline{X}..)$$

where $\hat{Y}_i$ is the adjusted mean of the ith variety, $\overline{Y}_{i.}$ is the observed (unadjusted) mean of the ith variety, $\overline{X}_{i.}$ is the mean plant stand of the ith variety and $\overline{X}..$ is the overall mean of the varieties. Another analysis of covariance table is constructed using the adjusted means as:

Item	d.f.	SS
Variety (adjusted)	$n − 1$	
Error (adjusted)	$(n − 1)(r − 1) − 1$	

The error adjusted sum of squares is calculated as

$$SS_{Ey} - (SS_{Exy})^2/SS_{Ex}$$

and variety + error adjusted sum of squares as

$$SS_y - (SS_{xy})^2/SS_x$$

From these the variety adjusted sum of squares is obtained as

Variety + error adjusted SS − Error adjusted SS

If the variation in plant stand is due to genetic reasons there is no use of analysis of covariance technique. Covariance adjustment of yield assumes a linear relationship but many of these associations are not linear throughout the range from 0 to 100%. At lower and upper ends it is curvilinear. Thus if the plant stand ranges from 25 to 75% use of covariance

4.5 Correlation and Regression Analysis

4.5.1 Correlation coefficient

The covariance is a measure of the joint variation of two or more variables. Their product moment is symbolized as $\mu 11$ (Pearson, 1895) or as their covariance symbolized by cov (Fisher, 1931). Considering two variables x and y, the cov (x, y) is the average product of the deviation of the two variables from their mean i.e.

$$\text{Cov } xy = \Sigma (x - \bar{x})(y - \bar{y})/n - 1$$

The covariance may be positive or negative and it is symmetrical in x and y and the variables need not be specified as dependent and independent. In the study of simultaneous variation of the two variables x and y we are particularly interested in the kind of relationship between the two variables (or the bivariate normal distribution of x and y) which is linear or nearly so. This relationship can be studied through correlation coefficient which measures the association between two variables and is calculated as

$$r_{xy} = \text{Cov } (x, y)/\sqrt{\text{Var}(x)\text{ Var }(y)}$$

where r_{xy} is the estimate of sample correlation coefficient. The population correlation coefficient is designated as p. Var (x) and Var (y) are the variances of characters x and y, respectively. r_{xy} is thus a numerical measurement of the degree of linear association between the deviations of x and y. It is sometimes called the 'total correlation' between the two variables as it includes all the known and unknown agents that make the two variables correlated but it does not necessarily imply any causal relationship. Many consequences of Mendelian inheritance and quantitative inheritance can be expressed in terms of correlation between relatives as correlation coefficient estimates the degree of closeness of linear relationship between x and y. The correlation coefficient $r(x, y)$ has the same sign as the Cov (x, y) of the two variables. Further, r_{xy} is independent of the change of origin of reference and scale (the unit of measurement). When x and y are independent variables then r_{xy} is 0 as Cov $(x, y) = 0.0$, i.e. they are uncorrelated, but then we can have situation when the variables may be uncorrelated but dependent. The limits for the correlation coefficient are as follows:

$$-1 \leq r \leq 1$$

The standard error of correlation coefficient is approximately $\dfrac{1 - r^2}{\sqrt{n - 1}}$. The distribution of r in small samples is far from normal because of such limitations. The degree of freedom for testing the correlation coefficient is $n - 2$. The degree of freedom in correlation study should be greater than 30.

4.5.1.1 Relationship between correlation coefficient and regression coefficient

Unlike the correlation coefficient, the regression coefficient is not symmetrical, i.e. $b_{yx} \neq b_{xy}$ as

$$b_{yx} = \text{Cov } xy/\sigma_x^2 \text{ and } b_{xy} = \text{Cov } xy/\sigma_y^2$$

$b_{yx} = b_{xy}$ only when $\sigma_x^2 = \sigma_y^2$. Also, there is no such limitation as in case of correlation coefficient. The test of reality of a relation between two variables is best studied by the significance of regression coefficient. The regression coefficient is related to correlation coefficient in the following manner.

$$b_{yx} = r_{xy} \, \sigma y/\sigma x$$

Similarly
$$b_{xy} = r_{xy} \, \sigma x/\sigma y$$

and
$$r_{xy} = \sqrt{b_{yx} b_{xy}}$$

i.e. the correlation coefficient is the geometric mean between the regression coefficients and $r_{xy} = b_{xy}$ if $\sigma_x^2 = \sigma_x^2$. Since σx and σy are also positive b_{yx}, b_{xy}, Cov xy and r have the same sign. The regression coefficients (b's) are independent of the change of origin but not of the scale. It is based on the reason that mean changes with the change in the origin and scale of measurement while the variance changes only with the change in the scale of measurement. A correlation coefficient is always a property of the population as well as of the variables and derives its usefulness of this. The regression coefficients of a casually dependent variable on a variable factor may, in certain cases, reveal essentially the same contributions of the later per unit of change in widely different populations in which the correlation coefficients differ correspondingly (Wright, 1968).

4.5.1.2 Coefficient of determination

The proportion of the variance of either of the two variables that is determined directly or indirectly by the other can be measured by the squared correlation coefficient (r^2). It can be shown that

$$\sigma_y^2 = b_{yx}^2 \sigma_x^2 + \sigma_{yy}^2$$
$$= r_{xy}^2 \sigma_y^2 + \sigma_{yx}^2$$

or
$$\sigma_{yx}^2 = \sigma_y^2 (1 - r_{xy}^2)$$

where σ_{yx}^2 is the average variance of deviation from the regression. σ_{yx}^2 can sometimes be used to denote the fraction of the variance of y that is determined by x and thus

$$\sigma_{yx}^2 = r_{xy}^2 \sigma_y^2$$

∴
$$r_{xy}^2 = \sigma_{yx}^2 / \sigma_y^2$$

In terms of sum of squares r^2 = regression SS/SS_y and $1 - r^2$ = remainder SS/SS_y. Since σ_{yx}^2 and σ_y^2 are positive $r^2 \leq 1.0$. Thus the departure of the value of r^2 from unity will be a measure of departure from the relationship between the variables from linearity.

4.5.2 Regression analysis

4.5.2.1 Linear regression model

The linear relationship between two variables x and y can be examined by studying regression and correlation. The linear regression model showing the causal relationship between the two variables takes the form as:

$$y = a + bx$$

where y is the dependent variable and x is the independent variable. This formula shows the straight line relationship. a is the intercept and b is the slope, gradient or the rate of change of y per unit change in x. Here x is the source of variation in y. In practice, we can have a number of lines or families say n ($i = 1, 2 \ldots n$) and two characters x and y have been scored on individuals of these lines.

Lines or families	Characters record	
$l = 1$	x_i	y_i
.	x_1	y_1
.	.	.
.	.	.
.	.	.
n	x_n	y_n

Thus we can have a set of observations as

$$y_1 = \alpha + \beta_{x1} = e_1$$
.
.
.
$$y_i = + \beta_{xn} = e_n$$

which takes the form as
$$y_i = \alpha + \beta_{xi}$$

where α and β are the true intercept and shape of the straight line. The regression line (β) needs not be straight but we have assumed that a straight line is often chosen as an approximation when it fits reasonably well over a range of x considered even though the true form is known to be non-linear. Any values of these quantities that we obtain from data therefore, are to be regarded as sample estimate of these population parameters.

In other words, $a = \hat{\alpha}$ and $b = \hat{\beta}$. In this study the aim is to estimate a and b.

4.5.2.2 Least squares solution

The value of a and b are chosen so as to minimize $\Sigma(y - \hat{y})^2$ and i.e. $\Sigma [y_i - a - b(x - \bar{x})]^2$. Theory shows that the resulting estimates a and b are unbiased and have the smallest standard error. The least square equation or normal equation is

$$\Sigma a + b \Sigma (x_1 - \bar{x}) = \Sigma y_i \qquad \ldots (1)$$
$$a \Sigma(x, -\bar{x}) + b \Sigma (x_i - \bar{x})^2 = \Sigma y_i (x_i - \bar{x}) \qquad \ldots (2)$$

Equation (1) reduces to
$$na = \Sigma y_i$$

$$\therefore \quad a = \frac{\Sigma y_i}{n} = \bar{y}$$

and equation (2) reduces to

$$b\ SSx = SPxy$$

$$\therefore \quad b = SPxy/SSx$$

The ANOVA in regression analysis takes the form given in Table 4.9.

Table 4.9 ANOVA in regression analysis

Item	d.f.
Regression	1
Remainder	$n - 2$
Total	$n - 2$

The various sum of squares are calculated as follows

$$\text{Total } SS = \Sigma y_i^2 - \frac{(\Sigma y_i^2)}{n}$$

$$\text{Regression } SS = (SPxy)^2/SSx$$

where

$$SP(xy) = \Sigma x_i y_i - (\Sigma x_i\ \Sigma y_i)/n$$

and

$$SSx = \Sigma x_i^2 - (\Sigma xi)^2/n$$

$$\text{Remainder } SS = \text{Total } SS - \text{Regression } SS$$

Whether the straight line is appropriate for measuring regression is tested by F test.

$$F(1, n - 2\ df) = \frac{\text{Regression } MS}{\text{Remainder } MS}$$

Deviation from regression is also tested by conducting F test as

$$F = \frac{\text{Remainder } MS}{\text{Error } MS}$$

Error MS used here is obtained from the general analysis of variance table.

For testing whether b differs significantly from 0.0 we will have to carry out t test as

$$t = \frac{b}{SE(b)} \quad \text{where } SE(b) = \sqrt{\frac{\text{Remainder } MS}{SSx}}$$

Thus through regression analysis we can examine: (i) whether y does depend on x; (ii) make prediction of y from x and here one can extrapolate; (iii) examine the shape of regression curve; (iv) measure of the error in Y after adjustments have been made for the effect of a related variable x as in the analysis of covariance and (v) examine the cause and effect relationship between variables (path coefficient analysis). The problem with regression analysis is that the x's are subject to error.

The adequacy of linear model can be determined by (i) the magnitude of r^2 where r is the correlation coefficient between two variables and (ii) lack of fit test applied. The linear regression equation takes the form as

$$Y = a + B(X - \bar{X})$$

Where the intercept $a = \bar{Y} - B\bar{X}$

4.5.3 Joint regression analysis

The difference between two regression coefficients say b_1 and b_2 can be due to chance or can be due to real difference. Joint regression analysis can be used to assess the homogeneity of any number of regression coefficients. When a set of genotypes is tested over a number of environments, for each genotype we will have a regression coefficient which we will see in the chapter on $G \times E$ interaction and there we would like to carry out joint regression analysis in order to examine whether or not the response of different genotypes can be explained with the help of single joint linear regression coefficient. In the joint regression analysis the ANOVA takes the form given in Table 4.10.

Table 4.10 ANOVA in joint regression analysis

Item	d.f.
Joint regression	1
Heterogeneity of regression	$s - 1$
Remainder	$s(t - 2)$

s = number of genotypes, t = number of environments ($j = 1 \ldots t$)

The sum of squares due to different items will be calculated as follows:

$$\text{Joint regression } SS = (\sum_j SPxy)^2 / \sum_j SSx$$

$$\text{Heterogeneity of regression} = \sum_j \text{Regression } SS - \text{Joint regression } SS$$

Remainder $SS = \Sigma$ Remainder ss

Significant joint regression item shows that the joint regression coefficient is not zero and there exists an overall relationship between x and y.

Joint regression coefficient is calculated as:

$$b = \sum_j SPxy / \sum_j SSx$$

with $SE(b) = \sqrt{\text{Joint reminder } MS / \sum_j SSx}$

4.5.4 Multiple Regression

The simple linear regression model can be extended to deal with linear relations among more than two variables. The multiple regression model takes the form as:

$$Y = a + b_1 X_1 + b_2 X_2 + \ldots + b_n X_n$$

Where $b_1, b_2 \ldots b_n$ are the regression coefficients and $X_1, X_2, \ldots X_n$ are the independent variables. The general normal equation in case of multiple regression takes the following form.

$$b_1 \Sigma X_1^2 + b_2 \Sigma X_1 X_2 + \ldots + b_n \Sigma X_1 X_n = \Sigma X_1 Y$$
.
.
.
$$b_1 \Sigma X_n X_1 + b_2 \Sigma X_n X_2 + \ldots + b_n \Sigma X_n^2 = \Sigma X_n Y$$

In matrix notations this equation can be written as

$$\underset{\sim}{S} \hat{\underset{\sim}{b}} = \underset{\sim}{R}$$

where

$$\hat{\underset{\sim}{b}} = \begin{pmatrix} \hat{b}_1 \\ \hat{b}_2 \\ \vdots \\ \hat{b}_n \end{pmatrix}$$

$$\underset{\sim}{S} = \begin{pmatrix} X_1^2 & X_1 X_2 & X_1 X_n \\ X_1 X_n & \ldots & X_n^2 \end{pmatrix}$$

$$\underset{\sim}{R} = \begin{pmatrix} X_1 Y \\ X_2 Y \\ \vdots \\ X_n Y \end{pmatrix}$$

The solution for $\hat{\underset{\sim}{b}}$ is $\hat{\underset{\sim}{b}} = \underset{\sim}{S}^{-1} \underset{\sim}{R}$.

4.5.5 Partial correlation and regression

If the aim is to study the correlation between two variables, say between x_1 and y, holding the third variable say x_2, constant the correlation coefficient between x_1 and y is referred to as partial correlation coefficient and is calculated as:

$$rx_{1y} \cdot x_2 = \frac{rx_{1y} - rx_{2x2} \cdot r_y x_2}{\sqrt{(1 - r^2 x_1 x_2)(1 - r^2 y x_2)}}$$

where $r_{x_1 x_2}$ is the correlation coefficient between x_1 and x_2 and r_{yx_2} is the correlation coefficient between y and x_2. Here the correlation of the third variable x_2 to variables x_1 and y is 1.0, i.e. $rx_1(x_2) \cdot y_{(x2)} = 1.0$. In case of correlation between y and two variable x_1 and x_2, the linear model takes the from as $y = a + b_1 x_1 + b_2 x_2$, where b_1 and b_2 are infact partially regression coefficients. b_1 measures the regression of between y and x_1 at constant x_2 and b_2 measures the regression between y and x_2 keeping the x_1 constant.

4.5.6 Multiple correlation

It is the correlation between the dependent variable y and a set of independent variables $(x_1, x_2, \ldots x_n)$ and is calculated as

$$R_y \cdot x_1 x_2 \ldots x_n = \sqrt{1 - \frac{\sigma^2_{y \cdot x_1 x_2 \ldots x_n}}{\sigma^2_y}}$$

where $R_y \cdot x_1 x_2 \ldots x_n$ is the multiple correlation coefficient. Here the coefficient of determination, $R^2_y \cdot x_1 x_2 \ldots x_n$ is calculated as:

$$R^2_y \cdot x_1 x_2 \ldots x_n = 1 - \frac{\sigma^2_y \cdot x_1 x_2 \ldots x_n}{\sigma^2_y}$$

In the ANOVA of a multiple regression analysis R^2 is the fraction of $S.S.$ due to regression and $1 - R^2$ is sum of squares due to deviation from the regression.

4.6 Orthogonal Polynomial

When the linear regression is inadequate to describe the relationship over the entire range of X then the polymodels (quadratic, cubic or high degree polynomials) can be used to describe the relation.

The quadratic and cubic models take the following form:

$$Y = \beta_0 + \beta_1 X + \beta_2 X^2 \text{ (quadratic)}$$
$$Y = \beta_0 + \beta_1 X + \beta_2 X^2 + \beta_3 X^3 \text{ (cubic)}$$

Orthogonal polynomials are equations such that all are pairwise uncorrelated or orthogonal. The expected value of Y can then be calculated following the general equation

$$\hat{Y} = Y + b_1 \lambda_1 \varepsilon_1 + b_2 \lambda_2 \varepsilon_2 + b_3 \lambda_3 \varepsilon_3 \ldots$$

where $\varepsilon_1, \varepsilon_2, \varepsilon_3 \ldots$ are chosen to be orthogonal polynomial which for equally spaced X's are defined as

$$\varepsilon_1 = \frac{X_i - \overline{X}}{d}$$

$$\varepsilon_2 = \left[\left(\frac{X_i - \overline{X}}{d}\right)^2 - \frac{n^2 - 1}{12}\right]$$

$$\varepsilon_{k+1} = \varepsilon_n \varepsilon_1 \frac{k^2(n^2 - k^2)}{4(4k^2 - 1)} \varepsilon_{k-1}$$

where k is the order of polynomial, d is the increment between successive X's. Thus an independent computation of any contribution according to the degree of independent variable and an independent test of contribution can be made. The sum of squares between n observations is thus partitioned into $n - 1$ items, each having (1) degree of freedom and each of which corresponds to a regression component of characteristic order in X. The first sum of squares removed by linear regression involves ε_1, the second involves ε_2 (quadratic), the third involves ε_3 (cubic) and so on. Assuming four levels of X's (X_1, X_2, X_3 and X_4) with four Y's (Y_1, Y_2, Y_3 and Y_4) with four replications ($r = 4$) the ANOVA takes the form given below.

Item	d.f.	SS	MS	F
Treatments	$3 = n - 1$			
Linear	1		MSL	MSL/MS$_E$
Quadratic	1		MS$_Q$	MS$_Q$/MS$_E$
Cubic	1		MS$_C$	MS$_C$/MS$_E$
Error	9		MS$_E$	
Total	15			

Even if there are more than four levels of X, usually only linear and quadratic and sometimes cubic responses are of interest. For four levels of X and equally spaced treatments, the coefficients and divisor for orthogonal comparisons in regression are T_1, T_2, T_3 and T_4. They are the totals of Y_1, Y_2, Y_3 and Y_4, respectively over blocks, r. The coefficients for comparison for different levels of X can be found as

	Treatment totals				
Comparison	T_1	T_2	T_3	T_4	λ
Linear	-3	-1	$+1$	$+3$	2
Quadratic	$+1$	-1	-1	$+1$	2
Cubic	-1	$+3$	-3	$+1$	10/3

Now if the coefficients of T_1, T_2, T_3 and T_4, are designated as $K_1 K_2 K_3$ and K_4, respectively then the sum of squares is calculated as follows:

Sum of squares due to any item

$$= \frac{(K_1 T_1 + K_2 T_2 + K_3 T_3 + K_4 T_4)^2}{r \sum_{i=1}^{4} K_i}$$

where r is the number of replications. The value of K_1, K_2, K_3 and K_4 will differ according to comparison as shown above.

Significant linear mean squares will indicate linear relationship between the two variables. Quadratic mean squares will show a non-linear relationship (parabolic curve) between variables and the significant cubic mean squares shows deviation from quadratic effects. In case only linear and quadratic effects show fitting the data then the final model will take the form

$$Y = \overline{Y} + b_1 \lambda_1 \varepsilon_1 + b_2 \lambda_2 \varepsilon_2$$

The values of λ_1 and λ_2 are 2 and 2, respectively. The values of b's are calculated as

$$(K_1 T_1 + K_1 T_2 + K_3 T_3 + K_4 T_4)/r \sum_{i=1}^{4} K_i^2$$

4.7 Non-Parametric Statistics

When we are studying genotypes for viral infection or insects infestation we can rank the genotype if we have a limited number in the experiment and we can analyse the ranks rather than the percentage damage for extracting meaningful information more quickly and easily. Also, if we are studying character like structure or flavour of food products we can use ranks instead of measuring the actual amount of chemical responsible. In case of study of genotype × environment interaction we can rank the genotypes on the basis of performance and analyse the ranks rather than yield for isolating stable genotypes(s). We can also study the correlation between characters over locations. Besides ranks, the non-parametric statistics includes counts or the signs of difference for paired observation. The advantages and use of non-parametric statistics in the study of genotypes × environment interaction are discussed in Chapter 10.

4.8 Multivariate Analysis

Multivariate distribution

It is the joint distribution of two of more random variables. Let $X_1, X_2 ... Xp$ are p variables. Then

$$\underline{X} = \begin{pmatrix} X_1 \\ X_2 \\ \vdots \\ X_p \end{pmatrix}$$

where X is the mean of a random sample of n observations of vector which is distributed with mean $E(X) = \mu$ and variance (covariance) or dispersion matrix

$$\underline{\Sigma} = \{\sigma_{ij} = \Sigma(X_i - \mu_i)(X_j - \mu_j)\}$$

of rank p and is said to be p variate multivariate normal if its probability density function is

$$f(X) = \frac{1}{(2\pi)^{p/2} |\underline{\Sigma}|^{1/2}} \exp\left\{-\frac{1}{2}(x-\mu)^T\right\}$$

if $p = 1$, $\frac{1}{\sqrt{2\pi}\sigma} e^{-1/2 \left(\frac{(x-\mu)^2}{\sigma}\right)}$

where $\underline{\Sigma}$ is generalized variance and there are p means, p variances and $\frac{1}{2}p(p-1)$ covariances. Through multivariate analysis we can compare means and variances of k samples from k populations. We can also study the relationship between p variables. Further, we can classify individuals into different categories on the basis of p variables.

The different multivariate analyses that are being used are principal component analysis, factor analysis, canonical analysis and discrimination and classification analysis.

4.8.1 Principal component analysis

This analysis seeks to explain variance by linear function of original random variables. Suppose there are p variables $(X_1, X_2, X_3, ... X_p)$

$$\begin{array}{ccc} \underline{Y} & = & \underline{T} & \underline{X} \\ p \times 1 & & p \times p & p \times 1 \end{array}$$

By a suitable choice of $\underline{T}$ we can transform X by rotational axes to new variables which are uncorrelated. As variance $(\underline{X}) = \Sigma$ which is positive and symmetric, there exists some $\underline{T}$ orthogonal correlation matrix such that

$$T' \underline{\Sigma} T = \underline{\Lambda} = \begin{bmatrix} \lambda_1 & 0 & 0 \\ 0 & \lambda_2 & 0 \\ 0 & 0 & \lambda_p \end{bmatrix}$$

where T' is the transpose of $\underline{T}$.

Let $\underline{Y} = T^1(\underline{X} - \mu)$ where $\mu = E(\underline{X})$

Variance $(\underline{Y}) = T'(\underline{X}) T = \underline{\Lambda}$

Thus $Y_1, Y_2, ... Y_p$ are independent functions and are called principal components and variance of Y's become

Var $(Y_i) = \lambda_i$ (eigen value)

$\lambda_1, \lambda_2, ... \lambda_p$ are given by solution

$$(\Sigma - \lambda I)\underline{t} = 0$$

where $\underline{t}$ is eigen vector. Suppose $\lambda_1 \geq 2 \geq \lambda_p > 0$, first few λ's are large and the rest are negligible. We hope that the variability of X is contained in just a few dimensions of $Y_1, Y_2,...$ where Y_1 has the greatest variance, Y_2 has the next highest possible variance, etc. The per cent of the total variance explained by Y_1 is

$$\frac{100\lambda_1}{\Sigma \lambda_i}$$

and per cent of the total variance explained by Y_1 and Y_2 is

$$\frac{100(\lambda_1 + \lambda_2)}{\Sigma \lambda i}$$

We stop when this reaches a suitably high proportion. Thus we decide to use first $q < p$ principal components. The component scores are also calculated. For data point X_i we have the scores as

$$Y_{i_1} = \underline{t}'_1 (X_i - X)$$

$$\cdot$$
$$\cdot$$
$$\cdot$$

$$Y_{iq} = \underline{t}'_q (X_i - X)$$

4.8.2 Factor analysis

This analysis seeks to explain the variance in terms of hypothetical factors. For random variables $\underline{X}MN(\mu, \Sigma)$, the m factor model takes the form as

$$\underline{X} = \underline{\mu} + \underline{\underline{\Lambda}} f + \varepsilon$$

where $f_1, f_2, \ldots f_n$ are the hypothetical factors, $\underline{\underline{\Lambda}} = \{\lambda_{ij}\}$ matrix of loadings of X's on $\underline{F}$'s and $\underline{\varepsilon}$ is the specific error variance with $\underline{\varepsilon} \sim N(\underline{O}, \underline{\psi})$ and independent of f.

$$\underline{\psi} = \begin{bmatrix} \psi_1 & 0 \\ 0 & \psi_p \end{bmatrix}$$

So, ε_i specific to X_i, $\underline{\Sigma} = \underline{\underline{\Lambda}} \underline{\underline{\Lambda}}' + \underline{\psi}$

Variance of $X_i = \sigma_{ii} = \sum_{j=1}^{m} \lambda_{ij}^2$

(communalities)

$+ \psi_i^2$ (specific variance)

Here we want to know (i) the loadings of $\underline{X}$ on each factor and (ii) the partitioning of var X into communal and specific. As in principal component analysis we normally use correlation matrix, R instead of covariances, S.

$$\underline{R} = \underline{\underline{\Lambda}} \underline{\underline{\Lambda}}' + \underline{\psi}$$

If $\underline{\underline{\Lambda}}$ does not appear to have useful interpretation we can rotate to $\underline{\underline{\Lambda}} \underline{T}$ where $\underline{T}$ is an arbitrary orthogonal rotation.

$$\underline{\Sigma} = \underline{\underline{\Lambda}} \underline{\underline{\Lambda}}' + \underline{\psi} = (\underline{\underline{\Lambda}}' \underline{T})(\underline{\underline{\Lambda}}' \underline{T}) + \underline{\psi}$$

Varimax criterion has been used to maximize 'Factor simplicity' which means at getting a simple structure with variables loadings heavily on just a few factors and very little on the rest. This is equal to variance of squared loadings

$$Sk = \frac{1}{p} \left[\sum_{j=1} (\lambda_{jk})^4 - \left(\frac{1}{p} \Sigma \lambda_{jk}\right)^2 \right]$$

The factor analysis is a principal component solution when the specific variance is zero.

4.8.3 Canonical analysis

Here we may hope to summerise all useful correlations between X_1 and X_2 set of variables in terms of just a few canonical correlations and variables.

$$\begin{pmatrix} X_1 \\ X_2 \end{pmatrix}_{p_2 \times 1}^{p_1 \times 1} = \begin{bmatrix} \Sigma 11 & \Sigma 12 \\ \Sigma 21 & \Sigma 22 \end{bmatrix}$$

$p_1 \times p_2$ matrix

We seek a dimensionality reduction (rather like we did with principal components) by defining linear function of X_1 and linear function of X_2 whose correlations explain all the covariances. We seek $U = \alpha'_1 X_1$ and $V = \beta' X_2$. We find α, β such that corr $(U, V) = \lambda$ is maximised. α and β are correlation eigen vectors. Canonical correlation λ between U and V is given by max λ where, λ^2 is max α. The $\lambda_1, \lambda_2, \ldots \lambda_{p1}$ are given by the solution:

$$|\lambda^2 \underline{I} - \underline{\Sigma}_{11}^{-1} \underline{\Sigma}_{12} \underline{\Sigma}_{22}^{-1} \underline{\Sigma}_{21}| = 0$$

There are thus p_1 pairs of canonical variables

$U_1 V_1$ with correlation λ_1

$\cdot$
$\cdot$
$\cdot$

$U_{p1} V_{p1}$ with correlation λ_p

and here $\lambda_1 > \lambda_2 \ldots \lambda_{p1}$.

4.8.4 Discrimination and classification

If p character $\underline{X}$ has been scored on individuals then on the basis of character $\underline{X}$ how can we classify individuals into one of the $\underline{K}$ populations.

4.8.4.1 Case of two populations

The aim of the discriminate analysis of Fisher (1936) is to choose the linear function, $l^1 X$

$$Z = l_1 X_1 + l_2 X_2 + l_3 X_3 + l_4 X_4 = \underline{\mathbf{I}}^1 \underline{\mathbf{X}}$$

which gives maximum discrimination between the two populations say I and II. X_1, X_2, X_3 and X_4 are the four characters scored on individuals of the two populations. The vector $\underline{\mathbf{I}} = \underline{\mathbf{S}}^{-1} \underline{\mathbf{d}}$ where

$$\underline{\mathbf{S}} = \frac{1}{n_1 + n_2 + -2} (\underline{\mathbf{S}}_I + \underline{\mathbf{S}}_{II})$$

$$\underline{\mathbf{d}} = \underline{\mathbf{X}}_I - \underline{\mathbf{X}}_{II}$$

$\underline{\mathbf{S}}_I$ and $\underline{\mathbf{S}}_{II}$ are the matrices of sum of squares and products about the mean for the two samples and $\underline{\mathbf{X}}_I$ and $\underline{\mathbf{X}}_{II}$ are the two observed means and n_1 and n_2 are the sample sizes. For the population the vector $\underline{\mathbf{I}}$ is an estimate of $\underline{\lambda} = \underline{\Sigma}^{-1} \underline{\delta}$. In the sample the vector $\underline{\mathbf{I}}$ is estimated as

$$\underline{\mathbf{I}} + \underline{\mathbf{S}}^{-1} \underline{\mathbf{d}}$$

The vector $\underline{\mathbf{I}}$ maximuses the variance ratio for between and within sample variation which is proportional to $(\underline{\mathbf{I}}^1 \underline{\mathbf{d}})2/(\underline{\mathbf{I}}^1 \underline{\mathbf{S}} \underline{\mathbf{I}})$, Now, the distance between the two population is

$$D^2 = \underline{\mathbf{d}}^{-1} \underline{\mathbf{S}}^{-1} \underline{\mathbf{d}}$$

Discriminate analysis would most clearly distinguish between the inter population relationship.

4.8.4.2 Discrimination between K populations

A measure of distance between two populations can be defined in the sense of Rao (1952) as a decreasing function of the probability of misclassification between populations. The D^2-statistics of Mahalanobis (1936) of estimating genetic distance (Mahalanobis distance) between population is discussed in Chapter 11. There is a link between discriminate analysis and canonical analysis. It can be shown that 1st canonical variable $\underline{\alpha}' \underline{X}$ is the same as 1st discriminate function (corresponds to λ. max, λ. is an eigen vector of $\underline{W}^{-1} B$) and thus in principle discriminant analysis can be done by canonical analysis programme. β is the between sample variance matrix and $\underline{W}$ is within sample variance matrix.

4.8.5 Principal co-ordinate analysis

Suppose we have p variables and n objects (genotypes or environments). We see similarity between the two objects. Given n objects association matrix between that objects

$$\underline{\mathbf{A}} = \{aij\} \equiv aij$$ some measure of similarity between objects, It seeks to represent the objects as points in space with some metric (not necessarily Euclidean) such that similar objects are close in space and dissimilar are far apart. A distance metric from objects is defined as

$$d_{ij}^2 = a_{ii} + a_{jj} - 2_{aij}$$

If A is asymmetric, $d_{ij}^2 = a_{ii} + a_{jj} - a_{ij} - a_{ji}$. The matrix $\underline{\alpha}$ is formed which is defined as

$$a_{ij} = a_{ij} - a_{i\cdot} - a_{\cdot j} + a_{\cdot\cdot}$$

where a_i = row means; a_j = column means and $a_{\cdot\cdot}$ = grand mean and $\alpha_{i\cdot} = 0$ and $\alpha_{j\cdot} = 0$, etc. Now eigen values $(\lambda_1 \ldots \lambda_n)$ and vector $(C_1 \ldots C'_n)$ of $\underline{\alpha}$ are calculated as

$$\underline{\alpha} = \lambda \, \underline{C}_1 \underline{C}'_1 + \lambda_2 \underline{C}_2 \underline{C}'_2 + \ldots \lambda_n \underline{C}_n \underline{C}'_n$$

Eigen vector is rescaled so that

$$\underline{C}'_i \underline{C}_i = \lambda_i$$

Then, $\underline{\alpha} = \underline{C}_1 \underline{C}'_i + \underline{C}_2 \underline{C}'_2 + \ldots \underline{C}_n \underline{C}'_n$

Points $Q_1, \ldots Q_n$ for objects are defined. Taking point Q_i, the ith row coordinates $(C_{11}, C_{12} \, C_{1n})$ are points in Euclidean space with distances apart = d_{ij}^2 = distance between objects as shown below:

	C_1	$C_2 \ldots$	C_n
Q_1	C_{11}	C_{12}	C_{1n}
Q_n	C_{n1}	C_{n2}	C_{nn}
	$\overline{C}_1$	$\overline{C}_2$	$\overline{C}_n$

We get the approximation to the distance given by

$$d_{ij}^2 = \alpha_{ii} + \alpha_{jj} - 2\alpha_{ij} = \sum_{k=1}^{n} (C_{ik} - C_{jk})^2$$

$$= \Delta^2(Q_i, Q_j) = \text{Euclidean distance}$$

As the origin of scaling is arbitrary we have choice of α to ensure that the mean co-ordinates $\overline{C}_1 = \overline{C}_2 = \overline{C}_n = 0$ and the centre of the map is at origin. The n dimensions is reduced to $n - 1$ as $\alpha_{i.} = \alpha.j = 0$

$$\Delta^2(Q_i, Q_j) = \text{Euclidean distance}$$

Total distance = $\sum_{i<j} d_{ij}^2 = n \sum_{i=1}^{n} \lambda_i$

If $\lambda_1 > \lambda_2 ... \lambda_r > \lambda_{n-1}$, we can stop at λ_r where the $\Delta^2(Q_j^r, Q_j^r) \simeq d_{ij}^2$. Thus we decide to use first r dimensions (corresponding to the biggest eigen value). It is in this sense that it is similar to principal component analysis.

4.8.5.1. Non-metric multidimensional scaling

Principal co-ordinates approximate true distances by distances in Euclidean space. In non-metric scaling we do not try to approximate but only preserve the rank order in whatever space we choose, i.e. given objects $O_1, O_2, ... O_n$ with distances d_{ij} (or assuming a_{ij}) we are trying to find co-ordinates X_i for each O_i in space of low dimensions m with some metric (not necessarily Euclidean)Δ. We apply rule such that as far as possible if $d_{ij} < d_{qr}$ then, $\Delta_{ij} < \Delta_{qr}$ for all i. j, q, r.

It is an iterative process. Start with an initial solution (e.g. principal coordinates) for particular m, the number of dimensions. Rank the true order distances as $d_{i_1 j_1} < d_{i_2 j_2} ... < d_{i_N j_N}$. For initial solution Δ_{ij} is computed and if, Δ_{ij} obeys the above rank order distances then find the best monotonic curve. If not then seek the best monotonic approximation to Δ say Δ' which does satisfy the following: .

$$\hat{\Delta}_{i_1 j_1} < \hat{\Delta}_{i_2 j_2} ... \hat{\Delta}_{i_m j_m}$$

Chose $\hat{\Delta}$ to mini mise stress coefficient

$$S = \sum_{i<j} \frac{(\Delta_{ij} - \hat{\Delta}_{ij})^2}{\Sigma \Delta_{ij}^2}$$

The resulting maps are often similar to principal coordinates map.

Procrustes rotation (Gower, 1975) is used to examine the robustness of any observed structure. Given points $\underline{Y}$ (environmental map) and $\underline{X}$ (varietal map)

$$\underline{Y} = a + b\underline{T} + \underline{X}$$

where a = origin; b = scale; $\underline{T}$ = rotation matrix (orthogonal in the sense that $\underline{TT'} = \underline{I}$). We choose scale, choose origin and rotate $\underline{X}$ to fit $\underline{Y}$ and these are done by least square.

This analysis is similar to principal component which is a $p \times p$ matrix reduced to small dimension, r, variable space (r) in the sense that the column of this matrix $\underline{X}$ becomes the row of the $\underline{R}$ matrix in principal coordinate analysis.

4.9 Growth Curves

Population growth can be described by the following curves.

1. Exponential growth curve
2. Logistic growth curve

The curve is represented by the equation

$$X_t = X_0 e^{rt}$$

where X_t is the population size at time t, X_0 is the population size at time 0, e is a constant (2.71828), r the rate of increase and t the time interval. The unit of time can be minutes, days, months, years, etc. This equation shows that the number of organism (population size) will continue to increase indefinitely, i.e., without limit and always faster and faster as shown in Figure 4.1. The increase in population size is at a compound interest rate. In other words, the rate of increase at any time is proportional to the size already attained which can be mathematically shown by the equation as

$$\frac{dX}{dt} = rX = (b - d)X$$

where t is the time, $r = (b - d)$ where b is the birth rate and d is the death rate and r is a constant and is

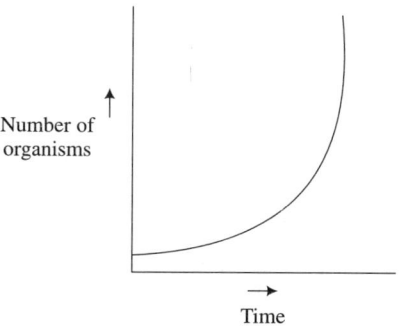

Fig. 4.1 Exponential growth curve showing number of organisms increasing toward infinity

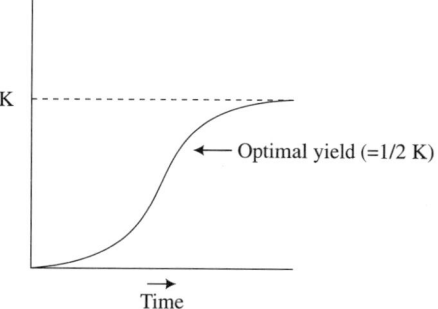

Fig. 4.2 Logistic growth curve. K is the carrying capacity of environment.

called the intrinsic rate of increase or the Malthusian parameter. $\dfrac{dX}{dt}$ is the rate of increase of X with the t and x is the number of individuals in the population at a given moment. The growth of natural population, the spread of epidemic or the growth of bacterial populations follow the exponential distribution. The exponential distribution arises in situation in which events occur at random (i.e. chance of an event occuring in a short time does not depend on how many events have already occured) in time. Sometimes r can be negative as in case of decay of emission from a radioactive element and the population size will decline exponentially. The logarithm of the exponential equation takes the form of a linear regression equation ($y = a + \beta x$).

It is assumed that the population live in an ideal environment having enough space and resources, free from competitors, diseases and insects pest and so forth but in reality it is not. Any factor can be limiting and it will limit further growth and so the exponential growth of a population cannot continue for ever. Thus various forces start operating against the increase and bring back the population to what we call a stable or equilibrium population size and this is how the population growth is regulated. This type of growth curve which is S-shaped as shown in Figure 4.2 is called a logistic growth curve. It is also called a sigmoid curve.

The growth equation takes the form as

$$\frac{dX}{dt} = r\frac{(K-X)}{K} = r\left(1 - \frac{X}{K}\right)$$

where K represents the upper limit of the population size and is, infact, the stable or equilibrium population size. The population will increase exponentially so long as X is less than K. At the equilibrium point, i.e. when $X = K$, $\dfrac{dX}{dt} = 0$, i.e. population ceases to grow. In other words, at equilibrium the population size X fluctuates up and down around some average X value. The number of individuals (X) at which $\dfrac{dX}{dt} = 0$ is called the carrying capacity of the environment. However, the moment X is greater than K (it can happen if there is drought or in case of crop disease, the crop may come to maturity) the population size is decreased. But when X is near zero, the dX/dt becomes close to rX and the growth is nearly exponential. Thus the population growth follows the exponential growth equation only under special circumstances and for a short period. If a growth character say plant height which is a metric character is plotted against number of days the resulting curve will be a sigmoid curve and if it is changed to a log scale, the resulting curve will look like as shown in Figure 4.3. Which shows that the growth period can be divided into the following three phases:

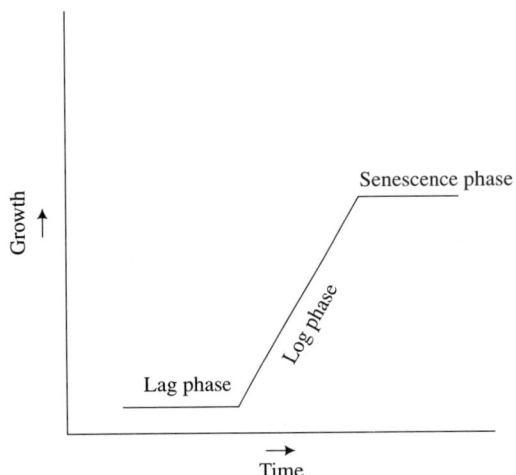

Fig. 4.3 Sigmoid curve in log scale.

(i) Lag phase of growth which indicates the slow phase of growth and is the period before the establishment of seedling in the field.
(ii) Log phase of growth which is the period of rapid increase of growth. The log phase of growth is frequently called the phase of exponential growth.
(iii) Senescence phase of growth which is the period of reduced size increase.

Time required for a specific change in the number of population.

The equation can be rewritten as

$$\frac{dX}{X} + \frac{dX}{K-X} = r\,dt$$

which upon integration gives the value of t, the time required to change the population number from X_0 to Xt.

$$t = \frac{1}{r}\ln\frac{X_t(K-K_0)}{(K-X_t)X_0}$$

But t from logistic equation can be calculated as

$$t = \frac{1}{r}\ln\frac{X_t}{X_0}$$

Thus we see that the time required for a specific change in the number of population is inversely proportional to r, the intrinsic rate of increase or the Malthusian parameter.

From the above equation the maximum rate of growth of the population in a particular environment can be estimated by differentiating the equation with respect to N which becomes

$$\frac{d}{dX}\left(\frac{dX}{dt}\right) = r\left(1 - \frac{2X}{K}\right)$$

After setting the derivative equal to zero, the population size that provides the maximum rate of growth of population is obtained as $X = \frac{K}{2}$. That is when $X = K/2$ the shape of the growth curve is steep. In other words, the rate of increase is greatest. This is also the greatest rate at which individuals can be removed without further reducing the populations size.

4.9.1 Competition

Suppose there are two species A and B competing with each other in a particular environment. Let α be the competition coefficient of species B with respect to species A and β be the competition coefficient of species A in competition with species B. The logistic equations for species A and B will be

$$\frac{dX_1}{dt} = r_1 X_1 \frac{(K_1 - X_1 - \alpha X_2)}{K_1}$$

and

$$\frac{dX_2}{dt} = r_2 X_2 \frac{(K_2 - X_2 - \beta X_1)}{K_2},$$

respectively where X_1, and X_2 are the numbers of individuals of species A and B, respectively and K_1, is the carrying capacity of environment when species A is present alone and K_2 is the carrying capacity of the environment with respect to species B when species A is not present. The carrying capacity K_1, is reduced by αX_2 in the presence of species B whereas the carrying capacity of the environment for species B, K_2, is reduced by βX_1, in the presence of species A. At equilibrium, each species will show zero growth. Thus

$$\frac{dX_1}{dt} = r_2 X_1 \left(\frac{K - X_1 - \alpha X_2}{K_1}\right) = 0 \text{ for species } A$$

and

$$\frac{dX_2}{dt} = r_2 X_2 \left(\frac{K_2 - X_2 - \beta X_1}{K_2} \right) = 0 \text{ for species } B$$

These equations yield the value of X_1, and X_2 as

$$X_1 = K_1 - \alpha X_2$$

and
$$X_2 = K_2 - \beta X_1$$

When the zero growth curve of species A falls outside the zero growth curve of species B, i.e. when species A is increasing and species B is decreasing their joint abundance will change in time leading to the growth of species A at the expense of species B and this will lead ultimately to the extinction of species B. Considering only one species occupying the environment at a time, the equilibrium condition will be $dX_1/dt = 0$ for species A and here $X_1 = K_1$ and $dX_2/dt = 0$ for species B and here $X_2 = K_2$

References

Fisher, R.A. 1946. Statistical Methods for Research Workers (10th edition) Oliver and Boyd, Edinburgh.

Fisher, R.A. and Yates, F. 1963. Statistical Tables. 6th edn. Longman, London.

Kempthorne, O. 1969. An Introduction to Genetics Statistics. Iowa State Univ. Press, Ames, Iowa.

Mardia, K.Y., Kent, J.D. and Bibby, J.M. 1979. Multivariate analysis. Academic Press, London.

Mather, K. 1940. Statistical Analysis in Biology (2nd edition), Methuen, London.

Morrison, D.F. 1976. Multivariate Statistical Methods. McGraw Hill, Tokyo.

Seal, H. 1964. Multivariate Statistical Analysis for Biologist. Methuen, London.

Searle, S. 1971. Linear Models, John Wiley & Sons, New York.

Senedecor, G.W and Cochran, WG. 1967. Statistical Methods. Iowa State University Press, U.S.A.

Steel, R.G.D. and Torrie, J.H. 1980. Principles and Procedures of Statistics, McGraw-Hill, Kogakusha Ltd.

Wilson, E.O. and Bossert, W.H. 1971. A Primer of Population Biology. Sinauer Associates, Inc. Publishers, Sunderland, Masschusetts.

Wright, S. 1968. Evolutions and Genetics of Populations Vol. I. The University of Chicago Press, Chicago.

5

Analysis of Means

The phenotypic value or effect (*P*) of an individual for a character can be considered as equal to the sum of genetic effects (*G*) and environmental effects (*E*) which are producing the characteristics of the individual.

Assuming no genotype × environment interaction

$$P = G + E$$

The environment could be physical or genetical in nature. There are different approaches to the estimation of the genetic effects. One approach is based on principles developed by Fisher, Immes and Tedin (1932), Mather (1949) and Mather and Jinks (1971). The underlying simple model in Mather and Jink's approach is shown in Figure 5.1. Considering one locus with two alleles (*A* and *a*) there will be three genotypes *AA*, *Aa* and *aa*. The genotype *Aa* is obtained as a result of crossing between the two parents $P_1(AA)$ and $P_2(aa)$. When these three genotypes are raised in an environment say Environment 1 and phenotypic values are recorded, *m* denotes the mean of the two pure breeding lines and is the origin from which all genetic effects are defined as deviations. This *m* represents the effects of common genes and the common environment that the genotypes share. When the same genotypes are grown in another environment say Environment 2 there is a change in the phenotypic values of these genotypes and hence there will be a shift in *m*, as shown in Figure 5.1(a-b). Now *aa* genotype deviates from *m* by an amount called – *da*, the additive effect and *AA* deviates from *m* by + *da* which are, in fact, the effect of decreasing allele, *a* and increasing allele, *A*, respectively. The two homozygotes (*AA* and *aa*) differ by a quantity, 2*da*. *Aa* genotype deviates from *m* by an amount called *ha*, the dominance effect. This *ha* can take sign positive or negative depending upon

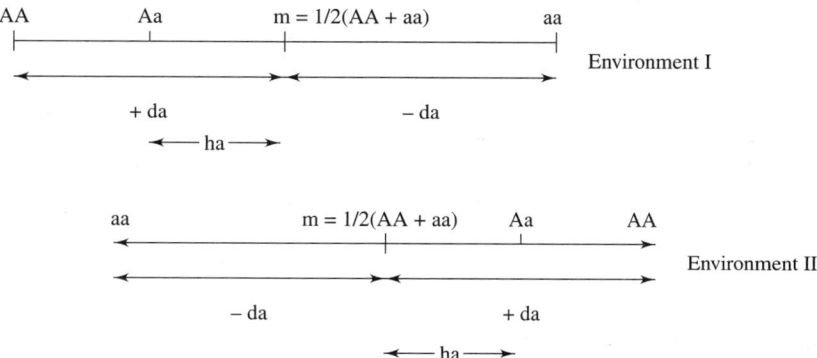

Fig. 5.1 (a-b) *a*, shows the *m*, the mean and additive effect (*da*) and dominance effect (*ha*) 1b. shows the shift in *m* as a result of change in the environment.

5.2 Biometrical Genetics–Analysis of Quantitative Variation

whether A is dominant to a or vice versa whereas the additive effects are always positive.

Extending this model to a polygenic case (k loci) the expectations of means of two pure breeding lines and their F_1 will be as follows:

$$\overline{P}_1 = m + [d]$$
$$\overline{P}_2 = m - [d]$$
$$\overline{F}_1 = m + [h]$$

where $\quad [d] = \Sigma da + db + ... + dk$
and $\quad [h] = \Sigma ha + hb + ... + hk$

These expectations will hold so long as P_1 contains all increasing alleles and P_2 contains all decreasing alleles. But when P_1 and P_2 contain both increasing as well as decreasing alleles then $[d]$ becomes

$$[d] = r_d [d]$$

where r_d is the coefficient of dispersion of alleles and is equal to

$$r_d = \sum_{i=1}^{k} di - 2 \sum_{i=1}^{k'} di$$

where k is the number of loci by which these two parents differ and k' is the number of loci at which P_1 contains the decreasing alleles whereas P_2 contains the increasing alleles. rd = 1.0 when $k' = 0$, i.e. P_1 contains all increasing while P_2 contains half increasing plus half decreasing alleles.

From the two pure lines, F_2, F_3... etc. and backcross generations can be produced. As the genotypic structure of these generations can be written in terms of P_1, P_2 and F_1 genotypes, for example F_2 generation population is made up of individuals of P_1, P_2 and F_1 genotypes; BC_1 generation population comprises of individuals of P_1 and F_1 genotypes and BC_2 generation is made up of P_2 and F_2 genotypes and so on, the expectations of means of these generations can be written in terms of three parameters namely, $[d]$ and $[h]$ and are as follows:

$$\overline{F}_2 = m + 1/2[h]$$
$$\overline{BC}_1 = m + 1/2[d] + 1/2 [h]$$
$$\overline{BC}_2 = m - 1/2[d] + 1/2 [h]$$

The general formula for writing down the expectation of means of any generation is as follows.

$$\overline{G} = m + (u - v) [d] + \beta[h]$$

where $\overline{G}$ is the mean of a generation. For one locus 2 alleles system, u is the frequency of increasing allele (A), v is the frequency of decreasing allele (a) and β is the frequency of heterozygote. In polygenic case u becomes the frequency of increasing alleles at all loci and $1 - u = v$. This formula shows that the expected generation means are all linear function of three parameters.

5.1 Scaling Tests

The above formula of expectation of means will work only when we make sure that no other sources of variation such as non-allelic interaction, genotype × environment interaction, sex linkage or maternal effect are present. To establish which of the sources is contributing to the variation at a significant level Mather (1949) devised scaling tests. A scaling test in essence is a simple relationship between the means and variances of related families which holds only if a specific source of variation is absent. With six generations (P_1, P_2, F_1, BC_1 and BC_2) the quantities A, B and C in the A, B and C scaling tests, respectively are as follows:

$$A = 2 \overline{BC}_1 - \overline{P}_1 - \overline{F}_1$$
$$B = 2 \overline{BC}_2 - \overline{P}_2 - \overline{F}_1$$
$$C = 4 \overline{F}_2 - 2\overline{F}_1 - \overline{P}_1 - \overline{P}_2$$

If we have generated and evaluated F_3 families then one can conduct D scaling test as D = $4\overline{F}3 - 2\overline{F}_2 - \overline{P}_1 - \overline{P}_2$.

Statistical tests are then carried out to test the null hypothesis that specific source of variation is not contributing at a significant level. The quantities A, B and C differing significantly from zero will invalidate the null hypothesis that non-allelic interaction is present. The principle of scaling tests is to detect deviation from additivity of gene action by combining generation means in such a way that the expected total is zero if the character is controlled by additive and dominance gene action. Nonadditivity is usually ascribed to epistasis, though other genetic causes (e.g. residual heterozygosity) could be responsible for failure of the scaling tests.

$$F_n = 1/4(P_1 + P_2) + 12F_{n-1}$$

5.2 Estimation of Parameters

The scaling criteria are correlated in the sense that P_1, P_2 and F_1 appear in all tests. To overcome this problem, Cavalli (1952) developed a joint scaling test for testing the adequacy of additive-dominance model and estimating the components of variation with their errors. The Cavalli procedure works even if the number of generations exceeds six as mentioned above. In fact, the scaling tests of Cavalli can be adapted to any generation and any type of inheritance. The unknown parameters m, $[d]$ and $[h]$ are estimated by weighted least square (Mather and Jinks, 1971).

The general linear regression model to account for the variation in a quantitative trait in terms of mean, additive and dominance effects of genes in generations following a cross between two pure breeding lines is as follows:

$$Y = b_1X_1 + b_2X_2 + b_3X_3$$

where Y is the observed mean of any generation, b_1, b_2 and b_3 are regression coefficients and represent m, $[d]$ and $[h]$, respectively and X_1, X_2 and X_3 are the coefficients of m, $[d]$ and $[h]$, respectively.

The assumptions underlying this model is that the Y's or the generation means are independently and normally distributed with similar variance. A brief outline of the weighted least square method of model fitting in terms of matrices is as follows. The different generations, their respective observed means, expected means and weights are given in Table 5.1.

Table 5.1 Generations with observed means expected means and weights

Generations	Observed mean	Expected mean (Coefficients of parameter)			Weights
	(Y_i)	(m)	$[d]$	$[h]$	(W)
$\bar{P}_1$	Y_1	1.0	1.0	0.0	W_1
$\bar{P}_2$	Y_2	1.0	−1.0	0.0	W_2
$\bar{F}_1$	Y_3	1.0	0.0	1.0	W_3
etc.					

The weight is calculated as the reciprocal of generation variance. In matrix form the Y_s' of different generations are expressed as

$$\underset{\sim}{X} = \begin{pmatrix} Y_1 \\ Y_2 \\ Y_3 \end{pmatrix},$$

the coefficients of m, $[d]$ and $[h]$ for different generation as

$$\underset{\sim}{A} = \begin{pmatrix} 1.0 & 1.0 & 0.0 \\ 1.0 & -1.0 & 0.0 \\ 1.0 & 0.0 & 1.0 \end{pmatrix},$$

the weights of the respective generations as

$$\underset{\sim}{W} = \begin{pmatrix} W_1 & 0.0 & 0.0 \\ 0.0 & W_2 & 0.0 \\ 0.0 & 0.0 & W_3 \end{pmatrix}$$

and the unknown parameters m, $[d]$ and $[h]$ as

$$\hat{\underset{\sim}{\theta}} = \begin{pmatrix} m \\ [d] \\ [h] \end{pmatrix}$$

We require values of m, $[d]$ and $[h]$ which minimize

$$Q = \Sigma W_i(Y_i - \varepsilon Y_i)^2$$

$$\underset{\sim}{Q} = (\underset{\sim}{X} - \underset{\sim}{A}\hat{\underset{\sim}{\theta}})^1 \underset{\sim}{W} (\underset{\sim}{X} - \underset{\sim}{A}\hat{\underset{\sim}{\theta}})$$

Now upon differentiation

$$\frac{\delta\theta}{\delta\hat{\theta}} = 2(\underset{\sim}{X} - \underset{\sim}{A}\hat{\theta})^1 \underset{\sim}{W}(-\underset{\sim}{A}) = \underset{\sim}{0}$$

$$-1/2 \frac{\delta\theta}{\delta\hat{\theta}} = (\underset{\sim}{X} - \underset{\sim}{A}\hat{\underset{\sim}{\theta}}) \underset{\sim}{W} \underset{\sim}{A} = \underset{\sim}{0}$$

$$\frac{\delta \underset{\sim}{u}v}{\delta\hat{\theta}} = \underset{\sim}{u}\frac{\delta \underset{\sim}{v}}{\delta\hat{\theta}} + \frac{\delta \underset{\sim}{u}}{\delta\hat{\theta}} \underset{\sim}{v} \text{ (for quadratic form)}$$

$$\frac{\delta \underset{\sim}{u}^1 \underset{\sim}{u}}{\delta\hat{\theta}} = 2\underset{\sim}{u}^1 \frac{\delta \underset{\sim}{u}}{\delta\hat{\theta}}$$

Thus
$$X'WA = \hat{\theta}'A' WA = 0$$
So
$$\hat{\theta}'A' WA = X'WA$$

Now transposing both sides

$$A' WA\hat{\theta} = A'WX \quad NBW = W'$$

Therefore, $\hat{\theta} = (A'WA)^{-1} A'WA$

and $\chi^2 = (X - A\hat{\theta})^{-1} W (X - A\hat{\theta})$

The χ^2 tests the adequacy of the model.

Mather and Jinks' method is based on the assumption that the populations have non-homogeneous variances. In theory the environmental component is same for all the populations such as P_1, P_2, F_1, F_2 and backcross generations and only the genetic components differ. But as the stability of response to the environment is a property of the genotype and/or the level of heterozygosity (Allard and Bradshaw, 1964) even though the source of variation is environmental and not genetic (Falconer, 1981) the environmental variances of the different generations populations will be increased in different amounts depending upon their sensitivity to the particular environment in which they are grown. Thus the variances of the populations will not be homogeneous. Weighted least squares is needed because the variances are not equal and when the variances are equal there would not be need for weighted least squares.

To ensure that the amount of information provided by each generation is the same, the variances of the observed generation means must be constant over all generations (Jinks and Perkins, 1968). This can be achieved by making the numbers of individuals in each generation proportional to its expected variance. The expected variances can themselves be obtained from the estimates of additive, dominance and environmental contribution to the within generation variances averaged over all previous experiments involving the same pair of inbred lines. This merely requires that for each generation the estimated additive, dominance and environmental components are summed in proportions in which they are expected to contribute to within generation variances (Mather, 1949). The relative number of individuals in each generation, arrived at in this way, can be converted into absolute numbers once the total number of plants which can be grown simultaneously is known. Although the aim of the experimental design is to equalize the amount of information provided by each generation this will be only approximately achieved in practice. It will, therefore, still be necessary to weight each generation mean by its observed amount of information to obtain maximum likelihood estimates of the parameters and a χ^2 test of the goodness of fit of the models. Hence it is essential that the estimates of the variances of the generation means from which the amounts of information are obtained as reciprocals (Is = $1/\sigma_{\bar{Y}}^2 = n/\sigma^2$ are based on as many degrees of freedom as possible. Since the number of plants in each generation is fixed by the space available, this can only be achieved by using single plant randomisation instead of the more usual randomisation of single plots containing a number of plants belonging to the same family.

From the above analysis conclusions can be drawn about the types of gene action involved in the determination of a quantitative trait. While interpreting the results one must keep in mind that the relative importance of the various types of gene action can not be inferred from the relative magnitudes of the parameters for example a relatively low value of [d] may result from either little or no additive genetic difference between the parents or dispersion of increasing and decreasing alleles between parents. Similarly, a relatively low value of [h] may be due to either little or no dominance or ambidirectional dominance. At this stage an inadequate model will indicate either the presence of non-allelic interaction or a simple two parameter model.

5.3 Epistasis or Non-allelic Interaction

Non-allelic interaction is of three kinds (Hayman and Mather, 1955; Van der Veen 1959; Mather, 1967). Considering two loci with 2 alleles each *A-a*, *B-b*, these interactions are as follows:

1. *i* type: It is the homozygote-homozygote interaction. Interactions between *A* and *B* loci in the genotypes *AABB*, *AAbb*, *aaBB* and *aabb* will be denoted as *iab*.

2. j type: It is the interaction between homozygote and heterozygote loci. It can be denoted as $j\ a/b$ or b/a. Letter before the stroke denotes homozygote at that locus and the letter after the stroke indicates heterozygote at that locus. The interactions in case of $AABb$ and $AaBB$ will be denoted as ja/b and jb/a, respectively.

3. l type: It is the heterozygote-heterozygote interaction and in case of $AaBb$ interaction will be denoted as lab.

There are three principal ways for specification of non-allelic interaction in case of two loci (Van der Veen, 1959).

(a) The F_∞-metric or Pureline-metric (Hayman 1954a; Smith and Robson 1959; Dempester 1956). $M(\theta a, \theta b) = m + \theta a da + \theta b db + (1 - \theta a^2)ha + (1 - \theta b^2)hb + \theta a \theta b iab + \theta a(1 - \theta b^2)\ ja/b + \theta b(1 - \theta a^2)\ jb/a + (1 - \theta a^2)(1 - \theta b^2)\ lab$.

(b) The Mixed-metric. (Hayman 1954a; Hayman and Mather 1955; Opsahl 1956: Jinks and Jones 1958). $M(\theta a, \theta b) = m + \theta a da + \theta b db + (1 - \theta a^2)ha + (1 - \theta b^2)hb + \theta a \theta b iab + \theta a(1/2 - \theta b^2)\ ja/b + \theta b(1/2 - \theta a^2)\ jb/a + (1/2 - \theta a^2)(1 - \theta b^2)\ lab$.

(c) The F_2-metric (Hayman (1955, 1958; Kempthorne 1957). $M(\theta a, \theta b) = m + \theta a da + \theta b db + (1/2 - \theta a^2)\ ha + (1/2 - \theta b^2)\ hb + \theta a \theta b\ iab + \theta a(1/2 - \theta b^2)\ ja/b + \theta b(1/2 - \theta b^2)\ jb/a + (1/2 - \theta a^2)(1/2 - \theta b)^2\ lab$.

m denotes the origin and the θ notation represents the phases AA, Aa and aa by the variable θa which takes the value $1, 0$ and -1, respectively. The genotypic value of an individual of genotype say $AABB$ will be

$$m + da + db + iab \text{ in } F_\infty \text{ metric}$$

$$m + da + db + iab - 1/2\ jab - 1/2\ jba + 1/4\ lab \text{ in mixed-metric}$$

and $\quad m + da + db - 1/2\ ha - 1/2\ hb + iab$
$$- 1/2j\ ab - 1/2\ jba + 1/4\ lab \text{ in } F_2\text{-metric.}$$

These expectations in case of polygenes take the following form:

$$m + [d] + [i] \text{ in } F_\infty \text{ metric}$$

$$m + [d] + [i] - 1/2\ [j] + 1/4\ [l] \text{ in mixed metric}$$

and $\quad m + [d] - 1/2\ [h] + [i] - 1/2\ [j] + 1/4\ [l]$
$$\text{in } F_2\text{-metric.}$$

The coefficients of interaction parameters, i, j and l in the F_∞-metric and in the F_2 metric are obtained by multiplication of coefficients attached to the single locus effects. The F_∞ metric differs from F_∞-metric in that the F_∞ generation mean (i.e. mean of all the pure breeding lines in F_∞ generation derived from a cross of two pure breeding lines) equals m in F_∞-metric whereas the F_2 generation mean coincides with the origin m in the F_2-metric. Although none of the metrics gives better or more information about gene action and interaction than others but the F_∞-metric is preferred for a variety of reasons. First, the coefficients of parameters are uniquely associated with homozygosity and heterozygosity at the loci. Secondly, the description of heterosis and maternal effects in terms of parameters becomes much more simpler which we shall see in later chapters. Thirdly, the F_∞-metric leads to a simpler and symmetrical conditions and description of digenic interactions.

Following F_∞-metric the expectations of means of generations following cross of two pure breeding lines can be written in the presence of digenic interactions as follows:

Generations	Expectation of means
P_1	$m + [d] + [i]$
P_2	$m - [d] + [i]$
F_1	$m - [h] + [l]$
F_2	$m - 1/2\ [h] + 1/4\ [l]$
etc.	

$$F_n = m + (1/2)^{n-1}\ [h] + (1/2)^{2(n-1)}[1]$$

And the expectation for the mean of any generation in the presence of digenic and trigenic interactions takes the form as

$$\overline{G} = m + X[d] + Y[h] + X^2[iab] + Y^2\ [lab]$$
$$+ X^3\ [iabc] + X^2Y\ [ab/c] + XY^2\ [a/bc]$$
$$+ Y^3\ [labc]$$

where X and Y are coefficients of $[d]$ and $[h]$, respectively; jab/c is the interaction between two homozygous and one heterozygous loci, ja/bc is the interaction between a homozygous and two heterozygous loci.

Following this above general formula the expectation of P_1, P_2 and F_1 means in the presence of trigenic interactions will be as follows:

$$\overline{P}_1 = m + [d] + [iab] + [iabc]$$
$$\overline{P}_2 = m - [d] + [iab] - [iabc]$$
and $$\overline{F}_1 = m + [h] + [lab] + [labc]$$

MODEL FITTING

The digenic model contains six components m, $[d]$, $[h]$, $[i]$, $[j]$ and $[l]$ and thus requires atleast seven generations to have estimates of these components and also to test the goodness of fit of the model whereas the trigenic model contains ten components and atleast eleven generations are needed to provide estimate of these components and to test the adequacy of the model. In case of failure of model containing digenic interaction parameters one should use a revised model incorporating particular forms of digenic interaction. Failure of the model at this stage could result from presence of higher order interaction like trigenic and other higher order interactions or linkage between pairs of interacting loci. The aim of model fitting is to determine the simplest model of gene action and interaction that will account for the observed variation.

Again, the parental lines may have increasing as well as decreasing alleles and thus to simplify the expectation in dispersion crosses in addition to r_d, the coefficient of dispersion of genes, r_i and r_j terms are added which are analogous to r_d and where

$$r_i = \frac{\overset{1/2k(k-1)}{\Sigma i} - \overset{k'(k-1)}{2\Sigma i}}{1/2\, k(k-1)}$$

The parameters $[i]$ $[j]$ and $[l]$ now take the form as:

$$[i] = r_i \overset{1/2\, k(k-1)}{\Sigma i}$$

$$[j] = r_j \overset{k(k-1)}{\Sigma j}$$

$$[l] = \overset{1/2\, k(k-1)}{\Sigma l}$$

As $iab = iba$ and $lab = lba$, i and j components are summed over $1/2k(k-1)$ loci whereas j is summed over $k(k-1)$ loci because $jab \neq jba$. The value of r_i ranges from 1 to -1. $r_i = 0$ when half of the pairs are associated, half dispersed and $r = -1$ when all pairs are dispersed. This shows that the real magnitudes and signs of $[d]$, $[i]$ and $[j]$ are therefore revealed only when the alleles of like effects are completely associated $(r = 1)$ in the parents. The parameters $[i]$ $[j]$ and $[j]$ show that the parameter $[i]$ can be of opposite sign to Σi and it will depend upon the sign of r_i whereas the signs of $[h]$ and $[l]$ will not be opposite to Σhi and Σli, respectively. The value of r, coefficient of dispersion, will never be -1 as with interactions involving 3 or more loci all pairs could never be dispersed.

All the classical non-allelic interactions (discussed in Chapter 2) can be identified from the relative magnitude and sign of d, h, i, j and l parameters at individual loci level (Hayman and Mather, 1955) but the same cannot be extended to the components level in polygenic case. Of the five components only $[h]$ and $[l]$ are uninfluenced by the distribution of alleles between the parents and therefore only $[h]$ and $[l]$ reflect the net direction of gene action and interaction and thus these two components can classify the predominant types of epistasis. When both $[h]$ and $[l]$ components are positive and are of higher magnitude it will indicate mainly complementary interaction between increasing genes whereas if both have negative sign and are of lower magnitude, it reflects mainly complementary interaction between dominant decreasing genes. When the component $[h]$ is positive and $[l]$ is negative, it shows duplicate interaction between increasing genes whereas negative $[h]$ and positive $[l]$ show duplicate interaction between decreasing genes.

5.4 Estimation of Coefficient of Dispersion

From the F_1 of the initial two pure breeding lines a large number of pure breeding lines (F_∞) are extracted through selfing and from this sample two pure lines are selected which represent the highest (P_H) and lowest (P_L) scoring parents for a character and thus expected to have at most of the loci increasing and decreasing alleles for which the original two parents showed heterosis in F_1. Thus, we have two sets of parents, the original P_1 and P_2 and the other P_H and P_L. From these two sets F_1, F_2, BC_1, and BC_2 families are developed and the models are then fitted to two sets separately and jointly. For the first set the parameters are m_1, $[d_1]$, $[h_1]$, $[i_1]$, $[j_1]$ and $[l_1]$ and for the second set are m_2, $[d_2]$, $[h_2]$, $[i_2]$, $[j_2]$ and $[l_2]$. Now we know that m, h and l are independent of r_d and r_i so the estimates of these three components will not differ between sets whereas d, i and j estimates which depend on the coefficient of dispersion may differ significantly between sets. Furthermore, if they do differ it will be because $r_{d2} > r_{d1}$, $r_{i2} > r_{i1}$ and $r_{j2} > r_{j1}$ and $r_{d2} = r_{i2} = r_{j2} \simeq 1$ and therefore we will find

$$[d_2] > [d_1], [i_2] > [i_1] \text{ and } [j_2] - [j_1]$$

Thus, the dispersion will lead to the under estimation of additive effects, $[d]$ and interaction between homozygous loci $[i]$ may not be detected at all. Now assuming that $m_1 = m_2 = m$, $h_1 = h_2 = [h]$ and $l_1 = l_2 = [l]$ then fitting a nine parameter model m, d_1, d_2, h, i_1, i_2 j_1, j_2 and l should be adequate which will confirm that P_H and P_L differs at the same loci for the same allele as P_1 and P_2. And if $[d_1] > [d_2]$ it further confirms that the same alleles are more associafed in P_H and P_L in comparison to in P_1 and P_2 and, therefore, d_1, i_2 and j_2 are relatively less affected by the dispersion and thus reflect the magnitude and sign of Σd, Σi and Σj more precisely. Further assuming that the association is complete ($r_d = 1$ or $r_i = 1$ and $r_j - 1.0$) the estimates of m, h, l, i_2, and j_2 will represent Σd, Σi , Σj and Σh and Σl and the coefficient of dispersion then can be estimated as:

$$r_{d1} = \frac{[d_1]}{[d_2]}$$

and

$$r_{i1} = \frac{[i_1]}{[i_2]}$$

This analysis (Jayasekara and Jinks, 1976) and (Pooni and Jinks, 1981) will give misleading results when genes are tightly linked and are still in dispersion phase in the F_∞ families. But then the probability of gene being in dispersion phase could be minimized by random mating the F_2, for few generations before extracting the F_∞ families by selfing.

5.5 Concept of Average Effect of Gene

The estimates of parameters provide an indication of the relative importance of the various kinds of gene effects affecting the total genetic variation of a plant character. But as the parameters are sum of genetic effects over loci (which can be either positive or negative), the failure to detect a particular gene effect may be due either to the absence of that effect or to the canceling of effects. Hence although these parameters provide useful information, the relative magnitude of additive, dominance and epistatic parameters estimated from generation mean analysis may not have relation to the relative importance of the various types of gene action in the F_2 population (Eberhart, 1964). But if the parameters are defined as deviation from the F_2 generation, the base population, significant estimates of genetic effects do indicate that the genetic effects of the type must be present for atleast one locus in the F_2 population and hence will contribute to the variation in F_2 population but the failure to detect certain genetic effects cannot be used as the evidence to conclude that such effects are not contributing to the variation among F_2 individuals because of canceling of positive and negative effects.

In case of random mating populations it is the gene and not the genotype that is transmitted from parent to offspring. This is where Fisher (1918, 1941) introduced the concept of average effect of gene. It is the average effect of genes of parents that will determine the mean genotypic value of its progeny. Let us now see how we can estimate the average effect of a gene. Assuming one locus with two alleles (A_1, A_2) the F_2 population will have individuals of genotypes A_1A_1, A_1A_2 and A_2A_2 with frequencies u^2, $2uv$ and v^2, respectively where u and v are the frequencies of genes A_1 and A_2, respectively. The genotypic values of these genotypes are $+d$, h and $-d$, respectively. Now either A_1 gene can be substituted by A_2 gene at random in the population or the reverse. Considering the former case that A_1 is substituted in A_1A_1 and A_1A_2 genotypes whose frequencies are $v^2/u^2 + uv$ and $uv/u^2 + uv$ i.e. u and v, respectively the genotypic value of A_1A_1 will change from d to h and of A_1A_2 from h to $-d$. The average effect of gene substitution (α) will then be

$$\alpha = u(d-h) + v(h+d)$$
$$= d + (v-u)h$$

Similarly, if A_2 is substituted in A_2A_2 and A_1A_2 genotypes at frequency v and u, respectively, the genotypic value of A_2A_2 will change from $-d$ to h and of A_1A_2 from h to $+d$. The average effect of gene substitution is, therefore,

$$\alpha = u(d-h) + v(h+d)$$
$$= d + (v-u)h$$

One can arrive at the similar derivation of α considering $\alpha = \alpha_1 - \alpha_2$ where α_1 and α_2 are the average effect of the gene A_1 and A_2 respectively. α_1 and α_2 are derived as the average deviation from the population mean due to the substitution of one gene by its allele in each genotype and are given in Table 5.2. The population mean is $d(u-v) + 2uvh$.

Table 5.2 Average effect of gene A_1 and A_2

Gene substitution	Frequency of genotypes in which gene substitution take place			Progeny value	Average effect of a gene
	A_1A_1	A_1A_2	A_2A_2		
A_1	u	v	–	$ud + vh$	$\alpha_1^* =$
A_2	–	u	v	$-vd + uh$	$\alpha_2^* =$

*$\alpha_1 = ud + uh - [d(v-u) + 2\,uvh] = v[d + (v-u)h]$
$\alpha_2 = -vd + uh - [d(v-u) + 2uvh] = -u[d + (v-u)h]$

The average effect of gene substitution (α) becomes

$$\alpha = \alpha_1 - \alpha_2$$
$$= v[d + (v-u)h] + u[d + (v-u)h]$$
$$= d + (v-u)h$$

Sum of the average effect of genes of an individual is called its breeding value (Falconer, 1960) and the variation in breeding values is attributed to the additive effect of genes. Breeding value of an individual is measured experimentally as mean of its progeny and thus is a property of an individual or a population of individuals. The breeding values of genotypes A_1, A_1, A_1A_2 and A_2A_2 will then be $2\alpha_1$, $\alpha_1 + \alpha_2$ and $2\alpha_2$, respectively and which in terms of α will be equal to $2v\alpha$, $(v-u)\alpha$ and $-2u\alpha$, respectively. The differences between breeding values and genotypic values are called dominant deviations. The genotypic values of A_1A_1, A_1A_2 and A_2A_2 can be derived as $2v(d-uh)$, $d(v-u) + h(1-2uv)$ and $-2u(d+vh)$, respectively which equals $2v(\alpha - vh)$, $(v-u)\alpha\ 2uvh$ and $-2u(\alpha + uh)$, respectively in terms of α, the average effect of gene substitution. Thus the dominant deviations for the genotypes A_1A_1, A_1A_2 and A_2A_2 will be $-2u^2h$, $2\,uvh$ and $-2u^2\,h$, respectively.

α can also be calculated as $\alpha = b_{xy}/\sigma_x^2$ where x is the number of alleles and y is the genotypic value (see Table 5.3).

Table 5.3 Showing genotype frequency, genotypic value, mean and number of alleles of three types of individuals in an F_2 population.

Genotype	AA	Aa	aa	Population mean
Frequency	u^2 (1/4)	$2uv$ (1/2)	v^2 (1/4)	
Genotypic value (Y)	D	h	$-d$	
Genotypic mean	$2v\,(d-uh)$	$d\,(v-u) + h\,(1-2uv)$	$-2u\,(d+vh)$	0
No. of alleles (X)	2	1	0	1
No. of alleles expressed as deviation	1	0	−1	0
Heterozygosity	0	1	0	½
Heterozygosity expressed as deviation	−1/2	1/2	−1/2	0

In fact a and d (Mather's d and h) were defined in tems of linear regression of genotypic value on the number of genes or heterozygosity (Fisher, 1918, 1941). The regression of genotypic value on number of gene, b_{GM} = Cov GM/Var M = Σgenotypic value × frequency × No. of alleles expressed as deviation/Variance of no.of alleles expressed as deviation (= Σ (No. of alleles expressed as deviation)^2 × (frequency) = ½ (v+u) d = d when u+v = 1.0. Regression of genotypic value on levels of heterozygosity, bGH = CovGH/Var H = 1/4h/ ¼ = h.

5.6 Other Models of Estimation of Genetic Effects

Anderson and Kempthorne (1954), Cockerham (1954, 1980) and Eberhart and Gardner (1966) proposed general models for detection and estimation of additive, dominance and epistatic effects. The factorial gene model of Anderson and Kampthorne is based on factorial model used in the design of experiment. The six parameters in the model are K_2, E, F, G, L and M. K_2 represents mean effect, E and F measure non-epistatic effects whereas G, L and M measure epistatic effects. Hayman (1958) presented a general factorial model where the generation mean is expressed as

$$\text{Mean} = m + \alpha d + \beta h + \alpha^2 i + 2\alpha \beta j + \beta^2 l$$

The genetic parameters d, h, i, j and l are sums of the gene parameters of Hayman and Mather (1953). Considering the above model the expectation of means of two pure breeding homozygous lines and their descendants are as follows:

$$\overline{P}_1 = m + d - 1/2h + i - j + 1/4l$$
$$\overline{P}_2 = m - d - 1/2h + i + j + 1/4l$$
$$\overline{F}_1 = m + 1/2h + 1/4l$$
$$\overline{F}_2 = m$$
$$\overline{B}_1 = m + 1/2d + 1/4i$$
$$\overline{B}_2 = m - 1/2d + 1/4i$$

Gamble (1962a) proposed a model similar to the above model. Thus all parameters m, d, h, j and l can be estimated directly from the means of above generations. Hayman and Mather's parameters are related to that of Anderson and Kempthorne as given below:

$$K_2 = m$$
$$E = 1/2h$$
$$F = d - 1/2h$$
$$G = 1/4l$$
$$L = j - 1/2l$$
$$M = i - j + 1/4l$$

Anderson and Kempthorne's model contains interaction between genes by which the parents differ and genes which are identical in the parents. Such epistasis must be accounted for in the model particularly when several crosses with some parents are in common are crossed because it is there the genes for which one pair of parents are identical may have alleles in other parents. In the analysis of means of two pure lines, F_1's F_2's and backcross generations such interactions may be incorporated and ignored as long as attention is confined to genetic material derived from the two parents. The factorial model of Anderson and Kempthorn is complicated even considering a 2 loci with 2 alleles. The parameters, F combining additive and dominance effects and L and M measuring pooled interaction are not easy to interpret.

Gardner and Eberhart (1966) and Eberhart and Gardner (1966) presented a general model for estimation of genetic effects including additive x additive epistasis using varieties and their crosses. The expectations of varietal means, selfs and the derived generation populations assuming additive x additive epistasis are as follows:

Y_k (kth open pollinated) variety
$$= \mu + a_k + aa_k + d_k$$

$$Y_k^s \text{ (Self)} = \mu + a_k + aa_k + 1/2\, d_k$$

$$Y_{kk'}(F_1) = \mu + 1/2(a_k + aa_k + a_{k'} + aa_{k'} + d_k + d_{k'})$$

$$F_2 = \mu + 1/2(a_k + aa_k + a_{k'} + a_{k'}) + 1/4\,(d_k + d_{k'}) + 1/4h_{kk'} + aa_{kk'}$$

where μ is the mean, a_k or $a_{k'}$ is the additive genetic effects in variety k or k' and aa_k or $aa_{k'}$ is the corresponding additive x additive interaction, d_k or $d_{k'}$ is a measure of inbreeding depression in variety k or k' and $h_{kk'}$ measures the heterosis and this parameter is due to differences in gene frequencies in k and k' varieties and to dominance and if additive $\times$ additive type interaction is present it will contribute to $h_{kk'}$.

In case of pure breeding homozygous varieties d_k becomes zero. These above equations for the means of the populations can be generalized by incorporating f_i, the coefficient of inbreeding (Carbonell et al. 1983). This is done by multiplying all d and h terms by $1 - f_i$, and the resulting expression can be derived by including f_i, in the expression for the genotypic frequencies. The population means after one generation of selfing can be obtained by simply substituting $f_i = 1/2$ in the appropriate equation.

The genetic architecture of the trait is known using standard least square procedure. The sum of squares for each item in the ANOVA is obtained by subtracting for the reduced model from that for the full model. The mean squares due to deviation from regression for the full model provides an F test for testing the adequacy of the model. If the model is inadequate a more complex model is required but then all the parameters in the model are biased. The parameters in the model are not orthogonal and in the presence of additive × additive interaction both additive and additive × additive effects are confounded. Further, inclusion of additive × dominance and dominance × dominance interactions and linkage in this model will result into many parameters to be estimated which would make the prediction of mean of any generation impossible. However, the prediction formulae for single and double crosses work in the absence of epistasis. The parameters in the model are related to Hayman and Mather's parameters in the absence of additive × dominance and dominance × dominance epistasis as follows:

$$m = \mu + 1/2\ h_{kk'} + aa_{kk'}$$

$$d = a_k + aa_k = -(a_{k'} + aa_{k'})$$

$$h = h_{kk'}$$

and $\quad i = -aa_{kk'}$

Cockerham (1980) presented a model based on factorial model of gene effects and is descriptive of the parental sources of genes.

$$G = \mu + \Sigma\ \alpha_i A_i + \Sigma\ \delta_{ij} D_{ij} + (\Sigma\ \alpha_i A_i)^2$$
$$+ (\Sigma\ \alpha_i A_i)(\Sigma \delta_{ij} D_{ij}) + (\Sigma \delta_{ij} D_{ij})^2 + \ldots$$

Further,

$$(\Sigma\ \alpha_i A_i)^2 = \sum_i \alpha_i^2\ (AA)_{ii} + 2 \sum_{i<j} \alpha_i \alpha_j\ (AA)_{ij}$$

where $\dfrac{\alpha_i}{2}$ is the proportion of genes in the entry from ith parent and $\alpha_i = 2\delta_{ii} + \delta_{ij}$, Ai is the sum of additive effects for the genes from ith parent, $\delta_{ij}(\delta_{jj})$ is the proportion of genotypes having alleles from parents i and j and D_{ij} is the sum of dominance effects.

The coefficients of δ will change depending upon whether the parents are pure breeding homozygous lines or open pollinated varieties and thus change with the level of inbreeding. The expectations of means of any generation following cross of two or more parents can be written and the parameters in the model can be estimated by least square. In case of homozygous lines $\delta_{ij} = 1$ and $\alpha_i = \alpha_j = \delta_{ij} = 1$ and thus the expectation of means of P_1, P_2, F_1 and F_2 populations will be as follows:

$$P_1 = u + 2A_1 + D_{11} + 4(AA)_{11}$$
$$+ 2(AD)_{1(11)} + \ldots$$

$$P_2 = u + 2A_2 + D_{22} + 4(AA)_{22}$$
$$+ 2(AD)_{2(22)} + \ldots$$

$$F_1 = u + A_1 + A_2 + D_{12}$$
$$+ (AA)_{11} + 2(AA)_{12} + (AA)_{22}$$
$$+ AD_{1(12)} + AD_{2(12)} + \ldots$$

$$F_2 = u + A_1 + A_2 + \frac{D_{11}}{4} + \frac{D_{12}}{2}\frac{D_{22}}{4}$$
$$+ (AA)_{11} + 2(AA)_{12} + (AA)_{22} +$$

$$\frac{(AD)_{1(11)}}{4} + \frac{AD_{1(12)}}{2} + \frac{(AD)_{1(22)}}{4} + \frac{(AD)_{2(11)}}{4}$$

$$+ \frac{(AD)_{2(12)}}{2} + \frac{(AD)_{2(22)}}{4} + \ldots$$

In this model as well as models of Anderson and Kempthorne, and Falconer there appears a dominance component in the expectation of pure breeding homozygous lines which looks unconvincing. Also, in these models unlike Mather's there is no simple way of including linkage parameters in the presence of non-allelic interaction and getting the estimates.

5.7 Detection of Epistasis by Other Methods

Epistasis can be detected by studying the relationship between the mean performance of individual and the level of its heterozygosity (Kempthorne, 1957) and the linear relationship between the two would indicate epistasis to be either negligible or undetectable. A linear relationship does not mean absence of epistatic effects because the net epistatic effects could be zero because of cancellation of positive and negative epistatic effects. As there is relation between hybrid vigour and inbreeding depression so the epistasis can also be detected by studing the relationship between the mean performance of individual and the different levels of homozygosity (i.e. inbreeding) and a curvilinear relationship would indicate presence of epistasis and the linear relationship would indicate additive gene actions with some level of dominance.

Epistatic effects when estimated from two parents and their derived generations may be confounded with genotype × environment interaction effects which result from the differential response of the generation under different levels of inbreeding to their environment. If single, three way and double crosses when progenies are non-inbred are used for estimation (Rawlings and Cockerham, 1962a, b), the possibility of confounding of epistasis with genotype × environment interaction is eliminated but then as we can see that the main effects and interaction effects are not completely free from other higher order gene effects unlike the generation mean analysis.

The underlying model is

$$G_{ij} = \mu + A_i + B_j + (AB)_{ij}$$

where A_i and B_j are the effects of the parents A_i and B_j, respectively and $(AB)_{ij}$ is the effect of the single cross progeny. The design effects have the following expectations in terms of genetic effects (Cockerham, 1980).

$$A_i = A_i + (AA)_{ii} + (AAA)_{iii} + ...$$
$$B_j = A_j + (AA)_{jj} + (AAA)_{jjj} + ...$$
$$(AB)_{ij} = D_{ij} + 2(AA)_{ij} + (AD)_{i(ij)}$$
$$+ (AD)_{j(ij)} + DD(ij)(ij)$$
$$+ 3(AAA)_{iij} + 3(AAA)_{ijj} + ...$$

In case of three parents A_i, B_j and C_k and three-way crosses, $A(BC)$ the model takes the form as

$$G_{i(jk)} = \mu + A_i + B_j \text{ and } C_k + (AB)_{ij}$$
$$+ AC)_{ik} + (BC)_{jk} + (ABC)_{iik}$$

The design effects have the following expectations.

$$A_i = A_i + (AA)_{ii} + (AAA)_{iii} + ...$$
$$B_j = 1/2\, A_j + 1/4\, (AA)_{jj} + 1/8\, (AAA)_{jjj} + ...$$

$$(AB)_{ij} = 1/2\, D_{ij} + 2/2\, (AA)_{ij}$$
$$+ 1/2\, (AD)_{i(ij)} + 1/4\, (AD)_{j(ij)}$$
$$+ \frac{1}{4}(DD)_{(ij)(ij)} + \frac{3}{2}(AAA)_{iij}$$
$$+ \frac{3}{4}(AAA)_{ijj} + ...$$
$$(BC)_{jk} = 2/4\, (AA)_{(jk)} + 2/8\, (AAA)_{ijk} + ...$$
$$+ 3/8\, (AAA)_{jkk} + ...$$
$$(ABC)_{ijk} = 1/4\, (AD)_{k(ij)}\ 1/4\, (AD)_{(ijk)}$$
$$= 2/4\, (DD)_{(ij)(ik)} + 6/4\, (AAA)_{ijk} + ...$$

For C_k and $(AC)_{ik}$ is substituted for j in B_j and $(AB)_{ij}$, respectively.

With four parents A_i, B_j, C_k and D_l and the double crosses, $(AB)(CD)$ the model is

$$G_{(ij)(kl)} = u + A_i + B_j + C_k + D_l + (AB)_{ij}$$
$$+ (AC)_{ik} + (AD)_{il} + (BC)_{jk} + (BD)_{jl}$$
$$+ (CD)_{kl} + (ABC)_{ijk} + (ABD)_{ijl}$$
$$+ (ACD)_{ikl} + (BCD)_{jkl}$$
$$+ (ABCD)_{ijkl}$$

with $A_i = 1/2 A_i + 1/4\, (AA)_{ii} + 1/8\, (AAA)_{iii} + ...$
$$(AB)_{ij} = 2/4\, (AA)_{ij} + 3/8\, (AAA)_{iij}$$
$$+ 3/8\, (AAA)_{ijj} + ...$$
$$(AC)_{ik} = 1/4\, D_{ik} + 2/4\, (AA)_{ik}$$
$$+ 1/8 (AD)_{i(ik)} + 1/8\, (AD)_{k(ik)}$$
$$+ 1/16\, (DD)_{(ik)(ik)} + 3/8\, (AAA)_{iik}$$
$$+ 3/8\, (AAA)_{ikk} + ...$$
$$(ACD)_{ikl} = 1/8\, (AD)_{k(il)} + 1/8\, (AD)_{l(ik)}$$
$$+ 2/16\, (DD)_{(ik)(il)}$$
$$+ 6/8\, (AAA)_{ikl} + ...$$
$$(ABCD)_{ijkl} = 2/16\, (DD)_{(ik)(jl)} + 2/16\, (DD)_{(il)(jk)}$$
$$+ D(AAA) + ...$$

Thus it can be seen that in case of single crosses the elimination of A and B effects eliminates all additive and some additive types of epistatic effects, leaving in the deviations (AB) all of the dominance and most of the epistatic effects. The analysis of single crosses thus provides test for dominance and epistasis whereas the analysis of three-way crosses provides tests for additive × dominance (AD) and higher order effects and that of double crosses for dominance × dominance (DD) and higher order effects.

Bauman (1959) introduced a test for epistasis using three-way crosses. With three parents A, B and C there will be three single crosses, $A \times B$, $B \times C$ and $A \times C$. In the absence of epistasis

$$\frac{\overline{(A \times C)} + \overline{(B \times C)}}{2} = \overline{(A \times B) \times C}$$

Thus the quantity,

$$\left[\frac{\overline{(A \times C)} + \overline{(B \times C)}}{2} - \overline{(A \times B) \times C}\right]$$

deviating significantly from zero will be a test of presence of epistasis. In the above example the parent C is called tester. Actually any of these parents can be used as tester. Thus there will be three three-way crosses and a combined test for epistasis with three degrees of freedom can be constructed. However, epistatic variance cannot be estimated by this procedure.

So far we have seen the gene effects determined exactly by the genetic entries in the experiment and more explicitly the phenotype is found as the algebraic sum of all the parameters associated with the genotypes in question. In the next chapter we will see that the quadratic functions (sum of squares or mean squares) provide a means of estimating the variances of these gene effects. Means are subjected to much lower error variance than variances and covariance and thus their estimates and comparisons involving these estimates are more precise as compared to variances and co variances and their comparisons for a given number of individuals recorded.

References

Allard, R. W. and Bradshaw, A.D. 1964. Implications of genotype-environment interaction in applied plant breeding. Crop Sci., **4**: 503–508.

Anderson, V.L. and Kempthorne, O. 1954. A model for study of quantitative inheritance. Genetics, **39(6)**: 883–898.

Bauman, L.F. 1959. Evidence of non-allelic gene interaction in determining yield, ear height and kernel row number in corn. Agron, J., **31**: 531–534.

Cavalli, L.L. 1952. An analysis of linkage in quantitative inheritance. Quantitative Inheritance (ed. E.C.**R. R**ieve and C.H. Waddington) pp. 135–144, HMSO, London.

Cockerham, C.C. 1954. An extension of the concept of partitioning of hereditary variance for analysis of covariance among relatives when epistasis is present. Genetics, **19**: 859–882.

Cockerham, 1956. Analysis of quantitative gene action. Brookhaven Symp. Quant. Biol., **9**: 53–68.

Cockerham, C.C. 1981. Random and fixed effects in plant genetics. Theor. Appl. Genet., **58**: 119–131.

Dempester, E.R. 1956. Some genetic problems in controlled populations. Proc. 3rd Berkely Symp. Math. Stas. and Prob. 4. (ed. J. Neyman) 23–40.

Eberhart, S.A. 1964. Theoretical relations among single, three-way and double cross hybrids. Biometrics, **20(3)**: 522–539.

Eberhart, S.A. and Gardner, C.O. 1966. A general model for genetic effect. Biometrics, **22(4)**: 864–881.

Falconer, D.S. 1989. Introduction to Quantitative Genetics. Longman. Burnt Mill.

Fisher, R.A. 1918. The correlations between relatives on the supposition of Mendelian inheritance. Trans. Roy Soc. Edin., **52**: 399–433.

Fisher, R.A., Immer, F.R. and Tedin. O. 1932. The genetical interpretation of statistics of the third degree in the study of quantitative inheritance. Genetics, **17**: 107–124.

Gale, J.S., Mather, K. and Jinks, J.L. 1977. Joint Scaling Tests. Heredity, **38(1)**: 47–51.

Gamble, E.E. 1962a. Gene effects in corn (Zea mays L.) Separation and relative importance of gene effects for yield. Canadian J. Plant Sci., **42**: 339–348.

Gardner, C.O. and Eberhart, S.A. 1966. Analysis and interpretation of the varietal cross diallel and related populations. Biometrics, **22(2)**: 439–452.

Hallauer, A.R. and Miranda, Fo, J.B. 1988. Quantitative genetics in maize breeding. Iowa state Univ. Press Ames. Iowa, USA.

Hartl, D.H. 2000. A primer of Population Genetics. Third Edition. Sinauer Associates, Inc. Publishers, Sunderland, Massachusetts, U.S.A.

Hayman, B.l. 1954a. The analysis of variance of diallel tables. Biometrics, **10**: 235–244.

Hayman, B.I 1958b. The separation of epistatic from additive and dominance variation in generation means Heredity, **12**: 371–390.

Hayman, B.l. and Mather, K. 1955. The description of gene interaction in continuous variation. Biometrics **10**: 69–82.

Jayasekara, N.E.M. and Jinks, J.L. 1976. Effects of gemne dispersion on estimate of components of generation means and variances. Heredity, **36**: 31–40.

Jinks, J.L. 1979. The Biometrical approach to Quantitative variation. Quantitative Genetic Variation: Academic Press, New York. pp. 81–107.

Jinks, J.L. and Jones, R.M. 1958. Estimates of components of heterosis. Genetics, **43**: 223–234.

Jinks, J.L., Perkins, J.M. and Pooni, H.S. 1973. The incidence of epistasis in normal and extreme environments, Heredity, **31(2)**: 263–269.

Kearsey, M.J. 1993. Biometrical genetics in breeding. In: Plant Breeding. Hayward. M.D.; Bose mark, N.O. and Romagos a, I. (eds). Chapman and Hall, London, pp. 163–193.

Kearsey, M.J. and Pooni. H.S. 1996. The Genetical Analysis of Quantitative Traits. Chapman and Hall, London.

Kempthorne, O. 1969. An Introduction to Genetical Statistics. Wiley. New York.

Mather, K. 1949. Biometrical Genetics. Methuen, London.

Mather, K. 1967. Complementary and duplicate gene interactions in biometrical genetics. Heredity, **22**: 97–103.

Mather, K. 1971. On Biometrical Genetics. Heredity, **26(3)**: 349–363.

Mather, K. 1979. Historical Review. Quantitative variation and Polygenic system. Quantitative Genetic Variation. Academic Press, New York.

Mather, K. and Jinks, J.L. 1982. Biometrical Genetics. 3rd edn. Chapman and Hall. London.

Opsahl, B. 1956. The discrimination of interaction and linkage in continuous variation. Biometrics, **12**: 415–432.

Pooni, H.S. and Jinks, J.L. 1981. The true nature of non-allelic interaction in *Nicotiana rustica* revealed by association crosses. Heredity, **42**: 41–48.

Stuber, C. W. and Cockerham, C.C. 1966. Gene effects and variances in hybrid populations. Genetics, **54**: 1279–1286.

Tan, W. Y. 1974. The approximate overall test for epistatic effects in biometrical genetics. Biometrics, **30**: 697–703.

Thompson. J. N. and Thoday, J. M. 1975(eds) 1979. Quantitative Genetic variation.

Rowe, K.E. and Alexander, W. L. 1980. Computations for estimating the genetic parameters in joint-scaling tests. Crop Sci., 20: 10-110.

Van der Veen, J.H. 1959. Test of non-allelic interaction and linkage for quantitative characters in generations derived from two diploid pure lines. Genetics, **12**: 415–432.

6

Analysis of Variance and Covariance

6.1 Partitioning of Variance

Variances and covariances (or correlation) are second degree statistics. The statistical variance components can be interpreted as we shall see in the following example in terms of genetic variance components. This is often achieved through the intermediary of covariances among relatives as these covariances can be expressed as linear combination of both types of variance components. If we look at the F_2 generation population of say n, individuals from a cross between two pure breeding lines from a replicated experiment, the phenotypic variance, $\hat{\sigma}_P^2$ (or VF_2) will be based on $n - 1$ degrees of freedom and will encompass variance due to difference in the genotypes of the individuals (genotypic variance, $\hat{\sigma}_G^2$) plus the environmental variance ($\hat{\sigma}_E^2$).

$$\hat{\sigma}_P^2 \text{ (or } VF_2) = \hat{\sigma}_G^2 + \hat{\sigma}_E^2$$

Or

$$V_P = V_G + V_E$$

Fisher (1918) first further partitioned the total genotypic variance ($\hat{\sigma}_G^2$) into the additive genetic variance ($\hat{\sigma}_A^2$), dominance variance ($\hat{\sigma}_D^2$) and the interaction variance ($\hat{\sigma}_{Int}^2$) components. Thus

$$\hat{\sigma}_P^2 = \hat{\sigma}_A^2 + \hat{\sigma}_D^2 + \hat{\sigma}_{Int}^2 + \hat{\sigma}_E^2$$

assuming no genotype × environment interaction.

In Mather's terminology D, H, I and E denote $\hat{\sigma}_A^2$, $\hat{\sigma}_D^2$, $\hat{\sigma}_{Int}^2$ and $\hat{\sigma}_E^2$, respectively.

6.2 Estimation of Variances in Generations Derived from Cross of two Pure Breeding Lines

Let us now see how in generations derived from cross of two parents we can calculate the variance components assuming no linkage and no epistasis. Again considering one locus with two alleles (A, a) the genotypes of the two parents (P_1, P_2), their F_1 and F_2 would be as given in Table 6.1

Table 6.1 Shows the genotype of P_1, P_2, F_1 and F_2 population along with frequency and genotypic value of individuals of different genotypes in F_2

Parental generation	$P_1(AA)$	x	$P_2(aa)$
F_1		Aa	
F_2 (genotype)	AA	Aa	aa
Frequency	1/4	1/2	1/4
Genotypic value	d	h	−d

All variations within non-segregating generations such as P_1, P_2 and F_1 will be a measure of environmental variance (E). The genotypic variance will be calculated as

$$\sigma_G^2 = 1/4\, d^2 + 1/2\, h^2 + 1/4(-d)^2 - (1/2h)^2$$
$$= 1/2\, d^2 + 1/4\, h^2$$

and considering multi-loci

$$\sigma_G^2 = 1/2\, \Sigma d^2 + 1/4\, \Sigma h^2$$
$$= 1/2\, D + 1/4\, H$$

where $D = \Sigma d^2$ and $H = \Sigma h^2$ and are additive and dominance variance components, respectively.

6.2 Biometrical Genetics–Analysis of Quantitative Variation

The phenotypic variance (σ_P^2) thus becomes

$$\sigma_P^2 = (VF_2) = 1/2 D + 1/2 H + E$$

From the F_2 population we can produce different generation populations through selfing, sibmating or random mating and the total variance in any generation derived by selfing, sibmating or random mating will take the form as

$$VF_n = (1 - \beta_n) D + \beta_n (1 - \beta_n) H + E$$

where β_n is the frequency of heterozygotes and $1 - \beta_n$ equals the frequency of homozygotes in nth generation.

Having known the genotypes, its frequency and genotypic values of individuals the variances in backcross generations can also be calculated on the same above line and found as

$$\sigma_G^2 (BC_1) = 1/4 D + 1/4 H - 1/2 F$$
$$\sigma_G^2 (BC_2) = 1/4 D + 1/4 H + 1/2 F$$

where $F = \Sigma\, dh$ and it takes sign +ve or –ve depending upon whether alleles from P_1 parent are in excess or not.

When one studies the advancing generations like F_3, F_4, F_5, etc. from a cross between two pure breeding lines one encounters a nested or hierarchical structure where a family in a generation can be traced to a single F_2 plant as shown in Figure 6.1 In the nth generation there will be $n - 1$ possible hierarchies or ranks, $r(r = 1$ to $n - 1)$ based upon common parents ($r = n - 1$), grandparents ($r = n - 2$), great grand parent ($r = n - 3$) etc., where r represents the round of recombination. Thus each generation is yielding a number of independent variances among which the total variance within that generation may be partitioned. For example, the total variance in F_3 generation can be partitioned into two types or ranks of variances, namely, the variance of true F_3 family means ($\sigma_1^2 F_3$), the rank 1 variance and the mean variance within F_3 families ($\sigma_2^2 F_3$), the rank 2 variance. Likewise in the F_4 generation there are three types of variances (1) the variance of the means of those lines that can be traced back to a single F_2 grandparent ($\sigma_1^2 F_4$) (2) Variance of means of these lines that can go back to the F_3 parents ($\sigma_2^2 F_4$) and (3) average within families variance ($\sigma_2^3 F_4$). The total variance of F_4 families contains variances of three ranks and thus the F_n generation families will have variances of $n - 1$ ranks. The expectations of $\sigma_2^2 F_3$ in terms of the heritable and non-heritable components of variations are given as

$$\sigma_1^2 F_3 = \frac{1}{2} D + \frac{1}{16} H$$

$$\sigma_2^2 = F_3 = \frac{1}{4} D + \frac{1}{8} H$$

and the total variance of F_3 families ($\sigma^2 F_3$) equals

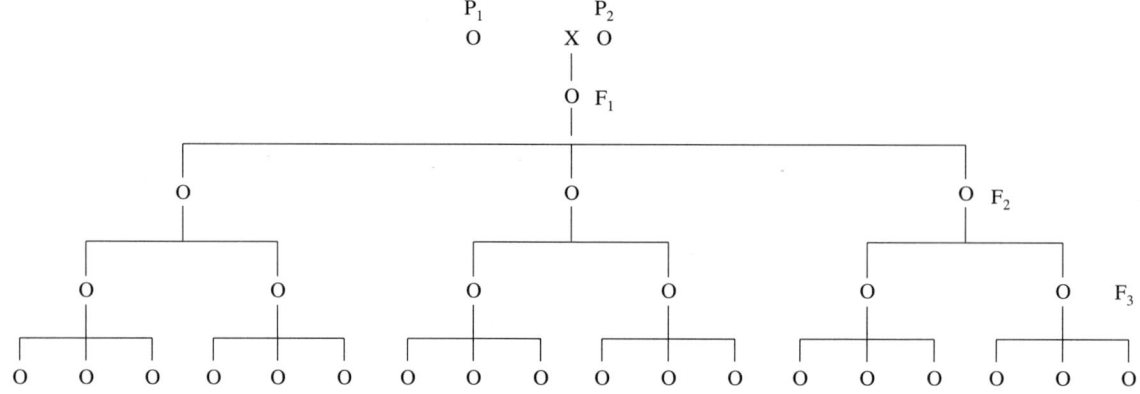

Fig. 6.1 Diagramatic representation of a nested design arising from selfing of an individual obtained from the cross of two parents, P_1 and P_2. Individuals are shown by 0's and all individuals under a broken line are families.

$$(\sigma^2 F_3) = \sigma_1^2 F_3 + \sigma_2^2 F_3$$
$$= 3/4\,D + 3/16\,H$$

The total variance in anyone generation is

$$VF_n = (1 - \beta_n)D + \beta_n(1 - \beta_n)H + E$$

where β_n is the frequency of heterozygotes and $1 - \beta_n$ equals the frequency of homozygotes, and the variances of different ranks are calculated as

$$V_r F_n = (1/2)^r D + (1/2)^{2n-r-1} H$$

where n takes a value from 2 to ∞ and r takes a value from 1 to $n - 1$

When the aim is to estimate the additive (D), dominance (H) and environment (E) components of variation then one can just raise basic generations such as P_1, P_2, F_1, F_2, BC_1 and BC_2 and from these generation variances estimates of D, H and E can be obtained as

$$E = VP_1 = VP_2 = VF_1 \simeq (VP_1 + VP_2 + VF_1)/3$$
$$H = 4VBC_1 + 4VBC_2 - 4VF_2 - 4E$$
$$D = 4VF_2 - 2VBC_1 - 2VBC_2$$

From the estimates of genetical components of variation one can estimate the dominance ratio as

$$\text{Dominance ratio} = \sqrt{\frac{H}{D}}$$

Dominance ratio greater than 1 shows over dominance, less than 1 shows partial dominance and dominance ratio equal to 1 indicates complete dominance.

6.3 Covariance/Correlation

The resemblance between relatives say for example between parents and offspring, is reflected in similarities of expression of quantitative traits. This resemblance is measured by correlation and thus by the covariance. The degree of resemblance expected provides basis for estimating genetic variance components. Different systematic designs as we shall see later will produce different types of relatives. Then using appropriate experimental design and subsequently statistical analysis statistical variance components are estimated. Genetic interpretation of these designs components are facilitated by translating them in covariance among relatives. Theoretical considerations of these genetic causes for resemblance between relatives permit the translation of these covariances into function of genetic variance (additive, dominance, and epistatic variance).

Let us consider a case of covariance between means of F_3 families with its parental means in F_2 generation. The genotypic values of F_3 families and their parents in F_2 generation are given in Table 6.2.

Table 6.2 Genotypic values of F_2 individuals and F_3 families

F_2	AA	Aa	aa
	$+d$	h	$-d$
	↓	↓	↓
F_3	AA	AA Aa aa	aa
	$+d$	$1/2h$	$-d$

The cov $\overline{F_3}\,\overline{F_2}$ (or W_1/F_{23} designated by Mather and Jinks, 1982) is calculated as

$$\text{Cov}\,\overline{F_3}\,\overline{F_2} = 1/4(d \cdot d) + 1/2(1/2h \cdot h)$$
$$+ 1/4(-d \cdot -d) - 1/2h \cdot 1/4h$$
$$= 1/2d^2 + 1/8h^2$$

which in polygenic case becomes

$$= 1/2 \Sigma d^2 + 1/8 \Sigma h^2$$
$$= 1/2D + 1/8H$$

With F_4 generation present together with F_2 and F_3 generations two kinds of covariances are possible (i) those between F_3 family means and means of F_4 families originating from the same F_2 grandparent, denoted as $W_1 F_{34}$ and (ii) that between F_3 individual means and means of their offspring families in F_4 generation denoted as $W_2 F_{34}$. There would be three kinds of covariances in the F_5 generation and four types of covariances in the F_6 generation. Thus like variances of different ranks we have covariances of different ranks and which can further be translated into genetical components of variation. The general formula for the expectation of covariances of rank j between generations F_{n-1} and F_n is

$$W_j F_{n(n-1)} = (1/2)^r D + (1/2)^{2n-r-2} H$$

Similar to what we have seen in case of means the different components of variation can be estimated through weighted least square method from the variances and co-variances of above mentioned different generations.

Since covariances do not have an environmental component in their expectations (Mather, 1949) they are potentially more useful than variances. A common drawback of the use of covariances is that they frequently involve the use of two non-contemporary generations, e.g. WF_2/F_3 and if the measurements are made in different seasons then the resulting statistics will be distorted by genotype x environment interaction. Therefore, the parents must be measured at the same time and under the same condition, as its progeny.

6.4 Estimation of Variances in Random Mating Population

When our interest lies in estimating genetical components of variation in a random mating population, one of the following mating designs is used. The term mating design refers to the system of mating used to develop progeny.

1. Biparental mating design (BIP)
2. Augmented biparental mating design (ABIP)
3. North Carolina mating designs.
   (a) North Carolina mating design I (NCM I)
   (b) North Carolina mating design II (NCM II)
   (c) North Carolina mating design III (NCM III)
4. Triple Test Cross (TTC)
5. Diallel
6. Partial diallel

The diallel and partial diallel mating designs are discussed in a separate chapter.

These mating designs generate different types of families and thereby involve the calculation of the variances and covariances of relatives of different kinds which are further translated in terms of genetic and environmental components of variation under the following assumptions:

1. Parents are random members of genetic population
2. The genetic population is assumed to be at random mating equilibrium.
3. Normal diploid behaviour at meiosis and showing solely Mendelian inheritance.
4. No environmental correlation among progenies.
5. Progenies are not inbred and can be considered random members of the same non-inbred population.
6. Linkage equilibrium.
7. Experimental error is normally and independently distributed.

6.5 Development of Progenies

1. Biparental Progenies (BIPs): In this mating design. $2n$ plants in a randomly mating population are randomly selected and crossed in pairs to produce n full sib progeny families. Now if say $2n = 50$ then the number of BIPs would be 25 and the BIPs would be made in the way shown in Figure 6.2. The randomly selected plants can be numbered from 1 to 50 and plant number 1 is crossed to plant number 2, plant number 3 is crossed to plant number 4 and so on. BIP permits the testing of a large part of the population. This mating design was developed by Mather (1949).

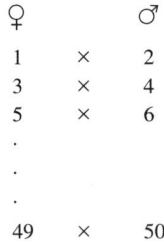

Fig. 6.2 Development of bi-parental progenies.

2. Augmented BIPs: In this mating design in addition to the BIPs the selfed families are developed by selfing the parents.

3. North Carolina Mating design I: Comstock and Robinson (1952) developed NCM I, NCM II and NCM III designs. These designs produce half-sib and full-sib families. NCM I is a hierarchical

or nested design where say n_1 males are randomly selected and each is crossed to a different set of say n_2 females such that no female is involved in more than one mating and thus $n_1 \times n_2$ full-sib progeny families as shown in Figure 6.3 are produced. For each of the n_1 males there are n_2 number of half-sib families.

♀ \ ♂	1	2	3	4 . . . n_1
	1	1′		
	2	2′		
	3	3′		
	.	.		
	.	.		
	.	.		
	n_2	n'_2		

Fig. 6.3 Development of progenies in NCM I design.

This design is especially suited for estimating variances in animal breeding where the sex ratio is not equal and in plants with multiflowers. This design allows us to test a large number of populations.

4. North Carolina Mating Design II: This design differs from NCMI in the sense that each of the n_1 males is crossed to the same set of n_2 females as shown in Figure 6.4 and thus is a cross-classified or factorial design.

♀ \ ♂	1	2	3	4 . . . n_1
1	×	×		×
2	×	×		×
3	×	×		×
4	×	×		×
.				
.				
.				
n_2	×	×		×

Fig. 6.4 Development of half-sib and full-sib families in NCMII.

5. North Carolina Mating Design III: In this mating design n randomly selected plants in the F_2 population derived from a cross of two pure breeding lines are used as male and each of the n males is crossed to the two pure line parents (P_1 and P_2) which are used as females and are called testers. This design is thus a backcross design and it generates $2n$ families.

6. Triple Test Cross: This design is an extension of NCM III in which F_1 of the two pure breeding parents (P_1 and P_2) is added as third tester and used as female along with their parents P_1 and P_2 as shown in Figure 6.5 and thereby generates a total of $3n$ progeny families. The progenies involving P_1, P_2 and F_1 as mother are called L_1, L_2 and L_3 families, respectively.

♀ \ ♂	1	2	3	. . .	n_1
$P_1(L_1)$	L_{1i}				L_{1n}
$P_2(L_2)$	L_{2i}				L_{2n}
$F_1(L_3)$	L_{3i}				L_{3n}

Fig. 6.5 Development of progeny families in TTC.

6.6 Estimation of Components of Variation

I. BIPs

The progenies developed in the different designs are raised in randomized block design in field and a number of quantitative traits are recorded and analysed. The ANOVA in BIPs takes the form as given in Table 6.3.

Table 6.3 ANOVA in BIPs

Item	*d.f.*	*M.S.*	*E.M.S.*
Replication	$r-1$		
Between Families	$n-1$	MS_b	$\sigma_W^2 + r\sigma_b^2$
Within families	$n(r-1)$	MS_W	σ_W^2
Total	$nr-1$		

Where r is the number of replications

$\sigma_b^2 = 1/4DR + 1/16HR$

$\sigma_w^2 = 1/4DR + 3/16HR + E_1$

and Total variation = $1/2DR + ¼\ HR + E_1$

The significant between families mean squares (MS_b) shows that σ_b^2 is greater than zero and that there is presence of genetic variation in the character under study. The intra-class correlation (t) value will provide an approximate estimate of half the heritability in broad sense. Also, the regression coefficient (regressing progeny means on parental or mid parental means) will provide an estimate of narrow sense heritability

II. Augmented Biparental Progenies

This is an extension of Biparental progenies described by Kearsey, 1970 and includes families derived by selfing the parents. In this mating design a sample of $2n$ plants are selected and paired off at random. Suppose the jth pair (where $j = 1 \ldots n$) parents are P_{j1} and P_{j2}. This P_{j1} and P_{j2} do not necessarily represent the two extremes of a character as supposed in the case of generation mean analysis. The parents P_{j1} and P_{j2} will produce four full-sib families like $P_{j1} \times P_{j1}$, $P_{j1} \times P_{j2}$ (F_1), $P_{j2} \times P_{j1}$ (RF_1) and $P_{j2} \times P_{j2}$ of which $P_{j1} \times P_{j1}$ and $P_{j2} \times P_{j2}$ are selfed families. The following comparisons then can be made between the four families.

Comparison	$P_{j1} \times P_{j1}$	$P_{j1} \times P_{j2}$	$P_{j2} \times P_{j1}$	$P_{j2} \times P_{j2}$
C_{j1}	1	0	0	1
C_{j2}	0	+1	−1	0
C_{j3}	−1	+1	+1	−1

The comparison C_{j1} is a function of d, the additive genetic effects and measures the variation of selfed family. The item $\sum_{j=1}^{n} C_{j1}^2/2r$ will be a sum of squares with n d.f. testing of the variation of the selfed families (additive genetic variation, Dikinson and Jinks, 1956). The comparison C_{j2} is a measure of the reciprocal differences and in the absence of such effects should be zero for all sets and thus $\sum_{j=1}^{n} C_{j2}^2/2r$ will be a sum of squares with n d.f. testing the reciprocal differences. The comparison C_{j3} is always a function of h, the dominance effects and thus $\sum_{j=1}^{n} C_{j3}^2/4r$ is a sum of squares measuring dominance at n d.f. This sum of squares is further partitioned into sum of squares for deviations of the C_{j3} around their own mean and the correction factor. The former is the conventional dominance component whereas the correction factor is a measure of directional dominance. The expectations of these comparisons in terms of additive and dominance parameters can be seen considering a set of parents derived from a population in linkage equilibrium, by inbreeding without selection, to same degree f, the coefficient inbreeding. With one locus and two alleles A and a the frequency of three genotypes are

AA	Aa	aa
$u^2 + uvf$	$2uv(1-f)$	$v^2 + uvf$
α	β	ψ

The genetic values of the selfed and crossed families will then be as given in Table 6.4.

The ANOVA in this design takes the form as given in Table 6.5.

Table 6.4 Genetic values of progeny family means in ABIP

Frequency	Parental		Genotypes	Genetic values of		
				Progeny	Family	Means
α^2	P_{j1}		P_{j2}	$P_{j1} \times P_{j1}$	$P_{j2} \times P_{j2}$	F_1 & RF_1
		AA × AA		d	d	d
$2\alpha\beta$		AA × Aa		d	1/2h	1/2 (d + h)
$2\alpha\gamma$		AA × aa		−d	−d	h
β^2		Aa × Aa		1/2h	1/2h	1/2h
$2\beta\gamma$		Aa × aa		1/2h	−d	1/2 (−d + h)
γ^2		aa × aa		−d	−d	−d

III. NCMI

The ANOVA in this design will take the form as given in Table 6.6.

In this design significant between males mean squares (MS_m) shows that σ_m^2 is greater than zero and there is presence of additive genetic variation in the character under investigation. Like BIPs. intraclass correlation coefficient can be calculated which provides estimate of approximately one quarter of the heritability in narrow sense.

Table 6.5 ANOVA in Augmented BIPs

Item	d.f.	M.S.	E.M.S.
Replication	$r-1$		
Between families within sets	$3n$		
P_1 vs P_2 (C_{j1})			
F_1 vs RF_1 (C_{j2})			
$P_1 + P_2$ vs $F_1 + RF_1$ (C_{j3})	n		
Directional dominance	1		
Dominance deviation	$n-1$	MS_2	$\sigma_1^2 + 4r\sigma_2^2$
Within families	$4n(r-1)$	MS_1	σ_1^2

Table 6.6 ANOVA in NCMI

Item	d.f.	M.S.	E.M.S.
Between replication	$r-1$		
Between males	(n_1-1)	MS_m	$\sigma_w^2 + r\sigma_{f/m}^2 + rn_2\sigma_m^2$
Between females within males	$n_1(n_2-1)$	$MS_{f/m}$	$\sigma_w^2 + r\sigma_{f/m}^2$
within families	$n_1 n_2 (r-1)$	MS_w	σ_w^2
Total	$n_1 n_2 r - 1$		

where r is the no. of replication

$$\sigma_w^2 = \tfrac{1}{4} DR + 3/16 HR + E_1$$
$$\sigma_m^2 = 1/8 DR$$
$$\sigma_{f/m}^2 = 1/8 DR + 1/16 HR$$

Total variation $= 1/2 DR + 1/4 HR + E_1$

IV. NCM II

The ANOVA of NCM II is given in Table 6.7.

Table 6.7 ANOVA in NCM II

Item	d.f.	M.S.	E.M.S.
Between replication	$r-1$		
Between males	$n_1 - 1$	MS_m	$\sigma_w^2 + r\sigma_{f \times m}^2 + n_2 r\sigma_m^2$
Between females	$n_2 - 1$	MS_f	$\sigma_w^2 + r\sigma_{m \times f}^2 + n_1 r\sigma_f^2$
Males × females	$(n_1 - 1) \times (n_2 \times 1)$	$MS_{f \times m}$	$\sigma_w^2 + r\sigma_{f \times m}^2$
Within families	$n_1 n_2 (r-1)$	MS_w	σ_w^2
Total	$n_1 n_2 r - 1$		

Where r is the number of replication

$$\sigma_w^2 = \tfrac{1}{4} DR + 3/16 HR + E_1$$
$$\sigma_m^2 = \sigma_f^2 = 1/8 DR$$
$$\sigma_{m \times f}^2 = 1/16 HR$$

and Total variation $= 1/2DR + 1/4HR + E_1$

In this design significant mean squares of the item between males (MS_m) and between females (MS_f) show that σ_m^2 and σ_f^2 are greater than zero and they both show the presence of additive genetic variation for the character under investigation. Significant interaction mean squares ($MS_{m \times f}$) shows the presence of dominance variation for the above character. With $n_1 = n_2$ significant variance ratio, MS_m/MS_f, shows the presence of material effects.

6.7 Expectations of σ_w^2, σ_b^2, σ_m^2 (or σ_f^2), $\sigma_{f/m}^2$ and $\sigma_{f \times m}^2$

Considering one locus with two alleles A and a the three genotypes AA, Aa and aa will be in the random mating population with frequency u^2, $2uv$ and v^2, respectively where u is the frequency of allele A and v is the frequency of allele, a. If in this population plants are randomly selected and crossed the genotype, frequency, genetic value and the genetic variance of the progeny families produced can be worked out as given in Table 6.8.

Now the mean variance within families σ_w^2 equals

6.8 Biometrical Genetics–Analysis of Quantitative Variation

Table 6.8 Genotypes, frequency, genotypic value and genotypic variance of random mating progenies

♀ \ ♂	Genotype Frequency	AA u^2	Aa $2uv$	aa v^2	Family mean
		1	2	3	4	5
Genotype AA	Frequency u^2	AA u^4 d 0	AA $2u^3v$ $1/2(d+h)$ $1/4(d-h)^2$	Aa u^2v^2 h 0	Aa $uv + vh$
Aa	$2uv$	AA Aa $2u^3v$ $1/2(d+h)$ $1/4(d-h)^2$	AA Aa aa $4u^2v^2$ $1/2\,h$ $1/2d^2 + 1/4h^2$	Aa aa $2uv^3$ $1/2(-d+h)$ $1/4(d+h)^2$	$1/2(u-v)d + 1/2h$
aa	v^2	Aa u^2v^2 h 0	Aa aa $2uv^3$ $1/2(-d+h)$ $1/4(d+h)^2$	aa v^4 $-d$ 0	$-vd + uh$
Family mean		$v + vh$	$1/2(u-v)d + 1/2h$	$-vd + uh$	$(u-v)d + 2uvd$

Variance of parents – Covariance of full-sibs

$$\sigma_w^2 = u^4 \cdot 0 + 4u^3v \cdot 1/4(d-h)^2 + 2u^2v^2 \cdot 0$$
$$+ 4u^2v^2 \cdot (1/2d^2 + 1/4h^2)$$
$$+ 4uv^3 \cdot 1/4(d+h)^2 + v^4 0$$
$$= uv[d + h(v-u)]^2 + 3u^2v^2h^2$$
$$= 1/4 DR + 3/16\, HR + E_1$$

considering polygenic case.

The between families variance (σ_b^2) becomes

σ_b^2 = Covariance of full-sibs

$$= u^4(d)^2 + 4u^3v \cdot (1/2d + 1/2h)^2$$
$$+ 2u^2v^2 \cdot (h)^2$$
$$+ 4u^2v^2 \cdot (1/2h)^2 + 4uv^3(-1/2d + 1/2h)^2$$
$$+ v^4(d)^2$$

.
.
$$= uv[d + h(v-u)]^2 + u^2v^2h^2$$
$$= 1/4 DR + 1/16 HR$$

The between males (σ_m^2) or females (σ_f^2) variance is covariance of half-sibs and equals

$\sigma_m^2 = \sigma_f^2$ = Covariance of half-sibs

$$u^2 \cdot (uv + vh)^2 + 2uv\,[1/2(u-v)$$
$$+ 1/2h]^2 + v^2(-vd + uh)^2$$

.
.
.
$$= 1/2\, uv\,[d + h(v-u)]^2$$
$$= 1/8\, DR$$

The variance due to interaction between males and females ($\sigma_{m \times f}^2$) becomes

$$\sigma_{m \times f}^2 = \text{Covariance } FS - 2\,\text{Cov } HS$$
$$= \sigma_b^2 - \sigma_m^2 - \sigma_f^2$$
$$= 1/16\, HR$$

The Total variance in a random mating population

$$= \sigma_w^2 + \sigma_b^2$$
$$= uv\,[d + h(v-u)]^2 + 3u^2v^2h^2$$
$$+ uv\,[d + h(v-u)]^2 + u^2v^2h^2$$
$$= 2uv\,[d + h(v-u)]^2 + 4u^2v^2h^2$$
$$= 1/2 DR + 1/4 HR + E$$

where $DR = \Sigma\, 4uv\,[d + h(v-u)]^2$

and $HR = \Sigma 16u^2v^2h^2$

The total variance (V_p) random mating population following Falconer (1960) is

$$V_p = V_A + V_D + V_E$$

where V_A (additive genetic variance)

$$= \text{variance of breeding value}$$
$$= a^2[4v^2u^2 + (v-u)^2 \cdot 2uv + 4u^2v^2]$$
$$\vdots$$
$$\vdots$$
$$= \Sigma\, 2uv\,[d + h(v-u)]^2$$

V_D (dominance genetic variance)

$$= \text{Variance of dominance deviation}$$
$$= h^2(\,4v^2u^2 + 8u^3v^3 + 4u^4v^2)$$
$$= \Sigma\, 4uvh^2$$

and V_E is the environmental variance

The genotypic variance (V_G)
$$= V_A + V_D$$
$$= \Sigma\, 2uv\,[d + h(v-u)]^2 + \Sigma\, 4uvh^2$$

Relationship between gene frequency, the degree of dominance and the amount of genetic variance:

The contribution of loci to the total genetic variance depends on the gene frequency and degree of dominance. The additive genetic variance depends not only on a, additive genetic effect but also on d, dominance effect. Further both variance components also depend on the allele frequencies. In case of particial dominance the genetic variance and additive genetic variance are maximum at gene frequency of 0.5 but the dominance variance is zero. In case of complete dominance ($d = a$), the genetic variance is maximum when $f(v) = 0.7$ or $f(u)$ 0.3 and so is additive genetic variance but dominance variance is maximum at $f(v) = 0.5$. In case of over dominance ($a = 0$) the additive genetic variance can equal zero at $u = v = 0.5$ even though the genetic variance is nonzero. Futher, genetic variance is maximum at $f(v) = 0.5$ and dominance variance is also maximum at this frequency $f(v) = 0.5$)). Thus we see that the genes contribute much more variance when at intermediate gene frequency then when at either high or low frequency. Recessive genes at low frequency in particular contribute very little variance.

6.7.1 Covariance between relatives in random mating population

The covariances between relatives in a random mating population can be expressed in terms of genetical components of variation as follows:

Covariance offspring – Parent $\text{Cov}(O, P) = 1/2 V_A$
Covariance half-sib, Cov (H.S.) $= 1/4\, V_A$
and Covariance full-sib, Cov (F.S.)
$$= 1/2 V_A + 1/4 V_D$$

The Cov (H.S.) and Cov (F.S.) will change with the level of inbreeding, f in the population (Cockerham, 1954 and Kempthorne, 1957) as

$$\text{Cov (H.S.)} = \left(\frac{1+f}{4}\right) V_A$$

and $\text{Cov (F.S.)} = \left(\dfrac{1+f}{2}\right) V_A + \left(\dfrac{1+f}{2}\right)^2 V_D$

Thus when $f = 1.0$, i.e. the parents are completely inbred. Cov (H.S.) is equal to the covariance of parent offspring in the original random mating population and Cov (F.S.) is the genotypic variance in the original population.

The expectations of these variances and covariances will vary with the levels of inbreeding. The total variation among the progenies will, however, not change and remain as

$$= 1/2\; DR + 1/4\; HR$$

or $\quad 1/2(1+f)\,D + 1/2(1-f)\,H_1 - 1/4(1-f)^2\,H_2$
$\quad - 1/2(1-f)\,F$

where f is the coefficient of inbreeding and D_1, H_1, H_2 and F are the diallel parameters with

$$DR = D + H_1 - H_2 - F.$$

The expectations of σ_m^2 (or σ_f^2), σ_{flm}^2, $\sigma_{m\times f}^2$, σ_b^2, and σ_w^2 will be as follows:

$$\sigma_m^2 = \sigma_f^2 = 1/8(1+f)\,DR$$
$$1/8(1+f)\,D + 1/8(1+f)\,H_1$$
$$- 1/8(1+f)\,H_2 - 1/8(1+f)\,F$$

or
$$\sigma_{f\times m}^2 = 1/16(1+f)^2\,HR$$
$$1/16(1+f)^2\,H_2$$

$$\sigma_{flm}^2 = 1/8(1+f)\,DR + 1/16(1+f)^2\,HR$$
$$\sigma_b^2 = 1/4(1+f)\,DR + 1/16(1+f)^2\,HR$$

or
$$1/4(1+f)\,DR + 1/4(1+f)H_1$$
$$- 1/16(3-f)\,(1+f)\,H_2 - 1/4(1+f)f$$

and $\quad \sigma_w^2 = 1/4(1-f)DR + 1/16(1-f)\,(3+f)$
$\quad HR + E$

or
$$1/4(1-f)D + 1/4(1-f)H_1 - 1/16(1-f)^2 H_2 - 1/4(1-f)F + E$$

6.6 NCM III and TTC

6.6.1 NCM III

The ANOVA of the NCM III is given in Table 6.9. In this design both additive (D) and dominance (H) components of variations are estimated with equal precision. The dominance ratio ($\sqrt{H/D}$) can be calculated and assuming no linkage, test for overdominance can be conducted by F test using mean squares due to tester $\times F_2$ interaction and F_2 males. Significant F coupled with dominance ratio greater than 1.0 would indicate a case of over dominance for the character under study.

Table 6.9 ANOVA of NCM III

Item	d.f.	M.S.	E.M.S.
Between testers	1	MS_t	$\sigma_w^2 + r\sigma_{t\times m}^2 + nr K_D^2$
Between F_2 males	$n-1$	MS_m	$\sigma_w^2 + 2r\sigma_m^2$
Testers $\times F_2$ Interaction	$n-1$	$MS_{t\times m}$	$\sigma_w^2 + r\sigma_{t\times m}^2$
Within families	$2n(r-1)$	MS_w	σ_w^2

Where r is the number of replication,

$$\sigma_w^2 = 1/8D + 1/8H + E$$
$$\sigma_{t\times m}^2 = 1/8H$$
$$\sigma_m^2 = 1/8D$$

The expectations of σ_m^2, $\sigma_{t\times m}^2$ and σ_w^2 can be worked out as given as Table 6.10.

From Table 6.10 it can be shown that

$$\sigma_m^2 = \text{Variance of } (\bar{L}_1 + \bar{L}_2)/2 = 1/8d^2$$
$$= 1/8D$$

$$\sigma_{t\times m}^2 = \text{Variance of } (\bar{L}_1 - \bar{L}_2) = 1/8h^2$$
$$= 1/8H$$

and $\sigma_w^2 = $ the within families variance,
$$= 1/8d^2 + 1/8h^2 + E$$
$$= 1/8D + 1/8H + E$$

This design does not provide a test for epistasis so the additive and dominance components will be biased in the presence of undetected epistasis. So is the case with all other mating designs discussed above. The estimated additive and dominance variances will contain epistatic variance when present.

6.8.2 TTC (Triple Test Cross)

Kearsey and Jinks (1968) provided unambiguous general test for epistasis. They extended the NCM III by adding one more tester, the F_1 of the two parents P_1 and P_2, making the total number of tester to 3 as the name of the design suggests and thus enabling to conduct a test for epistasis. This design is most

Table 6.10 Genotypes and genotypic values of progeny families in NCM III and TTC

Genotype Frequency	F_2 Population			Family mean
	AA	Aa	aa	
	1/4	1/2	1/4	
$\bar{L}_1 (P_1)$ AA	AA	Aa aa	Aa	$1/2(d+h)$
	d	$1/2d + 1/2h$	h	
$\bar{L}_2 (P_2)$ Aa	Aa	Aa aa	aa	$1/2(-d+h)$
	h	$-1/2d + 1/2h$	$-d$	
$\bar{L}_3 (F_1)$ Aa (in TTC)	AA Aa	AA Aa aa	Aa aa	$1/2h$
	$1/2d + 1/2h$	$1/2h$	$-1/2h + 1/2h$	
$(\bar{L}_1 + \bar{L}_2)/2$	$1/2(d+h)$	$1/2h$	$1/2(-d+h)$	$1/2h$
$(\bar{L}_1 - \bar{L}_2)/2$	$1/2(d-h)$	$1/2d$	$1/2(d+h)$	$1/2d$

flexible in that it can be applied to any population with any level of inbreeding with any gene frequency and degree of linkage disequilibrium or gene correlations. In the absence of epistasis this design provides a more efficient estimate of additive and dominance components.

The following three comparisons can be made between the progeny families.

Comparison	$\bar{L}_{1i}$	$\bar{L}_{2i}$	$\bar{L}_{3i}$
C_1	1	1	−2
C_2	1	−1	0
C_3	1	1	1

L_{1i} L_{2i} and L_{3i} are the ith families means where $i = 1 \ldots n$.

The comparison C_1 is a measure of epistasis whereas C_3 and C_2 are measures of additive and dominance variation, respectively. The sum of squares due to C_1 $\left(= \sum_{i=1}^{n} \bar{L}_{1i} + \bar{L}_{2i} + \bar{L}_{3i} \right)^2 / \sigma_{nr}$ comparison providing an overall test of epistasis ($\Sigma i^2 + \Sigma j^2 + \Sigma l^2$) at n d.f. can be partitioned into one degree of freedom for corrector factor which provides a test for additive × additive interaction and $n - 1$ degrees of freedom for corrected mean squares which detects the presence of additive × dominance and dominance × dominance epistasis. These can be seen from the expectations of $L_1 + L_2 - 2L_3$ worked out for each of the 9 possible genotypes in respect of genes $A - a$ and $B - b$ assuming digenic interaction as given in Table 6.11. The genotypes of L_1, L_2 and L_3 are AABB, aabb and AaBb, respectively.

Table 6.11 Expectation of $L_1 + L_2 - 2L_3$ in case of digenic interaction

Genotype	$L_1 + L_2 - 2L_3$		
	i	j	l
AABB	1/2	−1	1/2
AABb	1/2	−1/2	–
AAbb	1/2	–	−1/2
AaBB	1/2	−1/2	–
AaBb	1/2	–	–
Aabb	1/2	1/2	–
aaBB	1/2	–	−1/2
aaBb	1/2	1/2	–
aabb	1/2	1	1/2

One can find [i] term in all sets whereas the [j] [l] components vary. The correction factor with one d.f. thus takes [i] out of the C epistasis sum of squares Further more, it can be seen that in the $L_1 + L_2 - 2L_3$ item the additive (d) and dominance (h) contribution of the loci which L_1 and L_2 have in common do not appear and thus this test of epistasis is independent of the degree of inbreeding, gene frequencies, gene correlation, etc. The analysis of additive and dominance effects is similar to that of Comstock and Robinson (1952) i.e. by obtaining the variances of sum and differences of L_{1i} and L_{2i} except that $\sigma^2_{m \times l}$ in TTC is half of the $\sigma^2_{m \times t}$ of NCM III as shown in Table 6.12

Table 6.12 Variances of sum and differences of L_{1i} and L_{2i}

Item	d.f.	E.M.S.
Sums ($\bar{L}_{1i} + \bar{L}_{2i}$)	$n - 1$	$\sigma^2_w + 2r\sigma^2_m$ *
Difference ($\bar{L}_{1i} - \bar{L}_{2i}$)	$n - 1$	$\sigma^2_w + 2r\sigma^2_{ml}$
Error	$n(r - 1)$	σ^2_w

* In case of Sums ($\bar{L}_{1i} + \bar{L}_{2i} + \bar{L}_{3i}$) the expected mean square will be $\sigma^2_w + 3r \sigma^2_m$.

This change is made simply to facilitate comparison between additive and dominance components. The efficiency of $\sigma^2_{m \times l}$ to estimate dominance depends on the gene distribution in L_1 and L_2 tester lines and in the population. The absolute value of σ^2_{ml} is as much affected by dominance effects as by gene correlation in parents and linkage disequilibrium in population. In case of L_1 and L_2 being extreme high and low selected lines, it would have high degree of gene association but even in this situation σ^2_{ml}/σ^2_m will be a measure of average dominance as both components are affected to the same extent by excess coupling linkage. In the presence of linkage,

$$\sigma^2_m = 1/8 \ \Sigma d^2 + 1/4 \ \Sigma(1 - 2pjk)d_j d_k$$

and $$\sigma^2_{ml} = 1/8 \ \Sigma h^2 + 1/4 \ \Sigma(1 - 2pjk)h_j h_k$$

Thus in the presence of linkage

$$D = \Sigma d^2 \pm 2 \ \Sigma(1 - 2pjk)d_j d_k$$

and
$$H = \Sigma h^2 + 2\Sigma(1 - 2pjk)h_j h_k$$

where p_{jk} is the recombination frequency between It the jth and kth loci.

One further statistics that can be obtained in this design is the covariance of sum $(L_{1i} + L_{2i} + L_{3i})$ and difference $(L_{1i} + L_{2i})$ which has the expectation

$$\text{Cov sum/difference} = \sum_j \frac{1}{2} uvd_j h_j = -1/8\ F$$

where $F = \Sigma d_j h_j$.

The magnitude and sign of F provides information about the magnitude and direction of dominance. The test of significance of covariance is done by converting it into a correlation coefficient with $n - 3$ d.f.

F is calculated as

$$F = -\left[8\ \text{Cov}\left(\frac{L_{1i} + L_{2i} + L_{3i}}{3} \cdot \frac{L_{1i} + L_{2i}}{2}\right)\right]$$

or more correctly as

$$F = -8\ [\text{Cov}\ L_{3i} \cdot (L_{1i} + L_{2i})]$$

which is free from biases due to genotype × environment interaction (He Witt, 1980)

$$r\ \text{sum/difference} = \frac{\text{Cov sum/difference}}{\sqrt{V\ \text{sum} \times V\ \text{difference}}}$$

$$= \frac{\text{Cov}\ dh}{\sqrt{V_d \cdot V_h}} = \frac{F}{\sqrt{D \cdot H}}$$

The nearer that r sum/difference approaches a value of 1.0 the greater the directional element. Positive F indicates that increasing allele is dominant more often than the decreasing allele while the reverse is true if F is negative. Whether the dominance ratio is constant or variable, overall loci can be tested by estimating correlation between d_j, and h_j over loci. As the F denotes the sum of the product, $\Sigma\ d_j h_j$, the above correlation can be used for testing the consistency of dominance ratio over loci and has a maximum value of r, the coefficient of dispersion $(r = +1\ \text{or} -1)$ when the dominance ratio is constant at all loci. However, the problem with this dominance ratio is that if its value is less than 1 it does not mean that the variance there is not over-dominance at some loci as dominance less than 1 could be as a result of dispersion of dominance alleles between parents.

If F is not significant then either there is no dominance or the dominance is ambi-directional.

6.8.3 Consequences of inadequacy of testers in Triple Test Cross

In TTC additive, dominance and epistatic variances are detected by orthogonal partitioning of the family means and this results in independent estimates of D and H which is not possible with any other mating design. However, testers with extreme expression of each trait examined are required and if this requirement is not fulfilled then biased estimates of the variance are obtained. To overcome this problem modifications such as inclusion of F_1 between testers or extension of the range of generations with selfed or backcross families were developed.

If L_1 and L_2 difer only at $k_2 = k-k_1$ loci for the same alleles, the test for epistasis is no longer unambiguous and can spuriously detect non-allelic interaction when they may not exist but it still provides a test for epistasis and the adequacy of the testers simultaneously. The test of significance and the estimates of additive variance are biased to an extent related to the dominance and dominance x additive effects of the common loci while the significance and estimates of dominance variation are deflated as they reflect the dominance effects at the non-common loci only. The covariance of sums and differences is also underestimated for the same reason. The expectation for the variance of sum, $L_{1i} + L_{2i}(\sigma^2 m)$ assuming unequal gene frequency becomes

$$\sigma^2 m(L_{1i} + L_{2i}) = \Sigma\ uvd_i^2 + \Sigma\ uvh_i^2 - 2_{r1} \Sigma\ uvdi\ h_i$$

where coefficient of association $r_1 = (k_1 - 2k_1')/k'$

$$= \tfrac{1}{4} D_k + \tfrac{1}{4} H_{ki} - \tfrac{1}{2} F_{ki}$$

Where, of the k_1 loci at which there are common alleles in the testers and $(k_1 - k_1')$ increasing allele.

The expectation for the variance of differences $(L_1 - L_{2i})\ (\sigma^2\ ml)$ becomes

$$\sigma^2 ml = \Sigma\ uvh_i^2 = \tfrac{1}{4}\ H_{ki}$$

The expectation of covariance of $L_{1i} + L_{2i}$ and $L_{1i} - L_{2i}$ becomes

Cov sums/difference = $-r_2 \sum uvd_i h_i = -1/4 Fk_2$

Where $r_2 = (k_2 - 2k_2')/k_2$ where k_2' of the k_2 non-common alleles with increasing effect are present along with k_2-k_2' alleles of decreasing effects in L_1 and vice-versa in L_2 (Virk and Jinks, 1977).

6.9 Variants of TTC

As TTC is applicable to any population irrespective of its mating system, its gene and genotypic frequencies there would be many variants of TTC but only two versions of TTC will be briefly considered.

1. Where F_2 population derived from the two parents P_1 and P_2 is replaced by a population of inbred or pure breeding lines and no L_{3i} families are produced or a natural population is used in the place of F_2 population (Jinks, Perkins and Breese, 1969).

2. In self fertilising crops like peanut, lentil and others where it is cumbersome to make crosses besides crossing success is low and only few seeds are produced per cross, the family size would be too small to conduct a replicated experiment and in this situation selfed families of $L_{1i} + L_{2i}$ and L_{3i} can be raised instead (Pooni et al. 1980).

6.9.1 TTC with population of inbred lines

Here L_1 and L_2 are the two most extreme phenotypes selected from the population. For the ith inbred line (where $i = 1$ to n)

$$\overline{L}_{1i} + \overline{L}_{2i} - \overline{P}_i = u + \sum_j h_j$$

where $(u + \sum_j h_j)$ is constant for all n lines in the absence of epistasis and j is the number of loci at which L_1 and L_2 differ.

The variance of $L_{1i} + L_{2i} - P_i$ if significantly, greater than the error variance, indicates the presence of non-allelic interaction. Although under the condition $i_{ab} = lab$ and $jab = jba$, this test will fail to indicate epistasis when it is present. Also, if L_1 and L_2 are homozygous for the same allele at loci for which the inbred lines in the population carry different alleles, $\overline{L}_{1i} + \overline{L}_{2i} - \overline{P}_i$ will no longer be constant even in the absence of epistasis and thus will indicate epistasis when it is not there.

With linkage equilibrium and in the absence of epistasis,

$$\sigma_m^2 = \sum_j uv\, d_j^2 = 1/4D$$

and

$$\sigma_{ml}^2 = \sum_j uv\, h_j^2 = 1/4H$$

$$\sigma_w^2 = 0$$

In the presence of linkage

$$\sigma_m^2 = \sum_j uvd_j^2 + 2\sum_{jk} D_{jk}\, d_j d_k$$

$$\sigma_{ml}^2 = \sum_j uvh_j^2 \pm \sum_{jk} D_{jk}\, h_j h_k$$

The covariance of sum $(L_{1i} + L_{2i})$ on difference $(\overline{L}_{1i} - \overline{L}_{2i})$ has the expectation

$$\text{Cov sum/difference} = -\sum_j uvd_j h_j = -1/4F$$

provided there is no epistasis and correlated gene distribution.

In case of natural population, dominance will be detected only for loci at which the two parents differ and the expectations of σ_m^2 and a $\sigma_{m \times l}^2$ will take the form as

$$\sigma_{m \times l}^2 = 1/2 \sum_{i=1}^{k'} uvh^2$$

and

$$\sigma_m^2 = 1/2 \sum_{i=1}^{k} uvd^2 + 1/2 \sum_{i=k'}^{k} uvh^2$$

where k' is number of loci out of k loci at which the two parents differ but k is the number of loci at which the population is segregating.

6.9.2 Triple test cross (selfed families)

There are two versions of the TTC (selfed families). In one form selfs are produced where F_2 population has been used as male as is done in case of standard TTC families and in another form selfs are produced where $F\infty$ population has been used as male. Thus where L_{1i}, L_{2i} and L_{3i} are replaced by L_{1si}, L_{2si} and L_{3si}, the standard test for epistasis still applies. The variance component of $(\overline{L}_{1si} + \overline{L}_{2si} - \overline{L}_{3si})$ comparison provides on overall test of epistasis for n degrees of freedom which can be further partitioned as in case of standard *TTC* into one degree of freedom for the correction factor to provide a test of additive × additive interaction, the [*i*] component and ($n - 1$) degrees of freedom for the corrected mean squares which detects the presence of additive × dominance (Σj_{jk}^2) and dominance × dominance (Σl_{jk}^2) interaction together with some of their cross products.

In the absence of epistasis the comparisons ($\bar{L}_{1si} + \bar{L}_{2si}$) and ($\bar{L}_{1si} + \bar{L}_{2si} + \bar{L}_{3si}$) provide the estimates of the additive genetic variance and the σ_b^2 (or σ_m^2) for either of them (which are equal to one-fourth and one-ninth of the respective variances) estimates 1/8*D* for F_2 population and 1/4*D* for F_∞ population. Similarly σ_b^2 (or σ_m^2) for ($L_{1si} - L_{2si}$) detects and estimates the dominance component and it estimates 1/32*H* for F_2 population and 1/16*H* for F_∞ population which are one-quarter and one-half of the corresponding estimates from the normal triple test cross families, respectively.

In comparison to the normal triple test cross the coefficients for additive × dominance (*j*) and dominance × dominance (*l*) components in selfed families are smaller and thus this test is less efficient in detecting *j* and *l* types of interaction although both tests are equally efficient in detecting additive × additive type interaction (*i*). In the absence of epistasis, *D* is estimated with the same efficiency from the normal triple test cross families and selfed families. In other words the estimate of *D* from selfed families is as good as from normal triple test cross families. But in the presence of epistasis, the estimate of *D* from normal triple test cross families will be more affected by *j* type interaction than the estimates from the selfs. The dominance component (*H*) estimate from the selfs is much more biased in the presence of epistasis and is thus less reliable. The estimate of dominance component (*H*) from normal triple test cross families will be less affected by *i* and *l* types interaction in comparison to *H* from selfs.

6.10 Model Fitting to Mean Squares

The sum of squares in the ANOVAs of different mating designs are calculated in the way discussed in the Chapter 4.

Least squares estimates of variance components including epistatic component can be obtained by model fitting to either mean squares obtained in the different experiments of the same mating design or mean squares obtained from a combination of different mating designs such as NCMI and NCMII. The expectations of mean squares can be written in terms of parameters such as $D, H, I(\sigma_{AA}^2)$ and E. The variance components are estimated following weighted least square analysis (Mather & Jinks, 1971). Estimated variances are obtained by noting that each estimate is a linear function of mean squares and the means squares are independent.

6.11 Comparison of Efficiency of the Mating Designs

So far we have seen that the genetical architecture of metrical traits can be known using two broad methods, namely, analysis of generations derived from a cross between two lines parents (P_1 and P_2) such as P_1, P_2, F_1, F_2, BC_1 and BC_2 generations when the interest lies in just two or few genotypes and analysis of one of the several mating designs discussed above when the interest lies in a population of genotypes rather than just two genotypes. These two methods yield estimates of additive genetic (*D*), dominance (*H*) and environmental (*E*) variation. Now we can ask the following questions:

1. What should be the experimental size in order to have a precise estimate of components of variation and does this experimental size vary with the genetic control and the parameters (*D* and *H*)?

2. What are the relative efficiencies of these mating designs?

The problem was originally considered by Comstock and Robinson (1952) and later by Kearsey (1970, 1980). We will now consider each of these questions in turn.

1. Basic generations

The estimates of additive genetic variance (*D*), dominance variance (*H*) and their expected variances from variances of different generations as given in Table 6.13 would be as follows:

Table 8.13 Shows the different basic generations, their degrees of freedom, observed variances and the coefficients of components of variation

Generations	d.f.	Observed variance	Coefficients of component of variation		
			D	H	E
P_1, P_2 or F_1, F_2	n_1	VE	0.0	0.0	1.0
	n_2	VF_2	0.5	0.25	1.0
BC_1	n_3	VBC_1	0.25	0.25	1.0
BC_2	n_4	VBC_2	0.25	0.25	1.0

$$\hat{D} = 2[2VF_2 - VBC_1 - VBC_2]$$

$$\hat{H} = 4(VBC_2 + VBC_1 - VF_2 - VE)$$

and
$$V\hat{D} = 8\left(\frac{4VF_2^2}{n_2} + \frac{VBC_1^2}{n_3} + \frac{VBC_2^2}{n_4}\right) \quad (i)$$

$$V\hat{H} = 32\left(\frac{VBC_2^2}{n_3} + \frac{VBC_1^2}{n_4} + \frac{VBF_2^2}{n_2} + \frac{V_E^2}{n_1}\right) \quad (ii)$$

Now if N is the total degrees of freedom ($N = n_1 + n_2 + n_3 + n_4$) then eq. (i) can be rewritten as

$$V\hat{D} = \frac{8}{N}\left(\frac{4VF_2^2}{p_2} + \frac{VBC_1^2}{p_3} + \frac{VBC_2^2}{p_4}\right)$$

where p_i is the proportion of the total degrees of freedom with the ith generation such that $p_2 + p_3 + p_4 = 1.0$ and thus we can obtain value of p_i which minimises $V\hat{D}$ by differentiating Q with respect to p_2 and p_3 and are as follows:

$$p_2 = \frac{2VF_2}{2VF_2 + VBC_1 + VBC_2}$$

$$p_3 = \frac{VBC_1}{2VF_2 + VBC_1 + VBC_2} \quad (iii)$$

$$p_4 = \frac{2VBC_2}{2VF_2 + VBC_1 + VBC_2}$$

Substituting equation (iii) into (i) will result

$$V\hat{D} = \frac{8}{N}(2VF_2 + VBC_1 + VBC_2)^2$$

Thus the optimal proportions of different generations and $V\hat{D}$ for various genetical situations like different dominance ratio ($\sqrt{H/D}$), $F\sqrt{D/H}$, similar or different dominance ratio at different loci and different heritability estimates can be calculated. Similarly, the values of p_i which can minimise VH can be obtained as

$$p_1 = VE/VT$$
$$p_2 = VF_2/VT$$
$$p_3 = VBC_1/VT$$
$$p_4 = VBC_2/VT$$

where $VT = VE + VBC_1 + VBC_2 + VF_2$ and $V\hat{H}$ is thus estimated as

$$V\hat{H} = \frac{32}{N}(VF_2 + VBC_1 + VBC_2 + VE)^2$$

II. North Carolina Mating Design III

The additive (D) and dominance (H) components are estimated from the ANOVA given in Table 6.14 as follows:

Table 6.14 ANOVA of NCM III

Item	d.f.	M.S.	E.M.S.
Additive	$n - 1$	MS_A	$\sigma_w^2 + 2r\sigma_A^2$
Non-additive	$n - 1$	MS_D	$\sigma_w^2 + 2r\sigma_D^2$
Replication error	$2n(r - 1)$	MS_R	σ_w^2

where n is the number of F_2 individuals used as males and backcrossed to P_1 and P_2 and r is the family size.

$$\sigma_A^2 = 1/8 D$$
$$\sigma_D^2 = 1/8$$

and $\sigma_w^2 = 1/8 D + 1/8 H + E$

The additive (D) and dominance (H) components alongwith their variances are estimated as

$$\hat{D} = \frac{8(MS_A - MS_R)}{2r}$$

$$\hat{H} = \frac{8(MS_D - MS_R)}{2r}$$

and
$$V\hat{D} = \frac{32}{r^2}\left(\frac{MS_A^2}{n-1} + \frac{MS_R^2}{2n(r-1)}\right)$$

$$V\hat{H} = \frac{32}{r^2}\left(\frac{MS_D^2}{n-1} + \frac{MS_R^2}{2n(r-1)}\right)$$

The estimated variance of a mean squares say X based on K degrees of freedom is equal to $2X^2/(K+2)$, the true variance being $2E(X^2)/K$. Since the variance of mean squares varies with twice the square of the magnitude of its expectation and inversely with its number of degrees of freedom, the variance of the mean squares in the analysis varies greatly. This formula works well when gene frequencies are equal and errors of the model are all normally and independently distributed.

The standard error of the estimates of variance components is computed according to the formula:

$$2/c^2 \sum M_i^2/d_i + 2$$

Where M_i is the ith mean squares in the function, d_i is the degrees of freedom with the ith mean squares and c is the divisor of the function of mean squares (Moll et al., 1960).

Assuming that n is very large and thus substituting K for $n-1$, the values of r necessary to minimize $V\hat{D}$ and $V\hat{H}$ can be estimated. And since

$$D = 2h^2 n V F_2$$

and
$$H = 2\left(\frac{h}{d}\right)^2 h^2 n V F_2$$

the experimental sizes ND, NH required in order to have estimate of D and H greater than twice their standard errors, can be found as follows:

$$ND = \frac{3.8416\ NVD}{4h_n^4\ VF_2^2}$$

and
$$NH = \frac{3.8416\ NVH}{4(h/d)^2\ h_n^2\ VF_2^2}$$

where 3.8146 is the squares of the normal deviate for 5 per cent probability.

Kearsey (1980) having outlined the above procedures demonstrated that for optimum estimate the number of individuals in each generation be adjusted in a way so as to make the variance of every generation mean the same–a strategy proposed by Jinks and Perkins (1960), although it assumes some prior knowledge of the expected variances of the various generations. In the absence of any knowledge of the likely genetic control of the characters the number of individuals in non-segregating generations, F_2's and Backcross generations should be in the ratio of 20 : 30 : 50, a proportion suggested for an intermediate level of heritability and dominance. Increase in backcross individuals in comparison to F_2 individuals over the said ratio will drastically reduce efficiency of the design. As the design involved selfing which reduces heterozygosity and thus is not ideal for estimating dominance. In order to have precise estimate of additive genetic variance (D), it requires more F_2's and backcrosses while it is reverse for dominance component. When the potence ratio $\left(\frac{h}{d}\right)$ is high the number of BC_2 should be more in comparison to BC_1 among the backcross proportion. The proportion of non-segregating generations decline sharply with highly heritable characters. Experiments with NCM III demonstrated that the minimum experimental size remains relatively constant and low in the range $K = 20$ to 40 but increases rapidly if K is below 20. Thus at least 20 F_2 individuals should be sampled and backcrossed in NCM III irrespective of the genetic properties of the population and the parameters to be estimated. The number of replication should increase with the decrease in heritability estimate. Between these two designs the minimum experimental size required to detect additive or dominance component at 5 per cent level of significance is only 1/5th of the total experimental size for estimating additive genetic variance (D) and only 1/17 for dominance (H) at low heritability but at higher heritabilities NCM III is 2 to 17 times as efficient as basic generation design for estimating additive and dominance components. For estimating additive and dominance components NCM III does not require more than 600 and 10,000 completely randomized individuals, respectively. The experimental size in basic generation design varies greatly with the genetic models and parameters to be estimated and is not a practicable method of estimating dominance although one can successfully estimate additive variance with moderately sized heritabilities.

Keassey (1970) compared the efficiency of ABIP's, NCM I, NCM II, Diallel and NCM III by determining minimum experimental size required by these designs to detect dominance variation assuming that the additive variance is easily estimated.

The genetic components of variance from all above mentioned multiple mating designs can be expressed using four components, D_1, H_1, H_2 and F_1 as follows:

$$D_1 = 4Kuvd^2$$
$$H_1 = 4Kuvd^2b^2$$
$$H_2 = 16Ku^2v^2d^2b^2$$
$$F = 8Kuv(u-v)\,d^2b$$

where K is the number of loci controlling the character, u and v are the frequency of increasing and decreasing allele, respectively and is constant, d and h are the additive and dominance effects of gene, respectively and is constant and h/d is the dominance ratio, b.

Assuming that the total phenotypic variance being unity additive genetic variance (D_1) can be expressed as function of heritability in narrow sense.

$$h_n^2 = 1/2\,DR = 2\,Kuvd^2\,[1 + b(u-v)]^2$$

and thus

$$Kuvd^2\,(D_1) = \frac{h_n^2}{2}[1 + b(u-v)]^2$$

Similarly, H_1, H_2 and F components can also be expressed as function of heritability and dominance ratio. The environmental variance is calculated as

$$E = 1 - 1/2DR - 1/4HR = 1 - h_n^2 - 1/4H_2$$

The above parameters computed then can be combined to give values of EMS as shown in ANOVA for NCM II given in Table 6.15 which will in turn yield expected mean squares values (MS_2 and MS_1) for a given family size, r. Significant MS_2 when tested against MS_1 will demonstrate presence of dominance for a character and thus detection of dominance will depend on the ratio, σ_1^2/σ_2^2 where the magnitudes of these are functions of various genetic and environmental components of variation.

Table 6.15 ANOVA in NCM II

Item	d.f.	M.S.	E.M.S.
Between males	$n-1$		
Between females	$n-1$		
Males × females	$(n-1)(n-1)$	MS_2	$\sigma_1^2 + r\sigma_2^2$
Within families	$n^2(r-1)$	MS_1	σ_1^2

Now the objectives would be to find out:

1. What is that value of n, for a given family size and genetic situations which will result in MS_2 being significant at some level (α_1) on a certain proportion of occasions (α_2)?

2. Which combination of n and r minimises the total experimental size?

Let df_1 and df_2 be the degrees of freedom for MS_1 and MS_2, respectively and further let

$$\frac{E(MS_2)}{E(MS_1)}\;(\text{expected}) = K$$

and

$$\frac{MS_2}{MS_1}\;(\text{observed}) = K^\circ$$

We want K° to be significant at some level (α_1) on a certain proportion of occasions (α_2). Let the values of α_1 and α_2 be 0.05 and 0.95, respectively. The table value of $F(F\alpha_1(1))$ is calculated on the assumption that $K = 1$ and is attached with it some probability. If K is greater than unity, the value of F corresponding to α_i will be obtained by multiplying the tabulated value by K. Let this value be called $F\alpha_i(K)$. Now if $F \geq F\alpha_1(1)$, the null hypothesis of no dominance is rejected and we want this to happen on α_2 of occasions. Again, if $F(K)$ is the probability density of F, given a value of $K > 1$ we want F such that

$$\int_F^\infty F(K)\,d_F(K) = d_2$$

Supposing this value of F be called $F\alpha_2(K)$, we want that

$$F\alpha_1(1) = F\alpha_2(K)$$

But we know that

$$F\alpha(K) = KF\alpha_2(1)$$

and hence we want

$$F\alpha_1(1) = KF\alpha_2(1)$$

or
$$K = \frac{F\alpha_1(1)}{F\alpha_2(1)} \quad \text{(i)}$$

F can be replaced by χ^2 for large df_1 and then

$$K = \frac{\chi^2 \alpha_1}{\chi^2 \alpha_2} \quad \text{(ii)}$$

In NCM II

$$K = \frac{\sigma_1^2 + rt\sigma_2^2}{\sigma_1^2} = 1 + \frac{r\sigma_2^2}{\sigma_1^2} \quad \text{(iii)}$$

as $t = 1$ in NCM II and it takes the values of 2 in diallel and NCM III and 4 in ABIP's.

Comstock and Robinson (1952) suggested a method for calculation of n for a given value of r. As df_1 and df_2 are function of r, for large df_1 and df_2,

$$\sigma_z^2 = 1/2 \left(\frac{1}{df_1} + \frac{1}{df_2} \right) \quad \text{(iv)}$$

Now
$$K = \frac{F0.05(1)}{F_{0.95}(1)} = \frac{e_{0.05}^{2z}}{e_{0.05}^{2z}}$$

and
$$Z_{0.95} \simeq -Z\,0.05 \quad \text{(v)}$$

Hence
$$\log e^K = 4Z\,0.05$$

i.e.
$$Z\,0.05 = \log e^{K/4} \quad \text{(vi)}$$

Now putting
$$\frac{Z_{0.05}}{\sigma^2} = C_{0.1} \text{ (Normal deviate)}$$

$$= 1.644854$$

We can obtain

$$\sigma_z^2 = \left(\frac{\log e^K}{4 \times 1.644854} \right)^2$$

Now df_1 and df_2 are functions of n and r only and thus for a given r, n can be calculated to a close approximation. In case of NCM II

$$df_1 = n^2(r - 1)$$

$$df_2 = (n - 1)^2$$

For large n we may substitute $(n - 1)^2$ by n^2 so that

$$2\sigma_z^2 \simeq \frac{1}{n^2(r - 1)} + \frac{1}{n^2} = \frac{r}{n^2(r - 1)}$$

and
$$n \simeq \sqrt{\frac{r}{2\sigma_z^2(r - 1)}}$$

The value of n which need not be an integer will be slightly too small and hence the integer immediately above this approximation is substituted in

$$\frac{1}{n^2(r - 1)} + \frac{1}{(n - 1)^2}$$

and this value is compared with the old value of σ_z^2. If the new σ_z^2 is greater than the old σ_z^2 n is incremented by units of 1 until such a time that the newly calculated σ_z^2 is less than old one. As n is inversely related to r, with larger df_2 being sufficiently small for the validity of the approximation in equation (v) is doubtful and in such cases n is adjusted to satisfy the criteria of equations (i) and (ii).

Results from the simulation study using the above procedure showed that the modal values of n obtained from NC II, Diallel, NC III and ABIP's were 5, 8, 12 and 12, respectively. Further, all mating designs needed larger experimental size (upto 5000 randomized individuals) to detect dominance variatiom for moderate dominance values and heritabilities which is rather impracticable if a breeder or geneticist is working with many populations simnultaneously which they do. NCM III was found less sensitive to changes in gene frequency and inbreeding than other designs. The diallel and NCM III designs differ little in their efficiency but were found very sensitive to changes in gene frequency and inbreeding. These designs require a considerably smaller experimental size than does ABIP's at intermediate gene frequency But at extreme gene frequency ABIP's require less individuals than others. The NCM III invariably requires fewer individuals than all other designs These above observations were made under the assumption of polygenic control of a character with no epistasis and no linkage disequilibrium which is hardly met in real population and thus the above estimated experimental sizes are underestimated ones

All these point to that in many real experiments there is likely to be a higher chance of failure to detect dominance variation when it is present.

Since the sensitivity of these variance approaches to detect dominance is offten low, the estimated dominance component ($\hat{H}$) will be associated with large error and thus, these methods are not generally useful for detecting and estimating $\hat{H}$ in randomly mating population. Effect of inbreeding individuals prior to generating random mating progenies on total experimental size has markedly improved the sensitivity of the experiments but then it is possible that the natural selection during inbreeding process may distort the genotypic frequencies so as to make the estimated variance component unrepresentative of the base population. Thus the different parameters obtained from real experiments seem to have limited utility. In view of these problems it is desirable to adopt a more sensitive approach, the analysis of basic generations population derived from a cross of two pure breeding lines, to estimate these parameters. Selfed families are generally easier to produce than half-sib families and in the absence of dominance yield an estimate of additive genetic variance (D) with four fold greater power.

Where a prior knowledge of the relative magnitudes of genetic variance components is lacking, Pepper (1983) proposed two methods — a minimax strategy and a Bayesian strategy for choosing the number of parents for a given number of crosses in hierarchial (NCM I) and factorial (NCM II) mating designs. The underlying principle is to minimize either the maximum (Mini max strategy) or average sampling variance of the genetical variance components (Bayesian strategy). The sampling variances Var. ($\hat{\sigma}_A^2$ or D) and Var. ($\hat{\sigma}_D^2$ or H) are quadratic functions of the expected mean squares, i.e. the variances are functionally related to σ_w^2, σ_A^2 and σ_D^2 and σ_w^2 being constant for a given genetic variance and for design of equivalent size, are defined as the non-negative orthant of two dimensional space. He showed that in NCM l one should make 2 to 4 full-sib families per half-sib family regardless of the number of crosses used. The maximum and average values of sampling variances increases considerably than the optimal values if the number of full-sib/half-sib exceeds 4. NC II with equal number of males and females are optimal for simultaneous estimation of additive and dominance variance. Failure to achieve optimality with factorial design is less than with nested design. For same experimental size, factorial design is superior to nested design, a conclusion reached by previous workers using different procedures. The aim of genetical design is to estimate the relative magnitudes of quantitative genetic parameters and generally covariances among relatives are used for partitioning the genotypic variance. The number of covariances available from the mating design determines the maximum number of estimable genetic parameters. The mating design used above controls only the parents of the individual measured. These types of mating design include two co-variances among relatives (i). Cov (*HS*) and (ii) Cov (*FS*). The one factor mating designs like BIP's (full-sib families) or half-sib families or polycross progeny are sufficient to detect the presence of genetic variability. The two factors mating designs like NCM I, NCM II, NCl III, diallel or partial diallel can separate additive and dominance variance assuming no epistasis whereas triallel and quadriallel can separate additive × addition interaction which we shall see in the chapter on diallel If the assumption of no epistasis is invalid then estimates obtained will be biased. If mating systems are used that allow control of ancestors in more than one generations, the number of covariances among relatives increases and the number of estimables parameters also increases.

The genetic variances do suffer from large sampling errors and are generally affected by linkage and non-allelic interaction which we shall see in later chapter. It is also difficult to obtain enough relatives (covariances) to adequately estimate the different kinds of genetic variance particularly if the epistatic variances are present. Nevertheless, variances are the most positive measurers of dominance and epistatic gene action for characters showing quantitative variation. Reliable estimates of geneic variances with small standard error can only obtained from large experiments repeated over an adequate sampling of environment as well shall see in Chapter 8 on genotype × environment interaction.

6.11.1 Selection of a particular mating design

Selection of a particular mating design depends on the mating system of crop species. With sugarcane diallel cross is a difficult proposition because some varieties seem to be incompatible when crossed and some vaieties are male sterile. The NCM1 maximizes the number of parents used for a given number of crosses and thus should be suitable for sugarcane. Although NCM2 is the most suitable design for estimating variance but wherever there is presence of incompatibility one should make a number of sets of crosses, each set having only a small number of parents as shown below (Table 6.16)

Table 6.16

Items	d.f.	Mean sum of squares	Expected mean sum of squares
Sets	S–1		
Replication in set	S(r–1)		
Males in sets	S(f–1)	MS_{ms}	$\sigma_e^2 + r\sigma_{mf}^2 + rf\sigma_m^2$
Females in sets	S(f–1)	MS_{fs}	$\sigma_e^2 + r\sigma_{mf}^2 + rf\sigma_f^2$
Males x females in sets	S(m–1)(f–1)	MS_{mxfxs}	$\sigma_e^2 + r\sigma_{mf}^2$
Error	S(mf–1)(r–1)	MS_e	σ_e^2
Total	Smfr–1		

Similarly, instead of single diallel, multiple diallel, each involving a smaller number of parents can be used.

6.12 Uses of Estimates of Components of Variation

Estimates of components of variation will help in:

(i) deciding the type of breeding method to be used for exploiting a particular type variation.
(ii) deciding the type of selection to be applied
(iii) deciding when the selection should be applied and terminated.
(iv) determining the speed of selection.
(v) deciding the selection limit.
(vi) understanding the nature of gene action for quantitative traits.
(vii) identifying the superior population or genotype and
(viii) deciding the type of variety to be the most appropriate goal.

6.13 Negative Estimates of Variance Components

By definition, a variance component is always positive. However, there is nothing intrinsic to the analysis of variance to prevent negative estimates of variance component. Negative estimates of variance can occur if the assumptions underlying the analysis are incorrect. Negative estimates can arise because of the following reasons:

1. Inadequate genetic or statistical models.
2. Inadequate sampling of reference population.
3. Sampling error in estimation.
4. Poor experimental design.

Negative estimate can arise due to assortative mating. With split-split design negative estimates of variance could arise when the correlation between subplots in the same whole plot is less than the correlation between subplots in whole plot.

6.14 Comparison of Variability

When the objective is to compare the variability of different characters in one population or between populations for the same character we require the estimate of the coefficient of variability for that character(s). As the variances could be phenotypic, genotypic or environmental so we can have phenotypic, genotypic or environmental coefficient of variability and these will be estimated as follows:

Phenotypic coefficient of variability (P.C.V.)

$$= \frac{\sqrt{\sigma_{px}^2}}{\overline{X}}$$

Genotypic coefficient of variability (G.C.V.)

$$= \frac{\sqrt{\sigma_{bx}^2}}{\overline{X}}$$

Environmental coefficient of variability (E.C.V.)

$$= \frac{\sqrt{\sigma_{wx}^2}}{\overline{X}}$$

where, σ_{px}^2, σ_{bx}^2 and σ_{wx}^2 are the phenotypic, genotypic and environmental variances with respect to character, X and $\overline{X}$ is the mean of population for character X.

In estimation of phenotypic, genotypic and environmental variance with n lines or families and r number of replications analysis of variance will take the form as given in Table 6.17 for character X.

Table 6.17 Expectation of between families and within families variances

Item	d.f.	S.S.	M.S.	E.M.S.
Replication	$r-1$			
Between lines	$n-1$	SS_L	MS_L	$\sigma_{wx}^2 + r\sigma_{bx}^2$
Within lines	$(n-1) \times (r-1)$	SS_W	MS_W	σ_{wx}^2

The environmental variance is calculated as

$$\sigma_{wx}^2 = MS_w$$

The genetic variance is calculated as

$$\sigma_{bx}^2 = \frac{MS_L - MS_w}{r}$$

The phenotypic variance then becomes

$$\sigma_{px}^2 = \sigma_{wx}^2 + \sigma_{bx}^2$$

and $\overline{X}$ is the mean of n lines or families for the character X.

This measure of variability does not say anything about the range of variability within a particular population.

6.15 Partitioning of Covariance/Correlation

The covariance between characters, say between X and Y can be partitioned as

$$\text{Cov}(X, Y) = G(X, Y) + E(X, Y)$$

Phenotypic covariance, Genotypic variance and Environmental covariances. Further the genotypic covariance $G(X, Y)$ can be partitioned into additive, dominance and interaction covariance as

$$G(X, Y) = D(X, Y) + H(X, Y) + I(X, Y)$$

These covariances are estimated from the analysis of covariance as given in the analysis of covariance Table 6.18.

Table 6.18 ANOVA for covariance between characters X and Y

Item	d.f.	MS	E.M.S.
Between families	$n-1$	SP_{xy}	$\sigma_{Wxy} + r\sigma_{bxy}$
Within families	$(n-1)(r-1)$	SP_{Exy}	σ_{Wxy}

The environmental covariance is

$$\sigma_{Wxy} = SP_{Exy}$$

The genetic covariance is calculated as

$$\sigma_{bxy} = (SP_{xy} - SP_{Exy})/r$$

The phenotypic covariance then becomes

$$\sigma_{Pxy} = \sigma_{Wxy} + \sigma_{bxy}$$

σ_{bxy} is the additive genetic covariance, $D(X, Y)$ if the lines or families are population of inbred/pure breeding lines.

6.15.1 Partitioning of correlation coefficient

Panse (1940) developed the concept that phenotypic correlation involved two components, one genetic and the other environmental which can be calculated as follows.

r_p (Phenotypic correlation)

$$= \sigma_{pxy}/\sqrt{\sigma_{px}^2\, \sigma_{py}^2}$$

r_g (Genotypic correlation)

$$= \sigma_{bxy}/\sqrt{\sigma_{bx}^2\, \sigma_{by}^2}$$

and r_e (Environmental correlation)

$$= \frac{\sigma_{wxy}}{\sqrt{\sigma_{wx}^2 \sigma_{wy}^2}}$$

where σ_{py}^2, σ_{by}^2 are σ_{wy}^2 are the phenotypic, genotypic and environmental variances of the character Y, respectively.

The relationship between phenotypic, genotypic and environmental correlation is as follows:

$$r_p = \frac{\sigma_{pxy}}{\sigma_{px} \cdot \sigma_{py}} = \frac{\sigma_{bxy} + \sigma_{wxy}}{\sigma_{px} \cdot \sigma_{py}}$$

$$\vdots$$

$$= h_x \cdot h_y r_g + e_x e_y r_e$$

where h_x and h_y are the square root of hertability estimate for character x and y, respectively and e_x, and e_y are the environmental standard deviation of character x and y, respectively. This shows that $r_p \neq r_g + r_e$. The value of r_g can be greater than r_e and may differ in sign. Similarly, r_g can be greater than r_p, the phenotypic correlation coefficient and may differ in sign. The value of correlation coefficient can fall outside the normal range of –1 to +1 when covariance is over estimated whereas variance is under estimated.

6.15.2 Calculation of additive genetic correlation

The genetical correlation r_g (D) between two traits x and y in the population of inbreds/pure breeding lines will be

$$r_g (D) = D_{xy}/\sqrt{D_x D_y}$$

where D_{xy} is the additive genetic covariance between two traits x and y and D_x and D_y are their additive genetic variances. The correlation coefficient seems to be the most useful parameter for supplementing the means and standard deviation of normally distributed variables in describing bivariate and multivariate distribution, Correlation has played an important part in biometrical genetics because many of the consequences of Mendelian inheritance and later developments from it are expressed conveniently in terms of correlation between related plants or animals which can be seen later in the chapter.

6.15.3 Standard error of correlation coefficient

Robertson (1959b) gave a formula for estimating the variances of genetic correlation in terms of genetic correlation and heritability of the two traits using Reeve's (1955) formula as

$$\sigma_{rg}^2 = \frac{(1 - r_g^2)^2}{N - 1} \left[\frac{2}{h_x^2 h_y^2} + \frac{1}{2} \right]$$

where h_x^2 is the heritability of the trait, X

h_y^2 is the heritability of the trait, Y

N is the sample size.

However, Van Vyleck and Henderson (1961) showed that the precision of the estimated genetic correlation will be poor with a sample size of $N < 1000$ pairs and especially for traits with low heritability.

Robertson (1959b) gave formula for estimating the variances of genetic correlation in terms of genetic correlation and heritability of the two traits as

$$\sigma_{r_g}^2 = \frac{(1 - r_g^2)}{2} \cdot \frac{\sigma^2 h_1}{h_1^2} \cdot \frac{\sigma_{h2}^2}{h_1^2}$$

The variance of genetic correlation will in general be small when the S.E.'s of the heritability estimates are small. In general, in any correlation $r = \frac{\text{Cov}(x, y)}{\sqrt{\sigma_x^2 \sigma_y^2}}$ there appears three components, Cov (x y), σ_x^2 and σ_y^2 Now if the variances of the three components are $V(\sigma_x^2)$, $V(\sigma_y^2)$ and $V(\sigma_{xy}^2)$, and their covariances are Cov $(\sigma_{xy}^2, \sigma_x^2)$, Cov $(\sigma_{xy}^2, \sigma_y^2)$ and Cov (σ_x^2, σ_y^2), then the variance of the correlation is

$$\sigma_r^2 = r^2 \left[\frac{V(\sigma_{xy})}{\sigma_{xy}^2} + \frac{V(\sigma_x^2)}{4(\sigma_x^2)^2} + \frac{V(\sigma_y^2)}{4(\sigma_y^2)^2} \right.$$

$$\left. + \frac{\text{Cov}(\sigma_{xy}, \sigma_x^2)}{\sigma_{xy} \cdot \sigma_x^2} + \frac{\text{Cov}(\sigma_{xy}, \sigma_y^2)}{\sigma_{xy} \cdot \sigma_y^2} \right.$$

$$+ \frac{\text{Cov}(\sigma_x^2, \sigma_y^2)}{2\sigma_x^2, \sigma_y^2}\Bigg]$$

The three variances and three covariances appearing in the formulas can be calculated separately. This formula (Mode & Robinson (1959); Tallis (1959) and Scheinberg (1966)) can be used to estimate, variances of all correlations i.e. phenotypic, genotypic and environmental correlations.

6.16 Intra-Class Correlation

When it is desired to study the correlation between some characteristics of members of one or more families, intraclass correlation rather than the ordinary correlation coefficient is used. Here the pairs of the variables, say x and y, are really paired sample of the same variate. The test consists of say n unrelated families with uniform intra-family relationship sampled from a population, in which the observed test performance, P_{ij} of the ith individual in the jth family can be expressed as the sum of the two independent random components.

$$P_{ij} = b_i + w_{ij}$$

where b_i's are families means normally distributed with mean zero and variance σ_b^2 and the term w_{ij}'s are deviations normally distributed with zero mean and variance σ_w^2. The total phenotypic variance (σ_p^2) equals

$$\sigma_p^2 = \sigma_b^2 + \sigma_w^2$$

and the intra-class correlation coefficient, t, is estimated as

$$t = \sigma_b^2/\sigma_p^2$$

Thus, t measures the relative importance of factors producing variation between groups/families to the total variation and in this sense it measures the heritability. The estimates of σ_b^2, σ_e^2, σ_p^2 ($= \sigma_b^2 + \sigma_e^2$) are estimated from the analysis of variance as shown in Table 6.19.

Table 6.19 ANOVA for estimation of σ_b^2 and σ_w^2

Item	d.f.	M.S.	Expected mean square
Replication	$r-1$		
Between families	$(n-1)$	MS_b	$\sigma_w^2 + r\sigma_b^2$
With families	$(n-1)(r-1)$	MS_w	σ_w^2

$$\sigma_b^2 = \frac{MS_b - MS_w}{r}$$

The value of t would be 1.0 if σ_w^2 is zero and in this case the total variation will be solely due to σ_b^2. If there is no variation between family means (i.e. $\sigma_b^2 = 0.0$) then t will be negative. The value of t will approach zero, when σ_w^2 and σ_b^2 are of the same order.

6.17 Repeatability

When multiple measurements in individuals are obtained for characters in a population, then the concept of repeatability quantifies the correlation between measurements on an individual. Repeatability equals the intra-class correlation coefficient giving the proportion of total phenotypic variance which is due to differences among individuals. Such differences may represent long term environmental effects and/or genetic variation.

In case of F_2 generation population (the mixed population)

$$VP = VG + VE + 2\,\text{Cov}\,GE$$

The repeatability $(r) = \dfrac{V_G + V_{E\times G}}{V_p}$

Consequently, repeatability gives an upper limit to heritability of traits such as leaf shape or seed weights, fruit shape, seed content, sugar content (different canes from the same clone) which are expressed repeatedly in a plant's life time. Unlike estimates of heritability derived from field collected material sib-ship estimation of repeatability does not require assumptions about the breeding systems, levels of inbreeding or maternal effect in population although it is necessary that genetic control of that trait does not vary from one measurement to another.

Repeatability indicates the gain in accuracy to be expected from multiple measurements. Reduction of environmental variance is the chief advantage to be gained from multiple measurements.

References

Becker, W.A. 1967. Manual of Procedures in Quantitative Genetics. Second edition. Washington State University.

Cockerham, C.C. 1961. Implications of genetic variances in a hybrid breeding program. Crop. Sci., **1**: 47–52.

Comstock, R. 1996. Quantitative Genetics with Special Reference to Plants and Animal Breeding. Iowa State University Press, Ames.

Comstock, R.E. 1978. Quantitative genetics in maize breeding. In: D. Walden (ed.) Maize Genetics & Breeding. pp. 191–206. Wiley, New York.

Comstock, R.E and Robinson, H.F. 1952. The components of genetic variance in populations of biparental progenies and their use in estimating the average degree of dominanace. Biometrics. 4: 254-266.

Comstock, R.E. and Robinson, H.E 1952. Estimation of average dominance of genes. In: Gowen, J.W. (ed), Heterosis, Iowa State College Press, Ames, Iowa, pp. 494–516.

Dickinson, A.G. and Jinks, J.L. 1956. A generalized analysis of diallel crosses. Genetics, **41**: 65–78.

Falconer, D.S. and Mackay, T. 1996. Introduction to Quantitative Genetics. Longman, Essex, U.K.

Gardner. C.O. 1963. Estimation of genetic parameters in cross fertilizing plants and their implications in plant breeding. In: Statistical genetics and plant breeding. Hanson, W.O. and Robinson, H.F. (eds). National Academy of Sciences, Washington, pp. 225–252.

Griffing, B. 1975. Use of doubled-haploids in recurrent selection methods. Theo. Appl. Genet., **46**: 367–386.

Hallauer, A.R. and Miranda, FO, J.B. 1988. Quantitative Genetics in maize Breeding. Iowa State Univ. Press, Ames, Iowa, USA.

Hayman, B.1. 1960. Maximum likelihood method of estimation of genetic components of variation. Biometrics, **16**: 369–381.

Hartl, D. H. and Clark, A.c.1997. Principles of Population Genetics. 3rd Ed. Sinauer Associates, Sunderland, M.A.

Hewitt, J.K. 1980. A note on the test for the direction of dominance in the triple test cross in the presence of genotypes – environment interaction. Heredity, **45**: 293–295.

Jinks, J.L. 1979. The biometrical approach to quantitative variation. Quantitative Genetic Variation. Academic Press, New York, pp. 81–107.

Jinks, J.L. and Perkins, J.M. 1968. The detection of linked epistatic genes for a metrical trait. Heredity, **24**: 465–575.

Jinks, J.L., Perkins, J.M. and Breese, E.L. 1969. A general method of detecting additive, dominance and epistatic variation for metrical traits. 2. Application to inbred lines. Heredity, **24**: 45–57.

Jinks, J.L. and Virk, D.S. 1977. A modified triple test cross analysis to test and allow for inadequate testers. Heredity, **39**: 165–170.

Kearsey, MJ. 1970. Experimental sizes for detecting dominance variation. Heredity, **25**: 529–542.

Kearsey, MJ. 1980. The efficiency of North Carolina Experiment 3. and the selfing backcrossing series for estimating additive and dominance variation. Heredity, **45**: 73–82.

Kearsey, M.J. 1993. Biometrical genetics in breeding In: Plant Breeding. Hayward, M.D.; Bosemark, N.O. and Romagosa, I. (eds). Chapman & Hall. pp. 163–183.

Kearsey, M.J. and Jinks, J.L. 1968. A general method of detecting additive, dominance and epistatic variation for metrical traits. I. Theory, Heredity, **23**: 403–409.

Kearsey, M.J. and Pooni, H.S. 1996. The Genetical Analysis of Quantitative traits. Chapman & Hall, London.

Kempthorne, O. 1969. An Introduction to Genetics Statistics. Iowa State Univ. Press, Ames, Iowa.

Mather, K. 1979. Historical Review: Quantitative variation and polygenic system. Quantitative Genetic Variation. Academic Press, New York.

Mather, K. and Jinks, J.L. 1982. Biometrical Genetics. 3rd edn. Chapman and Hall, London.

Mather, K. and Vines, A. 1952. The inheritance of height and flowering time in a cross of *Nicotiana rustica*. Quantitative Inheritance. H.M.S.O., 45–80.

Mather, W.B. 1964; Principles of Quantitative Genetics. Burgess Publishing Company, Minneapolis, Minnesota.

Matzinger, D.F. and Cockerham, C.C. 1963. Simultaneous selfing and partial diallel test crossing. 1. Estimation of genetic and environmental parameters. Crop sci., **3**: 309–314.

Mode, C.J. and Robinson, H.F. 1959. Pleiotropism and the genetic variance and covariance. Biometrics, **15**: 518–537.

Ooigen, J.W. Van. 1986. Distribution of estimates of genetic variance components in self fertilizing crops. In:

Biometrics in Plant Breeding. Proceedings of the 6th Meeting of the Eucarpia Section. Biometrics in Plant Breeding. University of Birmingham, U.K. pp. 59–69.

Panse, V.G. 1940a. The application of genetics to plant breeding 2. The inheritance of quantitative characters and plant breeding J. Genet., **40**: 283–302.

Panse, V.G. 1940b. A statistical study of quantitative inheritance. Ann. Eugenics, **10**: 76–103.

Pepper, W.D. 1983. Choosing plant mating design allocations to estimate genetic variance components in the absence of prior knowledge of the relative magnitudes. Biometrics, **39**: 511–524.

Pooni, H.S. and Jinks, J.L. 1976. The efficiency and optimal size of triple test cross designs for detecting epistatic variation. Heredity, 36(2): 215–227.

Pooni, H.S., Jinks, J.L. and Pooni, G.S. 1980. A general method for detection and estimation of additive, dominance and epistatic variation for metrical traits. 4. Triple test cross and inbred line analysis. Heredity, **41**: 83–92.

Pooni, H.S., Jinks, J.L. and Pooni, G.S. 1980. A general method for detection and estimation of additive, dominance and epistatic variation for metrical traits 4. Triple test cross for normal families and their selfs. Heredity, **44**: 177–192.

Robertson, A. 1959b. The sampling variance of the correlation coefficient. Biometrics, **15**: 469–485.

Roff, D.A. 1997. Evolutionary Quantitative Genetics. Chapman and Hall, New York. Moll, R.H., Robinson, H. F. and Cockerham, C.C. 1960. Geneticvariability in an advanced generation of a cross of two open-pollinated varieties of corn. Agronomy J., 52; 171-173.

Scheinberg, E. 1960. Note: The sampling variance of correlation coefficients estimated in genetic experiments. Biometrics, **22**: 187–191.

Shaw, R.G. 1987. Maximum likelihood approaches applied to quantitative genetics of natural populations. Evolution, **41**: 812–826.

Tallis, G.M. 1959. Sampling errors of genetic correlation coefficients calculated from analyses of variance and covariance. Anat. Jul. of Statist., **1**: 35–43.

Van Vleck, L.D. and Henderson, C.R. 1961. Empirical sampling estimates of genetic correlations, Biometrics, **17**: 356–371.

Virk, D.S. and Jinks, J.L. 1977. The consequences of using inadequate testers in the simplified triple test cross. Heredity, **30**: 237–251.

Weber, W.E. 1982. Estimation of genetic variance components under linkage and sampling error in self fertilizing species. Agronomie, **2**: 201–212

Weir, B.S., Coekerham, C.C. and Reynolds, J. 1980. The effect of linkage and linkage disequilibrium on the covariances of non inbred relatives. Heredity, **45**: 351–359.

Wright, S.1969. Evolution and the Genetics of Populations. Vol. 2. The theory of Gene Frequencies.The University of Chicago Press. Chicago and London.

7

Diallel Analyses

A diallel cross is a set of all possible matings between several genotypes. The genotypes could be individuals, clones, homozygous lines, etc. (Hayman 1954).

A set of lines, say n, will generate a total of n^2 crosses, $\left(\frac{1}{2}(n-1) F_1\text{'s} + \frac{1}{2}(n-1) \text{ reciprocal } F_1\text{'s} + n \text{ selfs}\right)$.

The data recorded on the diallel cross raised in a replicated experiment can be analysed in the following ways:

1. Full diallel (F_1's, RF_1's, selfs)
2. Half diallel (F_1's, selfs)
3. Full diallel excluding self
4. Half diallel excluding self
5. Partial diallel

7.1 Full Diallel Analysis

The linear statistical model (Hayman, 1954a) which is an extension of Yates (1947) model for testing additive and maternal effects is as follows:

$$Y_{rs} = m + j_r + j_s + l + l_r + l_s + r_s + k_r - k_s + k_{rs}$$

and $Y_{rr} = m + 2j_r - (n-1) l - (n-2) l_r$

where Y_{rs} is the single $(r \times s)$ off-diagonal observation for each cell of diallel table (i.e. observation on individual resulting from a cross between the rth line as female and sth line as male), Y_{rr} the rth diagonal observation, m the general mean, j_r genetic effect of the rth line, j_{rs} the dominance deviation for the $r \times s$ reciprocal sum, $2k$, the difference between the effects of rth parental line used as female parent and as male parent, and $2 k_{rs}$ is the remaining discrepancy in the $(r \times s)$ reciprocal differences (i.e. the k_{rs} represents interactions between parental effects and own genotype of the progeny effects)

with $\sum_r j_r = 0$, $\sum_r k_r = 0$, $\sum_r k_{rs} = 0$ and $k_{sr} = -k_{rs}$

The j_{rs} is further subdivided as

$$j_{rs} = l + l_r + l_s + l_{rs}$$

where l is the mean dominance deviation,

l_r is the further dominance deviation due to rth parent,

l_{rs} is the residual dominance deviation in the $r \times s$ reciprocal sum

with $\sum_r l_r = 0$, $\sum_r l_{rs} = 0$ and $\sum_s l_{rs} = 0$

The various items in this analysis along with sum of squares are given in Table 7.1.

Assuming no maternal effects expectations of different items appeared in the ANOVA are as follows.

Item	Expectation
a	$2n \Sigma u_i v_i (d_i - h_i w_i)^2 + \sigma_e^2$
b	$8 \Sigma u_i^2 v_i^2 h_i^2 + \sigma_e^2$
b_1	$4n^2 (\Sigma u_i v_i h_i)^2/(n-1) + \sigma_e^2$
b_2	$4n \Sigma u_i v_i w_i^2 h_i^2/(n-2) + \sigma_e^2$
c	σ_e^2
d	σ_e^2

Thus significant mean squares due to a item shows the presence of additive genetic variation in

Table 7.1 ANOVA in Hayman's analysis

Item	d.f.	Sum of squares
a	$n-1$	$\Sigma(Y_{r.} + Y_{.r})^2/2n - 2Y^2.../n^2$
b	$\frac{1}{2}(n-1)$	$\Sigma(Y_{rs} + Y_{sr})^2/4 - \Sigma(Y_{r.} + Y_{.r})^2/2n + Y^2.../n^2$
b_1	1	$(Y... - ny.)^2/n^2(n-1)$
b_2	$n-1$	$\Sigma(Y_{r.} + Y_{.r} - nY_r)^2/n(n-2) - (2y... - ny.)^2/n^2(n-2)$
b_3	$\frac{1}{2}n(n-3)$	$\Sigma(Y_{rs} + Y_{sr})^2/4 - \Sigma Y_r^2 - (Y_{r.} + Y_{.r} - 2Y_r)^2/2(n-2) + (Y...-Y..)^2/(n-1)(n-2)$
c	$n-1$	$\Sigma(Y_{r.} + Y_{.r})^2/2n$
d	$\frac{1}{2}(n-1)(n-2)$	$\Sigma(Y_{rs} - Y_{sr})^2/4 - (Y_{r.} - Y_{.r})^2/2n$
error	$(n^2-1)(r-1)$	
Total	n^2-1	

where Y_{rs} is the score of the cross between rth female and sth male where $r = 1$ to n and $s = 1$ to n in an $n \times n$ diallel,
  Y_{sr} is the score of the reciprocal cross,
  $Y_{r.}$ is the sum of the rth array,
  $Y_{.s}$ is the sum of the sth array,
  $Y..$ is the sum of all the entries in the diallel table, and
  $Y.$ is the sum of scores of all the parental families.

the material under investigation. Significant b item indicates the presence of dominance component of variation. Significant b_1 shows directional dominance whereas significant b_2 item implies asymmetry in gene distribution and b_3 item shows the presence of the residual non-additive variation which is infact identical with the specific combining ability of Griffing (1956). In breeding jargon estimation of items a and b amounts to estimation of general combining ability and specific combining ability, respectively. The item a estimates (primarily though exclusively not) additive variance while item b estimates dominance. The c and d items indicate variations due to general reciprocal differences and specific reciprocal differences, respectively and in the absence of any cytoplasmic or maternal effects both provide estimate of environmental component of variation. Finally, σ_e^2 signifies the error component.

Hyman's analysis has been criticised on two grounds. Firstly, the analysis appears to have been calculated using a progressive fitting of the unknown parameters and thus a truely non-orthogonal analysis of variance is made orthogonal with the component of sum of squares summing-up to the total sum of squares (Kempthorne, 1956). Secondly, the representation of the maternal effects in the model has been questioned (Wearden, 1964; Topham, 1966) and the model has been shown to be representing a reciprocal effects model than a maternal effects model in strict sense. Walter and Gale (1977) proposed a method of removing the effects of the non-orthogonality from the full diallel table. They derived Hyman's analysis of variance following reciprocal effects as well as maternal effects and estimated different parameters following least squares. They showed that apart from S.S. $(\hat{jr})$, all sum of squares are precisely those given by Hayman. Hayman's (a) sum of square represents the sum of squares due to genetic effects only in the special case of no dominance. Indeed it was long recognised that the correct sum of squares for detecting genetic effects following Hayman's model is

$$\sum_r Y_{rr}^2 - (\sum_r Y_{rr})^2/n$$

which is identical with $4 \sum_r \hat{jr}^2$ of Walter and Gale's model and which is the sum of squares of diagonal entries. That is why the Y_{rr} in case of their reciprocal model is

$$Y_{rr} = m + 2j_r$$

which is different than Hayman's model.

In case of maternal affects the model takes the form as:

$$Y_{rr} = m + 2j_r + k_r$$

and $Y_{rs} = m + j_r + j_s + l + l_s + l_r + l_{rs} + kr \ (r \neq s)$

As k_{rs} term does not appear so the model assumes absence of interaction between maternal affect and own genotype effect. In this model Hayman's (d) mean squares estimates σ_e^2, the error term and thus provides a test for the adequacy of the maternal effect model. The only sum of squares affected by the change of model is due to S.S. $(\hat{j}_r)$ which tests for the genic effects (a), free from all possible maternal affects. But this test must not be used if there is significant d mean squares as it indicates the presence of maternal effect x own genotype interaction. Thus a more realistic model would be one which involves, in addition to maternal affect, interaction between maternal and own genotype effect. But then with diallel it would be difficult to ascertain whether the crosses show dominance deviation or there is interaction between maternal effect and own genotype effect. If however item d is not significant, the maternal effect model is biologically more appropriate than reciprocal effects model. Hayman's analysis is thus appropriate for detecting the various effects for which the analysis was devised.

7.2 Half Diallel Analysis

If reciprocal differences are assumed absent and only one of each pair of reciprocal crosses is raised then this half-diallel data can be analysed following Hayman's model (Morley-Jones, 1965). The sum of squares. corresponding to a, b_1, b_2 and b_3 can be obtained. The general ANOVA in half-diallel analysis will take the form as given in Table 7.2.

Table 7.2 General ANOVA in half-diallel analysis

Item	d.f.
Treatments (Parents + Crosses (F_1's))	$\frac{1}{2}n(n + 1) - 1$
Parents	$n - 1$
Crosses (F_1's)	$\frac{1}{2}n(n - 1) - 1$
Parents vs F_1's	1
Error	$\left(\frac{1}{2}n(n + 1) - 1\right)(r - 1)$

Following Hayman's model the ANOVA will take the form as given in Table 7.3.

Table 7.3 ANOVA using Hayman's analysis

Item	d.f.
a	$n - 1$
b_1	1
b_2	$n - 1$
b_3	$\frac{1}{2}n(n - 3)$
Total	$\frac{1}{2}n(n + 1) - 1$
Error	$\left(\frac{1}{2}n(n + 1) - 1\right)(r - 1)$

The item 'a' tests the additive variation whereas b_1, b_2 and b_3 test the three non-additive components described in Hayman's complete diallel analysis. All these items are tested against error. As the parent's selfed are included in this design so non-additivity is measured as deviation of the F_1's from the mid parent. Graphical analysis which is discussed in later section, can be carried out with the half-diallel data and if b_3 is significant in the above analysis then the interacting parent(s) can be isolated. As this design permits the estimation of additive genetic variance, D, hence a dominance ratio can be calculated which is not biased by the unequal gene frequencies. The comparison parents vs crosses (F_1's) provides test of unidirectional dominance which is similar to that of b_1. The half-diallel thus allows the test for non-allelic interaction and estimates additivity and dominance unambiguously.

The half diallel table by Morley-Jones can be criticized on the grounds as that of Hayman's full diallel table. Walters and Morton (1978) extended the method of Walters and Gale (1977) to the half-diallel table and showed that the correction of the mean squares for additive effects to allow for the partial confounding of the additive and dominance parameters is no longer necessary since each group of parameters (a, b_1 and b_2) in this method can be tested independently.

7.3 Genetical Structure of Quantitative Traits

The investigation of genetical properties of a group of homozygous lines using diallel cross method

7.4 Biometrical Genetics–Analysis of Quantitative Variation

has been presented in a series of papers (Jinks and Hayman, 1953; Hayman, 1954; and Jinks 1954). The analysis is based on the following assumptions:

1. Diploid segregation
2. No difference between reciprocal crosses
3. No non-allelic interaction
4. No multiple allelism
5. Homozygous parents
6. Genes independently distributed between the parents.

Failure of one of these assumptions could be identified by its characteristic effects on analysis of $W_r - V_r$.

7.3.1 Wr – Vr analysis

Considering one locus with two allele A and a there would be two pure breeding homozygous genotypes, AA and aa. If u and v are the frequencies of the genotypes AA and aa, respectively then the 2×2 diallel table will take the form as given in Table 7.4.

Table 7.4 Genotypes, their frequencis and genotypic values of the families in two parents (AA and aa) diallel

♀ \ ♂	Genotype Frequency Genotypic value	AA u d	aa v $-d$	Mean
AA		AA d	Aa h	$ud + vh$
aa		Aa h	aa $-d$	$uh - vd$
		$ud + vh$	$uh - vd$	$(u-v)d$ $+ 2uvh$

The variance of each row or the array variance (V_{ri} where $i = 2$ in the present case) and the covariance of the family means within array with the phenotypes of their respective non-recurrent parents (W_{ri}) can be calculated as follows:

$$V_{rAA} = ud^2 - vh^2 - (ud + vh)^2$$

$$= uv(d-h)^2$$
$$V_{raa} = uh^2 + vd^2 - (uh - vd)^2$$

.
.
.

$$= uv(d+h)^2$$
$$W_{rAA} = u \cdot d \cdot d + v \cdot h(-d) - (ud + vh)(u-v)d$$

.
.
.

$$= 2uvd(d-h)$$
$$W_{raa} = u \cdot d \cdot h + v \cdot h \cdot (-d) - (uh - vd) \times (u-v)d$$

.
.
.

$$= 2uvd(d+h)$$

The summary of the estimates of the above statistics is presented in Table 7.5.

Thus $W_r - V_r$ is constant over arrays and equals

$$W_r - V_r = uv(d-h)(d+h)$$
$$= uv(d^2 - h^2)$$
$$= 1/4 D - 1/4 H_1$$

When parents and their F_1's are raised in a replicated experiment and if mean squares corresponding to $W_r - V_r$ between arrays is not significant it shows no epistasis. The regression of covariance (W_r) on variance (V_r) will give a straight line of unit slope i.e.

$$b_{Wr/Vr} = 1$$

If interalia, two assumptions, namely, no epistasis and no linkage hold i.e., in case of additive-dominance model all the points will be dispersed along a line of unit slope with the dominant parents towards the origin and the recessive parents towards a limiting parabola described by $W_r = \sqrt{V_p^2 V_r}$.

Thus the order of parents along the regression line indicates the relative proportion of dominants to recessives in the corresponding parents and the points with lower values of W_r and V_r have greater

Table 7.5 Estimates of W_r, V_r, $W_r - V_r$ and $W_r + V_r$

Array	Frequency	W_r	V_r	$W_r - V_r$	$W_r + V_r$
AA	u	$2uvd(d-h)$	$uv(d-h)^2$	$uv(d-h)(d+h)$	$uv(d-h)(3d-h)$
Aa	v	$2uvd(d+h)$	$uv(d+h)^2$	$uv(d-h)(d+h)$	$uv(d+h)(3d+h)$.

proportion of dominants whereas the points with higher values of W_r and V_r have greater proportion of recessives. The position of the intercept tells about the degree of dominance. If it is below the origin overdominance is indicated whereas if it is above the origin then it shows partial dominance. If it passes through the origin then it shows complete dominance.

The expected relationships between W_r and V_r for different relative values of D and H_1 are given in Figure 7.1. The regression of W_r on V_r provides a geometric representation of the degree of dominance and also permits the separation of true dominance from spurious dominance caused by the non-allelic interaction.

The mean squares corresponding to $W_r + V_r$ between arrays if significant shows the presence of dominance. In case of no dominance there is no significant regression, i.e. the variances and covariances of all arrays are identical, within experimental error, being estimate of a single point where $W_r/V_r = 2$. In other words, in the absence of dominance the arrays would be clustered at random around the mid-point of the line. If the heritability for a particular quantitative character is very low or there is no genetic variation for that character then all the array points in the $W_r - V_r$ graph will form a straight line running parallel to V_r but due to sampling variation it could swing in either direction making b negative. Further, significant correlation between $(W_r + V_r)$ and parental means indicates mean directional dominance and the not significant correlation shows ambidirectional dominance.

7.3.1.1 Effects of interaction and gene association and dispersion on $W_r - V_r$ graph

Failure of either of the assumptions (no epistasis and no linkage) leads to departure from the rectilinear relationship between W_r and V_r and $W_r - V_r$ is not constant. Coughtrey and Mather (1970) showed the effects of interaction and of gene association and dispersion on W_r/V_r graph for different gene frequencies (Figure 7.2).

I. Effects of non-allelic interaction

Considering two loci (A, B) with two alleles at each locus and assuming $da = db = ha = hb$ and $iab = jab = jba = lab = \theta da$, where θ specifies the type and strength of the interaction, the statistics $W_r - V_r$ and $W_r + V_r$ can be derived for the four parental arrays, namely, $AABB, AAbb, aaBB$ and $aabb$. θda is positive for complementary and negative for the duplicate interaction and it has a range from 0 (no interaction)

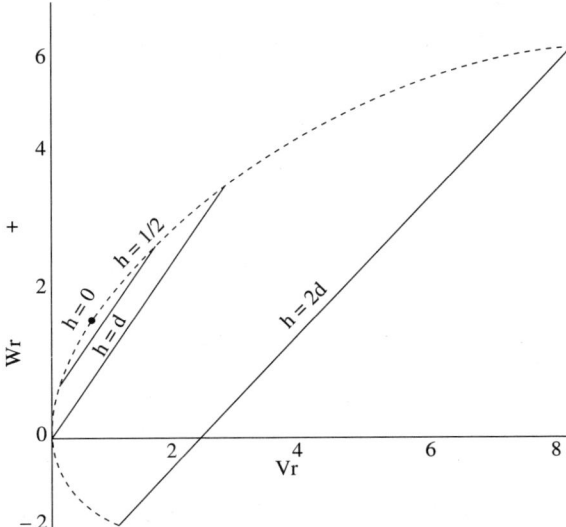

Fig. 7.1 The theoretical regression of W_r on V_r for various degrees of dominance i.e. h/d ratio. The curve (broken line) joins the points of the arrays whose common parents contain all the dominance or all the recessive alleles (Jinks, 1954).

7.6 Biometrical Genetics–Analysis of Quantitative Variation

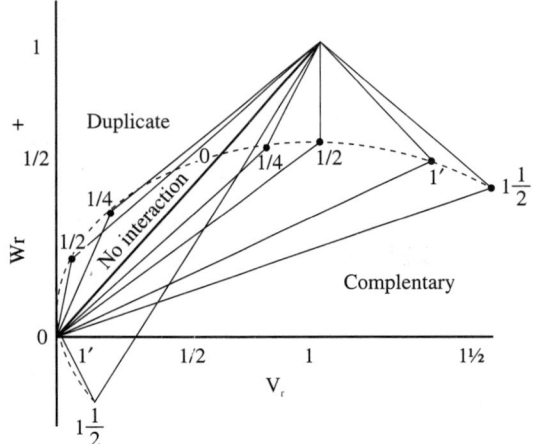

Fig. 7.2 The effect of complementary and duplicate interactions between two gene pairs on the $W_r - V_r$ graph (adapted from Mather, 1966).

to 1 (full interaction). It can be shown that $W_r - V_r$ is constant over arrays. Secondly, when there is complementary interaction i.e. $\theta da = + 1$, the value of W_r goes down and V_r value goes up and as a result $W_r - V_r$ is negative and thus there is over estimation of dominance. In otherwords, complementary interaction inflates H_1, the dominance components relative to D, the additive genetic variance component. The curve thus is concave upwards with b deviating from unity. The degree of curvature depends on the relative frequencies of dominant and recessive genes and the intensity of interaction (θda). For complete duplicate interaction i.e. when $\theta da = -1$ both W_r and V_r values are zero and thus there is a straightline relationship between W_r and V_r. When θda has a value of $-1 < \theta < 0$, W_r values go up whereas V_r values go down and as a result $W_r - V_r$ values are positive. There is thus underestimation of dominance, i.e. there is deflation of dominance ratio. The $W_r - V_r$ graph now is concave downwards with b deviating from unity. With i type interaction alone at equal gene frequencies the curve is concave upwards, however array points from parents containing genes in dispersion no longer lie on a line of unit slope. With l type interaction at equal gene frequencies the arrays points scatter along a line of unit slope whereas in case of j type interaction the scattering is always irregular. Thus, W_r, V_r regression can detect the complementary genes type interaction but not duplicate interaction (Mather 1967).

Interaction affecting a proportion of the members of an array will increase the variance of that array whereas the covariance will fall. This will result in moving the regression line in $W_r - V_r$ graph to right giving an apparent increase in dominance but its slope will fall below the expected value of 1. If the ratio of interacting to non-interacting members in the array is low the array involving interacting parents can be omitted from the $W_r - V_r$ analysis which then would lead to an improvement of the average slope and at the same time reduce the experimental error of the regression coefficient.

II. Effects of Association/Dispersion of Genes

Again considering 2 loci with 2 alleles at each locus and C_1, C_2, C_3 and C_4 being the proportion of gametes AB, aB, Ab and ab, respectively the W_r and V_r values of the four parental arrays namely $AABB$, $AAbb$, $aaBB$ and $aabb$ can be derived under the condition $C_1 C_4 \neq C_2 C_3$. Like epistasis. $W_r - V_r$ is not constant in the presence of linkage disequilibrium. When there is association phase linkage $W_r - V_r$ will be higher for arrays $AABB$ and $aabb$ whereas its values will be lower for arrays $AAbb$ and $aaBB$. The situation will be just the reverse when there is dispersion phage linkage. Thus association of genes will lead to a $W_r - V_r$ graph which is convex upwards whereas dispersion of genes will result in a curve which is concave upwards and in both cases b will deviate from unity. Association phage linkage under estimates the dominance whereas dispersion phage linkage over estimates the dominance because repulsion phage linkage leads to an inflation of dominance component relative to additive genetic component and thus leads to spurious overdominance. So both the effects of interaction and association/dispersion of genes on $W_r - V_r$ show that graphical analysis does not provide unambiguous test for the presence of epistasis or linkage when both are present.

7.4 Estimation of Genetical Components of Variation

Besides W_r and V_r other statistics that can be estimated from the diallel table are as follows:

V_p (Variance of the parents)
$$= ud^2 + vd^2 - [(u-v)d]^2$$
$$= 4u^2v^2d^2$$
$$= \frac{1}{4}D$$

$\overline{V}_r$ (Mean variance of the arrays)
$$= u[ud^2 + vh^2 - (ud + vh)^2]$$
$$- v[uh^2 + vd^2 - (uh - vd)^2]$$
.
.
.
.
$$= 1/4D + 1/4H_1 - 1/4F$$

$V_{\bar{r}}$ (Variance of the means of arrays)
$$= u(ud + vh)^2 + v(uh - vd)^2$$
$$- [(u-v)d + 2uvh]^2$$
.
.
.
.
$$= 1/4D + 1/4H_1 - 1/4H_2 - 1/4F$$

$\overline{W}_r$ (Mean covariance between the arrays and the parents)
$$= u \cdot 2uvd(d-h) - v \cdot 2uvd(d+h)$$
.
.
.
$$= 1/2D - 1/4F$$

From these above statistics and the mean scores of the parents and their F_1's the following genetical components of variation can be estimated as

$$D = V_p - E = \Sigma\, 4uvd^2$$
$$H_1 = 4\overline{V}_r + V_p - 4\overline{W}_r - \frac{3n-2}{n}E = \Sigma\, 4uvh^2$$
$$H_2 = 4\overline{V}_r - 4V_{\bar{r}} + \frac{2(n^2-1)}{n^2}E = \Sigma\, 16u^2v^2h^2$$
$$F = 2VP - 4\overline{W}_r - \frac{2(n-2)}{n}E$$
$$= \Sigma\, 8\,uv(u-v)\,dh$$

F indicates whether dominant or recessive alleles are more frequent in the parents, being positive if dominant alleles are in excess.

$$h^2 = 4(\overline{P} - \overline{F}_1)^2 - \frac{4(n-1)}{n^2}E = \Sigma\, 4uvh$$

From the estimates of the genetical components of variation the following information can be obtained.

The dominance ratio is calculated as

$$\sqrt{\frac{H_1}{D}}$$

If this ratio is greater than 1 it is a case of over-dominance, if it is less than 1 it indicates partial dominance whereas if it equals 1 it shows complete dominance.

The ratio of $H_2/4H_1$ provides an idea about the distribution of genes having positive and negative effects. It's value different than 0.25 shows asymmetrical distribution of genes.

The ratio $\dfrac{\sqrt{4DH_1 + F}}{\sqrt{4DH_1 - F}} = \dfrac{K_D}{K_R}$ provides information about the relative frequencies of dominant and recessive genes. Values ≥ 1 show that the dominant and recessive loci are not in equal frequency.

The ratio $\dfrac{h^2}{4H_1}$ tells us the number of genes determining a character which shows some degree of dominance and is thus different than Mather's number of effective factors.

The ratio being $\dfrac{1/2F}{\sqrt{D(H_1 - H_2)}}$ different than unity suggests that the ratio of h to d is not consistent over all loci:

Finally, the estimate of heritability (heritability in broad sense, h_b^2 as well as heritability in narrow sense, h_b^2 can be obtained as

$$h_b^2 = \frac{1/2D + 1/2H_1 - 1/4H_2 - 1/2F}{1/2D + 1/2H_1 - 1/4H_2 - 1/2F + E}$$

$$h_n^2 = \frac{1/2D + 1/2H_1 - 1/2H_2 - 1/2F}{1/2D + 1/2H_1 - 1/4H_2 - 1/2F + E}$$

When the additive-dominance model is not adequate i.e. when there is non-allelic interaction the estimates of D_1, H_1, H_2 and F will be biased and the dominance and other ratios will give misleading conclusions. The expectation of a, b, b_1, b_2, b_3, c and d of Hayman's analysis can be written in terms of the diallel parameters, namely, D, H_1, H_2, F and E as given in Table 7.6.

7.4.1 F_2 Diallel

Jinks (1956) worked out expectation for the F_2 and backcross generations of diallel cross. The expected statistics for the F_2 generation are of the same general form as those of the F_1 generation except that the contribution of h is halved by one generation of inbreeding. Thus the coefficients of H_1 and H_2 are now one quarter of the F_1 statistics while the coefficient of F_1 is halved. Like F_1 diallel, the four genetic parameters D, H_1, H_2 and F can be estimated from the F_2 diallel and the test of homogeneity of least squares estimates of these four parameters over generations will provide a further test of non-allelic interaction. As only the mean variances of F_2 families have linkage terms appeared in the expectation, the existence of heterogeneity after including this statistics will indicate the presence of linkage provided the heterogeneity did not exist before excluding the statistics, the mean variance of the F_2 families. Thus in the presence of non-allelic interaction it no longer unambiguously tests the presence of linkage.

Table 7.6 Relation between Hayman's parameter and diallel parameters

Hayman (1954)	Jinks and Hayman 1953	Mather (1949)
a	$\frac{1}{2}n(D - F + H_1 - H_2) + E$	$\frac{1}{2}nD + E$
b	$\frac{1}{2}H_2 + E$	$\frac{1}{2}H + E$
b_1	$\frac{1}{4}n^2h^2/(n-1) + E$	
b_2	$n(H_1 - H_2)/(n-2) + E$	
c	E	E
d	E	E

To detect genic interaction it is advisable to go for F_2 diallel rather than F_1 diallel. In the F_1 generation genic interaction is reflected only by the cross means while in the F_2 they can be detected also from the variance among individuals within crosses. The existence of dominance can be tested, by two-way analysis of variance (array × blocks) of the $(W_{rij} + V_{rij})$ values. Similarly, an analysis of the $(W_{rij} - V_{rij})$ values-and the joint regression analysis of W_{rij} on V_{rij} over blocks can be conducted. These are the standard tests for the model's adequacy. In the F_2-diallel the expectations of the different statistics take the form as

$$V_{\bar{P}} = D$$

$$\overline{W}_r = \frac{1}{2}D + \frac{1}{8}F + \frac{1}{116}E$$

$$V_r = \frac{1}{4}D + \frac{1}{16}H_1 - \frac{1}{8}F = \frac{1}{290}E$$

$$V_{\bar{r}} = \frac{1}{4}D + \frac{1}{16}H_1 - \frac{1}{16}H_2 - \frac{1}{8}F + \frac{1}{290}E$$

$$\overline{V}F_2 = \frac{1}{4}D + \frac{1}{8}H_1 + E$$

The parameters D_1, H_1, H_2, F and E can be obtained by least squares. Hayman (1958) and Jinks (1956) derived various statistics from F_2-diallel to demonstrate the overall importance of dominance and interaction relative to additive variance. The relationship between W_r and V_r described for the F_1 analysis also holds, i.e. $bW_r/V_r = 1$ in the absence of non-allelic interaction and non random distribution of alleles. However, the point of intersection of the regression line with W_r axis is no longer $W_r - V_r = \frac{1}{4}(D - H_1)$ but $\frac{1}{4}\left(D - \frac{1}{4}H_1\right)$. In the absence of non allelic interaction $W_r - V_r$ graph for the F_2 and backcross generations should be identical but this is not necessarily true in the presence of interaction. A further relationship exists between the regressions of W_r on V_r for the F_1 and F_2 (or backcross generations). If we draw a line for each array through the W_r/V_r co-ordinates for its F_1 and F_2 array points, they will converge on a common point of intersection whose own coordinates vary characteristically with the degree of dominance. The W_r co-ordinate for the point of intersection is independent of the dominance relations as it contains only terms in d being $\frac{1}{2}D$

$= \Sigma 2uvd^2$ whereas V_r co-ordinate does not vary with the degree of dominance and for one locus with 2 alleles case has the expectation

$$uv(d-h)^2 + \frac{uvdh - (3/4)uvh^2}{uvdh} 2uvdh$$

This expression has a maximum value of $1/4D$ when $h = 0$. The point of intersection for other generations, e.g. F_2 and F_3 can be obtained by halving the contribution of dominance in the above expression. Thus for nth generation inbreeding the coordinates of intersection will be

$$W_r = 2uvd^2$$

and

$$V_r = uv(d - (1/n)h)^2$$
$$+ \frac{(1/n)uv\,dh - (3/4)\,uv\,((1/n)h)^2}{(1/n)\,uvdh \times (2/n)\,uvdh}$$

when n is large, h reduces to zero and the V_r co-ordinate approaches the value of $\frac{1}{4}D = \Sigma uvd^2$. Therefore with a low initial degree of dominance or a large number of generations of selfing the point of intersection approaches the co-ordinates for $W_r - V_r$ points of an F_1 diallel showing no dominance but these relationships break down in the presence of non-allelic interaction.

7.5 Diallel Analysis with Heterozygous Parents

Dickinson and Jinks (1956) worked out the parameters for diallel cross when the parents are heterozygous or open pollinated varieties. Considering one locus with 2 alleles (A, a) with u and u, the allele frequencies of A and a, respectively and f_1 the inbreeding coefficient the genotypic frequencies of genotypes AA, Aa and aa will be as follows:

Genotype	Frequency
AA	$u^2 + uvf = \alpha$
Aa	$2uv(1-f) = \beta$
aa	$v^2 + uvf = \gamma$

Where $\alpha + \beta + \gamma = 1$.

Using these genotypes as parents in the diallel cross a variety of statistics can be calculated. Unlike homozygous diallel we have on the diagonal the selfs of the heterozygous parents which will show a fall in mean equal to $1/2h$ from the parental value. We will have thus variances of the actual heterozygous parents (V_{p_1}) as well as of self means which can be regarded as hypothetical parental population (V_{p_2}). The various statistics that can be calculated are as follows:

$$V_{P1} = \frac{1}{2}D_1 + \frac{1}{4}H_{111} - \frac{1}{4}F_{11}$$

$$V_{P2} = \frac{1}{2}D_1 + \frac{1}{16}H_{111} - \frac{1}{8}F_{11}$$

$$\overline{V}_r = \frac{1}{8}D_1 + \frac{1}{16}H_1 - \frac{1}{16}F_1$$

$$V_{\bar{r}} = \frac{1}{8}D_1 + \frac{1}{16}H_1 - \frac{1}{16}H_{11} - \frac{1}{16}F_1$$

$$\overline{W}_{P_1/r} = \frac{1}{4}D_1 + \frac{1}{8}H_{111} - \frac{1}{8}H_{iv} - \frac{1}{16}F_1 - \frac{1}{16}F_{11}$$

$$\overline{W}_{P_2/r} = \frac{1}{4}D_1 + \frac{1}{16}H_{111} - \frac{1}{16}H_{iv} - \frac{1}{16}F_1 - \frac{1}{32}F_{11}$$

$$W_{P_1/P_2} = \frac{1}{2}D_1 = \frac{1}{8}H_{111} - \frac{3}{16}F_1$$

From the above statistics the following genetic parameters are calculated as:

$$D_1 = 2V_{P_1} + 8V_{P_2} - 8W_{P_1/P_2}$$

$$H_1 = 4V_{P_1} + 16V_{P_2} + 16\overline{V}_r + 16\overline{W}_{P_1/r}$$
$$\quad - 32W_{P_2/r} - 16\overline{W}_{P_1/P_2}$$

$$H_{11} = 16\overline{V}_r - 16V_{\bar{r}}$$

$$H_{111} = 16V_{P_1} + 16V_{P_2} - 32W_{P_1/P_2}$$

$$H_{IV} = 8V_{P_1} + 16\overline{W}_{P_1/r} + 16\overline{W}_{P_2/r} - 8W_{P_1/P_2}$$

$$F_1 = 8V_{P_1} + 32V_{P_2} + 16\overline{W}_{P_1/r}$$
$$\quad - 32\overline{W}_{P_2/r} - 32W_{P_1/P_2}$$

$$F_{1I} = 16V_{P_1} + 32V_{P_2} - 48W_{P_1/P_2}$$

In case of random mating

$$D_1 = D_{11} = 4 \Sigma\, uvd^2$$
$$H_1 = H_{111} = 8 \Sigma\, uv(1 - 2uv)h^2$$
$$H_{11} = H_{IV} = 16 \Sigma\, u^2v^2h^2$$
$$F_1 = F_{11} = 16 \Sigma\, uv(u - v)\, dh$$

whereas in case of complete inbred populations

$$F_{11} = H_{111} = H_{IV} = 0$$

and all other components of variation reduce to those of diallel of inbreds, namely, D_1, H_1, H_2 and F. From these estimates the mean level of dominance is calculated as

$$\sqrt{\dfrac{\tfrac{1}{2}(H_1 + H_{IV})}{D}}$$

which equals h/d if the n loci have equal effects which is the exception rather then the rule. It can also be estimated from the ratio as

$$\sqrt{\dfrac{H_{11}}{D_1^2}} = \dfrac{h}{\sqrt{r}}\, d^2$$

The estimates of h/d and $\dfrac{h}{\sqrt{r}}\, d^2$ can be solved simultaneously for h and d if $r = 1$. This analysis also provides the estimates of coefficient of inbreeding or the degree of heterozygosity of loci showing dominance and of the allele frequency at such loci. Whether dominant or recessives are in excess can also be determined by the sign of F_1 and F_{11} and which can be further supported by the sign of the difference of $2\bar{P}_2 - \bar{P}_1$. As there is heterogeneity of $(W_r - V_r)$ over arrays due to heterozygous arrays, $W_r - V_r$ is not constant. The slope b deviates from unity even in the absence of non-allelic interaction. The heterozygosity of the parents will tend to under estimate dominance.

7.5.1 Identification of heterozygous parent(s)

If the set of parents used in the diallel includes both homozygous as well as heterozygous parents then $W_r - V_r$ graph will show these points to be distributed in a triangular pattern with the homozygous genotypes occupying the same relative positions on the line of unit slope as in homozygous diallel with the complete heterozygote falling at the apex of triangle. Three broad patterns of $W_r - V_r$ graph have been encountered in the outbreeding species: (1) the presence of a clear triangle-a conclusion that heterozygosity is present, (2) where amongst the set of selected plants the individual genotypes are homozygous for the genes by which they differ and (3) where possible heterozogosity is present but a line of unit slope may also be fitted.

If more than one of the assumptions under the diallel analysis fail greater difficulty can be encountered in separating the causes. If both heterozygosity and epistasis (complementary) are present one may eliminate the possible offending array and re-examine for heterozygosity if the aim is to find out the heterozygous lines in the set as in case of selection of parent (s) in the development of synthetics.

7.6 Combining Ability Analysis

Griffing (1956) suggested four ways or methods in which atleast $n(n-1)/2$ set of crosses can be obtained. These four methods are: (1) parents, F_1's and reciprocal F_1's, (2) parents and F_1's, (3) F_1's and reciprocal F_1's and (4) F_1's. Each of these methods can be analysed following fixed model (model I) and random model (model II), respectively and the expectation of mean squares will differ according to whether random or fixed model is used. The general model underlying combining ability analysis is

$$Y_{ij} = g_i + g_{ij} + s_{ij} + r_{ij}$$

where Y_{ij} is the mean of the F_1 resulting from crossing of the ith and the jth parents,

g_j is the general combining ability (GCA) effect of ith parent,

g_j is the general combining ability (GCA) effect of the jth parent,

s_{ij} is the specific combining ability (SCA) effect of the cross between ith and jth parents, and

r_{ij} is the reciprocal effect if reciprocal F_1's are included and maternal effects are present.

The general combining ability of the ith line in the diallel can be defined as the mean performance of the crosses having ith line as one of the parents. The specific combining ability of ith × jth cross in an array can be defined as the deviation in mean of this cross from the average mean of that array. The terms general combining ability and specific combining ability were coined by Sprague and Tatun (1942).

With fixed model the effects sum to zero. Thus

$$\Sigma g_i = 0, \Sigma g_j = 0, \Sigma r_{ij} = 0 \text{ and } \Sigma s_{ij} = 0$$

The genotypic effects are constant, i.e., the effects g_i, g_j, s_{ij} and r_{ij} are constants and the expectations of their mean squares are the constant squared. Thus in the usual sense they have no variance but one can compute an average of the squared effects and is called the variance of fixed effects. With random model, the expectations of g_i, g_j, s_{ij} and r_{ij} are all zero and the expectations of these terms squared are

$$Y(g_i^2) = E(g_j^2) = \sigma_g^2 = \text{variance due to GCA,}$$

$$E(s_{ij}^2) = \sigma_s^2 = \text{variance due to SGA,}$$

and $E(r_{ij}^2) = \sigma_r^2 = \text{variance due to reciprocal effects}$

Although general and specific combining ability effects and variances can be estimated in all eight cases, Griffing (1958) suggested that only a modified diallel crossing system in which selfs are not included (methods 3 and 4) can be used to estimate genetic parameters of the population from which the inbred parents are supposed to have been derived by selfing but without any kind selection and a random sample of such inbreds is involved in the diallel (Fisher).

7.6.1. Estimation of GCA and SCA effects

In case of method II and model I the analysis takes the form as

$$Y_{ijkl} = \mu + g_i + g_j + s_{ij} + b_k + (bg)_{jk} + e_{ijkl}$$

where u is the general mean

- g_i is the general combining ability effects of ith line
- g_j is the general combining effect jth line
- s_{ij} is the specific combining ability effect of $(i \times j)$th cross
- b_k is the kth block effect
- $(bg)_{ik}$ is the interaction between $(i \times j)$th genotype and kth block
- e_{ijkl} is the error.

The analysis of variance takes the form as given in Table 7.7.

The mean square for error is estimated as

$$M'_e = \frac{\text{Mean square of error (estimated from general analysis of variance table)}}{\text{Number of replication}}$$

General combining ability and specific combining ability effects are estimated as follows:

Table 7.7 ANOVA in combining ability analysis (Method II and Model I)

Item	d.f.	S.S.
General combining ability	$n - 1$	$S_g = \frac{1}{n+2} [\sum_{i=1}^{n} (X_{i.} + X_{ii})^2 - \frac{4}{n} X^2..]$
Specific combining ability	$\frac{n(n-1)}{2}$	$S_s = \sum_i^n \sum_j^n X_{ij}^2 - \frac{1}{n+2} \sum_{i=1}^{n} (X_{i.} + X_{ii})^2 + \frac{2}{(n+1)(n+2)} X^2..$
Error	m	

where X_{ij} is the mean value of $(i \times j)$th cross.
X_{ii} is the mean value of ith parent
$X_{i.}$ is the total of the array of ith parent
$X..$ is the grand total of $n(n-1)/2$ progenies and n parental lines.

$$g_i = \frac{1}{n+2}\left(X_{i\cdot} + X_{ii} - \frac{2}{n}X_{\cdot\cdot}\right)$$

$$s_{ij} = X_{ij} - \frac{1}{n+2}(X_{i\cdot} + X_{ii} + X_{j\cdot} + X_{jj})$$

$$+ \frac{2}{(n-1)(n+2)} X_{\cdot\cdot}$$

The variance of the difference between any two mean values is

$$\text{Var}(X_{ij} - X_{kl}) = 2\sigma_e^2 = 2\sigma_{M'_e}^2$$

The variances of effects and of differences between effects are estimated as follows:

$$\text{Var}(\hat{g}_i) = \frac{n-1}{n(n+2)}\sigma_e^2$$

$$\text{Var}(\hat{s}_{ij}) = \frac{n^2+n+2}{(n+1)(n+2)}\sigma_e^2 \quad (i \neq j)$$

$$\text{Var}(\hat{g}_i - \hat{g}_j) = \frac{2}{(n+2)}\sigma_e^2 \quad (i \neq j)$$

$$\text{Var}(\hat{s}_{ij} - \hat{s}_{ik}) = \frac{2(n+1)}{(n+2)}\sigma_e^2 \quad (i \neq j, k, j \neq k)$$

$$\text{Var}(\hat{s}_{ij} - \hat{s}_{kl}) = \frac{2}{(n+2)}\sigma_e^2 \quad (i \neq j, k, l; j \neq k, l; k \neq l)$$

7.6.2 Estimation of GCA and SCA variances

The ANOVAs in case of method III (F_1's and RF_1's) and method IV (F_1's) take the form as given Tables 7.8 and 7.9, respectively.

Table 7.8 ANOVA in case of method III and model II

Item	d.f.	M.S.	E.M.S.
General combining ability	$n-1$	Mg	$\sigma_e^2 + 2\sigma_s^2 + 2(n-2)\sigma_g^2$
Specific combining ability	$n(n-3)/2$	Ms	$\sigma_e^2 + 2\sigma_s^2$
Reciprocal	$n(n-1)/2$	Mr	$\sigma_e^2 + 2\sigma_r^2$
Error		Me	σ_e^2

Table 7.9 ANOVA in case of method IV and model II

Item	d.f.	M.S.	E.M.S.
General combining ability	n	Mg	$\sigma_e^2 + \sigma_s^2 + (n-2)\sigma_g^2$
Specific combining ability	$n(n-3)/2$	Ms	$\sigma_e^2 + \sigma_s^2$
Error		Me	σ_e^2

where, σ_g^2 = variance due to GCA = σ_{gca}^2
σ_s^2 = variance due to SCA = σ_{sca}^2

The population genetic variance can be described in terms of *gca* and *sca* variances which could further be partitioned into additive and non-additive components of variation.

Population genetic variance $(\sigma_G^2) = 2\sigma_{gca}^2 + \sigma_{sca}^2$ (Total genetic variance among single cross progeny)

where, $\sigma_{gca}^2 = 1/2\,\sigma_A^2 + 1/4\,\sigma_{AA}^2$

$$\sigma_{sca}^2 = \sigma_D^2 + \sigma_{AD}^2 + \sigma_{DD}^2$$

Thus *gca* variance contains not only additive genetic variance but also a part of additive x additive genetic variance whereas the *sca* variance includes all of the dominance and the remaining epistatic variance.

In Mather's notation,

$$\sigma_{gca}^2 = \frac{1}{4}DR$$

and

$$\sigma_{sca}^2 = \frac{1}{4}HR$$

There is thus no neat division of genetic variance. Both additive genetic and dominance components are confounded with epistatic variance and so additive genetic variance and dominance variance can be estimated only in the absence of epistasis. Epistasis affects the estimates of general and specific combining ability mean squares, variance and effects in an unpredictable manner. The *gca* effects of lines are due to additive and additive x additive epistasis in the presence of epistasis. Besides estimates of the components of variation other informations that can be obtained are as follows:

The heritability estimate can be obtained as

$$h^2(n) = 2\sigma^2_{gca}/(2\sigma^2_{gca} + \sigma^2_{sca} + \sigma^2_e)$$

Assuming equal gene frequency the average degree of dominance is calculated as

$$\text{Average degree of dominance} = \sqrt{\sigma^2_{sca}/\sigma^2_{gca}}$$

Baker (1978) showed the relative importance of gca and sca in predicting the performance of progeny by considering the ratio, $2\sigma^2_{gca}/2\sigma^2_{gca} + \sigma^2_{sca}$ The closer this ratio to unity, the better the predictability based on gca alone.

Finally, one must be aware of the fact that the estimates of gca and sca are relative to and dependent on the particular set of inbred lines included in the study. General combining ability was originally coined to describe the breeding value of an individual parent in hybrid combinations with other parents. The breeding value will be enhanced where a parent is homozygous for genes exhibiting dominance and epistasis in the favoured direction. This term in diallel analysis is used to designate differences between the arrays and so has ultimately assumed equivalance with additive (homozygous) genetic variance. In the same way sca has been equated with dominance and epistatic variance where it was originally coined to describe the special properties of an individual cross. It is important to realize that both gca and sca may be associated with dominance and epistasis. They refer to properties of individual as parents in particular cross combinations and not necessarily to distinguish between properties of the genes controlling the variation. In practice, interactions do occur and gca and sca effects can be high or low and positive or negative.

There are thus three main approaches to the analysis of diallel cross. These three approaches differ in three main ways:

I. The Material Under Investigation

Whether the material under investigation is analysed following Eisenhart's (1947) model I (Fixed model) or model II (random model). Hayman and Jinks studied the genetical parameters of fixed set of inbred lines and thus followed model I whereas Kempthorne (1956, 1958) followed random model and worked out parameters of random mating population from which inbred lines were derived by selfing without selection. Griffing (1956a) pointed that both models are practicable.

II. Statistical and Genetical Models of Gene Action

Here Hayman and Jink's approach is similar to Kempthorne's but differs from Griffing. Hayman and Jink's polygene model and Hayman's (1954a) five components of genetic variation, D, H_1, H_2, F and h^2 alongwith $W_r - V_r$ graph, approximately characterize the genetic situation of a fixed set of inbred lines. Griffing used the statistical concept of general and specific combining ability.

III. Methods of Estimation of Parameters

Here Hayman and Jink's approach differ from Kempthorne's and Griffing. Hayman (1954a) gave methods of estimating the five components from statistics involving both parents and off spring. Some of these statistics were additional to those. supplied by the ANOVA of diallel table given by Hayman's (1954b). Kempthorne (1956) and Griffing (1958) described five components of variation for which Kempthorne suggested the need of estimation procedures. The genetic components of Hayman and Jinks are interpretation of these components in terms of their particular model. Griffing (1956b), 1958) restricted his analysis to a modified diallel crossing system in which selfs are not included and the genetic variation could be expressed in terms of the variances of general and specific combining abilities. Hayman's parameters are related to those of Kempthorne and Griffing as can be seen below.

Hayman	Kempthorn	Griffing
μ_0	μ_1	μ_0
μ_1	μ	μ_0
σ^2_0	σ^2_1	σ^2_1
C_{01}	C	σ^2_s
σ^2_1	σ^2_G	$\sigma^2_{gca} + \sigma^2_{sca}$

7.14 Biometrical Genetics–Analysis of Quantitative Variation

| C_{11} | Cov(P,0) | σ^2_{gca} |

where in a $n \times n$ diallel
μ_0 is the mean of a diagonal variable,
μ_1 is the mean of non-diagonal variable,
σ^2_0 is the variance of a diagonal variable,
C_{01} is the covariance between diagonal variable and any other variable in the same array,
σ^2_1 is the variance of non-diagonal variable,
C_{11} is the covariance between non-diagonal variables in the same array (Covariance half sib), and $\mu = \mu I - \mu_0$

7.7 Partial Diallel

Kempthorne and Curnow (1961) proposed this mating design. In a diallel with the increase in the number of parents, the number of possible single crosses increases tremendously. With limited resources available it would not be possible to go for bigger size diallel but this problem can be solved by sampling only certain crosses from all possible crosses between n parents and thus reducing the total number of crosses considerably. In the partial diallel a breeder handles only $ns/2$ crosses where s is a whole number equal or greater than 2 and n and s cannot both be odd or even.

The following crosses are sampled.

line $1 \times$ line $k + 1, k + 2, \ldots, k + s$
line $2 \times$ line $k + 2, k + 3, \ldots, k + s + 1$
.
.
.
line $i \times$ line $k + i, k + i + 1, \ldots, k + i - 1 + s$
line $n \times$ line $k + n, k + n + 1, \ldots, k + n - 1 + s$

where $k = (n - 1 + s)/2$

All the numbers above n are reduced by multiplies of n so as to be kept between 1 and n. There is a analogy between this method of sampling in the diallel cross and the experimental design for blocks of two plots in which s blocks containing treatment number i also contains treatment numbered $k + i, k + i + 1$, ... $k + 2 - 1 + 3$. This circulant design was discussed by Kempthorne (1953). To every design for blocks of two plots in which each treatment occurs the same number of times, there corresponds a method of sampling the diallel cross in which such line is involved in the same number of crosses and conversely treatments i and j, occuring in the same block corresponding to sampling the cross $i \times j$. Any balanced incomplete block design corresponds to the complete diallel cross.

In the partial diallel cross the particular subset of crosses to be sampled is determined by the original numbering of the parents. Changing the order of the parents will result in yielding different subsets of crosses and in fact for n parents there will be $(P!)/2$ possible sets of any particular size.

The underlying model is

$$Y_{ij} = \mu + g_i + g_j + s_{ij}$$

where Y_{ij} is the observed mean of ith $\times$ jth cross, μ is the overall mean, g_i and g_j are the general combining abilities of the ith and jth parent, respectively, and s_{ij} is the specific combining ability of ith $\times$ jth cross.

The only way to calculate the gca sum of squares is to assume the absence of sca and the sca is then obtained by subtraction. In the absence of sca the expected value of observed Y_{ij} is

$$Y_{ij} = \mu + g_i + g_j$$

The g_i's are estimated by minimizing

$$X = \sum_{i,j} (Y_{ij} - \mu - g_i - g_j)^2$$

Now if R_i is the ith array total the normal equation takes the form

$$AG = Q$$

where A is the symmetric circulant matrix with the elements a_{ij} such that $a_{ii} = s$, the number of crosses per array and $a_{ij} = a_{ji} = 1$ if cross $i \times j$ is used and $a_{ij} = a_{ji} = 0$ otherwise, G is the column vector with elements g_i, and Q is the column vector with elements $Q_i = R_i - s\mu$.

Now if a_{ij} is the element in the ith row and jth column of A^{-1}, the gca sum of squares is $\sum_{ij} a_{ij} Q_i Q_j$ and the ANOVA takes the form as given in Table 7.10.

Table 7.10 Sum of squares and their expectations in partial diallel

Item	d.f.	E.M.S.
Replication	$r-1$	
General combining ability	$n-1$	$\sigma_e^2 + r\sigma_s^2 + [rs(n-2)/(n-1)]\sigma_g^2$
Specific combining	$ns/2 - 1$	$\sigma_e^2 + \sigma_s^2$
Error	$(r-1)(ns/2 - 1)$	σ_e^2
Total	$rns/2 - 1$	

The genetical interpretations of the variances are as follows.

When the parents are random mating progenies ($F = 0.0$),

$$\sigma_g^2 = \text{Cov (H.S.)} = (1/8)DR$$

$$\sigma_s^2 = \text{Cov (F.S.)} - 2\,\text{Cov (H.S.)} = (1/16)HR$$

and $\sigma_w^2 = V_p - \text{Cov (F.S.)} = (1/4)DR + (3/16)HR$

The expectations of Cov (H.S.) and Cov (F.S.) will change with the change of the degree of inbreeding in the diallel parental population in the manner given in chapter 6.

When the diallel parents are inbred lines, i.e. $F = 1.0$

$$\sigma_g^2 = (1/4)DR$$
$$\sigma_s^2 = (1/4)HR$$

and $\sigma_w^2 = E$

7.7.1 Number of crosses per parent in partial diallel

It is important for the breeder to know the size of s, the number of crosses per parent, necessary to provide efficient estimates of parameters. Result from a overall diallel is questionable in random model as the small sample of parents does not represent the population and the bigger diallels suffer from the practical problem of making crosses and their evaluation in a suitable field design. Such problem is not there in fixed model as the elite individuals are used for producing synthetics or parents of F_1 hybrid. In this context Bray (1971) compared partial diallels with the complete diallel which corresponds to method IV of Griffing (1956), each with the same number of parents. He took 20 different subsets of data from a 12-clone diallel of Lucern. The two full and partial diallels were compared for estimation of gca and sca variance components, estimation of heritability, detection of gca and sca, gca effects or selection for the best. He concluded that the use of a partial diallel of any size involves risks in that it samples only part of the potentially available data. Statistically estimates or conclusions may be very different than the actual values for the population of parents. The extent of any such error and the likelihood of it depend upon both the number of crosses sampled and the nature of the character under study. Characters for which parents show specific combining ability would seem to be particularly prone to misinterpretation. The effect of different number of crosses on the variance of the difference between gca effects for different number of parents which is calculated following Kempthorne and Curnow (1961) and is given below showed that it is not possible to state clearly an optimum size for the partial diallel although more than 8 or 10 crosses per parent would probably be unnecessary.

$$\text{Average } V(\hat{g}_i - \hat{g}_j) = 2\left(\frac{P_a^0}{P-1} - \frac{1}{2N(P-1)}\right) \times (\sigma_s^2 + \sigma_e^2/R)$$

where

P is the number of parents,
N is the number of crosses per parent,
a^0 is diagonal term in the inverse of cross matrix,
R is the number of replication

However, Murty et al. (1967) and Anand and Murty (1969) suggested half the number of parents. The number of crosses per parent varies with the character even within a population. How partial diallel will fare in efficiency compared with complete diallel each for the same number of crosses is another problem which needs to be investigated. Unlike full diallel, the degrees of freedom are more evenly partitioned between gca and sca mean squares in partial diallel and supposed to be estimated with equal precision. However, the increase in precision appears to be largely vitiated by decrease in the expected value of gca mean squares but the extent of this inferiority needs to be studied considering the economic and statistical advantages offered by this design. A comparison of partial diallel with other mating designs (Kearsey, 1970) has shown that partial diallel appears to yield no more information than the other designs while the crossing programme is complicated and analysis is computanously heavy.

7.8 Number of Parents, Family Size and Number of Replications in Diallel

Like other mating designs in order to optimize the resources it is essential to work out the number of parents to be included in the diallel, number of individuals per cross to be grown and the number of replications to be taken in the field design in order to obtain realistic indication as to the presence and the relative importance of main types of genetic variation. Hayman (1960) considered the problem of number of parents and concluded that a diallel cross of 10 parents would supply useful estimates of the genetic components of variation within a population. His conclusion is based on sampling theory approach to work out which value of n is needed to have useful significant estimate of the parameters in population of which any given diallel cross is a random sample. Having worked out the expectations of the statistics, μ_0, μ, σ_0^2, C_{01}, σ_1^2 and C_{11} he derived the unbiased estimators of these six parameters. The expected variances and covariances of the above six estimators were also calculated. He then showed that the ratio of the squares of each parameter and its sampling variance is expressed in n, the number of parents in the diallel which approximates χ^2 with one degree of freedom and thus the significance of the estimators of each parameter can be tested. Then by adjusting the values of n, it can be seen which value of n results in significant estimates of these parameters. With selected parents, the assumption that parents are a random sample from some equilibrium base population is completely invalid and estimation of variance components will not provide useful information. Cockerham (1963) found this number (10 parents) too conservative. Kempthorne and Curnow (1961) pointed out that a design that minimizes the variances of two estimates does not necessarily minimize the variance of their ratio and the results of Hayman must be viewed in this context when the ratios of genetic variances such as heritability and degree of dominance are of interest. Pederson (1971) used different values of variables (p, r and n) in his simulation study and got information on two parameters of the population, namely, heritability (for individuals, families) and degree of dominance. Only crosses among parental lines (method IV and model II) with f either 0 or 1 was considered. The ANOVA in this case takes the form given in Table 7.11.

Table 7.11 ANOVA in method IV and model II

Item	d.f.	E.M.S.
General combining ability	$p - 1$	$\sigma_E^2 + n\sigma_P^2$ $+ \frac{1}{4}nr(1+f)\sigma_D^2$ $+ \frac{1}{4}nr(p-2)$ $(1+f)\sigma_A^2$
Specific combining ability	$\frac{1}{2}p(p-1)$	$\sigma_E^2 + n\sigma_P^2$ $+ \frac{1}{4}nr(1+f)^2\sigma_D^2$
Error	$\frac{1}{2}(r-1)$ $\times (p-2)(p+1)$	$\sigma_E^2 + n\sigma_P^2$
Sampling error	$\frac{1}{2}r_p(p-1)(n-1)$	σ_E^2

where σ_P^2 is plot to plot environmental variance which is zero if all the individuals of a family are grown as a discrete unit within a replicate,
σ_D^2 is the dominance variance,
σ_A^2 is the additive genetic variance.

Within family variance (σ_w^2) is calculated as

$$\sigma_E^2 = \sigma_w^2 = \sigma_G^2 - \text{Cov(F.S.)}$$
$$= \sigma_w^2 + 1/2(1-f)\sigma_A^2 + 1/4(1+f)(1-f)\sigma_D^2$$

From these estimates heritability and degree of dominance are calculated. The sampling variance of heritability is calculated following Kempthorne (1957) as the variance of the ratio.

Var (x/y)

$$= \frac{1}{y^2}\left[\text{Var}(x) - 2x/y\, \text{Cov}(x,y) + \left(\frac{x}{y}\right)^2 \text{Var}(y)\right]$$

The most appropriate combination of p, r and n was defined as one which maximizes the quantity, i, the amount of information per individual.

$$i = I/\left[\frac{1}{2}p(p-1)\, r_n\right]$$

where $I = I$/variance (h^2 or d, the degree of dominance) To overcome the effect of plot effects on the results he used the relationship shown by Smith (1938).

Variance of plot means = $\frac{1}{n^b}$ (Variance of individuals).

Estimates of b range from 0.16 to 0.8 for various crops with an average of 0.5.

Thus, $\frac{\sigma_E^2}{n} + \sigma_P^2 = \frac{\sigma_E^2}{n^b}$ i.e. $\sigma_P^2 = \sigma_E^2 \left(\frac{1}{n^b} - \frac{1}{n}\right)$

He concluded from the results of heritability that 8 to 10 parents are preferred with few individuals per cross and with only 2 or 3 replicates. There is a practical problem of testing 2–3 individuals per cross in the field so he suggested that n of 10 to 20 is more realistic even though this will enevitably lead at a loss of efficiency. With less than 8 parents the estimator may be appreciably biased. Multiple diallels were generally found twice as efficient as the use of a single diallel and thus as many 8 or 10 parents diallel should be carried out as are feasible. Conclusion about p, n and r based on results from degree of dominance is less general. When the degree of dominance is zero the sampling variance can be minimized only by increasing one or all of the three variables. For higher degree of dominance, number of p around 16 with small r and large n is preferred. However, if both the heritability and the degree of dominance are to be considered it is the former which is of greater practical value and thus the number of parents should be around 10. Finally, an increase in f from 0 to 1 gave about 16-fold increase in the relative amount of information and this may be a means of improving the efficiency if the value of degree of dominance is expected to be small.

7.9 Reference Population in Diallel

Two reference populations have been used in the diallel analysis for which the genetic parameters (the components of variance) can be estimated.

1. The random mating population from which crosses are considered to be a random sample (Kempthorne, 1956);
2. The set of parents utilized in the diallel cross (Hayman, 1954b).

The two reference random mating populations are shown in Figure 7.3.

In case of former the set of n homozygous lines is assumed to be a random sample from the population of homozygous lines derived by inbreeding the original random mating population in linkage equilibrium without selection. If the sample size is large and random, the original reference population will be similar to the derived reference population in genetical structure. On the contrary if the sample size is small or the fixed set of parents in general will not fulfill the assumption of uncorrelated gene effects and in many cases there are many genes for a particular character then the number of parents included in the diallel and so the F_1's cannot be considered a sample from the original population or when the sample size is small the derived reference population will be different than the original reference population in gene frequencies because of random genetic drift and it is inbred with an inbreeding coefficient of $f = (1/2)p$ or $1(1/p)$ according to whether the parents are a sample of non-inbred parents or their inbred relatives. Thus these two types of reference populations (Kuehl et at. 1968) will have genetic implications over and above those imposed by statistical assumption of random or fixed effects on the estimates of genetical parameters (Griffing 1956b; Wearden 1964). But the derived reference random mating population which can be constituted by repeated cycles of random mating among parents will be a mere representative of the genes contained by a set of parents at hand than a random mating population and is more amenable to evaluation

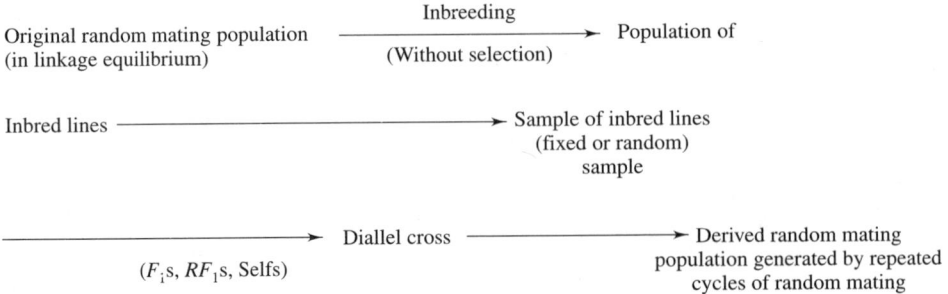

Fig. 7.3 Reference populations.

of biases and errors of inferences than a fixed set of parents and crosses. Wright called these two reference populations as ancestral and descendent reference populations.

As F_1's are a sample of crosses from the ancestral reference population the estimation of additive and dominance genetic variance must be based on the general and specific combining ability variance estimates derived from the F_1 data only. But as the selfs progenies have a mean inbreeding coefficient of 1/2 or 1 in case of non-inbred and inbred parents, respectively and constitute a 1/pth part of the full diallel table and the population is not in linkage equilibrium as self progenies have a higher probability of homozygosity at all loci then the F_1's so the σ_g^2, the gca variance has no simple interpretation in terms of additive component unless the parents are in linkage equilibrium and have similar coefficient of inbreeding. The estimation of dominance from σ_s^2, the sca variance is not without problem. The complete diallel (method I) analysis provides sca mean squares based on deviations of both F_1's and selfs progenies from the mean array effects and is equivalent to Hayman's b item and is used to estimate dominance. Considering polygenic and assuming ambidirectional dominance, the mean value on the diagonal is lowered because of the inbreeding depression than off it with a consequent inflation of the sca deviation and the estimate of dominance component derived from them. The unbiased estimate of dominance component is obtained only when F_1's vs. selfs contrast is removed from the sca sum of squares, thus using Hayman's $b - b_1$ items and thus is different than the use of b_3 alone which is equivalent to Griffing's method III

sca item for ancestral reference population. Further, if the parents are in linkage equilibrium the parameters of descendent reference population (F_1's. RF_1's, selfs) can be estimated more precisely than those of the ancestral reference population because in case of former no sampling of allele is involved.

7.10 Comparison with other Mating Designs

If only the aim is to have information on the detection and estimation of different forms of genetic variation the other available designs discussed in Chapter 6 will provide similar information with more precision for similar amount of efforts. In general, for non-inbred population under situation of extreme frequencies of the dominant allele and low heritability experimental sizes are less for the Augmented BIPs and NC III designs than for the diallel.

7.11 Assumptions under Diallel Analysis

Diallel analysis is open to criticism on both genetical and statistical grounds (Gilbert, 1958). He suggested that the various genetical assumptions underlying Hayman's analysis are difficult to be met in practice. The assumption that genes are independently distributed across parents is impossible to find. Frey (1976) pointed out that genes at k loci cannot be independent unless a minimum of 2^k parents are used in the diallel. However, this assumption is assured if parents are produced by random mating followed by non-selective inbreeding. The assumption of no epistasis does not hold at bio-chemical level where each gene gives rise to a gene product and there epistasis cannot be avoided. Inversion or other chromosomal devices that maintain heterozygosity in plants which is advantageous may violate the assumption of homozygous parents. Although the assumption of diploid segregation is justified how can yield be explained where endosperm which makes up the greater part of the yield as triploid.

The least squares estimation procedure applied to the analysis of F_2 and backcross generations is criticized on the grounds of statistical inefficiency but efficient estimation by maximum likelihood procedure is more troublesome. Further the two statistics W_r and V_r should be independently and normally distributed which they are not. Also, as there is overall heterogeneity in the values of $W_r + V_r$ statistics, its use to determine the dominance of parents needs to be justified. Finally, since W_r is statistically correlated with V_r (V_r should be tested for heterogeneity).

7.12 Varietal Diallel

When a diallel is made from a fixed set, say n, of open pollinated varieties and selfs are included mean, additive and dominance effects can be tested separately and a test for epistasis can be made. Hayman (1957) showed that in a full diallel of n parents after accounting for the mean, $(n-1)$ additive terms and $n(n-1)/2$ dominance terms there are $n(n-1)/2$ deviations to test for epistasis composed of all types. Infact when the parental selfs are not included then there are only $n(n-3)/2$ deviations for epistasis. Gardner and Eberhart (1966) proposed the following models where only parental varieties and their $n(n-1)/2$ crosses are grown in the experiment.

$$Y_{kk'} = \mu_v = \frac{1}{2}(v_k + v_{k'})$$

$$Y_{kk'} = \mu_v + \frac{1}{2}(v_k + v_{k'}) + \gamma \bar{h}$$

$$Y_{kk'} = \mu_v + \frac{1}{2}(v_k + v_{k'}) + \gamma(\bar{h} + h_k + h_{k'})$$

$$Y_{kk'} = \mu_v + \frac{1}{2}(v_k + v_{k'}) + \gamma(\bar{h} + h_k + h_{k'} + s_{kk'})$$

where, $Y_{kk'}$ is the observed mean of the cross between ith and jth varieties,

μ_v is the mean of all parental varieties and equals $\mu + \bar{d}$, where $\bar{d} = \sum_i d_i/n$, d_i being the cumulative contribution of heterozygous loci, and v_k or $v_{k'}$ is the variety effect and equals

$a_i + (d_i - \bar{d})$ where a_i, is the cumulative contribution of homozygous loci.

With $n \geq 4$, $h_{kk'}$ the heterosis parameter is further partitioned into three components as

$$h_{kk'} = \bar{h} + h_{k'} + h_k + s_{kk'}$$

where $\sum h_k = \sum s_{kk'} = 0$,

$\bar{h}$ is the average heterosis due to the particular set of varieties in the diallel,

h_k or $h_{k'}$ is the average heterosis due to the variety, k or k'.

$s_{kk'}$ is the specific heterosis in the cross of varieties k and k'.

In these models

$$Y = 0 \text{ when } k = k'$$
$$r = 1 \text{ when } k \neq k'$$

The ANOVA takes the form as given in Table 7.12.

Table 7.12 ANOVA in Varietal diallel analysis

Item	d.f.
Mean	1
Additive	$n-1$
Dominance	n
Heterosis ($h_{kk'}$)	$n(n-1)/2$
Average ($\bar{h}$)	1
Variety (h_k)	$(n-1)$
Specific ($s_{kk'}$)	$n(n-3)/2$
Epistasis	$n(n-1)/2$

Eberhart (1964) and Eberhart and Gardner (1966) showed that the joint analysis of single, three-way and double crosses from four parents resulted in fifteen degrees of freedom for epistasis, none of which involved epistatic effects of an order higher than additive × additive.

When the parental varieties are not included in the experiment the general combining ability takes the form as

$$g_k = 1/2 v_k + h_k$$

This shows that the gca effect depends not only on the variety effect but also on variety heterosis, i.e., on dominance effect. The sum of squares for crosses and its partition into general combining ability and specific combining ability may be done according to model I, method IV of Griffing (1956) or by the general least squares procedure.

7.13 Triallel Analysis

Assuming no epistasis diallel analysis as we have seen provides estimates of additive and dominance genetic variances. The triallel and quadrallel analyses proposed by Rawlings and Cockerham (1962a, 1962b) can separate additive × additive epistatic variance. The underlying model in triallel analysis is

$$Y_{i(jk)l} = \mu + r_i + G_{i(jk)} + e_{i(jk)l}$$

where the genotypic effect, $G_{i(jk)}$ can be defined as linear function of uncorrelated effects as

$$G_{i(jk)} = (g_i + g_j + g_k) + (S_{2ij} + S_{2ik} + S_{2jk})$$
$$+ S_{3ijk} + (O_{1i} + O_{1i(j)} + O_{1(k)})$$
$$+ (O_{2ai \cdot j} + O_{2ai \cdot k} + O_{2ajk})$$
$$+ (O_{2bi(j)} + O_{2bi(k)} + O_{3i \cdot jk})$$

where

g_i is the average effect of line i averaged overall orders,
S_{2ij} is the 2-line interaction effect of lines i and j appearing together averaged overall orders,
S_{3ijk} is the 3-line interaction effect of lines i, j and k appearing together averaged overall orders,
O_{1i} is the 1-line order effect of line i as parent,
$O_{1(i)}$ is the 1-line order effect of the line i as grandparent,
$O_{2ai \cdot j}$ is the 2-line order interaction effect of lines i and j averaged over two orders $i(j -)$ and $j (i -)$ interaction,
O_{2ajk} is the 2-line order interaction effect of grand parents j and k due to particular order $- (jk)$,
$O_{2bi\ (j)}$ is the 2-line order interaction effect of parent line i and grandparent line j due to particular order, $i(j)$,
$O_{3i \cdot jk}$ is the 3-line order interaction effect of parent line i and grandparent line j and k due to particular order $i(jk)$.

Average effects of lines (g) and interaction effects (S_2, S_3) are analogous in meaning to general combining ability and specific combining ability effects. respectively of diallel analysis whereas the order effects (O_1, O_{2a}, O_{2b} and O_3) items have appeared because the parents can be combined in different ways in three-way hybrids and because of the ancestory of lines—parental and grandparental, in crosses.

The various sum of squares are calculated as shown in Table 7.13.

The expectations of various statistical variances in terms of genetical components of variation are as follows.

$$\sigma_g^2 = \frac{2}{9}\sigma_A^2 + \frac{1}{16}\sigma_{AA}^2 + \frac{25}{1152}\sigma_{AAA}^2,$$

$$\sigma_{S_2}^2 = \frac{1}{9}\sigma_D^2 + \frac{25}{288}\sigma_{AD}^2 + \frac{1}{16}\sigma_{AD}^2 + \frac{1}{36}\sigma_{DD}^2$$
$$+ \frac{41}{1152}\sigma_{AAD}^2 + \frac{1}{64}\sigma_{ADD}^2 + \frac{1}{444}\sigma_{DDD}^2,$$

$$\sigma_{s3}^2 = \frac{1}{24}\sigma_{AD}^2 + \frac{1}{24}\sigma_{DD}^2 + \frac{3}{64}\sigma_{AAA}^2$$
$$+ \frac{5}{95}\sigma_{AAD}^2 + \frac{1}{24}\sigma_{ADD}^2 + \frac{1}{32}\sigma_{DDD}^2,$$

$$\sigma_{o1}^2 = \frac{1}{8}\sigma_A^2 + \frac{9}{64}\sigma_{AA}^2 + \frac{49}{512}\sigma_{AAA}^2,$$

$$\sigma_{o2a}^2 = \frac{1}{4}\sigma_D^2 + \frac{1}{32}\sigma_{AA}^2 + \frac{9}{64}\sigma_{AD}^2 + \frac{1}{16}\sigma_{DD}^2$$
$$+ \frac{3}{64}\sigma_{AAA}^2 + \frac{41}{512}\sigma_{AAD}^2 + \frac{9}{256}\sigma_{ADD}^2$$
$$+ \frac{1}{64}\sigma_{DDD}^2,$$

$$\sigma_{o2b}^2 = \frac{1}{64}\sigma_{AD}^2 + \frac{3}{256}\sigma_{AAA}^2 + \sigma_{AAD}^2$$
$$+ \frac{1}{256}\sigma_{ADD}^2,$$

and

Table 7.13 Analysis of variance of three-way hybrids

Item	d.f.	S.S.
Replication	$r - 1$	$R = 2/pp_1p_2 \frac{1}{r} \sum_{i=1}^{r} Y... - CF$
Hybrids	$3pC_3 - 1$	$H = \frac{1}{r} \sum_{i<j<k}^{pC_3} \sum^{3} Y_1^2(jk) - CF$
1-line general	p_1	$G = \frac{2}{3} rp_2p_3 \sum_i^p Y_i^2... - \frac{3p_1}{p_3} CF$
2-line specific	$pp_{3/2}$	$S_2 = \frac{1}{3} rp_4 \sum_{i<j}^{pC_2} Y_{ij}^2... - \frac{3p_2}{p_4} CF - \frac{2p_3C}{p_4} G$
3-line specific	$pp_1p_5/6$	$S_3 = \frac{1}{3} r \sum_{i<j<k}^{pC_3} Y_{ijk}^2 - CF - G - S_2$
1-line order	p_1	$O_1 = \frac{1}{3} r\, pp_2 \sum_i^p [2Y_{i(..)}.\, Y.(i.)]^2$
2-line order (a)	$pp_{3/2}$	$O_{2a} = \frac{1}{6} r\, pp_1 \sum_{i<j}^{pC_2} [Y_{i(j.)} + Y_j(j.) - 2_y.(ij)]^2 - \frac{p}{2p_1} O_1$
2-line order (b)	$p_1p_{2/2}$	$O_{2b} = \frac{1}{2} r p_3 \sum_{i<j}^{pC_2} [Y_{i(j.)} - Y_j(i.)]^2 - \frac{3p_2}{2p_3} O_1$
3-line order	$pp_2p_4/3$	$O_3 = \frac{1}{r} \sum_{i<j<k}^{pC_3} \sum^{3} Y_i^2(jk). - \frac{1}{3r} \sum_{i<j<k}^{pC_3} y_{ijk}^2 - O_1 - O_{2a} - O_{2b}$
Error	$(r - 1)(3pC_3 - 1)$	
MSR		
MSH		
MSG		$\sigma_e^2 + 3r\,\sigma_{s_3}^2 + 6rp_3\,\sigma_{s_2}^2 + (3rp_2p_3/2)\,\sigma_g^2$
MSS$_2$		$\sigma_e^2 + 3r\sigma_{s_3}^2 + 3rp_4\,\sigma_{s_2}^2$
MSS$_3$		$\sigma_e^2 + 3r\sigma_{s_3}^2$
MSO$_1$		$\sigma_e^2 + r\sigma_{o3}^2 + 3rp_2\,\sigma_{o2b}^2 + (r_p/3)\,\sigma_{o2a}^2 + (r_{pp2}/3)\,\sigma_{o1}^2$
MSO$_{2a}$		$\sigma_e^2 + r\sigma_{o3}^2 + (2_{rp1}/3)\,\sigma_{o2a}^2$
MSO$_{2b}$		$\sigma_e^2 + r\sigma_{o3}^2 + 2rp_3\,\sigma_{o2b}^2$
MSO$_3$		$\sigma_e^2 + r\sigma_{o3}^2$
MS$_E$		σ_e^2

where

$Y_{i(jk)}$ is the sum of hybrids $i(jk)$ over r replications,

Y_{ijk} is the sum of all hybrids with lines i, j and k in any order,

$Y_{i(j.)}$ is the sum of all hybrids with parent i and grandparent j,

$Y_{.(ij)}$ is the sum of all hybrids with grandparents i and j,

$Y_{ij..}$ is the sum of all hybrids with lines i and j any order,

$Y_{i(..)}$ is the sum of all hybrids with parent,

$Y_{.(i.)}$ is the sum of all hybrids with grandparent i,

$Y_{i...}$ is the sum of all hybrids with line i as either parents or grandparent,

$Y....$ is the grand total.

$$\sigma^2_{o3} = \frac{1}{32}\sigma^2_{AD} + \frac{1}{8}\sigma^2_{DD} + \frac{13}{256}\sigma^2_{AAD}$$
$$+ \frac{13}{128}\sigma^2_{ADD} + \frac{3}{32}\sigma^2_{DDD},$$

As we have more mean squares available in this mating design we can estimate genetic components of variance including additive × additive epistasis through least squares or weighted least squares or maximum likelihood. Both triallel and quadrallel mating designs provide the same type of genetic information but the larger coefficients of the genetic components of variance in the triallel design make far more reliable information and more powerful testing (Table 7.14). In these designs the development of progeny family requires more time and labour and the analysis is more complex in comparison to diallel mating design.

Table 7.14 Showing coefficients of components of genetic variance in single, three-way and double crosses developed using inbreds, calculated following Mather's notation

Types of crosses	Components of genetic variance					
	D	H	DXD	DXH	HXH	DXDXD
Single cross	1/4	1/4	1/4	1/8	1/16	1/8
Three-way cross	3/8	1/8	9/64	3/64	1/64	27/512
Double cross	1/4	1/16	1/16	1/64	1/256	1/64

7.13.1 Order of lines in three-way hybrids

Unlike single and double cross hybrids the average effects of lines in three-way hybrids do not account for all of the additive genetic effects not contained in the random error. The additive genetic effects contribute to the one-line order effects O_{1i} in three way hybrids. Thus it is apparent that the order is important in the selection of three-way hybrids even if all genetic variances are due to additive effects. If all genetic variances are additive, the optimum order for a set of three lines utilizes the line with the largest average effect as the single parent. This is unlike the double cross hybrids where order of lines is irrelevant if all genetic variances is additive.

References

Baker, R.J. 1978. Issues in diallel analysis. Crop Sci., **18:** 533–537.

Bray, R.A, 1971. Quantitative evaluation of the circulant partial diallel cross. Heredity, 27: 189–202.

Cockerham, C.C. 1963. Estimation of genetic variance. In: Hanson, W.D. and Robinson, H.F. (ed.) Statistical Genetics and Plant Breeding, Pub. 982, NAS, NRC, Washington, D,C., pp. 53–94.

Coughtrey, A. and Mather, K. 1970, Interaction and gene association and dispersion in diallel crosses where the gene frequencies are unequal. Heredity. **25:** 79–88.

Dickinson, A.G. and Jinks, J.L. 1956. A generalized analysis of diallel crosses. Genetics 41: 65–78,

Eberhart, S.A. 1964. Theoretical relations among single, threeway and double cross hybrids, Biometrics. **20:** 522–539.

Eberhart, S.A. and Gardner, C.O. 1966. A general model for genetic effects. Biometrics, 22: 864–871.

Eisenhart, C. 1947. The assumptions underlying the analysis of variance. Biometrics, **3:** 1–21.

Ferreira, P.E. 1988. A new look at Jinks-Hayman's method for estimation of genetical components in diallel crosses. Heredity, 60: 347–353.

Feyt, H. 1976. Etude Critique del' analyse des croisements dialleles aumoyen la simulation. Ann. A meliot. Plantes. **26:** 173–195.

Gardner, C.P. and Eberhart, S.A. 1966. Analysis and interpretation of the variety cross diallel and related populations. Biometrics, 22: 439–452.

Gilbert, N.F.G. 1958. Diallel cross in plant breeding. Heredity, **12:** 477–492.

Griffing, B.1956. A generalized treatment of the use of diallel crosses in quantitative inheritance. Heredity, **10:** 31–50.

Griffing, B. 1956. Concept of general and specific combining ability in relation to diallel crossing system. Aust. J. Biol. Sci., **9:** 462–493.

Griffing, B. 1958. Application of sampling variable in the identification of methods which yield unbiased estimates of genotypic variance components. Aust. J. Biol. Sci., **11:** 219–245.

Hallauer, A.R. and Miranda Fo, J.B. 1988. Quantitative Genetics in maize Breeding, Iowa State Univ. Press, Ames, Iowa, USA.

Hayman, B.I. 1954a. The analysis of variance of diallel tables. Biometrics, **10:** 235–244.

Hayman, B.I. 1954b. The theory and analysis of diallel crosses. Genetics, **39**: 789–809.

Hayman, B.I. 1957. Interaction, heterosis and diallel crosses. II Genetics, **42**: 336–355.

Hayman, B.I. 1958a. The theory and analysis of diallel crosses. II Genetics, **43**: 63–85.

Hayman, B.L. 1958b. The separation of epistatic from additive and dominance variation in generation means. Heredity, **12**: 371–390.

Hayman, B.I. 1960a. Maximum likelihood estimation of genetic components of variation, Biometrics, **16**: 369–381.

Hayman, B.I. 1960b. The separation of epistatic from additive and dominance variation in generation means. II. Genetica. **31**: 133–146.

Hayman, B.I. 1963. Notes on diallel cross theory. In: Hanson, W.D. and Rohinson, H.F. (eds.) Statistical Genetics and Plant Breeding NAS, NRC, Washington DC, pp. 571–578.

Hinklemann, K. 1975. Design of genetical experiments. In: A survey of statistical designs and linear models. J.N. Srivastava (ed.) North Holland.

Hinkelmann, K. 1976. Diallel and multicross designs: What do they achieve? Proc. Int. Conf. Quant. Genet. Pollack, E. Kempthome, O. and Bailey, T.B. (eds). pp. 659–676.

Jinks, J.L. 1954. The analysis of continuous variation in a diallel cross of *Nicotiana rustica* varieties, Genetics, **39**: 767–788.

Jinks, J.L. 1956. The F_2 and backcross generations from a set of diallel crosses. Heredity, **10**: 1–30.

Jinks, J.L. and Hayman, B.I. 1953. The analysis of diallel crosses. Heredity, **9**: 223–238.

Kearsey, MJ. and Pooni, H.S. 1996. The Genetic Analysis of Quantitative traits. Chapman and Hall, London.

Kempthorne, O. 1956. The theory of diallel cross. Genetics. **41**: 451–459.

Kempthorne, O, 1976. Status of quantitative genetic theory. Proc. Int. Conf. Quant. Genet. Pollack, E., Kempthorne, O. and Bailey, T.B. (eds). pp. 719–760.

Kempthorne, O. and Curnow, R.N. 1961. The partial diallel cross. Biometrics. **17**: 229–250.

Kuehl, R.O., Rawlings, J.O. and Cockerham, C.C. 1968. Reference populations for diallel experiments. Biometrics. **24**: 881–901.

Mather, K. 1949a. Biometrical Genetics (1st edn.), Methuen, London.

Mather, K. 1949b. The genetical theory of continuous variation. Proc. 8th Int. Congo Genetics, Hereditas, Suppl., Vol. pp. 376–401.

Mather, K. 1967b. Complementary and duplicate gene interactions in biometrical genetics. Heredity, **22**: 97–103.

Mather, K. and Jinks, J.L. 1982. Biometrical Genetics. 3rd edn. Chapman and Hall, London.

Matzinger, D.E and Kempthorne, O. 1956. The modified diallel table with partial inbreeding and interaction with the environment. Genetics, **41**: 822–833.

Mayo, O. 1980. The Theory of Plant Breeding. Claredon Press, Oxford.

Morley-Jones, R. 1965. The analysis of variance of diallel table. Heredity, **20**: 117–121.

Murty, B.R., Arunachalam, V. and Anand, I.J. 1967. Diallel and partial diallel analyses of some yield factors in *Linum usitatissimum*. Heredity, **22**: 35–41.

Nassar, R.E 1965. Effects of correlated gene distribution due to sampling on the diallel analysis. Genetics, **52**: 9–20.

Oakes, M.W. 1967. The analysis of diallel crosses of heterozygous or multiple allelic lines. Heredity, **22**: 83–95.

Pederson, D.G. 1971. The estimation of heritability and degree of dominance from a diallel cross. Heredity, **27**: 247–264.

Pooni, H.S., Jinks, J.L. and Singh, R.K. 1984. Methods of analysis and estimation of genetic parameters from a diallel set of crosses. Heredity, **52**: 243–252.

Rawlings, J.O. and Cockerham, C.C. 1962a. Analysis of double cross hybrid populations. Biometrics, **18**: 229–244.

Rawlings, J.O. and Cockerham, C.C. 1962b. Triallel analysis. Crop Sci., **2**: 228–231.

Smith, H.E. 1938. An empirical law describing heterogeneity in the yields of agricultural crops. J. Agric. Sci. **28**: 1–23.

Sokol, M.J. and Baker, R.J. 1977. Evaluation of the assumptions required for the genetic interpretation of diallel experiments in self pollinating crops. Can. J. Plant Sci., **57**: 1185–1191.

Sprague, G.F. and Tatum, L.A. 1942. General versus specific combining ability in single crosses of corn. J. Amer. Soc. Agron., **34**: 923–932.

Topham, P.B. 1966. Diallel analysis involving maternal and maternal interactions effects. Heredity. **21**: 665–674.

Walters, D.E. and Gale, J.S. 1977. A note on the Hayman analysis of variance for a full diallel table. Heredity. **38:** 401–407.

Walters, D.E. and Morton, J.R. 1978. On the analysis of variance of half diallel table. Biometrics. **34:** 91–94.

Wearden, S. 1964. Alternative analyses of the diallel cross. Heredity. **19:** 669–680.

Wright, A.J. 1985. Diallel designs, analysis and reference population. Heredity, **54:** 307–311.

8

Genotype × Environment Interaction

8.1 Definition

In Chapter 5 we saw that the phenotypic value (P) of an individual in an environment is equal to genotypic value (G) of that individual and environmental value (E), assuming no genotype × environment interaction. Now in the presence of genotype × environment interaction, the phenotypic value of an individual becomes

$$P = G + G \times E + E$$

where $G \times E$ is the genotype × environment interaction effect. Further we saw in Chapter 5 that the genotypic value can be partitioned into a general mean, additive effect (d), dominance effect (h), assuming no non-allelic interaction and thus this $g \times e$ interaction term can be further partitioned into additive effects × environment interaction ($d \times E$), dominance effects × environment interaction ($h \times E$) as follows:

$$P = m + [d] + [d] \times E + E$$

or

$$m + [h] + [h] \times E + E$$

or

$$m + [d] + [h] + [d] \times E + [h] \times E + E$$

depending upon the genotype of that individual. In the presence of non-allelic interaction, additive × additive (i), additive × dominance (j) and dominance × dominance (I) type interaction effects will further interact with the environment and these terms will appear in the expectation of mean of an individual raised in an environment.

At variance level we saw in Chapter 6 that the phenotypic variance of an individual equals its genotypic variance plus the environmental variance assuming no $g \times e$ interaction. Now considering genotype × environment interaction the phenotypic variance of an individual becomes

$$V_p = V_G + V_{G \times E} + E$$

where $V_{G \times E}$ is the genotype × environment interaction variance. Again, the genotype × environment interaction variance, $V_{G \times E}$ can be further partitioned into additive genetic × environment interaction variance $V_{A \times E}$, dominance genetic × environment $V_{D \times E}$ and genic interaction × environment interaction variance $V_{Int \times E}$. Thus the phenotypic variance will take the form as

$$V_p = V_A + V_D + V_{A \times E} + V_{D \times E} + V_{Int \times E} + E$$

Thus we can see how the mean and variance of an individual is affected by the presence of $g \times e$ interaction and also we can see how the different parameters such as dominance ratio, heritability, etc. will be affected by $g \times e$ interaction. We can also visualize the discrepancy between the expected and the observed level of achievement in selection programme based on parameters obtained without taking into account the genotype × environment interaction.

Genotype × environment interaction is said to occur when the different genotypes respond differently to environmental changes as can be seen in Figures 8.1 (a, b and c) where the two genotypes (*A. B*) are shown to be interacting for yield in three different

8.2 Biometrical Genetics–Analysis of Quantitative Variation

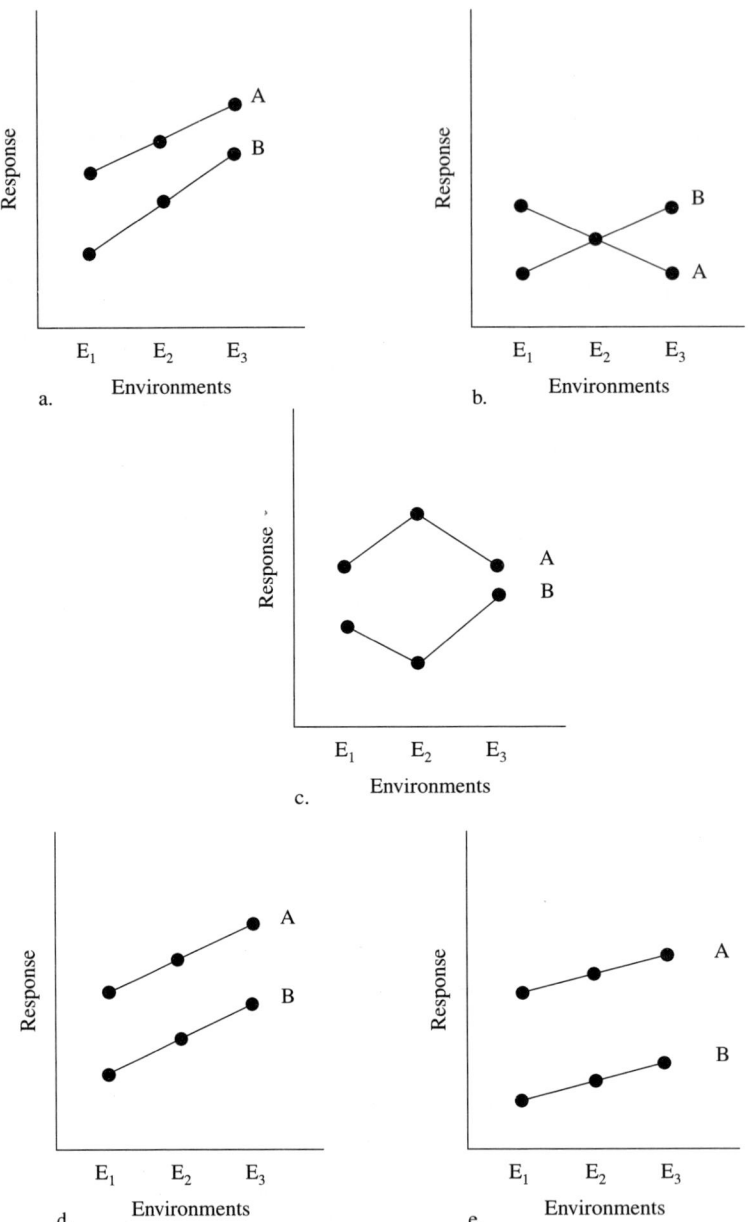

Fig. 8.1 (a-e) Figures (a, b, c) show genotype × environment interaction. Figures (d, e) show no genotype × environment interaction. A. and B, are two genotypes.

environments (E_1, E_2, and E_3). The two genotypes respond similarly to the change in the environment as shown in Figures 8.1 (d, e) and thus they do not show genotype × environment interaction.

Genotype × environment interactions vary with the material tested and the sites chosen for testing.

8.2 ANOVA for Genotype × Environment Interaction

The unbiased estimate of genetic and $g \times e$ interaction variances can be obtained from the expected mean squares from the ANOVA of the replicated experiment involving a number of genotypes tested at different locations or in different years or both. The precise form the expected mean squares takes as we have seen in Chapter 6, will depend on the kind of model (random, fixed or mixed model) the analysis follows. In case of random model both genotypes and environments (locations, years, etc.) selected for the experiment represent a random sample from their respective populations whereas both are fixed samples in case of fixed model. Genotypes are a random sample and the environment fixed or vice versa in case of mixed model. The expected means squares in case of random fixed and mixed models are given in Table 8.1.

Although not a very sensitive test significant $MS_{G \times E}$ indicates the presence of genotype × environment interaction. This is because of the reason that the large number of degrees of freedom for $g \times e$ interaction item makes the interaction mean squares not-significant in *F-test* even when the interaction sum of squares is large. Thus even if interaction mean squares is not significant, the worker should not stop from carrying out further analysis of $g \times e$ interaction as one is interested in studying the underlying stability structure of individual genotype. However, if the entire interaction sum of squares can be described by a single l-df interaction contrast (Single df interaction contrast can arise in a case where one or several genotypes or environments are expected to behave differently from the others perhaps because some are controls) or single- df for non-additivity (a case we shall see later in which all regression lines in a Finlay–Wilkinson's diagram intersect at a single point which is not met very often in practice and where interaction can be explained as a constant times the matrix of products of genotype × environment main (additive) effects, Tukey, 1949) causing this source's, mean squares to equal the interaction sum of squares, it is in this case that if $g \times e$ interaction for 1 df is not significant it will not be worthwhile to partition the interaction. The two misunderstandings about testing for interaction are: first that the test of interaction should not be regarded as independent but rather preliminary (Bancroft 1964). Consequently if genotypes and environments are tested at 0.05 level of significance, the interaction should be tested at the level of 0.25 instead of the usual 0.05. Second, this test of interaction lacks power

Table 8.1 ANOVA for $g \times e$ interaction

Item	d.f	M.S.	E. M. S. Random	Fixed	Mixed
Genotypes	$(S-1)$	MS_G	$\sigma_e^2 + r\sigma_{g \times e}^2 + r \times t\sigma_G^2$	$\sigma_e^2 + rt\sigma_G^2$	$\sigma_e^2 + r\sigma_{g \times e}^2 + rt\sigma_G^2$
Environments	$(t-1)$	MS_E	$\sigma_e^2 + r\sigma_{g \times e}^2 + r \times S\sigma_e^2$	$\sigma_e^2 + r\sigma_E^2$	$\sigma^2 + rS\sigma_E^2$
Genotype × Environment	$(S-1) \times (t-1)$	$MS_{G \times E}$	$\sigma_e^2 + r\sigma_{g \times e}^2$	$\sigma_e^2 + r\sigma_{g \times e}^2$	$\sigma_e^2 + r\sigma_{g \times e}^2$
Error	$St(r-1)$	MS_E	σ_e^2	σ_e^2	σ_e^2

where S is the number of genotype

t is the number of environments

r is the number of replications

σ_G^2 is the genotype variance

σ_E^2 is the environment variance

σ_e^2 is the error variance

$\sigma_{g \times e}^2$ is the genotype × interaction variance

Table 8.2 ANOVA for g × e interaction when environment involves locations and years

Item	d. f.	M.S.	E. M. S
Genotype	$(S-1)$	MS_G	$\sigma_e^2 + r\sigma_{GLY}^2 + rY\sigma_{GL}^2 + rL\sigma_{GY}^2 + rLY\sigma_G^2$
Genotype × Year	$(S-1) \times (Y-1)$	$MS_{G \times Y}$	$\sigma_e^2 + r\sigma_{GLY}^2 + rL\sigma_{GL}^2$
Genotype × Location	$(S-1) \times (L-1)$	$MS_{G \times L}$	$\sigma_e^2 + r\sigma_{GLY}^2 + rY\sigma_{GL}^2$
Genotype × Location × Year	$(S-1) \times (Y-1)$	$MS_{G \times L \times Y}$	$\sigma_e^2 + r\sigma_{GLY}^2$
	$(L-1)$		
Error	$(r-1)SLY$	MS_E	σ_e^2

where

s is the number of genotypes

Y is the number of years

L is the number of Locations

r is the number of replications

σ_G^2 is the genotype variance

σ_{GY}^2 is the genotype × year interaction variance

σ_{GL}^2 is the genotype × location interaction variance

σ_{GLY}^2 is the genotype × location × year interaction variance

σ_e^2 is the error variance

as has been pointed out by some workers (Freeman, 1973; Cox, 1984). The use of standard ANOVA is subject to the restrictions generally applicable to linear models such as homogeneity of error variances and additivity of effects. Thus the larger the trial, the less successful such methods will be in detection of detection of genotype × environment interaction.

When the environment includes locations and years the ANOVA will take the form as given in Table 8.2

The genotype × location × year interaction is called the second order interaction. Matzinger (1963) concluded that the second order interactions were prevalent in cotton, soyabean and tobacco. The second order interaction tends to be important in peanut because of its indeterminate nature. The genotype × environment interactions are similar in many autogamous crops. The ANOVA table shows that the g × e interaction may result in change in genotypic value and genetic variance components from location to location and from year to year.

8.3 Types of Environment

The environments are of two kinds: (i).Macroenvironment and (ii), Microenvironment. The microenvironment refers to environmental heterogeneities within a single macroenvironment, single plot or even a single plant which are uncontrolled and usually attributed to error variation in statistical analysis. On the other hand, macroenvironment refers to climatic, edaphic and management conditions which itself are made up of a number of factors (like different day length, temperature, humidity, rainfall, different levels of nutrients such as N, P, K or other micronutrients, different types of soils, number of irrigation, different dates of planting and plant densities and thus includes predictable as well as unpredictable environments) wherein treatments in general can be cross-classified against the genotypes (Jinks, 1976);

1. An unresponsive genotype which shows little or no change in performance in comparison with other genotypes to an improvement in the environment such as the addition of fertilizers, water, heat, etc.

2. A tolerant or well buffered genotype which displays little or no change in performance in comparison with other genotypes to a worsening of the environment such as excessive application or shortage of nutrients, water, heat, etc.

3. A stable or reliable genotypes which displays little or no change in performance compared with other genotypes to environmental changes which are largely beyond our control such as differences between seasons within an agrocimatic region, etc. In short, stability can be characterized by reduced variation across years. Thus variation of a genotype across years measures stability over time which is the only component of variation or risk which is relevant to a farmer contemplating of adoption of a variety in a given location.

4. A highly adaptive or flexible genotype is one which displays little or no change in performance compared with other genotypes when grown in different agro-climatic regions. Adaptibility can thus be defined as reduced variation across locations.

Similarly genotypes can be classified into the following. two categories according to the kind of response they show to this kind of variation.

1. Homeostasis: A genotype, all plants of which are more alike in size, shape, in days to flowering and in days to physiological maturity than other genotypes, is described as stable, uniform or homeostatic and the phenomenon is called homeostasis. In other words homeostasis is the ability of the genotype to stabilize itself in different environments.

2. Developmental stability: A genotype in which all repetitive structures on the same plant reach the same physiological or morphological stage over a shorter period of time or at maturity more alike in 'their dimension and quality than other genotype is said to display developmental stability.

The regulation of phenotype during development of whole plant or individual organs in response to environmental influence is defined as plasticity. General adaptibility can also be attributed to variation in phenotypic plasticity.

Variation within non-segregating generations such as P_1, P_2, F_1 has been used as a measure of developmental stability in that a lower variance signifies a greater ability to cope with the vagaries of the environment. For outbreeding species F_1 appears to be more stable than its inbreds whereas F_1 is no more stable than its parents in inbreeding species.

8.4 Approaches to Analysis of $g \times e$ interaction

The two approaches to the analysis of $g \times e$ interaction are as follows:

1. Statistical
2. Genetical

Statistical analysis provides information regarding mean performance of a genotype for a particular trait averaged over all environments and variation in performance over these environments. The later is subjected to statistical reduction, transformation or partitioning and these in turn form the basis of various methods of $g \times e$ interaction analysis as given below:

(a) Parametric Approach

1. Univariate analysis
   (a) Regression analysis (Yates and Cochran, 1938; Finlay and Wilkinson, 1963; Eberhart and Russell, 1966; Perkins and Jinks 1968).
   (b) Stability variance analysis

2. Multivariate analysis
   1. Principal Component analysis
   2. Factor analysis
   3. Canonical components analysis
   4. Cluster analysis
   5. Biplot analysis

(b) Non-parametric Approach

8.4.1 Regression analysis

8.4.1.1 Perkins and Jinks model

Of the various regression analyses we will discuss Perkins and Jinks analysis first and then develop relationship with other analyses. Perkins and Jinks (1968) proposed the following model.

$$Y_{ij} \quad = \mu + d_i \quad + (1 + \beta_i)e_j + \delta_{ij}$$

(Mean yield of ith genotype in jth environment) — Mean performance — Environmental sensitivity

This model has been developed from yet another general linear model which is

$$Y_{ijk} = \mu + d_i + e_j + g_{ij} + e_{ijk}$$

where Y_{ijk} is the yield of kth replicate of the ith genotype in the jth environment, μ, is the general mean, d_i is the deviation of the ith genotype from mean. It could be $[d]$ or $[h]$ depending upon whether that particular genotype is a pure breeding line or F_1, e_j is the deviation of the jth environment from the mean, g_{ij} is the interaction effect like d_i. It could be g_{dj} or g_{hj} depending upon the genotype, and e_{ijk} is the random environmental variable of the kth replicate.

$i = 1 \ldots S$ = number of genotypes,
$j = 1 \ldots t$ = number of environments,
and $k = 1 \ldots r$ = number of replications.

The assumptions are

$$\sum_{i=1}^{S} d_i = 0, \sum_{j=1}^{t} e_j = 0, \sum_{i \text{ or } j}^{S \text{ or } t} g_{ij} = 0$$

and e_{ijk} has a expected mean of 0 and variance σ^2. As g_{ij} is regressed on e_j

$$g_{ij} = \beta_i e_j + \delta_{ij}$$

where β_i is a linear regression coefficient for the ith genotype and δ_{ij}, the deviation or non-linear component. When S genotypes are grown in t environments, the mean performance data can be arranged in the fashion given in Table 8.3.

Table 8.3 Mean performance of S genotypes in t environments

	$j = 1, 2, 3$		t	Sum	Mean
Genotype					
$i = 1$	Y_{11}		Y_{1t}	$Y_{1.}$	$Y_{1.}/t = \mu + d_i$
2					
3					
S	Y_{Si}		Y_{St}	$Y_{S.}$	$Y_{S.}/t = \mu + d_S$
Sum $Y.1$			$Y.t$	$Y..$	$Y..$
Mean	$Y._{1/s} = \mu + e_1$	$Y.t/s$		$Y../st = \mu$	
	$= \mu + et$				

Here, $\mu + e_i - \mu \ldots \mu + et - \mu$ are the e_j's (X's) and Y_{ij} forms thus for each genotype a regression analysis is conducted and the ANOVA takes the form as given in Table 8.4.

Table 8.4 ANOVA for individual genotype

Item	d.f.	S.S.
Regression	1	$\sum_{j=1}^{t}(Y_{ij}e_j)^2 / \sum_{j=1}^{t} e_j^2$
Remainder	$S - 2$	$\sum_{j=1}^{t} \delta_{ij}^2$

The regression coefficient $(1 + \beta_i)$ becomes

$$\sum_{j=1}^{t} Y_{ij} e_j / \sum_{j=1}^{t} e_j^2$$

and the regression sum of squares becomes

$$\frac{\sum_{j=1}^{t}(Y_{ij}e_j)^2}{\sum_{j=1}^{t} e_j^2} = (1+\beta)^2 \sum_{j=1}^{t} e_j^2$$

which shows that only a small portion of regression mean squares measures β_i.

When S number of regression analyses are combined into a joint regression analysis, the total regression sum of squares then becomes

$$\sum_{i=1}^{S} (\text{regression S.S.}) S = \sum_{i=1}^{S}(1+\beta i)^2 \sum_{j=1}^{t} e_j^2$$

and since $\sum_{i=1}^{S} \beta_i = 0$ this becomes

$$S \sum_{j=1}^{t} e_j^2 + \sum_{j=1}^{t} \beta_i^2 \sum_{j=1}^{t} e_j^2$$

where $S \sum_{j=1}^{t} e_j^2$ is the joint regression sum of squares which in this analysis equals the environmental sum of squares. $\sum_{i=1}^{S} \beta_i^2 \sum_{j=1}^{t} e_j^2$ is the heterogeneity between regression sum of squares and the total remainder sum of squares becomes

$$\text{Total remainder } SS = \sum_{i,j=1}^{St} \delta_{ij}^2$$

The ANOVA for the joint regression analysis will take the form as given in Table 8.5.

The genotype × environment interaction is thus partitioned into heterogeneity between regressions and remainder. Genotype × environment interactions can be said to be present if either the heterogeneity between regressions mean squares, the remainder mean squares or both are significant. The heterogeneity mean squares if significant means that some of the β_i's are significantly positive and others significantly negative since $\sum_{i}^{s} \beta_i = 0$.

Table 8.5 ANOVA for joint regression analysis

Item	d.f.	M.S
Genotype	$(S-1)$	$t \Sigma(di)^2/(S-1)$
Environments	$(t-1)$	$S \sum_{j=1}^{t} e_j^2 /(t-1)$
Genotype × Environment interaction	$(t-1) \times (S-1)$	
Heterogeneity between regressions	$(S-1)$	$\sum_{i=1}^{s} \beta_i^2 \sum_{j=1}^{t} e_j^2 (S-1)$
Remainder	$(S-1) \times (t-2)$	$\sum_{i,j=1}^{St} \delta_{ij}^2 (S-1)(t-2)$
Error	$St(r-1)$	σ_e^2 *

* σ_e^2 is the error mean squares from the within genotype, within environment variation averaged overall genotypes and environments.

If the heterogeneity mean squares alone is significant we can predict the performance of each line across environments within the limits of sampling error. If the remainder mean squares alone is significant, it shows either no relationship or no sample relationship between the genotype environment interactions and the environmental values and hence no prediction can be made. If both items are significant then the practical utility of any predictions will depend upon the relative magnitudes of the two mean squares and if the heterogeneity mean squares is larger than the remainder mean squares then the predictions of the genotype × environment interactions based on linear regression will still have considerable practical value. The not significant heterogeneity mean squares does not rule out the possibility that the regression of $(d_i + g_{ij})$ on e_j for some of genotypes taken individualy may be highly significant when tested against their remainder mean squares and thus predictions can be made for these particular genotypes. In this analysis we can also construct a ANOVA with the items namely, joint regression, heterogeneity of regression and remainder and their significant joint regression mean squares will indicate that β_i is not zero.

According to this model a genotype with an average sensitivity will have a β_i value of zero and such a genotype will show no genotype × environment interaction. A genotype with $\beta_i > 0.0$ will be unusually sensitive to the environment and such a genotype will be undesirable as its performance will show an above average variation between environments. But such a genotype may be useful if it is specifically recommended for better environments. Finally, a genotype with β_i significantly negative and more specifically when $\beta_i = -1.0$ will be relatively indifferent to variation in the environments and is desirable as its performance has maintained overall environments including the poorer environments.

The overall efficiency of joint regression analysis in accounting for genotype × environment interaction is measured by the linear proportion of variance accounted for by the linear regression (Frip and caten, 1971).

8.4.1.2 Measurement of environment

We can have either physical or biological measurement of the environment but the best measure of the combined effects of all physical factors constituting the environment will be provided by the organism itself, although the environmental effect will no longer be measured without error. To have independent estimate of environment it would be better to divide the replicate of each genotype into two groups, using one group to measure the interaction and the other to measure the environment or estimate the environment from the replicate of the same material in a closely related and contemporary environment. Secondly one can use one or more standard varieties to assess the environment. For example, parental genotypes can be used as standards in relation to any generation derived by crosses between them. Knight (1970) while discussing the disadvantages of biologically determined environmental index suggested that the environmental index is more accurately assessed with a large number of entries. However, whether one makes independent or dependent assessment of environment the items, heterogeneity in regression and remainder will not be affected and the ranking of genotypes on the basis of linear regression coefficient is not going to alter.

8.4.1.3 Relationship with statistics of Yates and Cochran or Finlay and Wilkinson

Perkins and Jinks' parameters are related to Yates and Cochran's or Finlay and Wilkinson's in the way shown in Table 8.6. Perkins and Jinks' analysis is similar to that of Yates and Cochran and of Finlay and Wilkinson in that they are all joint regression analysis i.e. genotype × environment interaction is regressed on the additive environmental effects.

Table 8.6 Relationship between Perkins and Jinks's and Yates and Cochran's or Finlay and Wilkinson's statistics

Parameter	Yates and Cochran or Finlay and Wilkinson's statistics	Perkins and Jinks statistics	Relationship
Slope	b	β	$b - 1 = \beta$
Genetic value	$\bar{Y}$	d	$\bar{Y} - \mu = d$
Grand mean	$\bar{\bar{Y}}$	μ	$\bar{\bar{Y}} = \mu$
Environment	x	e	$x - \mu/e$

Finlay and Wilkinson (1963) characterized the stability of genotypes on the basis of regression coefficient (b) values in the following manner:

$b = 1$ average stability
$b > 1$ below average stability
$b < 1$ above average stability
$b = 0$ absolute stability

These above four categories of b can be associated with either high yield or low yield. Infact, bi and yield are often positively correlated (Eberhart and Russell, 1966). A genotype with high mean yield and b equals 1, will have a general adaptability between associated with low yield it can be said to be poorly adapted to all the environments. Regression coefficient greater than 1 signifies genotype with higher sensitivity to environmental changes and greater specificity of adaptibility to high yielding environments. Regression coefficients less than 1 characterize the genotypes with low sensitivity to environmental changes and therefore higher specificity of adaptibility to low-yielding environments.

8.4.1.4 Eberhert and Russell's model

The model takes the form as

$$Y_{ij} = \mu_i + \beta_i I_j + \delta_{ij}$$

where

Y_{ij} is the mean of ith genotype in the jth environment ($i = 1, 2, \ldots S; j = 1, 2, \ldots t$),
μ_i is the ith genotype mean overall environments,
β_i is the regression coefficient,
I_j is the environmental index,

and δ_{ij} is the deviation from regression

While constructing the model they added together the environmental sum of squares and genotype × environment interaction sum of squares and then repartitioned this total variation within genotypes into: (1) a linear component between environments (Environment, linear) with 1 degree of freedom, (2) a linear component of genotype × environment interaction (Genotype × Environment, linear) with ($t - 1$) degrees of freedom, (3) deviation from regression for each of the S genotypes with ($t - 2$) degrees of freedom and (4) the pooled deviation. Freeman and Perkins (1971) pointed out that the sum of squares for the linear component between environment (Environment, linear) with 1 degree of freedom is the same as the total sum of squares for environments with ($t - 1$) degrees of freedom and which is questionable. The ANOVA takes the form as given in Table 8.7.

The environmental index, I_j is calculated as the mean of all genotypes at the jth environment minus the grand mean.

$$I_j = (\sum_i Y_{ij}/S) - (\sum_i \sum_j Y_{ij}/St)$$

Table 8.7 Analysis of variance in stability parameter analysis

Item	d.f.	S.S.	M.S.
Genotype	$(S-1)$	$1/t \sum_i Y_i^2 - CF$	MS_1
Environments + Genotype × Environment	$S(t-1)$	$\sum_i \sum_j Y_{ij}^2 - \sum_i Y_{i.}^2/t$	
Environment(linear) I		$1/S (\sum_j Y.j\, I_j)^2 / \sum_j I_j^2$	
Genotype × Environment (linear)	$(S-1)$	$\sum_i [(\sum_j Y_{ij} I_j)^2/ \sum_j I_j^2] -$ Environment (linear) S.S.	MS_2
Pooled deviations	$S(t-2)$	$\sum_i \sum_j \delta_{ij}^2$	MS_3
Genotype 1	$(t-2)$	$\left\{(\sum_i Y_{ij})^2 - \dfrac{(Y_{1.})^2}{t}\right\} - (\sum_j Y_{ij} I_j)^2/\sum_j I_j^2 = \sum_j \delta_{ij}^2$	
.	.	.	
.	.	.	
.	.	.	
Genotype S	$(t-2)$	$\left[\sum_j Y_{sj}^2 - \dfrac{(Ys.)^2}{t}\right] - [(\sum_j Ys_j I_j)^2] = \sum_j \delta_{sj}^2$	
Pooled error	$t(r-1)(S-1)$		

and $\sum_j I_j = 0$

The deviation from regression, $S^2 d_i$ is obtained as

$$S_{d_i}^2 = [\sum_j \delta_{ij}^2 / (t-2)] - S_e^2/r$$

where S_r^2/r is the estimate of the pooled error or the variance of a genotype mean at the jth environment. The regression coefficient, b is estimated as

$$b_i = \sum Y_{ij} I_j / \sum_j I_j^2$$

and SE(b) is calculated as

$$SE\,(b) = \sqrt{\left[\sum_j \delta_{ij}^2 / (t-2)\right] / \left(\sum_j I_\alpha^2\right)}$$

The significant genotype × environment (linear) mean square (MS_2) when tested against pooled deviations mean square (MS_3) shows that the genotypes differ for their regression on the environmental index. The significance of deviations from regression for each genotype is tested against pooled error.

They used two parameters, namely, regression coefficient and deviation from regression to characterize the stability of a genotype. A stable genotype was defined as one with $b = 1$ and $Sd^2 = 0$. Breese (1969) strongly advocated the use of this second parameter of stability. He suggested that the term stability now be reserved to describe measurements of unpredictable irregularities (which corresponds to deviation of mean squares from regression) in the response to environments. But Witcombe and Whittington (1970) noted that this δ_{ij} is predictable when environmental changes which are not accounted for in the analysis are known.

The use of remainder mean squares as a measure of stability is not valid in certain circumstances. For example, if some disease resistant entry is tested along with other susceptible entries in a multilocational trial and if disease is present then for resistant variety the S_d^2 will be larger and significant even though it is superior in stability. However in the mean standard deviation (which will be discussed later in the section) a resistant entry may tend to have a lower standard deviation. Eberhart and Russell subtracts a constant which is a measure of inexplicable residual variation from the remainder mean squares. The value thus obtained termed S_d^2 obviously bears a sample relationship to the remainder mean squares. Further this remainder mean squares is not independent from the regression coefficient.

8.4.1.5 Genotypic stability parameters

Tai (1971) proposed a method of partitioning the genotype × environment interaction on the principle of structural relationship which overcomes the problem of regressing Y's on X's which are not independent. The genotype × environment interaction effect of ith genotype is partitioned into two components namely, α_i which measures the linear response to the environmental effect and λ_i which is the deviation from linear response in terms of error variance and are called genotypic stability parameters. These parameters are estimated as follows:

$$\alpha_i = \frac{MS_t}{MS_t - MS_e}(b-1)$$

$$\lambda_i = \frac{S}{(S-1)} \frac{(t-2)}{(t-1)} \frac{S^2 d}{MSe/P} - \lambda\left(\frac{b-1}{s-1}\right)\frac{MSb}{MS_e}$$

where b and S^2d are the phenotypic stability parameters of Eberhart and Russell and MSt, MS_b and MS_e are mean squares due to environments, replicates within environments and error deviates, respectively. He distinguishes between the linear component of interaction and the additive effects of the environment so that his regression coefficients have a mean of zero as do those of Perkins and Jinks, rather than of 1 as do those of Eberhart and Russell. Also, whereas Eberhart and Russell subtracted a pooled error estimate from the non-linear component of the interaction in order to get a value of 0 for S^2d for a stable variety, Tai divided the non-linear interaction term by the pooled error estimate so that the equivalent value of his parameter is 1. Thus a genotype with $\alpha = 0$ and $\lambda = 1$ is considered having average stability and a variety is stable when $\alpha = -1$ and $\lambda = 1$. This method thus uses an essentially similar technique to Ebertart and Russell except that maximum likelihood estimation of a structural relationship has been used.

8.4.1.6 Stability statistics of individual genotype

When only a small fraction of the genotype × environment interaction sum of squares can be attributed to heterogeneity among the regressions (Baker, 1969; Byth, Eiseman and De Lacy, 1976) the characterization of genotypes on the basis of regression coefficient may not be very effective and then it might be of greater interest to partition genotype × environment interaction to stability variance statistics assignable to each genotype.

Wrick (1962, 1964) proposed ecovalence, the contribution of a genotype to the interaction sum of squares in a two way analysis of variance for the measure of its instability and is computed as follows and which was later modified by Kang and Miller (1984) as W_i mean squares.

$$W_i = \Sigma Y_{ij}^2 - \left(\frac{2}{S}\right)\Sigma_j Y_{ij} Y_{.j}$$

$$+ \left(\frac{1}{S^2}\right)\Sigma_j Y_{.j}^2 - \left(\frac{1}{t}\right)\left(Y_{i\cdot} - \frac{Y_{..}^2}{S}\right)^2$$

$$= \Sigma_j (\bar{Y}_y - \bar{Y}_{i\cdot} - \bar{Y}_{\cdot j} + \bar{Y}_{..})^2$$

However, no means of testing the significance of this parameter was proposed. Freeman and Perkins (1971) criticised this partitioning on the grounds that given S genotypes and t environments consideration of the alternative forms of the interaction sum of squares in terms of totals shows that there is no way of dividing it into S groups. Also, the degrees of freedom for interactions are $(t-1)(S-1)$ and this number is not in general divisible by S.

Shukla (1972) developed an unbiased estimate of stability variance of the ith genotype ($\hat{\sigma}_i^2$) and proposed a criterion for testing the significance of $\hat{\sigma}_i^2$ to determine whether or not a genotype is stable. The stability variance of ith genotype is defined as the variance over environments of $(g_{ij} + \bar{e}_{ij})$ appearing in Perkins and Jinks model,

$$Y_{ij} = \mu + d_i + e_j + g_{ij} + e_{ijk}$$

Thus, $g_{ij} + \bar{e}_{ij} = V_{ij}$

where $\bar{e}_{ij} = \sum_{k=1}^{r} e_{ijk}$ and $E(\bar{e}_{ij}) = 0$

Under the assumptions

$$E(V_{ij}) = 0, V(V_{ij}) = \sigma_i^2$$
$$E(V_{ij}, V_{i'j'}) = 0 \text{ for } i \neq i' \text{ or } j \neq j';$$
$$V(g_{ij}) = \sigma'^2_i; E(g_{ij}, \bar{e}_{ij}) = 0$$
$$i = 1, 2, \ldots, S$$
$$V(\bar{e}_{ij}) = \sigma_0^2$$

Then $\sigma_t^2 = \sigma'^2_0 + \sigma_0^2$

where

σ_i^2 is the between environmental variance of the ith genotype and called stability variance of the ith genotype,

σ_0^2 is the within environmental variance,

and σ'^2_i is the interaction variance of the ith genotype.

A genotype is stable when

$\sigma_t^2 = \sigma_0^2$ and it is possible only when $\sigma'^2_i = 0$ i.e. no interaction variance.

The unbiased estimator of σ_i^2 denoted by $\hat{\sigma}_i^2$ is calculated as

$$\hat{\sigma}_i^2 = \frac{1}{(t-1)(S-1)(S-2)} [S(S-1)\sum_j (u_{ij} - \bar{u}_{i\cdot})^2$$
$$- \sum_i \sum_j (u_{ij} - \bar{u}_{i\cdot})^2$$

where $u_{ij} = \bar{y}_{ij} - \bar{Y}_{\cdot j}$ and $\bar{u}_{i\cdot} = \sum_i u_{ij}/t$

and $\sum_j (u_{ij} - u_{i\cdot})^2 = \sum_j (Y_{ij} - \bar{Y}_{i\cdot} - \bar{Y}_{\cdot j}' + Y..)^2$

Multipling each $\hat{\sigma}_i^2$ by $(S-1)(t-1)/S$, S components of genotype × environment interaction, one corresponding to each genotype can be obtained. These estimates are not statistically independent and they can be negative as they are differences of two sum of squares. The $\hat{\sigma}_0^2$ is calculated as

$$\hat{\sigma}_0^2 = \sum_i \sum_j \sum_k (Y_{ijk} - \bar{Y}_{ij})^2 / Str(r-1)$$

with $St(r-1)$ degrees of freedom.

This estimate of $\hat{\sigma}_0^2$ is analogous to pooled error of Perkins and Jinks. To test whether certain genotypes are stable or not a F test is carried out

$$F = \hat{\sigma}_i^2 / \hat{\sigma}_0^2$$

When $\hat{\sigma}_i^2$ is negative or less than $\hat{\sigma}_0^2$, $\hat{\sigma}_i^2$ may be taken equal to zero.

The present approach differs from Baker (1969) in that the former estimates $(S-2)(\sigma_i^2 + \bar{\sigma}^2)/S$ while the later estimates σ_i^2 where

$$\bar{\sigma}^2 = \frac{1}{S} \sum_{i=1}^{s} \sigma_i^2$$

The present method can be extended to use covariate (S) to remove its linear effect from genotype × environment interaction mean sum of squares. The remainder of genotype × environment interaction mean sum of squares (S_i^2) can be assigned to each cultivar and the significance is tested against, again $\hat{\sigma}_0^2$. The estimate of individual S_i^2 again, is analogous to Perkins and Jinks mean sum of squares, $\sum_j \delta_{ij}^2$ $(t-2)$ and Baker's (1969) deviation from regression sum of squares but the present approach has the advantage in that they are unbiased estimates of S_i^2 and the mean of $\hat{S}_i^2$ is the same as the mean of sum of squares of departure from regression, (Balance) the term he used.

$$\hat{S}_i^2 = \frac{S}{(S-2)(t-2)} \left[S_i - \sum_i \frac{S_i}{S(S-1)} \right]$$

Where $S_i = \sum_{j=1}^{t} (u_{ij} - \bar{u}_{i\cdot} - b_{izj})^2$

$$\hat{b}_i = \sum_j \frac{(u_{ij} - \bar{u}_{i\cdot}) Z_j}{\sum_j Z_j^2}$$

and Z_j is a measure of some characteristics of jth environment.

The definition of stability here is similar to Baker's and Eberhart and Russell's. A significant departure of regression coefficient from zero will be indicated by a relatively high stability variance but a regression coefficient of zero need not mean that the particular genotype is stable. A zero regression coefficient can result if there is no linear relationship between genotype mean and environmental mean, yet the stability variance (σ_i^2) may be greater than σ_0^2. Tai's method differs from the present in the sense that he considered the model under certain side conditions on the interaction. According to Tai's definition of stability one should have $b_i' = -1$ and $S_i^2 = \sigma_0^2$ but his definition of average stability coincides with the present concept of stability.

Kang et al. (1987) found that $\sum_i (u_{ij} - \bar{u}_{i\cdot})^2$ in the formula for calculating $\hat{\sigma}_f^2$ is equal to $\hat{W}_i$ sum of squares and that the remainder in the equation is a constant for a given body of data. They suggested that Shukla's statistics $\hat{\sigma}_i^2$ and $\hat{S}_i^2$ be preferred over Wrick's $\hat{W}_i$ in $G \times E$ interaction study.

Plaisted and Peterson (1959) proposed a similar parameter to that of Wrick. In an experiment of S genotypes and t environments ANOVA is constructed for every pair of genotypes and variance due to $g \times e$ interaction is estimated and this is repeated for all pairs, $(S-1)S$ of genotypes. The mean of the estimated variance components of $G \times E$ interaction for all pairs of genotypes that included the ith genotype is the stability variance of that ith genotype (θ_i) and this is estimated as follows:

$$\theta_i = \frac{S}{2(S-1)(t-1)} \sum_{j=1}^{t} (Y_{ij} - \bar{Y}_{i\cdot} - \bar{Y}_{\cdot j} + \bar{Y}_{\cdot\cdot})^2$$

$$+ \frac{\sum_{i=1}^{s}\sum_{j=1}^{t}(Y_{ij} - \bar{Y}_{i\cdot} - \bar{Y}_{\cdot j} + \bar{Y}_{\cdot\cdot})^2}{2(S-1)(t-1)}$$

Plaisted (1960) proposed a method in which ith genotype is deleted from the entire body of data and the $g \times e$ interaction variance from this subset is the stability index for the ith genotype

$$\theta_{(i)} = \frac{-S}{(S-1)(S-2)(t-1)}$$

$$\times \sum_{j=1}^{s}(Y_{ij} - \bar{Y}_{i\cdot} + \bar{Y}_{\cdot j} + \bar{Y}_{\cdot\cdot})^2$$

$$+ \frac{\sum_{i=1}^{s}\sum_{i=1}^{t}(Y_{ij} - \bar{Y}_{i\cdot} - \bar{Y}_{\cdot j} + \bar{Y}_{\cdot\cdot})^2}{(S-2)(t-1)}$$

8.4.1.7 C.V. and other Statistics

In addition to these above methods the conventional statistics, the coefficient of variability of ith genotype (CV_i) calculated from the analysis of variance is a measure of stability of ith genotype.

$$CV_i = \frac{S_i}{\bar{Y}_{i\cdot}} \times 100$$

where S_i is the standard deviation of ith genotype and $\bar{Y}_{i\cdot}$ as the mean of the ith genotype over all environments.

The variance of ith genotype across environments (S_i^2) can also be used as a measure of stability.

$$S_i^2 = \sum_{j=1}^{t}(Y_{ij} - \bar{Y}_{i\cdot})^2/(t-1)$$

The CV statistics is not very suitable as a measure of stability because it has the normal limitation i.e. low mean yield is associated with high CV and high mean yield is associated with low CV. In other words the high yield environment is characterized by relatively low CV and low yield environment by relatively high CV.

Yau and Hamlin (1994) suggested relative yield as a measure of entry performance in variable environments and the variance of relative yield across sites as a stability measure. They showed that relative yield has three advantages: (i) conversion of the entry variance of yield across sites to a practical stability measure, (ii) equal weight given to each site when calculating entry means a across sites and, (iii) ease in comparing large number of entries tested in different experiments.

The yield of genotypes at any site is expressed as relative to the mean yield of that site.

$$RY_{ij} = (Y_{ij}/Y_{\cdot j}) \times 100$$

where RY denotes the relative yield; Y_{ij} is the yield of the ith genotype at jth location; $Y_{\cdot j}$ denotes the mean yield of jth locations. The mean relative yield ($RY_{r\cdot}$) is the average of individual relative yield across locations, i.e.

$$RY_{i\cdot} = (\sum_{j=1}^{n} RY_{ij})/n$$

where n is the number of locations. The variance (S_i^2) of the genotype relative yield is calculated as

$$S_i^2 = [\sum_{j=1}^{n} (RY_{ij} - RY_{i.})^2]/(n-1)$$

8.4.1.8 Comparison of stability statistics

We have seen nine statistics currently in use to characterize stability. We have also seen relation between different statistics offered by different workers including Witcombe and Whittington, 1970. Lin *et al.* (1985) compared all these statistics together after considering the basic structure of these statistics from which the different methods were derived. All these statistics given in Table 8.8 have been classified into four groups on the basis of the structure of these statistics (whether the statistic is based on deviation from the average genotypic effects ($Y_{ij} - Y_{i.}$) or on genotype × environment interaction effect ($Y_{ij} - \overline{Y}_{i.} - \overline{Y}_{.j} - \overline{Y}_{..}$) or whether the statistics are sum of squares or regression coefficient or deviation mean squares from regression). They have been further classified into 3 groups on the basis of concepts of stability they represent, namely, Type I (A genotype is considered stable if its among-environmental variance is small i.e. genotype is stable in absolute sense), Type II (A genotype is considered stable if its response to environment corresponds to the mean response of all the genotype in the trial) and Type III (A genotype is considered stable if the deviation mean squares is small).

The difference between two statistics in group A is a mere change of scale. For small CV ($CV < 20\%$). Variance $[\log_{(y)}]$ = Variance $(y)/[(\text{mean }(y)]^2$ the statistics will yield similar conclusion but when CV exceeds 20% the two statistics will yield different conclusion. All the statistics in group B are equivalent in the sense that their first items are proportional to each other and their second items are constant. However, θ_i, $\theta(1)$ and W_i^2 are merely index numbers whereas σ_i^2 is an unbiased estimate of variance of ith genotype. The two statistics in group C are

Table 8.8 Formula of the nine stability statistics

Group	Type	Formula
A	1	$S_i^2 = \sum_{j=1}^{t} (Y_{ij} - \overline{Y}_{i.})^2/(t-1)$
	1	$CV_i = S_i/\overline{Y}_{i.} \cdot \sqrt{\overline{Y}_{i.}} \times 100$
B	2	$\theta_i = \dfrac{S}{2(S-1)(t-1)} \sum_{j=1}^{t} (Y_{ij} - \overline{Y}_{i.} - \overline{Y}_{.j} + \overline{Y}_{..})^2 + \dfrac{SS(GE)^*}{2(S-1)(t-1)}$
	2	$\theta_{(i)} = \dfrac{-S}{(S-1)(S-2)(t-1)} \sum_{j=1}^{t} (Y_{ij} - \overline{Y}_{i.} - \overline{Y}_{.j} + \overline{Y}_{..})^2 + \dfrac{SS(GE)}{(S-2)(t-1)}$
	2	$W_i^2 = \sum_{j=1}^{2} (Y_{ij} - \overline{Y}_{i.} - \overline{Y}_{.j} + \overline{Y}_{..})^2$
	2	$\sigma_i^2 = \dfrac{S}{(S-2)(t-1)} \sum_{j=1}^{t} (Y_{ij} - \overline{Y}_{i.} - \overline{Y}_{.j} + \overline{Y}_{..})^2 - \dfrac{SS(GE)}{(S-1)(S-2)(t-1)}$
C	2	$b_i = \sum_{j=1}^{t} (Y_{ij} - \overline{Y}_{i.})(\overline{Y}_{.j} - \overline{Y}_{..})/\sum_{j=1}^{t} (\overline{Y}_{.j} - \overline{Y}_{..})^2$
	2	$\beta_i \sum_{j=1}^{t} (Y_{ij} - \overline{Y}_{i.} - \overline{Y}_{.j} + \overline{Y}_{..})(\overline{Y}_{.j} - \overline{Y}_{..}) / \sum_{j=1}^{t} (\overline{Y}_{.j} - \overline{Y}_{..})^2$
D	3	$\delta_i^2 = \dfrac{1}{(t-2)} [\sum_{j=1}^{t} (\overline{Y}_{ij} - \overline{Y}_{i.})^2 \beta_i^2 \sum_{j=1}^{t} (\overline{Y}_{.j} - \overline{Y}_{..})^2]$
		$\delta_i^2 = \dfrac{1}{(t-2)} [\sum_{j=1}^{t} (Y_{ij} - \overline{Y}_{i.} - \overline{Y}_{.j} + \overline{Y}_{..})^2 - \beta_i^2 \sum_{j=1}^{t} (\overline{Y}_{.j} - \overline{Y}_{..})^2]$

*$SS(GM) = \sum_{j=1}^{S} \sum_{j=1}^{t} (Y_{ij} - \overline{Y}_{-i} - \overline{Y}_{.j} - \overline{Y}_{..})^2$

equivalent and so are the two statistics in group D. The statistics of group C and D are to be preferred over those of group B because the former gives the shape of response of genotypes to changes in environment as well as its variation.

A breeder would like to go in for a variety giving high yield coupled with stability. Now among these three types of stability the stability in type I sense is analogous to homeostasis and it has been found to be associated with a relatively poor response and low yield in high yielding environment. Also, this is the kind of response a genotype shows to changes within one macroenvironment or to changes within a restricted geographical range. Thus, if the breeder's aim to breed a variety for wider adaptability, stability of this kind is not an important character to look for but when a variety is bred for specific environments or for a restricted geographical area, homeostasis can be considered a feature of the particular variety alongwith high yield.

The estimate of stability statistics for a genotype showing type II stability is a relative measure and depends on the genotypes included in the study. Thus inferences drawn to be confined to the test genotypes and should not be generalized unless the genotypes are a random sample of those grown in the area under test. A genotype, stable by definition is so only with respect to the other genotypes in the test. It may or may not remain stable if tested with another set of genotypes. Further, in case of the linear regression approach to genotype × environment interaction analysis Weisberg (1980) and Daniel and Wood (1980) showed that the fits are largely determined by one or two data points. This means that stability statistics of a variety may be unduly influenced by its performance in only one or two environments and to that extent may be seriously misleading (Westcott, 1986). Knight (1970) showed that the effect of logarithmic transformation is to minimize the genotypic differences at the high values and maximize the differences at the low values and thus transformation may induce linearity if differences between genotypes at higher values are greater than those below them. He also found that the exceptionally high yield values had a greater influence on the mean yield and the regression coefficient when calculated on an arithmetic scale was higher than when calculated on a logarithmic scale and as a result the mean yield and regression coefficient are often positively cor-related but correlation disappears when a logarithmic scale is used. Thus the underlying mathematical model relating the response of a genotype to environmental change changes with the change in scale and thus different inferences will be drawn.

Stability of a genotype judged by the parameter, the deviation mean squares from regression, has been questioned on the ground that the regression model being descriptive (i.e. where independent variable, the environmental index can't be measured prior to the experiment), the deviation mean squares from regression indicates not more than how good is the fit. A poor fit (i.e. small R_i^2 or large δ_i^2) or heterogeneous deviations mean squares should be taken to mean that the linear regression model is not adequate to estimate variability. That means this mean squares is not independent of the regression coefficient (Hardwick and Wood, 1972). Only in prediction model where independent variables are measured prior to the experiment, the deviation mean squares from regression may have deterministic property that can be associated with a genotype which the proponents assumed. Witcombe and Whittington (1971) examined a subset of six entries out of twenty five originally analysed and found five of the six entries to be aberrant in that the highest nitrogen levels were supra-optimal, the other entry showed a sustained yield response upto the highest nitrogen level. This later entry when analysed using Eberhart and Russell's model was shown to be stable and was having low deviation mean squares from regression. However, when the same subset of six entries was analysed again using the same model the same entry appeared to be unstable, thus having significant deviation mean squares. Westcott (1986) concluded from the results that a variety could have marked deviations from linear regression, not because it was inherently irregular but because it showed a different response pattern from the majority of the group with which it was being compared.

In general, group A statistics are useful if a breeder is concerned about stability of a genotype over a range of environments.

8.4.2 Analysis of non-linear genotype × environment interaction

Although there are numerous reports of linear relationship (Freeman, 1973; Hill 1975) between genotype and environment index there are many others where the relationship is shown to be non-linear. Witcombe and Whittington (1971) noted that every genotype has many different responses to environments, each specific to a particular type of variables and as environment comprises many variables, non-linear genotype × environment can be explained by these phenomena. Perkins and Jinks (1968, 1773) and Jinks and Pooni (1979) showed that a significant part of non-linearity results from thresholds in response of genotypes to environmental differences. Two phase and multiphase regressions are involved to explain away non-linearity in the genotypic response which is occasionally induced to response thresholds. By threshold response we mean that every genotype has an upper limit to its phenotypic expression beyond which further improvement in the environment will evoke no response. This property of phenotype will vary between genotypes and will be subject to genetic control. In the simplest case the response of a genotype to an improving environment can be represented by two intersecting straight lines or in other words by a pair of regression equations as follows which differ in means, a's and regression coefficients, b's.

$$Y = a_1 + b_1 x$$
$$Y = a_2 + b_2 x$$

In terms of biometrical genetical parameters the two intersecting linear regressions will be expressed as

$gd_j = b_{dej}$ in one subset of environments
 say e_j comprises X_1 to X_{10}

and $gd_j = b'_{dej}$ in another subset of environments,
 say e_j represents X_{11} to X_{20}.

Figures (8.2) and (8.3) show the two intersecting straight lines and two overlapping straight lines response of the genotypes, respectively.

Occurrences of threshold responses to increasing nutrients level were reported by Boyd et al. (*1976*). They noted that the intersecting straight lines is the model of response expected from Leibig's (1855) Laws of minimum and Blackman's subsequent "Limiting factors" (1905) which describe the joint effect of several limiting factors. In the present case genotype is the limiting factor leading to a constant phenotypic expression beyond a certain value of environmental index because in general there is no other measure of an environment's value than the average phenotype of all or some of the genotypes grown in that environment.

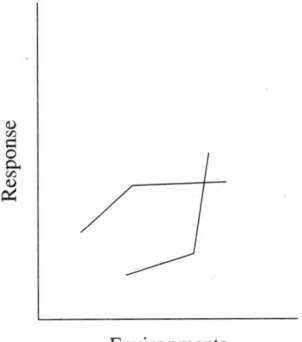

Fig. 8.2 **Two intersecting straight lines response of genotypes.**

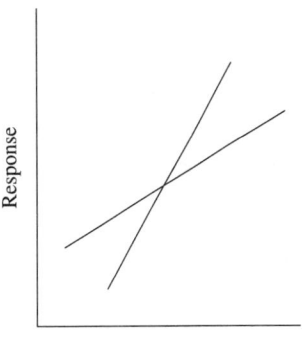

Fig. 8.3 **Two overlapping straight lines response of genotypes.**

We will now consider a simple case of two genotypes raised in a number of environments and see how the responses can be described when one of the two genotypes shows threshold response. Let us assume that A shows the linear response while B ceases to respond in above average environments. Now the response of B to environmental change that is, its regression on the environmental value, e_j will be completely described by two intersecting straight lines with a positive and zero slope in the poor and better environments, respectively. The average response of A and B will have a slope of one. The response curve of A and B will, therefore deviate symmetrically either side of this slope. At environmental value where B changes its slope from b_2 to 0, the slope of A must change from b, to $b_1 + b_2$ to keep its symmetry with the corresponding value of B. Thus the response of A will be completely described by two intersecting straight lines with a slope of b_1 in the poorer environments and $1 + \dfrac{b_2}{b_1}$ in the better environments.

These relationships for two genotypes can be extended to 3 or more genotypes considered in the study. The average response of all the genotypes must still be a straight line of unit slope and the responses of all the individuals genotypes must still be symmetrical around this value. Now if the first of S genotypes stops responding to an improvement in e_j, each of the remaining $S - 1$ genotypes must on an average increase their rates of response by

$$1 + \dfrac{b_1}{\sum\limits_{2}^{S} b_i}$$

If r genotype ($r = 1...S$) are showing the threshold response then each of the remaining $(S - r)$ genotypes will have increased their rates of response by

$$1 = \dfrac{\sum\limits_{i}^{r} b_i}{\sum\limits_{r+1}^{S} b_i}$$

Thus the remaining $(S - r)$ genotypes will on average show a steadily accelerating response to any further increase in e_j as r increases from S to $S - 1$.

We can have another situation in which some of the r genotypes may show a reduced rather than a zero response to improving environments and then the remaining $S - r$ genotypes will increase their response instead by

$$1 = \dfrac{\sum\limits_{i}^{r} C_i}{\sum\limits_{r+1}^{S} b_i}$$

where C_1 is the absolute change in the linear slope of the ith genotype.

These relationships will hold and a pair of intersecting straight lines will account for the response of each of the genotypes in study only if two or more genotypes reach its limit at the same e_j value. But if it is not so then it is unlikely that a pair of intersecting straight lines will account for response if any genotype that has not reached its limit. This is because then the responses of these genotypes will accelerate at each environmental value at which an additional genotype reaches its limit. In this situation, the responses of such genotypes are, therefore, more likely to fit a polynomial, for example, a quadratic curve $Y = a + bx + Cx^2$ and here $g_{dj} = b_{dej} + b'_{dej^2}$.

One can proceed by fitting linear, quadratic and two intersecting straight lines model. The procedure of fitting a two intersecting regression can be presented in a matrix form as follows (Pooni and Jinks, 1980).

$$Y_i = a_i D_i + a_2(1 - D_i) + \{bD_i + b_2(1 - D_i)\}X_i + U_{1i} D_1 + U_2(1 - D_i)$$

where a, a_2, b_1 and b_2 are the regression coefficients parameters, Y_i and X_i are the dependent and independent variable vectors; U_{1i} and U_{2i} are the remainder vectors for the two regressions and D_i a discrete vectors of coefficients with the value of 0 and 1, respectively below and above the changes over points between two sets of environments. A best fitting pairs of straight lines can be selected from all possible pairs (1) $Y \cdot Y_i$ and (2) $Y_{j+1} \cdot Y_s$ where i is given every value from 3 to $S - 3$. It would be better if all $S - 5$ possible pairs are examined while finding out the best pair of intersecting lines. A conservative test of significance is used by multiplying the observed probability (wherever $P < 0.05$) obtained from the normal test of significance by $(n-5)$ before deciding the level of significance.

A single straight line is rejected in favour of the best fitting pair if it resulted in a significant reduction of the residual mean squares and a significant difference between bI and bII. A single straight line is rejected in favour of a quadratic or high order polynomial only if this resulted in a significant reduction in the residual mean squares and the quadratic (or higher order) regression coefficient was significant.

In fitting two phase regressions, it is assumed that a complete switch over from one to the other takes place between the pair of environments which minimises the remainder mean squares. However, switch over between the two regressions may not be so abrupt as assumed and infact may take place gradually. This is likely when the response of a genotype is dependent on so many factors. In this situation, the two regressions will overlap and make a proportionate contribution to the phenotypic values over the whole range of environments. Thus D_i is replaced by the cumulative normal integral (Goldfeld and Quandt, 1972) from $-\infty$ to $(x_i - x_o)/\sigma$. Again, the regression ss is maximised by changing the reference environment x_o by environmental scores between $x(3)$ and $x(S-3)$ and the various parameters are estimated. Results from this overlapping lines can be compared with intersecting straight lines, each estimation process takes 4 degrees of freedom. Assuming $a_1 = a_2 = a$, the two regression coefficients b_1 and b_2 with the same a can be estimated and now it takes the form of a quadratic equation which can further be compared with a single linear regression and improvement over it can be seen.

Amongst these models the two intersecting straight lines were found to be the best fitting model for the largest number of genotypes. But there is still a significant remainder mean squares left which suggests that a more complex model is required. Verma et al. (1978) first claimed that the two intersecting straight lines provide a better model of genotype × environment interaction than a single straight line but they did not apply statistical criteria of the kind laid down above to distinguish between alternative models.

Mariani et al. (1983) proposed a three phase regression by dividing the whole range of environments into poor, average and good environments using confidence limits of the general mean.

The two regressions and multiple regressions approach suffers from the problem of choice of truncation point while dividing the environmental range and also the computational complications when each genotype is assumed to have its own response threshold.

8.4.3 Limitations with regression models

Limitations with linear model are that (1) X variate (the independent variate) is subject to error. Intervals along X-axis should be equally spaced so that the precise nature of relationship between Y, the dependent variate and X, can be determined and (2) genotype means (Y variate) contribute to and hence are not statistically independent of the environmental means (X variate) on which they are regressed.

Regression techniques to characterize genotype responses to the environments are an over simplification. Interactions are still occuring which are not identified because biological indices do not give information on the physical nature of the environment. Some workers have found that regression parameters account for a considerable amount of genotype × environment interaction while others have found that these parameters account for only a small amount of such interactions and secondly regression parameters have been found to vary considerably from trial to trial. The regression models have the disadvantages that all effects are expected to be linear or at least linear after transformation (of other known form such as quadratic) so that if several different response patterns are infact being assessed important interactions will be misinterpreted. However, univariate models may be expected to be helpful under the conditions that the test sites are representative of commercial growing conditions in their localities and the genotypes under test are infact widely disparate in genotypes.

8.5 Multivariate Analysis of Genotype × Environment Interaction

All these above univariate techniques in parametric approach deal with only the individual aspect of stability and they do not provide a complete picture of the response. For example, when we talk of response of a genotype it may be characterized as stable in type I sense but unstable in Type II sense or it may be judged as Type III unstable but stable in Type II sense, etc. and thus it is difficult to reconcile all of these assessments into a unified conclusion. The reason is that there is often more than one way in which responses differ. That is a genotype's response to environments is multivariate or in words of Hardwick and Wood (1972) there is variation between genotypes in more than one dimension. Multivariate analyses can throw light on relationship, interdependence and relative importance of characteristics involved and yield meaningful information. A series of univariate analysis carried out separately for each of the variables is, in general, not adequate as it ignores the correlation among variable. Through multivariate analysis the number of comparisons which must be made is reduced by grouping lines or varieties which respond similarly and then comparing such groups. This technique can also be applied to environments so that suitable environments for testing may be delinearised. The various multivariate analyses that are being used are 1. Principal component analysis 2. Factor analysis 3. Canonical analysis 4. Cluster analysis 5. Principal co-ordinate analysis 6. Biplot analysis and others. These analyses are discussed in Chapter 4.

8.5.1 Principal component analysis

Williams (1952) showed that the least square estimation of regression coefficients in the linear regression approach of genotype × environment interaction was equivalent to extracting the first principal component of the genotypic performance. The validity of the model then can be examined by extracting further principle components or more informally, by inspection of the residual correlation matrix after extracting the first component. While examining plant height of some inbred lines of *Nicotiana rustica*, Perkins and Jinks (1968) found that the deviation from regression could be divided into two groups on the basis of whether the genotypes showed positive or negative residual correlation with each other. Regression analysis within each group then showed a better fit to the data although significant deviations still existed. These groups coincided with a well established major gene difference, mop-head/ non-mophead. However, such a clear-cut biological explanation has seldom been observed in other studies.

Mendel (1971) considered the principal component approach further by using a multiplicative model (as has long been recognised, Fisher and Mackenzig, 1923). This method showed the number of dimensions necessary to contain the genotypic variation and gave estimates of the corresponding coefficients without however any prior knowledge of which factors these dimensions represented. Hardwick and Wood (1972) suggested that Mendel's method may prove particularly valuable when the deviations from regressions are substantial but environmental variables have been measured. Grafius and Keisling (1960) used factor analysis method to construct orthogonal vectors representing environmental effects and thus predicted genotypic response in terms of these factors.

8.5.2 Biplot method

If principal component analysis is applied to the residual (the interaction) from the additive ANOVA rather than to the original yield data to generate a multiplicative model, it becomes a AMMI (additive main effects and multiplicative interaction) model. The biplot or AMMI model is

$$Y_{ij} = \mu + \alpha_i + \beta_j + \sum_{n=1}^{N} \lambda_n \gamma_{in} \delta_{jn} + \theta_{ij} + e_{ijk}$$

where, Y_{ij} is the yield of ith genotype in jth environment,

μ is the grand mean,

α_i's ($i = 1 ... S$) are genotype mean deviation (the genotype mean – the grand mean),

β_j's ($j = 1 ... t$) are the environmental mean deviation,

λ_n is the eigen value of principal component analysis (*PCA*) axis *n*,
γ_{in} and δ_{jn} are the genotype and environment PCA scores for PCA axis *n*,
N is the number of PCA axes retained in the model,
θ_{ij} is the residual, and
e_{ijk} is the replicate error associated with *k*th replicate ($k = 1 \dots r$)

The least square fit for balanced data is obtained by first fitting the additive part by the ordinary analysis of variance and then analyzing the interaction (non-additive residual) by PCA (Gabriel, 1978). The number of *df* attached to each PCA is calculated by the simplest method (Gollob, 1968) which closely approximates other methods. PCA axis is assigned $(S + t - 1 - 2n)df$. Frequently the first one or few PCA axes describe most of the interaction sum of squares with relatively few degrees of freedom. There is automatic generation of a model sequence by including non to all of the interaction PCA axes. The biplot from AMMI identifies single *df* interaction, contrast, concurrence, and Finlay-Wilkinson's regression method of partitioning the interaction. It may identify other models or subcases as most appropriate for a given body of data. If only the additive or only the multiplicative portions of AMMI model are significant, then the ANOVA or PCA subcases are indicated. If the PCA axes are effective in concentrating the interaction *SS* into a few axes, then the interaction is probably highly complex. A general ANOVA test for interaction is then appropriate for diagonising the general means model. The ANOVA for the AMMI model is given Table 8.9.

Table 8.9 ANOVA for AMMI model

Item	df
Replication	$(r - 1)$
Genotype	$(s - 1)$
Environment	$(t - 1)$
Genotype × Environment interaction	$(s - 1)(t - 1)$
PCA 1	$(s + t - 3)$
PCA 2	$(s + t - 5)$
3	$(s + t - 7)$
Residual	$\{(S - 1)(t - 1) - df \text{ for } PCA\ 1$
Error	$+ df\ PCA\ 2 + \dots\}$

Gauch Jr. (1988) in a trial of 15 genotypes in 15 environments (sites and years) found that the best predictive model is AMMI with one interaction PCA axis. He also showed that Finlay-Wilkinson analysis accounted for only 9% of the interaction *SS* whereas the concurrence model captured 2%. Thus for these data the classical analysis of interaction were not effective. Assuming that the data = pattern + noise where the pattern pertains to inherent features of genotypes and environments and thus can be predicted whereas the noise relates to stochastic, uncontrolled usually unexplainable variability among replicates and is of no predictive value, one wants to find as much possible of the pattern while eliminating the maximum noise (Freeman, 1973). Appropriate multivariate models selectively recover pattern in their early *df*, and selectively recover noise in their late *df*. This has been demonstrated in simulated data by a variant of PCA called reciprocal averaging (also called correspondence analysis (Gauch, 1982)). Pattern involves sizeable correlations among matrix values whereas noise involves idiosyncratic deviations in individual matrix cells, so eigen analysis (as in correspondence analysis or PCA or AMMI) initially concentrates on the pattern. After most of the pattern is exhausted, at first eigen values extract noise at a somewhat accelerated rate by exploiting chance correlations in the noise, but this is soon exhausted and the remainder of the *SS* is extracted at a slow rate.

Kempton (1980) used this approach to form principal components biplot and also extended Finlay and Wilkinson's plot. He showed how the biplot gives the expected response of a genotype in an environment using the information from linear regression and principal component analysis. Ordinations followed by some form of pattern analysis will allow detection of either groups of small genotypes or groups of similar environments.

The difficulties encountered in the application of multivariate analysis in the study of genotype × environment interaction are statistical as well as interpretational (Seal, 1964; Gorden 1981 and Silvey, 1982). Besides multivariate techniques do not correct the defects that the linear regression analysis possesses. Hill (1975) noted that despite its imperfection, the linear regression approach has the twin merits of simplicity and biological significance attributes which are lacking in multivariate statistical approach.

The ordination techniques such as principal component analysis, principal coordinates analysis and factor analysis often present a large percentage of the original n dimensional variation in a few dimensions and thus may simplify interpretation of GXE structure. Complex relationships among environments or among genotypes can be adequately represented in a scatter diagram (Westcott, 1987). Alternatively, cluster analysis has been used to group locations that discriminate among genotypes in a similar manner or to summarize patterns of genotypic performance across environments. The combination of ordination and cluster analysis is termed pattern analysis. With biplot facility from AMMI analysis, both genotypes and environments occur on the same scattergram and inferences can be drawn about specific x environment combination. The biplot displays simultaneously the rows and the columns of a matrix.

AMMI model - The AMMI model integrates the usual additive analysis of variance for the additive effects with the principle components analysis for the multiplicative effects. The additive part of the AMMI model (μ, γ, δ) is estimated first with ANOVA and then the multiplicative part (λ, γ, δ) is estimated with the PCA to explain the pattern in GXE interaction (Gauch and Zobel, 1989). The additive part of the model is simply the genotypes mean plus the environment mean minus grand mean. The interaction part equals the genotype PCA score times the environment PCA score. The AMMI model differs from regression models in that the former partitions the sum of squares of GXE interaction into principal component axes whereas the later partitions the sum of squares of GXE interaction into joint, genotypic and environmental regressions. AMMI analysis provides a graphical representation (or biplot) to summarize information on main effects and interactions (PC1) of both genotypes and environments simultaneously.

The AMMI model removes residual or noise variation from GXE interaction. Various studies have demonstrated the effectiveness of the AMMI model in capturing and partitioning the sum of squares of GXE interaction in comparison to the linear regression models.

With S number of genotypes and t number of environments, the matrix representing GXE interaction sum of squares (g_{ij}) takes the following from:

$$\begin{matrix} g_{11} & g_{11} \cdots \cdots g_{1t} \\ g_{21} & g_{22} \cdots \cdots g_{2t} \\ g_{31} & g_{32} \cdots \cdots g_{3t} \\ . & . & . \\ . & . & . \\ . & . & . \\ g_{S1} & g_{S2} \cdots \cdots g_{St} \end{matrix}$$

Biplot display - The figure 8.4 shows main effects means on the abscissa and PCA1 values as ordinates. Genotypes or environments that appear almost on a perpendicular line have similar means and those that fall almost on a horizontal line have similar interactions parrterns. Genotypes (or environments) with large PCA1 scores (either positive or negative) have high interaction whereas genotypes (or environments) with PCA1 scores near zero have small interactions (Figure 8.5)

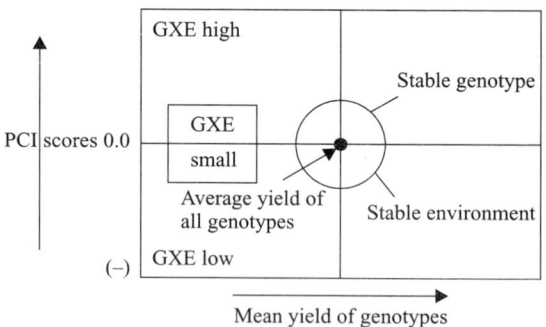

Fig. 8.4 Showing plotting of PC1 scores of genotypes and/or environments against mean yield.

8.5.3 GGE bilot

When GXE experiments are conducted at multiple locations across different countries, they fail to differentiate genotypes. In such conditions there is

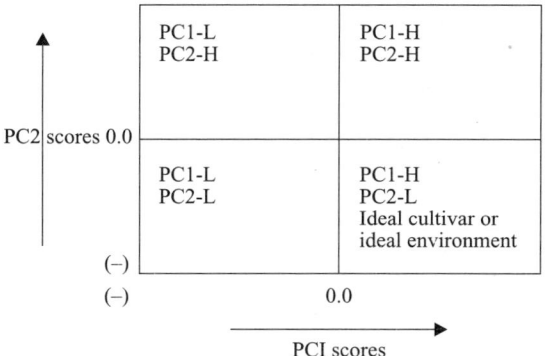

Fig. 8.5 Showing plotting of PC1 scores against PC2 scores for genotypes or environments. H = High; L = Low.

a need to divide the target region into two or more different mega-environments and this will lead to improvements in cultivar differentiation. The target region should be homogeneous. Taking a global prospective mega-environments have been defined as a broad, not necessarily contiguous area occurring in more than one country and frequently transcontinental, defined by similar biotic stresses, cropping system requirement, by volume of production (Braun, 1996). It can also be defined as a portion of crop species growing region with a homogeneous environment that causes some varieties to perform similarly (Gauch and Zobel, 1997). Polygo view of a GGE biplot indicates presence or absence of crossover GXE interactions involving the most responsive genotype and is suggestive of the existence or absence of different mega-environments among tested sites. The biplot is constructed by the first two PCs derived from subjecting the environment-centered yield data to singular value decomposition (Kroonenberg, 1997).

As the matrix is environment centered, the environment (column) means are zero. It effectively identifies the GXE interaction pattern in the data which involves identification of environment in which a genotype is performed best, selection of superior cultivars and test environment for a given mega environment. Ideal cultivars are the ones with a large PC1 score (high yielding ability and a small PC2 score (stability). Test environment is one with a large PC1 score (more discriminating of genotypes in terms of the genotypic main effect) and a low PC2 score (more representative of the overall environment). The is a near perfect correlation between genotypic PC1 scores and genotype main effects which is usually met for yearly but not for all database and so the alternative is to replace PC1 by regression of environment-centred yield on the genotype main effects. Vertex genotypes are the most responsive lying at longest distance from the origin (0, 0). Vertex genotype in each sector is the best genotype at sites whose markers fall into the respective sector. Position of genotype on the biplot is unique. The perpendicular lines are also unique.

GGE biplot is a methodology for graphical analysis of multienvironment trial data. GGE refers to the genotype main effect (G) plus the genotype x environment interaction (GE) and a biplot is a plot that simultaneously displays both the genotypes and the environments (or in more general terms, both the row and the column factors (Yan *et al*, 2000, 2001). Thus GGE is a biplot that display the GGE of multienvironment trial. GGE biplot simultaneously plots information regarding G and GXE interactions. GGE biplot is constructed by plotting the first two principal components (PC1 and PC2) derived from singular value decomposition of the environment-centered data. The singular value decomposition of an environment-centered matrix, say Y produces a principal component analysis of the covariance of the matrix Y. Models that decompose the environment-centered data are referred to as site-regression models or SREG and SRGE with two PCs is termed SREG2. By singular value decomposition, a GXE matrix of mean yield of s genotypes and in t environment can be approximated as the product of a genotype matrix and an environment matrix. Thus the yield of genotype i at environment j, Y_{ij} can be estimated as

$$Y_{ij} = \Sigma \lambda n \varepsilon n \, \eta n \quad (\lambda_i \geq \lambda_2 \geq \lambda_3 \geq \lambda_n)$$

Where n is the number of PCs needed to approximate the original data, with $r \leq \min(s, t)$; λn is the singular value of PCn, the square of which is the sum of squares explained by the PCn. ε_n and η_n are the ith genotype score and the jth environment score, respectively. For PC$_n$ the singular value decomposition allows the GXE interaction table of means to be displayed in a plot having s points for genotypes plus t points for the environments. Each genotype is represented by a point called a marker and is defined by the genotype's scores on all PCs and each environment is represented by a marker defined by the environment's scores on all PCs. Such a plot is termed a biplot as both the genotypes and environment are plotted in a single plot. Biplots can be multidimensional but two-dimensional biplot using only two PCs (the first and the second PC) are most common. In order to achieve symmetric scaling between the genotype scores and the environment scores, the above mentioned equation takes the following from.

$$Y_{ij} = \Sigma \lambda n \, \xi_n \, \eta_n$$

Scaling has the advantages in that PC1 and PC2 have the same units.

The general linear regression model for predicting the mean yield of ith genotype in jth environment takes the following form.

$$Y_{ij} = \mu + \alpha_i + \beta_j + \varphi_{ij}$$

Where μ is the grand, α_i is the main effect of the ith genotype, β_j is the main effect of the jth environment and φ_{ij} is the interaction between ith genotype and jth environment. Deletion of α_i and/ or β_j or all $\mu + \alpha_i + \beta_j$ allows variation to be explainable by the deleted term (s) to be absorbed into the φ_{ij} term. It is the matrix of the φ_{ij} values that is subjected to SVD. Subjecting the φ_{ij} to SVD results in AMMI whereas by deleting the α_i term and then subjecting the φ_{ij} to SVD results in SERG model.

$$\varphi_{ij} = Y_{ij} - \mu - \beta_j = \Sigma \, \xi_n^* \, \eta_n^* - \text{(Equation 2)}$$

Thus equation 2 characterizes a GGE biplot whereas the equations 1 characterizes a GE biplot. The GGE biplot can thus be used for summarizing the effects of genotype and GXE interactions and for identifying the mega-environments. The yield of a particular genotype at a particular location can be approximated by the product of the genotype PC1 score by the location PC1 score. Geometrically, this is the length of the location vector (the absolute distance from the plot origin to the marker of the location) multiplied by the length of the genotype vector (the absolute distance from the plot origin to the marker of the genotype) and by the *cosine* of the angle between them (Kroonenberg, 1995). How much of the G + GXE variability has been capture by GGE biplot axis can be know from the comparison of the partitioning of the total sum of squares (TSS =SS_G + SS_{GXE}) of GXE centered data matrix provided by singular value decomposition and the partitioning of this TSS provided by the analysis of variance technique.

8.6 Non-Parametric Approach

While using parametric techniques (for measuring stability) it is assumed that a quantitative character shows a normal distribution which may not be true for many characters. Also, parametric measures are relatively more sensitive to errors of measurements and addition or deletion of one or few observations causes great variation in the parametric stability measures. Thus, it is worthwhile to go for non-parametric measures for stability. Clustering of genotypes according to their response structure (Genotypes, Environments or both may be grouped to simplify the data, Byth, Eiseman and De Lacy, 1976) is one non-parametric approach and the other is stability measures based on ranks of genotypes in each environment.

8.6.1 Cluster analysis

Clustering allows subsets of genotypes to be identified by the characteristics of the separate groups although not directly in terms of stability. If a genotype well known for its performance and stability is included in the test, it will serve as an example for the other genotypes in the same subset. These genotypes may be regarded as having the overall characteristics of this variety. There are several methods of clustering genotypes (or environments) based on the similarity of response characteristics (Lin et al. 1986) and are summarised in Table 8.10. These methods can be classified into two groups depending upon whether a method uses similarity (or dissimilarity) agglomerative or algorithm (Lance and Williams, 1967) to group objects. Further there are two classes of similarity measure: unicriterion and multicriterion.

Table 8.10 A summary of the similarity coefficients between genotypes i and i'

Similarity measures

1. Unicriterion approach

Euclidean distance

$$d^2(A)_{ii'} = \sum_{j=1}^{t} (Y_{ij} - Y_{i'j})^2 \quad \text{(Hanson 1970); Mungomery et al. (1977) and Johnson (1977))}$$

$$d^2(B)_{ii'} = \sum_{j=1}^{t} [(Y_{ij} - \bar{Y}_j) - (Y_{ij} - \bar{Y}_{i'.})/t \quad \text{(Albou El-Fittouh et al. (1969))}$$

Standardized distance

$$d_s^2(A)_{ii'} = \sum_{j=1}^{t} \left(\frac{Y_{ij} - \bar{Y}_{i.}}{S_i^*} - \frac{Y_{i'j} - Y'_{i.}}{S_{i'}} \right)^2 \quad \text{Fox and Rosielle (1982)}$$

$$d_s^2(B)_{ii'} = \sum_{j=1}^{t} \left(\frac{Y_{ij} - \bar{Y}_{i.} - \bar{Y}_{.j} + \bar{Y}_{..}}{W_i^*} - \frac{Y_{ij} - \bar{Y}_{i'} + \bar{Y}_{.j} + \bar{Y}_{..}}{W_i'} \right)^2 \quad \text{Albou-El Fittouh et al. (1969)}$$

Dissimilarity index

$$D(A)_{ii'} = \frac{SSD^*(V_iV_{i'}) - SSD(V_i) - SSD(V_{i'})}{S_iSSD(V_i)} \quad \text{Lin and Thompson (1975)}$$

$$D(B)_{ii'} = \sum_{j=1}^{t} [(Y_{ij} - \bar{Y}_{i.}) - (Y_{i'j} - Y_{i'.})]/2(t-1)) \quad \text{Lin (1982)}$$

Correlation coefficient

$$r(A)_{ii'} = \frac{\sum_j (Y_{ij} - \bar{Y}_{i.})(Y_{ij} - \bar{Y}_{i.})}{[\sum_j (Y_{ij} - \bar{Y}_{i.})^2 \sum_j (Y'_{ij} - \bar{Y}_{i.})^2]^{1/2}} \quad \text{Guitard (1960)}$$

$$r(B)_{ii'} = \frac{\sum_j (Y_{ij} - \bar{Y}_{i.} - \bar{Y}_{.j} - \bar{Y}_{..})(Y_{ij} - \bar{Y}_{i.} - \bar{Y}_{.j} - \bar{Y}_{..})}{[\sum_j (Y_{ij} - \bar{Y}_{i.} - \bar{Y}_{.j} - \bar{Y}_{..})^2 \sum (Y_{i'j} - \bar{Y}_{i'.} - \bar{Y}_{.j} - \bar{Y}_{..})^2]^{1/2}} \quad \text{Habgood (1977)}$$

Multicriterion approach

Patten distance $d_{ii'}^2 = 2(1 - \cos ii')$ Lefkovitch (1985)

Frechet distance $\psi_{ii'}^2 = (Y_{i.} - Y_{i'.})^2 + (S_i + S_{i'})^2$

$SSD(V_iV_{i'})$ is the sum of squares of deviation from the joint regression, and $SSD(V_i)$ is the sum of squares of deviation from the individual regression.

The unicriterion approach of similarity can be grouped into four: (1) Euclidean Distance (d), (ii) Standardized distance (ds), (iii) Dissimilarity index (D), and (iv) Correlation coefficient (r). Each of these groups is further divided into two different types of index: Index (A), uses both genetic effects and genotype × environment interaction for measuring similarity and Index (B) indicates that the similarity is based on $g \times e$ interaction alone.

Hanson's (1970) parameter, the Euclidean distance originally defined as genotypic stability and correlation coefficients of Guitard (1960) and Habgood (1977) have not been used for clustering but genotypic pattern can be identified by inspection of the distance matrices.

Johnson's (1977) model combines both regression and multiplicative effects but because of small residuals, the distance approximately equals $d(A)_{ii'}$. Thompson's (1975) dissimilarity index $D(A)_{ii'}$ is based on the test statistics obtained from a joint regression analysis of a pair of genotypes. If the joint regression does not hold, Lin's (1982) method which assumes no specific model for $g \times e$ interaction, can be used.

Lefkovitch's (1980, 1982, 1985) multicriterion conditional clustering procedure uses a cluster algorithm that permits more than one measure of pair-wise relationship. His measures of dissimilarity of genotypes are the mean overall environments ($\overline{Y}_{i\cdot}$), the variance across environments (S_i^2) and among environments pattern distance ($d_{ii'}$). The procedure can be extended to use any additional measure of dissimilarity. The genotypes can be grouped using all measures either sequentially or simultaneously to decide on group homogeneity and thus this procedure provides an analytical means to investigate each measure separately.

The relationship among different similarity measures can be seen by expressing the indices (except $D(A)_{ii}$) in terms of $(S_i - S_i')^2$, $r(A)_{ii'}$ $[= \cos \theta_{ii'}]$ and $(Y_{i\cdot} - Y_{i'\cdot})$ as follows:

$$d^2(A)_{ii'} = t d^2(B)_{ii} + t(Y_{i\cdot} - Y_{i'\cdot})^2 \quad \text{(i)}$$

$$d^2(B)_{ii'} = [(S_i - S_{i'})^2 + 2(1 - r(A)_{ii}) S_i S_{i'}]$$
$$\times (t-1)/t \quad \text{(ii)}$$

$$d_s^2(A)_{i'i'} = 2(t-1)[1 - r(A)_{ii'}] \quad \text{(iii)}$$

$$d_s^2(B)_{ii'} = 2[1 - r(B)_{ii'}] \quad \text{(iv)}$$

Also

$$D(B)_{ii'} = d^2(B)_{ii'}/[2(t-1)/t] \quad \text{(v)}$$

and $\quad d_{ii'}^2 = d_s^2(A)_{ii'}/(t-1) \quad \text{(vi)}$

Lefkovitch's (1985) three measures of dissimilarity are combined in $d^2(A)_{ii}$ as shown by equations (i) and (ii). Equations (iii) and (iv) show that the standarized distance and the simple correlation coefficients are equivalent. Equation (v) shows that Lin's (1982) dissimilarity index is equivalent to $d^2(B)_{ii'}$ and equation (vi) shows that Lefkovitch's (1985) pattern distance is equivalent to $ds(A)_{ii'}$.

Furthermore, multicriterion approach of Letkovitch uses two criteria S_i^2 and $r(A)_{ii}$ instead of criterion, the $g \times e$ interaction mean squares, by $D(B)_{ii'}$. The mean and the $g \times e$ interaction mean squares are combined into one coefficient in $d^2(A)_{ii'}$ and $D(A)_{ii'}$ whereas they are kept separate in $D(B)_{ii'}$ and Lefkovitch's decomposition. Finally, the meaning of similarity differs in the following ways: for $d^2(A)_{ii'}$ and $D(A)_{ii}$ it means a complete equality of genotypic responses across environments; for $d^2(B)$ and $D(B)_{ii'}$ it means equality of all within location differences; and for $d_s^2(A)_{ii'}$, $r(A)_{ii'}$ and $r(B)_{ii'}$ it means the equality of all within location ratios. Problems of arbitrary choice of distance measures can only be resolved by trying several clustering procedures and if meaningful clusters exist, they should emerge on most procedures,

8.6.2 Ranks of genotypes

Several non-parametric methods proposed by Hühn (1979) are based on ranks of genotypes in each environment and use the concept of homeostasis as a measure of stability. Hühn (1987) used two statistics, the mean of the absolute rank differences of a genotype over the t environments and the variance among the ranks on t environments to measure the phenotypic stability and applied it to the data on 20 genotypes of winter wheat in 10 environments (locations). The general linear model of $g \times e$ interaction is reduced to

$$Y_{ijk} = \mu + d_i + e_{ijk}$$

if genotypes are ranked separately in each environment and thus environmental effects will have no influence on stability as defined by means and variances of the ranks which are given below:

$$S_i^{(1)} = 2 \sum_{j=1}^{t-1} \sum_{j'=j+1}^{t} |r_{ij} - r_{ij'}|/[t(t-1)]$$

and

$$S_i^{(2)} = \sum_{j=1}^{t} (r_{ij} - \overline{r}_{i\cdot})^2/(t-1)$$

where r_{ij} is the rank of the ith genotype in jth environment,

K is the number of genotypes ($i = 1 \ldots K$),

N is the number of environments ($j = 1 \ldots N$),

$$\overline{r}_{i\cdot} = \sum_{j=1}^{N} r_{ij}/t$$

For a genotype with maximum stability $S_i^{(1)} = 0$ and also zero variance indicates maximum stability (Figure 8.6). To test the null hypothesis that all genotypes have the same phenotypic stability, the following statistic is computed.

	Low yield, $b > 1.0$	High yield, $b > 1.0$
Mean $S_1^{(1)}$ Or variance $S_2^{(2)}$	3 Poorly adapted to all environments	2 Well adapted to favourable environments
	Low yield, $b < 1.0$	High yield, $b = 1.0$ (stable)
	4 Specific adaptation to poor environments	1 Well adapted to all environments

Mean yield

Fig. 8.6 Showing plotting of mean yield vs $S_1^{(1)}$ and vs $S_2^{(2)}$

$$S^m = \sum_{i=1}^{K} Z_i^{(m)} = \sum_{i=1}^{K} [S_i^{(m)} - [ES_i^m]]^2 / \text{var.}(S_i^{(m)}), \quad m = 1, 2,$$

which has approximately a chi-squared distribution with K degrees of freedom. This global test has a Type I error probability equal to the choosen level of significance. If the interest lies in knowing the stability of a particular genotype the Z_i statistic is computed which is

$$Z_i^{(m)} = (S_i^{(m)} - E[S_i^{(m)}])^2 / \text{Var}(S_i^{(m)}), \quad m = 1, 2,$$

and again it has approximately a chi-squared distribution with 1 degree of freedom but here the Type I error is with much higher probability. Testing of stability differences among genotypes is done using standard procedures of multiple comparison among observed $S^{(m)}$ values. The problem with the non-parametric methods is the occurrence of ties whereas the advantage of the non-parametric approach is that a genotype's response characteristics can be assessed qualitatively.

Even when $S_1^{(1)}$, $S_2^{(2)}$ and ranks are to be taken into account this method has got no superiority over linear regression analysis. There are problems with ranks. $S_1^{(1)}$ and $S_2^{(2)}$ are correlated. The correlations are medium to low. Considering the plots yield vs $S_1^{(1)}$ or $S_2^{(2)}$ different genotypes are selected and so no use of using ranks.

8.6.3 Study of environmental factors

We saw earlier that the environment can be classified into micro- and macro-environments. Also, the environment is made up of a large number of controllable and uncontrollable factors. When we raise a group of genotypes in a number of environments, the genotype × environment interactions will comprise genotype × controllable environmental factors effects plus genotype × uncontrollable environmental factor effects. The experiment should be designed in such a way as to provide information about the contribution of each of the environmental factors separately and also contribution of factors in factorial combinations. Several authors have used simple regression analysis to analyse the effect on yield of each individual factor in turn (Jones, 1979 and Beckette, 1982). Regression analyses provide satisfactory results only if one or few factors are showing large effects on yield and are independent in action but are no longer appropriate if several factors are having equally large effects or when several of them are inter correlated. In this case the obvious choice is to go for multiple regression analysis. Dowker's (1963) regression of yield on rainfall for two varieties of maize at four planting densities showed that the regression coefficients were very different. Various workers including Feyerhem and Paulsen (1982) on wheat, and Haun (1982) on maize have carried out multiple regression analysis using climatic variables with the aim of predicting yield of a crop rather than the yield of a particular genotype. Perkins (1972) found that differences between genotypes could be explained in terms of multiple regression on climatic factors. But how each of the climatic factors independently or jointly affect yield remains to be investigated. The situation gets complicated if the different environmental factors are showing interaction and the response is of quadratic, cubic or higher order polynomial nature. In this situation many factors will be needed to examine the adequacy of the fit of regression which is not a problem as the breeder is conducting trial over several seasons in different locations. Here we can also apply the various multivariate techniques described to find out an environmental factor or set of environmental factors which explains most of the genotype × environment interaction. Also, environments can be clustered using various clustering methods and each cluster of environments can be further characterized. By cluster analysis we are partitioning the heterogeneous area into homogeneous ones. Cluster analysis can be used to classify locations to minimize the within cluster genotype × environment interaction and thereby minimize the difference between clusters so as to obtain a number of different regions which could be evaluated separately in breeding. Principal component analysis can be used to assess regional and temporal variation.

However, large interactions of genotype with locations could still exist in a subregion and moreover, the genotype × year interaction can't be reduced by a subdivision of a cropping area. Shukla (1972b) found that three of his varieties were particularly affected by variations in the environmental variables but for his material there was no clear advantage of canonical analysis over regression analysis on environmental variables. However, Wood (1972) found a more logical explanation of the variation in the growth of grasses from different sites using canonical analysis than was obtained by any other technique.

It will not be out of context if we stress the point that we must be aware of the fact that biotic and abiotic stress which can vary over locations and seasons during the critical phages of crop growth, will affect the physiology and thus will be reflected in the differential response of genotypes.

We thus see that genotype × environment interaction may arise from combination of many factors simultaneously, uncontrollably and in unknown proportions. By inclusion of all of the many physical and biological factors in all combinations that a genotype might subsequently encounter the genotype × environment study would become unmanageable. Thus the aim would be, therefore, to reduce the number to manageable proportion by identifying those environmental factors which are critical in that they have major or limiting influence on performance or in other words to identify those factors which account for the greater amount of variation between genotypes. It would be worthwhile to subdivide the reduced number of environmental variables into those controllable and uncontrollable factors. While classifying a factor as uncontrollable or controllable one must not only see whether it is practical and possible to control a factor on commercial as well as experimental scale but also work out the economics of controlling that particular factor. Now for each case an optimum genotype can defined. In case of genotypes showing genotype × uncontrollable environmental factor interactions, an optimum genotype is one which displays maximum in-difference or minimum sensitivity to either variation and for genotypes showing genotype × controllable environmental factor interactions, an optimum genotype is one which displays maximum responsiveness or minimum sensitivity to their variation. If the two optima are under independent genetic control, they could be combined to give an optimum genotype whose performance would be reliable and predictable.

8.7 Genetical Analysis of Genotype × Environment Interactions

The study of genotype × environment interaction at the genetical level will aim at as follows:

1. The determination of the extent of the genetical control of genotype × environment interaction.

2. The identification of the kinds of gene action and interaction involved and to quantify their relative contributions in order to make prediction across generations.

3. The determination of the extent to which the same gene loci control mean performance and environmental sensitivity for the same and different characteristics of the phenotype.

4. The determination of the extent to which the same gene loci control the environmental sensitivities of the different characters in response to the same environmental variable and of the same characters to different environmental variables.

5. The translation of this information into formulae which predict the expected sensitivity of any character to any variable in any generation and the response and correlated response to selection for a particular level of sensitivity of any character to any

environmental variable. In this approach biometrical model fitting is done to specify the contribution of genetic, environmental and genotype × environment interaction to means and variances in terms of various parameters (Mather and Jones, 1958; Jones and Mather 1958; Jinks and Stevens, 1959).

8.7.1 Generations derived from cross of two pure breeding lines

1. Means

In the presence of genotype × environment interaction the expectation of means of two pure breeding lines and generations derived from their cross in the *j*th environment assuming additive-dominance model will be as follows (Bucio Alanis, 1966; Bucio Alanis and Hill, 1966).

$$\overline{P}_{1j} = m + [d] + e_j + g_{dj}$$

$$\overline{P}_{2j} = m - [d] + e_j - g_{dj}$$

$$\overline{F}_{1j} = m + [h] + e_j + g_{hj}$$

where e_j is the additive environmental effect of the *j*th environment (for average environment e_j value will be zero and for poor environment e_j will be negative and for good environment e_j value will be positive). g_{dj} and g_{hj} are the interaction of additive genetic component (d) and dominance component (h) with e_j, respectively and

$$\sum_{j=1}^{t} e_j = 0, \sum_{j=1}^{t} g_{dj} = 0 \text{ and } \sum_{j=1}^{t} g_{hj} = 0$$

In the *j*th environment with the inclusion of other generation (F_2, BC_1, BC_2, etc.) means with P_1, P_2 and F_1, the weighted least squares estimate of m, $[d]$, $[h]$, e_j, g_{dj} and g_{hj} parameters can be obtained and a test of goodness of fit of the model can be conducted (Mather and Jinks, 1971). For *t* environments there will be *t* number of e_j, g_{dj} and g_{hj} and thus the test of adequacy of the model can not be conducted. To overcome this problem of over parameterization the *t* number of means of P_1, P_2, F_1, etc. generations are regressed on the *t* values of $m + e_j$ (= $\overline{P}_{1j} + \overline{P}_{2j}/2$). As m, $[d]$ and $[h]$ are constant in the expectations so infact $e_j + g_{dj}$, $e_j - g_{hj}$, etc. are regressed on e_js. The regression coefficients (b_d, b_h) obtained thus summarise all the effects of *t* values of e_j's, g_d's, g_h's and are as follows:

$$b\overline{P}_{1j} = 1 + b_d$$

$$b\overline{P}_{2j} = 1 - b_d$$

$$b\overline{F}_{1j} = 1 + b_h$$

Thus the genotype × environment components are linear function of the additive environmental components i.e.

$$g_{dj} = b_d e_j$$

$$g_{hj} = b_h e_j$$

The expectation of the above generations in the *j*th environment then takes the following form.

$$\overline{P}_{ij} = m + [d] + (1 + b_d)e_j$$

$$\overline{P}_{2j} = m - [d] + (1 - b_d)e_j$$

$$\overline{F}_{ij} = m + [h] + (1 + b_h)e_j$$

For each particular generation (d) and b_d or $[h]$ and b_h are constant and only e_j will change with the change in environment. Again, after combining other generations means with those of P_1, P_2, and F_1, the weighted least square estimates of m, $[d]$, $[h]$, b_d, b_h and e_j can be obtained and the adequacy of the model can be tested. With the estimates of these parameters the mean of any generation, even if not included in the analysis can be predicted across environments. For example, the expected mean of *n*th generation of inbreeding in the *j*th environment will be

$$\overline{F}_{nj} = m + \beta_n [h] + (1 + \beta_n b_h) e_j$$

where β_h is the frequency of heterozygotes in the *n*th generation.

In the expectations of P_{1j}, P_{2j}, F_{1j} we find that b_d is similar to $[d]$ and b_h is similar to $[h]$ in both signs and coefficients. The reason is that b_d and b_h are the genotype × environment interaction equivalents of $[d]$ and $[h]$, respectively. In other words, they are the additive genetic and dominance components of environmental sensitivity. Thus in the presence of genotype × environment interaction the mean of any generation consists of two parts, namely, its mean performance specified by m, $[d]$ and $[h]$ and its environmental sensitivity which is a linear function of additive environmental effects e_j specified by b_d and b_h. In the presence of non-allelic interaction (i), (j) and (l) components will be added and the corresponding genotype × environment interaction components g_{ij}, g_{jj} and gl_j, will appear and further the regression coefficients b_i, b_j and b_l can be computed. Having known the relative contribution (magnitude as well as direction) of additive, dominance and epistatic effects, its interaction with the environments and the environmental sensitivities, the relative sensitivities of any generation or any population of known genetic constitution can be predicted provided those above parameters are estimated using environments which represent a random sample from the range of environments that a particular crop can encounter.

The genetical study provides us the following information.

1. For many of the characters of economic importance the β's are expected to be positive but negative β's will occur when the character under study is one of resistance to adverse environmental conditions such as disease resistance, drought resistance, etc.

2. When $\beta_d = \beta_h = 0$, $[d]$ and $[h]$ will not change over environments.

3. When $\beta_d \neq \beta_h \neq 0$, the relative values of $[d]$ and $[h]$ components will depend upon the particular environment and the sign of β's.

4. When $\beta_d > \beta_h$, $[d]$ is less stable over environments than is $[h]$, the dominance component.

5. When e_j is positive, the difference between $\overline{P}_1 - \overline{P}_2 = 2\,[d] + 2g_{dj}$ and when e_j is negative $\overline{P}_1 - \overline{P}_2 = 2\,[d] - 2g_{dj}$. The former will always be positive whereas the later can be positive or negative dependng upon the relative magnitudes of (d) and g_{dj}. Thus the difference between two pure breeding lines (P_1, P_2) is more easily detected when the general performance of line is better than the average i.e. when e_j is positive whereas the genotypes are less likely to be recognised when e_j is negative. Clearly, then where the relative positions of the lines change frequently, the different lines get selected for different environments as opposed to the situation where one line is selected for general use.

6. Where all the genotype × environment interactions are accounted for by the linear function of the environmental values (β's), these β's provide a simple measure of the relative abilities of the generation or genotype to increase or reduce their phenotypic expression in different environments. Thus the relative abilities can be assessed as the ratio of the β's, and $\beta P_1/\beta P_2$ represents the ability of the environment to alter the phenotypic expression of P_1 relative to that of P_2.

7. The potence ratio, $\dfrac{h}{d}$ in the jth environment when genotype × environment interaction is present will be estimated as

$$\frac{[h] + g_{hj}}{[d] + g_{dj}}$$

2. Variances

In the presence of genotype × environment interaction the expectations of the variances of P_1, P_2 and F_1 will be as follows:

$$VP_1 = E_1 + GD + 2\,W_{ejgdij}$$

$$VP_2 = E_1 + GD - 2\,W_{ejgdij}$$

$$VF_1 = E_1 + GH + 2\,W_{ejghij}$$

where E_1 is the environmental variance,

$GD = \Sigma gdij^2$, which is the sum of the interaction between additive genetic effect at the ith locus and the jth environment summed over all loci and environments,

and $GH = \Sigma ghij^2$, which is the corresponding component for the dominance genetic effects. W_{ejgdij} is the product of the additive environmental effect of the jth environment and the interaction of the later with the additive genetic affect at the ith locus summed over all loci and environments,

and W_{ejghij} is the corresponding component for the dominance genetic effect.

The genotype × interaction components of the variances of the parental and F_1 families differ more than those of the variance of any other families (Mather and Jins, 1971). Therefore, the parental and F_1 families provide the most sensitive comparison for the detection of genotype × environment interaction (Perkins and Jinks, 1970). The within family variances of P_1, P_2 F_1 unequally differ only in the presence of genotype × environment interaction, the test for which can be done using Bartlett's test of homogeneity of variances. This test holds irrespective of the presence of linkage and epistasis. The epistatic components of P_1, P_2 and F_1 are completely confounded with the additive and dominance components, respectively and their interactions with the environments are similarly confounded.

If there is no genotype × environment interaction i.e. when $GD = GH = g_d = g_h = 0$ and all the terms involving W are zero then $VP_1 = VP_2 = VF_1 = E_1$. In this situation a simple additive-dominance model using parameters D, H, F, and E_1 should be adequate. The failure of this simple model could be due to genotype × environment interaction or epistasis or linkage. We cannot fit a model incorporating epistasis as the number of parameters to be estimated is more than the number of statistics available for their solution.

The dominance ratio in the presence of genotype × environment interaction will be instead

$$\sqrt{\frac{H + GH}{D + GD}}$$

and this will change with the change in the environment. The other parameter such as heritability will also change as the additive and dominance components are confounded with the genotype × environment interaction effect. The estimates of D, H and F may differ over environments in the presence of genotype × environment interaction in one of the two ways. For example, if all kinds of gene actions are equally sensitive to environmental differences, the estimates of D, H and F will change at the same rate from one environment to another. If on other hand, the different kinds of gene actions are not equally sensitive to the environmental differences the estimates of D, H and F will change to different extents over the environments. Comstock pointed out that the relative sensitivities of these estimates depend on the relative magnitudes of the different sorts of genetic variances.

Table 8.11 Genotype × environment interaction in case of triple test cross families

Item	D. F.	E.M.S.
Epistasis (i type)	1	
Epistasis (i) × Environment	$(S - 1) \times 1$	
Epistasis (j, l type)	$n - 1$	
Epistasis (j, l type) × Environment	$(S - 1)(n - 1)$	
Error	$3nS(r - 1)$	
Sums ($\bar{L}_{1i} + \bar{L}_{2i} + \bar{L}_{3i}$)	$n - 1$	$\sigma_{e_1}^2 + 3r\sigma_{Sm}^2 + 3rS\sigma_m^2$
Sums × Environment	$(S - 1)(n - 1)$	$\sigma_{e_1}^2 + 3r\sigma_{Sm}^2$
Error	$3n(r - 1)$	$\sigma_{e_1}^2$
Differences ($\bar{L}_{1i} - \bar{L}_{2i}$)	$n - 1$	$\sigma_{e_2}^2 + 2r\sigma_{Sml}^2 + 2rS\sigma_{ml}^2$
Differences × Environment	$(S - 1)(n - 1)$	$\sigma_{e_2}^2 + 2r\sigma_{Sml}^2$
Error	$2nS(r - 1)$	$\sigma_{e_2}^2$

8.7.2 Multiple mating design

8.7.2.1 Triple test cross

Perkins and Jinks (1970) extended the triple test cross analysis to detect (1) interaction between the progeny genotype and micro-environmental effects and (2) interaction between progeny genotype and macro-environmental effects. Although testing the homogeneity of variances within families and correlation between means and variances are the two methods of detecting genotype × microenvironmental interaction, these are not applicable to triple test cross as the L_1 and L_2 families variances could be heterogeneous and means and variances of these families could be correlated in the absence of interaction. However, sums and differences for means and variances of L_1 and L_2 families are not correlated. The variances of the sums and differences are still expected to be heterogeneous in the absence of genotype × environment interaction but there is now no correlation between $1/2\ (\bar{L}_{1i} + \bar{L}_{2i})$ and $1/2\ (\sigma^2 L_{1i} + \sigma^2 L_{2s})$ and between $1/2\ (\bar{L}_{1i} - \bar{L}_{2i})$ and $1/2\ (\sigma^2 L_{1i} - \sigma^2 L_{2i})$. These covariances have the following expectations in the presence of interaction.

$$\text{Cov}\ \frac{1}{2}\ (\bar{L}_{1i} + \bar{L}_{2i}), \frac{1}{2}\ (\sigma^2 L_{1i} + \sigma^2 L_{2i})$$

$$= \frac{1}{4}\ dgde$$

$$\text{Cov}\ \frac{1}{2}\ (\bar{L}_{1i} - \bar{L}_{2i}) \cdot \frac{1}{2}\ (\sigma^2 L_{1i} - \sigma^2 L_{2i})$$

$$= -(1/8)\ hg^2d + (1/8)\ hg^2h + (1/4)\ hghe$$

Thus significance of either covariance shows the presence of interaction with microenvironment. The covariance sum clearly detects additive effects × micro environmental effects interaction and the covariance difference detects, although not exclusively, interaction with the dominance effects.

Genotype × Macro-environment interaction

When individuals of each of the L_{1i}, L_{2i} and L_{3i} progeny families of a triple test cross, where $i = 1$ to n are raised in each of S environments, the analysis of variance takes the form as given in Table 8.11.

In the absence of epistasis significant sums × environments and differences × environment items show the presence of interactions between the environments and the additive and dominance effects, respectively. Estimates of σ^2_{Sm} and σ^2_{Sml} provide measures of the above interactions which can be compared with the estimates of the additive and dominance components, σ^2_m and σ^2_{ml}. In case of standard triple test cross (the two pure breeding lines and their F_1 making L_1, L_2 and L_3 families with F_2 populations) the expectations of these statistical variances are:

$$\sigma^2_m = \frac{1}{8} D$$

$$\sigma^2_{ml} = \frac{1}{8} H$$

$$\sigma^2_{Sm} = \frac{1}{8} G_2 D$$

$$\sigma^2_{Sml} = \frac{1}{8} G_2 H$$

where $G_2 D = \Sigma g^2 d_{ij}$ and $G_2 H = \Sigma g^2 h_{ij}$ are additive × macro environment interaction and dominance × macro-environment interaction variances, respectively.

8.7.2.2 Diallel cross

If the entries from the diallel table are raised in a replicated experiment in a number of environments, we can test the presence of genotype-environment interaction using (1) the means and variances of the parents and (2) the variances and covariances of the arrays of the diallel table. In addition, the analysis will provide an assessment of the stability of additive, dominance and interaction components under different environmental conditions (Allard, 1956). In the absence of non-allelic interaction consistency of the additive component of variation can be detected unambiguously by the not significant selfs × environment interaction item in the analysis of variance table. In the presence of epistasis, the additive component of selfs is confounded with interaction. The analysis of the variances and covariances of the arrays of the diallel entries takes the form as given in Table 8.12.

Table 8.12 ANOVA of the diallel entries in genotype × environment interaction

Item	D.F.
Environment (Years)	$t-1$
Dominance	1
Dominance × Environment	$t-1$
Arrays	$n-1$
Arrays × Environments	$(t-1)(n-1)$
Dominance × Arrays	$n-1$
Dominance × Arrays × Environments	$2 \times t(n-1)$

The environment (or year) item is calculated from the difference between the sum of all $Wr + Vr$ overall arrays and blocks for each year. This item if significant detects the variation in $\overline{Wr} + \overline{Vr}$ i.e. in $D + H_1 - F$ over years but detects the variation in dominance if Wr and Vr are rescaled on dividing by VP, the variance of the parents in the absence of epistasis. Assuming epistasis this significant item shows the presence of dominance as well as epistasis. The dominance item is calculated from the difference between the sum of Wr and the sum of Vr over all arrays, blocks and all years ($\overline{Wr} - \overline{Vr}$). Assuming no epistasis, this item, if significant, detects the presence of mean dominance. In the presence of epistasis, $Wr - Vr$ item is not exclusively concerned with mean dominance but is also influenced by non-allelic interaction and its significance shows that either dominance is not complete i.e. $Wr - Vr \neq 0$ or that epistasis is involved or both. The significant dominance × environments item tests the consistency of mean dominance effects and/or epistatic effects. The arrays item is calculated from the sum of $\overline{Wr} + \overline{Vr}$ for each array over all blocks and environments. The significant arrays item indicates significant difference in dominance among parents in the absence of non-allelic interaction. When epistasis is present, the significant items shows either significant dominance and/or non-allelic interaction. The arrays × environment item again tests the stability of dominance component of variation. The significant dominance × arrays item indicates the presence of epistasis as we can see that the difference in $Wr - Vr$ for arrays can occur only when there is epistasis. The item environment × dominance × arrays again tests the stability of the non-allelic interaction over environments. Finally, when $h = d$ and i is positive i.e. with recessive epistasis or complementary genes, the ratio of $(\overline{Wr} - \overline{Vr})/(Wr - Vr)$ will be small but this ratio will be large when i is negative i.e. with dominant epistasis or duplicate gene action.

If the diallel material is raised in a number of environments over years then σ_G^2 or σ_S^2 the variance due to gca or sca at a single location becomes

$$\sigma_{gca}^2 = \sigma_g^2 + \sigma_{gl}^2 + \sigma_{gy}^2 + \sigma_{gly}^2$$

$$\sigma_{sca}^2 = \sigma_s^2 + \sigma_{sl}^2 + \sigma_{sy}^2 + \sigma_{sly}^2$$

where σ_g^2 or σ_s^2 is the variance due to gca or sca,

σ_{gl}^2 or σ_{sl}^2 is the interaction variance of gca or sca with location,

σ_{gy}^2 or σ_{sy}^2 as the interaction variance of gca or sca with year,

and σ_{gly}^2 or σ_{sly}^2 is the interaction variance of gca or sca with year and location.

In such experiments one must see whether there is any proportionality i.e.

$$\sigma_S^2/\sigma_G^2 = \sigma_s^2/\sigma_g^2 = \sigma_{sl}^2/\sigma_{gl}^2 = \sigma_{gy}^2/\sigma_{gy}^2 \text{ etc.}$$

and if this were true then the dominance ratio could be estimated much more reliably using estimates of σ_S^2/σ_G^2 rather than σ_s^2/σ_g^2. If pooled estimate of σ_G^2 is lower than estimate from a single environment this indicates large biases from interaction of general combining ability with years and locations. When pooled, σ_S^2 is greater than σ_S^2 at a single location. It shows that the specific combining ability is not interacting with location or year or both although on the average it does. The ANOVA in case of combining ability analysis over locations and years takes the form given in Table 8.13.

When the environment is not specified, the interaction of general combining ability and specific combining ability with the environment can be estimated from the ANOVA given in Table 8.14.

Table 8.13 Interactions of parents, F_1's, general combining abillity and specific combining abillity with environments in diallel analysis

Item	d.f
Replications	$r-1$
Locations	$(l-1)$
Years	$(y-1)$
Entries	$n(n+1)/2 - 1$
Parents	$n-1$
Crosses	$n(n-1)/2 - 1$
Parents vs crosses	1
GCA	$n-1$
SCA	$n(n-3)/2$
Entries × location	$((n(n+1)/2) - 1)(l-1)$
Entries × Year	$((n(n+1)/2) - 1)(y-1)$
Entries × location × year	$((n(n+1)/2) - 1)(l-1)(y-1)$
Parents × locations	$(n-1)(l-1)$
Parents × Years	$(n-1)(y-1)$
Parents × Years × location	$(n-1)(l-1)(y-1)$
Crosses × locations	$(((n(n-1)/2) - 1) - 1)(l-1)$
Crosses × Years	$((n(n-1)/2) - 1)(y-1)$
Crosses × Years location	$((n(n-1)/2) - 1)(l-1)(y-1)$
Parents vs crosses × locations	$l-1$
Parents vs crosses × years	$y-1$
Parents vs crosses × locations × Years	$(l-1)(y-1)$
GCA × location	$(n-1)(l-1)$
GCA × years	$(n-1)(y-1)$
GCA × location × years	$(n-1)(l-1)(y-1)$
SCA × location	$(n(n+3)/2)(l-1)$
SCA × years	$(n(n+3)/2)(y-1)$
SCA × locations × years	$(n(n+3)/2)(l-1)(y-1)$
Pooled error	$ly((r-1)((n(n+1)/2) - 1)$

The various sum of squares are calculated as

$$SS_t = \frac{2\sum_K X..k}{n(n+1)} - \frac{2X^2...}{n(n+1)t}$$

$$SS_{gl} = \frac{\sum_k \sum_i (X_{i,k} + X_{iik})^2}{n+2} - \frac{4\sum_k X^2..k}{(n(n+2))}$$

$$- \frac{\sum_i (Xi.. + Xii.)^2}{(n+2)t} + \frac{4X^2..}{(n(n+2)t}$$

$$SS_{sl} = \sum_k \sum_{i \le j} \sum X_{ijk}^2 - \frac{\sum_k \sum_i (Xi.k + X_{iik})^2}{n+2}$$

$$+ \frac{2\sum_k X^2..k}{(n+1)(n+2)} - \frac{\sum_{i \le j} X_{ij.}^2}{t} +$$

$$\frac{\sum_i (Xi.. + Xii)^2}{(n+2)t} - \frac{2X^2...}{(n+1)(n+2)t}$$

Thus we can see that if diallel experiment is repeated over environments, we can obtain an unbiased estimate of additive genetic variance (σ_A^2). The additive genetic variance estimated from one environment is biased by genotype × environment interaction and σ_A^2 estimates $\sigma_A^2 + \sigma_{A \times E}^2$.

8.7.3 Interaction between epistatic gene action and environment

Jinks et at. (1973) raised a set of 82 pure breeding lines derived from the cross of two pure breeding lines and showed that the occurrence and magnitude of epistatic component depends on the environment. In general it is greatest in one or both extremes of the environmental range and smallest in the average environments. The epistatic component is linearly related to biological or physical measure of environment in some cases. They further concluded that the same character may display two or more even three qualitatively distinct kinds of genetical architecture

Table 8.14 Expectations of mean sum of squares in case of parents and F_1's for modal I and II

Item	d.f.	S.S.	E.M.S. Model I	E.M.S. Model II
General combining ability (GCA)	$n-1$	SS_g	$\sigma_e^2 + \frac{(n+2)}{n-1} t \sum_i g_i^2$	$\sigma_e^2 + \sigma_{sl}^2 + (n+2)\sigma_{gl}^2 + t\sigma_s^2 + (n+2)t\sigma_g^2$
Specific combining ability (SCA)	$n(n-1)/2$	SS_s	$\sigma_e^2 + \frac{2t}{n(n-1)} \sum_i \sum_{\leq j} s_{ij}^2$	$\sigma_e^2 + \sigma_{sl}^2 + t\sigma_s^2$
Environment	$(t-1)$	SS_t	$\sigma_e^2 + \frac{n(n+1)}{2(t-1)} \sum_k t_k^2$	$\sigma_e^2 + \sigma_{sl}^2 + 2(n+1)\sigma_{gl}^2 + \frac{n(n+1)}{2}\sigma_t^2$
GCA × Environments	$(n-1)(t-1)$	SS_{gl}	$\sigma_e^2 + \frac{(n+2)}{(n-1)} \sum_k \sum_i (gl)^2 ik$	$\sigma_e^2 + \sigma_{sl}^2 (n+2)\sigma_{gl}^2$
SCA × Environment	$n(n-1)/2(t-1)$	SS_{sl}		$\sigma_e^2 + \sigma_{sl}^2$
Error	$t(n(n+1)/2 - 1)(r-1)$			σ_e^2

in different portions of the environmental range. These are the genetical architectures normally associated with characters which have been subjected to different types of selection (Mather, 1953; Breese and Mather, 1960; Kearsey and Kojima, 1967; Mather, 1967). For example, in average environments there is little or no evidence for epistasis for most characters and thus display the genetical architecture of a character with an intermediate optimum subject to stabilizing selection. The same characters have significant positive epistatic component in extremely high environments whereas significant negative epistatic component in extremely low environmental range and thus they display the genetical architecture of character whose optimum is towards the extreme of its phenotypic distribution and subject to directional selection.

8.8 Genetic Control of Environmental Sensitivity

8.8.1 Theories of environmental sensitivity

Lerner (1954) found that genotypes differed in their reliability or developmental stability in the same or over different environments and that in general F_1's were the more stable or reliable. He further concluded that the heterozygosity *per se* was the basis of developmental and genetical stability or homeostasis The second theory was proposed by Mather and Jink (1953). They showed that in so far as F_1's were superior, it was because of their gene content and not their heterozygosity *per se* and this explanation held at the macro, micro and intra individual level of environmental variation for a range of phenotypic characters. However, the general finding that mean of a genotype is positively correlated with variance when raised in a range of environments has confused the issue. This positive relationship can be attribute, to scalar or some other cause and it can be removed by transformation or change of scale but not without affecting the interpretations. As genotypes having high mean with low variance or low mean with variance have been selected (we shall see later) and thus relationship found broken or reversed, it shows that the scalar relationship accounts only for a small part of the total variation in environmental sensitivity Many workers have, however, divided the variance of genotype over environments by its mean over environments to get rid of this assumed relationship Thus the resulting quantity called the coefficient of variation, $CV \left(= \sqrt{\dfrac{\text{variance}}{\text{mean}}}\right)$, is thought providing a measure of environmental sensitivity. As F_1 which gives higher mean with low CV, it contributed to belief that homeostasis is a function of heterozygosity.

Comstock also contradicted this belief by demonstrating that if heterozygosity *per se* is involved then the ratio of bias (induced by $g \times e$ interaction to variance estimated) may conceivably be either higher or lower for dominance variance than for the additive genetic variance. He further noted that the inequality of this ratio might have a critical bearing on inference about the level of dominance based on the relative magnitude of estimates of dominance and additive genetic variance.

$$\dfrac{VD_E}{VD} \gtrless \dfrac{VH_E}{VH}$$

Where VD_E is variance due to interaction between additive genetic variance and environments,

VD is additive genetic variance,

VH_E is variance due to interaction between dominance genetic variance and environments,

and VH is dominance genetic variance.

He demonstrated that in case of over-dominance the upward bias in the estimates of dominance variance could be less than in additive genetic variance in a single environment i.e.

$$\dfrac{VDE}{VD} > \dfrac{VHE}{VH}$$

On the other hand, if genes are showing little dominance, the upward bias in the estimate of dominance could be more than in the additive genetic variance i.e.

$$\dfrac{VDE}{VD} < \dfrac{VHE}{VH}$$

However, the chance of misinterpretation as a consequence of disproportion in biases appears less likely than in the above case. Thus we see that the ratio of biases to variance estimated is not necessarily the same for all kinds of genetic variances. Then the interpretation assuming equal bias would underestimate the true level of dominance.

In deciding which is the best genotype, mean performance and environmental sensitivity must be considered together. The two parameters were found to be positively correlated. Jinks (1976) concluded from experiments on genotype × environment interaction that there exists a considerable genetical variation for environmental sensitivity to environmental variables of all kinds and this variation is at least as widespread as variation in mean performance. Genotypes with higher and lower than average environmental sensitivities can be selected in segregating population following hybridization and these selected differences can be fixed. Then there is a positive correlation between mean performance and environmental sensitivity. The positive correlation could be due to either pleiotropy or linkage. If pleiotropy is the possible cause of positive correlation then this relationship is difficult to break and this would lead to a scalar relationship. If linkage disequilibrium is the cause i.e. the relationship is due to past selection leading to an association between completely or partially independent genetical system controlling mean performance and environmental sensitivity, then this relationship is partially or completely breakable by segregation and recombination following hybridization between contrasting genotypes. These two phenomena are practically indistinguishable when the two functionally independent genetical systems are tightly linked. Mather and Caligari (1987) showed from the study of chromosome assay of genotypes with contrasting sensitivities that the genes controlling mean performance and those controlling sensitivity in respect of two environmental variables, temperature and density, were located on different chromosomes which demonstrates that independent genetical systems are determining mean and environmental sensitivity. Results from experiments with *Nicotiana rustica* and *Schizophyllum commune* have suggested that mean performance and environmental sensitivity are under partial independent genetical control (Jinks and Conolly 1973–75). Crosses between varieties in which mean performance and environmental sensitivity were correlated, yielded pure lines showing every combination of high and low mean performance with high and low environmental sensitivity. In this attempt single seed descent method proved useful and segregants were found in F_7 population. Thus we can adopt a similar strategy to produce a variety with a desired level of environmental sensitivity as we currently do for breeding for higher mean performance.

8.8.2 General or specific nature of genotypes involvement in environmental sensitivity

Chromosome assay in Drosophila provided the direct evidence that different genetical systems were involved in the sensitivity to different environmental factors such as temperature, density, etc. Assays of the performance of inbred lines, F_1's and subsequen generations in environments comprising all 16 combinations of 4 major nutrients N, P, K and Ca, plant density, temperature and seasons, sowing date confirmed the earlier finding that high sensitivity to one kind of environmental variable can often be combined in the same genotype with either high or low sensitivity to another kind of environmental variable.

8.8.3 Choice of selection environment and the effect of the chosen environment on the properties of the selected lines

Accepting that the genes controlling mean performance and those controlling environmental sensitivity to the specific environmental variable are independent even to limited extent, the following conclusions can be drawn about the effect on the selection lines obtained from the environment chosen for carrying out selection.

1. We have earlier noticed that the greater discrimination between genotypes is possible in better environment. This is true only if the narrow heritability does not fall inversely proportional to the phenotypic variance. Thus selection would be more successful when there is greater discrimination among genotypes. Also we know that the mean performance and environmental sensitivity are generally positively correlated so phenotypic variance will be greater in better environment i.e. environment leading to a higher mean performance. Also genotypes show relatively more variation in poorer environments as against better environments which probably occurs because whatever factors limit yield in different environments do not affect all genotypes equally.

2. The environment chosen to carry out selection for mean performance will itself determine in part the environmental sensitivity of the resulting selections.

The following are the findings from the study on selection in *Schizophyllum* raised in a number of environments (Jinks 1976). Different levels of temperature constituted the environment. Low temperature made up the poor environment and high temperature constituted the good environment.

1. Selection for high mean performance in good environment gives high performance combined with higher environmental sensitivity.

2. Selection for high mean performance in a poor environment gives high performance combined with low environmental sensitivity.

3. Selection for low mean performance in a good environment gives low performance with low environmental sensitivity.

4. Selection for low mean performance in a poor environment gives low performance with high environmental sensitivity.

5. Selection for high mean performance in a good environment or for low performance in a poor environment is more sensitive to environmental variation than selection for high mean performance in a poor environment or for a low mean performance in a good environment while in a poor environment selection for high mean performance in a poor environment or selection for low mean performance in a good environment is more sensitive to environmental variation than selection for high mean performance in a good environment or for low mean performance is a poor environment. These findings suggest that there is a genetical component to the environmental sensitivity that is specific to the

environment of selection. This specificity appears to be under control of recessive genes. Epistasis, linkage of interacting genes, reciprocal differences and the interactions contribute more frequently to the variability in mean performance than in environmental sensitivity.

6. Selections made on the basis of mean performance in the two different environments maintain their deviations in the direction of selection over extreme environments much better than selections made at a single environment. Hence to achieve a desirable average performance over a range of environments including an acceptable performance even in the most of these environments, selection must be based on average performance in two or more contrasting environments within this environmental range. Similar results from comparable selection experiments with mice were reported earlier (Falconer and Latyszenwski, 1952). Bateman (1971) concluded that the more widely adapted genotypes will generally be produced by selection in the less favourable environments. In other words, disruptive selection has been proved to be effective for combining yield, adaptation and stability.

References

Allard, R. W. 1956. The analysis of genetic environmental interactions by means of diallel crosses. Genetics, **41**: 305–318.

Baker, R.J. 1969. Genotype-environment interaction in yield of wheat. Can. J. Plant Sci., **49**: 743–75'.

Baker, R.J., 1988. Tests for cross over genotype-environment interactions. Cand. J. Plant. Sci., **68**: 405–410.

Bancroft, T.A. 1964. Analysis and inference for incompletely specified models involving the use of preliminary test (s) of significance. Biometrics, **20**: 427–442.

Basford, K., Kroonenburg, P.M.; De Lacy, I.H. and Lawrence, P.K 1990. Multi-attribute evaluation for regional cotton trials. Theor, Appl. Genet., **79**: 225–234.

Becker, H.C. 1981. Correlations among some statistical measures of phenotypic stability. Euphytica, **30**: 835–840.

Becker, R.J. and Leon, J. 1988. Stability analysis in plant breeding. Plant Breeding, **101**: 1–23.

Beckett, J.L. 1982. Variety x environment interactions in sugarbeet variety trials. J. Agric. Sci., **98**: 425–435.

Boerwinkle E. and Hallman, D.M. 1993. Genotype-by-environment interaction. It is a face of life. In: Genetics of Cellular, Individual, Family and Population Variability, Sing CF, Hanis, C.L. (eds.) Oxford, Oxford University Press.

Breese, E.L. 1969. The measurement and significance of genotype-environment interactions in grasses, Heredity, **24**: 27–44.

Breese, E.L. and Mather, K. 1960. The organisation of polygenic activity within chromosome in Drosophila, 2. Variability. Heredity, **14**: 375–400.

Brennan, P.S., Byth, D.E., Drake, D.W., Delacy, LH. and Butler, D.G. 1981. Determination of the location and number of test environments for a wheat cultivar evaluation program. Aust. J. Agric. Res., **32**: 189–201.

Brown, K.D., Sorrells, M.E. and Coffman, W.R. 1983. A method of classification and testing of environments. Crop Sci., **23**: 889–893.

Bucio Alanis, L. 1966. Environmental and genotype-environmental components of variability. I. Inbred lines. Heredity, **21**: 387–397.

Bucio Alanis, L. and Hill, J. 1966. Environmental and genotype-environmental components of variability, 2. Heterozygotes. Heredity, **21**: 399–403.

Byth, D.E., Eisemann, R.L. and De Lacy, 1.D. 1976. Two way pattern analysis of a large data set to evaluate genotypic adaptation. Heredity, **27**: 215–230.

Ceccarr"i, S. 1989. Wide adaptation. How wide? Euphytica. **40**: 197–205.

Ceccarelli, S. and Grano, S. 1991a. Selection environment and environmental sensitivity in barley. Euphytica, **57**: 157–167.

Ceccarelli, S., Nachit, M.M., Ferrara, G.O., Mekni, M.S., Tahir, M., Van Leur, J. and Srivastava, J.P. 1987. Breeding strategies for improving cereal yield and stability under drought. In: Drought Tolerance in Winter Cereals. Srivastava, J.P., Proceddu, E., Acevedo, E. and Verma. S. (eds). John Wiley. Chichester, pp. 101–104.

Comstock, R.E. and Moll, R.H. 1963. Genotype-environment interactions. In: Statistical Genetics and Plant Breeding. NAS, Washington pp. 164–196.

Connolly, V. and Jinks, J.L. 1975. The genetical architecture of general and specific environmental sensitivity. Heredity, **35**: 249–260.

Cox, D.R. 1984. Interaction. International Statistical Review, **52**: 1–31.

Crossa, J. 1990. Statistical analyses of multi location trials. Adv. Agron. **44**: 55–85.

Crossa, J. and Cornelius, P.L. 1997. Sites regression and shifted multipiicate model clustering of cultivar sites under heterogeneity of error vanance. Crop Sci., **37**: 406–415.

Crossa, J., Fox, P.N., Pfeiffer, W.H., Rajaram, S. and Gauch, H.G. 1991. AMMI adjustment for statistical analysis of an international wheat trial. Theor. Appl. Genet., **81**: 27–37.

Crossa, J., Gauch, H.G. and Zobel, RW. 1990. Additive main effects and multipiicative interaction analysis of two international maize cultivars trials. Crop Sci. **30**: 493–500.

Crossa, J., Gauch, H.G. Jr and Zoble, R.W. 1990. Additive main effects and multiplicative interaction analysis of two international maize cultivar trials. Crop. Sci., **30**: 493-500.

Crossa, J. et al. 1991. AMMI adjustment for statistical analysis of an international wheat yield trial. Theor. Appl. Genet., 81: 27-37.

Daniel, C. and Wood, F.S. 1980. Fitting equations to Data. Second Edition, Wiley, New York.

De Lacy, I.H., Eisemann, R.L. and Cooper, M. 1990. The importance of genotype by environment interaction in regional variety trial. In: Genotype by-Environment Interaction and Plant Breeding, Kang, M.S. (ed.) Louisiana State University, Baton, Rouge, pp. 287–300.

Dowker, B. D. 1963. Rainfall reliability and maize yields in Machakos district. E. Afr. Agric. dor. J., **28**: 134–138.

Eberhart, S.A. and Russell, W.A. 1966. Stability parameters for comparing varieties. Crop Sci., **6**: 36–40.

Eisemann, R.L., Cooper, M. and Woodruff, D.R. 1990. Beyond the analytical methodology-better interpretation and exploitation of genotype-by environment interaction in breeding. In: Genotype-by Environment Interaction and Plant Breeding, Kang, M.S. (ed.) Louisiana State University, Baton Rouge, pp. 108–117.

Falconer, D.S. 1989. Introduction to Quantitative Genetics, Longman, Burnt. Mill.

Falconer, D.S. and Latyszewski, M. 1952. The environment in relation to selection for size in mice. J. Genet., **51**: 67–80.

Feyerherm, A.M. and Paulsen, G.M. 1981. Development of wheat yield prediction model. Agron. J., **73**: 277–282.

Finlay, R.W. and Wilkinson, G.N. 1963. The analysis of adaptibility in a plant breeding programme. Aust. J. Agric. Res., **14**: 742–754.

Fisher, R.A. and Mackenzie, W.A. 1923. Studies in crop variation, 2. The manurial response of di fferent potatoes varieties. J. Agric. Sci., Camb., **13**: 311–320.

Flinn, J.C. and Garrity, D.P. 1989. Yield stability and modern rice technology. In: Variability in Grain Yields. Implications for Agricultural Research and Policy in Developing Countries. Anderson, J.R. and Hazell, P.B.R. (eds.), John Hopkins University Press, Baltimore, pp. 251–264.

Fox, P. . 1982b. Reference sets of genotypes and selection for yield in unpredictable environments. Crop Sci., **22**: 1171–75.

Fox, P.N. and Boyd, W.J.R. 1985. The nature of genotype × environment interactions for wheat yield in Western Australia. Field Crops Res., **11**: 387–398.

Fox, P.N. and Rosiellele, A.A. 1982a. Reducing the influence of environmental main effects on pattern analysis of plant breeding environments. Euphytica, **31**: 645–656.

Fox, P.N., Skovmand, B., Thompson, B.K., Braun, H.J. and Cormier, R. 1990. Yield and adaptation of hexaploid spring triticale. Euphytica, **47**: 57–64.

Freeman, G.H. 1973. Statistical methods for the analysis of genotype environment interactions. Heredity, **31**: 339–354.

Freeman, G.H. and Perkins, J.M. 1971. Environmental and genotype-environmental components of variability. 8. Relations between genotypes grown in different environments. Heredity, **27**: 15–23.

Fripp, Yvonne and Caten, C.E. 1971. Genotype-environment interactions in *Schizophyllum commune*. 1. Analysis and character. Heredity, **27**: 393–407.

Fripp, Yvonne and Caten, CE. 1973. Genotype-environment interactions in *Schizophyllum commune*. 3. The relation between mean expression and sensitivity to change in environment. Heredity, **30**: 441–449.

Gauch, H.G. 1982. Noise reduction by eigen vector ordinations. Ecology, **63**: 1643–1649.

Gauch, J.R. H.G. 1988. Model selection and validation for yield trials with interaction, Biometrics, **44**: 705–715.

Gauch, H.G. 1990. Full and reduced models for yield trials. Theor. Appl. Genet., **80**: 153–160.

Gauch, H.G. and Zobel, R.W. 1988. Predictive and postdictive success of statistical analysis of yield trials. Theor. Appl. Genet., **76**: 1–10.

Goldfeld, S.M. and Quandt, R.E. 1972. Non-linear methods in Econometrics. North Holland Publishing Co., Amsterdam.

Gollob, H.E 1968. A statistical model which combines features of factors analytic and analysis of variance technique. Psychometrike, **33**: 73–116.

Gorden, A.D. 1981. Classification. Chapman and Hall, London

Grafius, J.E. and Kiesling, R.L. 1960. The prediction of relative yields of different oat varieties based on known environmental variables. Agron. J., **52**: 396–399.

Guitard, A.A. 1960. The use of diallel correlations for determining the relative locational performance of varieties of barley. Can. J. Plant Sci., **40**: 645–659.

Habgood, R.M. 1977. Estimation of genetic diversity of self fertilizing cereal cultivars based on genotype-environment interaction. Euphytica, **26**: 485–489.

Hanson. W.D. 1970. Genotypic stability. Theor. Appl. Genet., **40**: 226–231.

Hardwick, R.C. and Wood, J.T. 1972. Regression methods for studying genotype-environment interactions. Heredity, **28**: 209–222.

Huehn, M. 1990. Nonparametric measures of phenotypic stability. Part 1: Theory. Euphytical, 47: 189-194.

Haun, J.R. 1982. Early prediction of corn yields from daily weather data and single predetermined seasonal constants. Agricultural Meteorology, **27**: 191–207.

Hill, J. 1975. Genotype-environment interactions-a challenge for the plant breeding. J. Agric. Sci., Camb. **85**: 477–493.

Hill, R.R. and Baylor, J.J. 1983. Genotype × environment interactions analysis in alfalfa. Crop Sci., **23**: 811–815.

Hühn, M. 1979. Beitrage zur Erfassung der phanotypischen Stabilitat, 1. Vorsechlag einiger auf Ranginformationen beruhenden Stabilitatsparameter, EDP in Medicine and Biology, **10**: 112–117.

Hühn, M. 1990. Non-parametric estimation and testing of genotype environment interactions by ranks. In: plant Genotypes by Environment Interaction and Plant Breeding, Kang, M.S. (ed.), Louisiana State University, Baton Rouge, pp. 69–93.

Jinks, J.L. and Connolly, V. 1973. Selection for specif and general response to environmental differences. Heredity, **30**: 33–40.

Jinks, J.L. and Pooni, H.S. 1979. Non-linear genotype-environment interactions arising from response thresholds, 1. Parents, F_1's and selections. Heredity, **43**: 57–70.

Jinks, J.L. and Stevens, J.M. 1959. The components variation among family means of diallel crosses. Joh Genetics., **44**: 297–308.

Jinks, J.L. and Conlolly, 1975. Determination of environment sensitivity of selection lines by the selection environment. Heredity, 34: 401-406.

Jinks, J.L. and Pooni, H.S. 1982. Determination of environment sensitivity of selction lines of Nicotiana rustica by the selection environment. Heredity, 49: 291-294.

Johnson, G.R. 1977. Analysis of genotypic similarity interms of mean yield and stability of environmental; response in a set of maize hybrids. Crop Sci., **17**: 827 842.

Jones, H.G. 1979. Effects of weather on spring barley yields in Britain. J. of Natl. lnst. of Agric. Bot., **15**: 24–33.

Jones, R.M. and Mather, K. 1958. Interaction of genotype- and environment in continuous variation, 2. Analysis. Biometrics, **14**: 489–498.

Kang, M.S. (ed.) 1990. Genotype by enviroment interaction and plant breeding. Deptt. of Agronomy, Louisiana State Univ. Agric. Centre. Baton Rouge, LA

Kearsey, MJ. and Kojima, K. 1967. The genetic architecture of body weight and egg hatchability in *Drosophila melanogaster*. Genetics, **56**: 23–27.

Kempton, R.A. 1984. The use of biplots in interpreting variety by environment interaction. J. of Agric. Sci., **103**: 113–135.

Knight, R.C. 1970. The measurement and interpretation of genotype-environment interactions. Euphytica, **19**: 225–235.

Lance, G.N. and Williams, W.I. 1967. A general theory of classificationary sorting strategies 1. Hierarchial systems. Computer J., **9**: 373–380.

Lerner, I.M. 1954. Genetic Homeostasis. Oliver and Boyd, Edinburgh

Lefkovitch, L.P. 1980. Conditional clustering. Biornetrics, **36**: 43–48.

Lefkovitch, L.P. 1982. Conditional clusters, musters and probability. Math. Biosci., **60**: 207–234.

Lefkovitch, L.P. 1985. Multicriteria clustering in genotype-environment interaction problems. Theor. Appl. Genet., **70**: 585–589.

Lin, C.S. 1982. Grouping genotypes by a cluster method directly related to genotype-environment interaction mean squares. Theor. Appl. Genet., **62**: 277–280.

Lin, C.S. and Binns, L.P. 1991. Genetic properties of four types of Stability parameters. Theor, Appl. Genet., **82**: 505–509.

Lin, C.S., Binns, M.R. and Lefkovitch, L.P. 1985. Stability ic analysis. Where do we stand? Crop Sci., **26**: 894–900.

Ludlow, M.M. and Muchow, R.C. 1989. A critical evaluation of traits for improving crop yields in water limited environments. Adv. Agron., 43: 107–153.

Magari, R., Kang, M.S. and Zhang, Y. 1997. Genotype by environment interaction for ear moisture loss rate in com. Crop. Sci., 37: 774 – 777.

Mather, K. 1953a. The genetical structure of populations. Symp. Soc. Exp. Biol., 7: 66 – 93.

Mather, K. 1953b. Genetical control of stability in development. Heredity, 7: 297 – 336.

Mather, K. and Morley Jones, R. 1958. Interaction of genotype and environment in continuous variation I. Description. Biometrics, 14: 343-359.

Matzinger, D.F. 1963. Experimental estimates of genetic parameters and their applications in self fertilizing plants. In: Statistical genetics and plant breeding, NAS-NRC 982, pp. 253 – 275.

Mendel, J. 1971. Principal component analysis of variance and data structure. Statist. Neerlandica, 26: 119 – 129.

Menz, K.M. 1980. A comparative analysis of wheat adaptation across international environments using stochastic dominance and pattern analysis, Field Crops Res., 3: 33 – 44.

Nachit, M.M. et al. 1992. Use of AMMI and linear regression models to analyze genotypw-environment interaction in durum wheat. Theor. Appl. Genet., 83: 597-601.

Nassar, R. and Huhn, M. 1987. Studies on estimation of phenotypic stability: Tests of significance for non- parametric measures of phenotypic stability, Biometrics, 43: 45 – 53.

Patel, J.D., Reinbergs, E., Mather, D.E., Choo, T.M. and In Sterling, J .D.E. 1987. Natural selection in a doubled-haploid mixture and a composite cross of barley, Crop Sci., 27: 474–479.

Perkins, J.M. 1972. The principal component analysis of genotype environment interactions and physical measurement of the environment. Heredity, 29: 51–70.

Perkins, J.M. and Jinks, J.L. 1968a. Environmental and genotype environmental components of variability 3. Multiple lines and crosses. Heredity, 23: 339–356.

Perkins, J.M. and Jinks, J.L. 1968b. Environmental and genotype environmental components of variability 4. Non-linear interactions for multiple inbred lines. Heredity, 23: 525–535.

Perkins, J.M. and Jinks, J.L. 1970. Detection and estimation of genotype-environmental, linkage and epistasis components of variation for a metrical trait. Heredity, 25: 157–177.

Perkins ..I.M. and Jinks J.L.. 1971. Analysis of genotype × environment interaction in Triple test cross data. Heredity, 26: 203–209.

Perkins, J.M. and Jinks, J.L. 1973. The assessment and specificity of environmental and genotype-environmental components of variability. Heredity, 30: 111–126.

Peterson, CJ. and Pfeiffer, W.H. 1989. International winter wheat evaluation. Relationship among test sites based on cultivar performance. Crop Sci., 29: 276–282.

Plaisted, R.L. and Peterson, L.C. 1959. A technique for evaluating the ability of selections to yield consistently in different locations or seasons. Am. Potato J., 36: 381–385.

Pooni, H.S. and Jinks, J.L. 1980. Non-linear genotype x environment interaction. Heredity, 45: 389–400.

Pooni, H.S., Jinks, J.L. and Tayasekara, N.E.M. 1978. An investigation of gene action and genotype x environment interactions in two crosses of *Nicotiana rustica* by triple test cross and inbred line analysis. Heredity, 41: 83–92.

Rathjen, AJ. and Pederson, D.G. 1986. Selection for improved grain yields in variable environments. DSIR Plant Breeding Symposium. Special Publication 5, NZ. Agronomy Society

Romagosa, I., Fox, P.N., Garcia del Moral, L.E, Romos, I.M., Garcia de Moral, B., Roca de Togores, E and Molina-Cano, J.L. 1993. Integration of statistical and physiological analyses and adoptation of near isogenic barley lines. Theor. Appl. Genet.,

Roseille, A.A. and Hamblin, 1. 1981. Theoretical aspects of selection for yield in stress and non-stress environments. Crop Sci., 21: 943–946.

Royo, C., Romagosa, I and Rodriguez, A. 1991. Comparative adaptation of triticale and spring wheat in Spain. In: Proc Second lnt. Triticale Symp., CIMMYT, Mexico, D.E, pp. 593–597.

Seal, H. 1964. Multivariate Statistical Analysis for Biologists. Methuen, London.

Shukla, G.K. 1972. Some statistical aspects of partitioning genotype-environmental components of variability. Heredity, 28: 237–245.

Silvey, V. 1982. Analysis of crop variety adaptation from performance trials in England and Wales. In: Proc. XIth Intl. Biometrics Conf., pp. 157–163.

Sprague, G.F. and Federer, WT. 1951. A comparison of variance components in com yield trials. II Error, year × variety, location x variety and variety components. Agron. J. 43: 535–541.

Tai, G.C.C. 1971. Genotype stability analysis and its application to potato regional trials. Crop Sci., **11**: 184–190.

Tuckey, J.W. 1949. One degree of freedom for non-additivity. Biometrics, **5**: 232–242.

Verma, M.M., Chahal *G.S.* and .. urty, B.R. 1978. Limitations of conventional regression analysis-a proposed modification. Theor. App. Genet., **53**: 89–91.

Weisberg, S. 1980. Applied Linear Regression. Wiley, New York.

Westocott, B. 1986. Some methods of analysing genotype-environment interaction. Heredity, **56**: 243–253.

Westocott, B. 1987. A method of assessing the yield stability of crop genotypes J. Agric. Sci., Camb., **108**: 267–274.

Williams, E.J. 1952. The interpretation of interaction in factorial experiments. Biometrics, **39**: 65–81.

Witcombe, J.R. and Whittington, WJ. 1971. A study of the genotype-environment interaction shown by germinating seeds of *Brassica napus*. Heredity, **26**: 397–411.

Wood, J.T. 1972. The multivariate approach to genotype-environment interactions. Paper presented to the Symposium on genotype × environment Interactions, Birmingham University, September 1972.

Wrick, G. 1962. Uber eine Methods zur Erfassung der okologischen streubreite in Feldversuchen. Z. Pflzucht., **47**: 92–96.

Yates. F. and Cochran. WG. 1938. The analysis of groups of experiments. 1. Agric. Sci., Camb., **28**: 556–580.

Yau, S.K. and Hamlin, J. 1994. Relative yield as a measure of entry performance in variable environments. Crop Sci., **34**: 813–817.

Yan, W. et al. 2001. Two types of GGE Biplots for analyzing multi-environment trial data. Crop Sci., 41: 656-663.

Yan, W. et al. 2000. Cultivar evaluation and mega-environment investigation based of GGE Biplot. Crop Sci., 40: 597-605.

Yan, W. and Rajcan, I. 2002. Biplot analysis of test sites and trait relations of soybean in Ontario. Crop sci., 42: 11-20.

Zobel, R.W, Wright, *M.l.* and Gauch, H.G. 1988. Statistical analysis of yield trial. Agron. J., **80**: 388–393.

9

Analysis of Reciprocal Differences

So far we have assumed that there is no reciprocal difference between the crosses, i.e., $\overline{F}_1 = R\overline{F}_1$. The reciprocal differences can arise because of maternal and/or cytoplasmic effects and sex linkage, the later being expressed as functions of gene values and their frequencies.

9.1 Maternal Effects

Once the phenomena, sex linkage and preferential segregations, have been ruled out, the reciprocal differences can arise due to a whole host of possible extra-nuclear maternal effects.

Maternal effects on quantitative characters include

(i) those effects originating from differences in maternal environment provided to the developing seed by the female parent. The differences in maternal environment include differences in nutritional factors contributing to the prenatal and post-natal environments, transmission of either antibodies or pathogens from dam to offspring and common external environments of the sibs determining the maternal behaviour pattern.

(ii) those originating from differences in cytoplasm.

(iii) its interactions with other effects like nuclear genetic effects and physical environmental effects.

The maternal effects can be transient or persistent in nature. The degree of transience observed in reciprocal difference from one generation to the next can be used to infer whether the reciprocal difference is due to mothering ability (maternal environment) and/or effects from physiological interactions between particular seed and pollen parent or due to the effects of cytoplasmic origin (extra-chromosomal factors). Since a reciprocal difference in maternal environment would be conditioned by the genotype of the maternal parent as well as the particular environment in which the female parent was raised, reciprocal difference of same size and magnitude would not be expected to occur in succeeding generations. Several facts suggest that the maternal effects are determined by the maternal genotype, possibly interacting with the cytoplasm rather than cytoplasmic factor *per se*. Such a factor is not self perpetuating and thus disappears unless replaced by the effect of an appropriate nuclear gene. Transient maternal effects which arise because of difference in maternal environment are prevalent in animal kingdom. In plants, transient maternal effects are due to differences especially in nutritional factors, in the environment of the embryo provided by the seed parent. If on the other hand the reciprocal difference is entirely due to cytoplasmic (extra-chromosomal) factors such as plasma genes, plasmids or plasmons, its intensity among progenies with a common cytoplasmic lineary will be the same regardless of the nuclear genetic composition of either parent. The consistency shows evidence both for self perpetuating and independent effect transmission. The degree to which this type of reciprocal difference occurs depends on cytoplasmic gene or other factors autonomous of nuclear genes. Persistent maternal (or paternal) effects arise through unequal contributions of cytoplasmic determinants from the female and male gametes. Male gametes are almost devoid of cytoplasm. Meiotic drive which involves differential segregation of whole chromosome as found in males of Drosophila and mice will contribute to the paternal variance. However, there is an unusual kind of reciprocal difference which while being transient persists over few generations. It affects quantitative characters and its magnitude is environmentally dependent. It has many similarities to the recently reported examples of maternally inherited differences in clonally reproducing plants (Beddows, Breese and Lewis, 1962).

Maternal effects can be separated from the cytoplasmic effects by simple test. Maternal effects can be estimated by comparing means of F_1 seed with those of selfed seed on the same female whereas cytoplasmic effects can be detected by comparing means of F_2 seeds born on F_1 plants from reciprocal crosses. Separation of cytoplasmic effects from effects of maternal genotype requires a crossing scheme in which reciprocal F_1's are crossed to a common male tester but even there the estimation of those effects requires the assumption of no nuclear × cytoplasmic interaction.

Now if the maternal effect is present then the aim would be to separate out genetically determined variation from maternally (or paternally) determined variations. The total variation in the presence of maternal effect may take the form (Wilham, 1963) as

$$\sigma_T^2 = \sigma_A^2 + \sigma_D^2 + \sigma_{Am}^2 + \sigma_{Dm}^2 + \sigma_{A,Am}^2$$
$$+ \sigma_{D,Dm}^2 + \sigma_{me}^2 + \sigma_w^2$$

where σ_{Am}^2 is the variance due to maternal additive effects,

σ_{Dm}^2 is the variance due to maternal dominance effects,

σ_{me}^2 is the variance due to maternal environment,

and σ_w^2 is the residual environmental variation.

Mather and Jinks (1971, 1982) considered non-allelic interaction for maternal effects on the pattern of epistasis for autosomal genes and a form of genotype × environment interaction i.e. maternal effect × progeny genotype or maternal environment × progeny genotype interactions which we shall see later in this chapter.

Reciprocal effects have been found to be insignificant in many species of plants (Cockerham, 1963). Mather and Jinks (1971) showed maternal effects in plants for seedling traits but which diminished with increasing age.

9.2 Estimation of Maternal Effects

9.2.1 Generations derived from the cross of two pure breeding lines

9.2.1.1 Generation means

The maternal effects, determined by the delayed effects of maternal (or paternal) genotypes on its progeny phenotype of the genotype AA, Aa and aa are $+dm$, $+hm$ and $-dm$, respectively, which is over and above the autosomal effects $+d$, $+h$ and $-d$ associated with these genotypes, respectively. Thus the expectations of different generations such as F_2, BC_1 and BC_2 can be worked out assuming additive-dominance model and the adequacy of the model can be tested using χ^2 (Mather and Jinks, 1971, 1982). In the event of additive-dominance model being adequate, significant difference between the reci-procals of F_1, BC_1 and BC_2 generations will test for the presence of reciprocal difference.

$$\overline{F}_1 - \overline{RF}_1 = 2[dm]$$
$$\overline{BC}_1 - \overline{RBC}_1 = [dm] - [hm]$$
$$\overline{BC}_2 - \overline{RBC}_2 = [dm] - [hm]$$

But a model in which the cytoplasmic effects are assumed to be transmitted unchanged by the maternal parent for an indeterminate number of generations, the comparisons $\overline{F}_1 - \overline{RF}_1$ and $\overline{RBC}_2 - \overline{BC}_2$ should be significant and should be equal in magnitude whereas the comparison $\overline{BC}_1 - \overline{RBC}_1$ should not be significant.

If a model with m, $[d]$, $[h]$, $[dm]$ and $[hm]$ parameters fails then the failure could be due to non-allelic interactions. The non-allelic interactions could be for autosomal genes or it could be for maternal effects $im(dm \times dm)$, $jm(dm \times hm)$ and $jm(hm \times$

hm). There could be another type of interaction i.e. maternal effect × progeny genotype. It is a form of genotype × environment interaction i.e. maternal environment × progeny genotype ($d.\ dm$, $h.\ dm$. or $h.\ hm$). If a model with m, $[d]$, $[h]$ $[dm]$, $[hm]$ and $[i]$, $[j]$ and $[l]$ parameters fails then the failure can be attributed to linkage of interacting genes or the model is inadequate for reciprocal differences. Analysis of P_1, F_1 and BC_2 families with P_1 as mother or analysis of P_2, F_1 and BC_2 families with P_2 as mother will indicate whether or not there is need of going for interaction of maternal effects model or inclusion of progeny genotype × maternal effect in the model.

When $\overline{F}_2 > \overline{F}_1$, it is a manifestation of maternal effects or when $\overline{RBC}_1\ (F_1 \times P_1) > \overline{BC}_1\ (P_1 \times F_1)$, it shows superiority of F_1 as mother. The superiority of F_2 is a property both of the progeny's own genotype and the genotype of the F_1 mother. To take into account the effect due to genotype of F_1 mother, f parameter can be added in the model as shown in the expectations of means of basic generations given below:

Generations	Expectations
P_1	$m + [d]$
P_2	$m - [d]$
F_1	$m + [h]$
F_2	$m + \frac{1}{2}[h] + [f]$
$P_1 \times F_1$	$m + \frac{1}{2}[d] + \frac{1}{2}[h]$
$F_1 \times P_1$	$m + \frac{1}{2}[d] + \frac{1}{2}[h] + [f]$
$P_2 \times F_2$	$m - \frac{1}{2}[d] + \frac{1}{2}[h]$
$F_1 \times P_2$	$m - \frac{1}{2}[d] + [h] + [f]$

When the objective is to separate out the effect of progeny's own genotype from those of maternal genotype, the expectations of the means of different generations will take the form (Barnes, 1968) as:

Generations	Expected mean
$P_1 \times P_2$	$m + [dp] + [dm]$
$P_2 \times P_2$	$m - [dp] - [dm]$
$P_1 \times P_2$	$m + [hp] + [dm]$
$P_2 \times P_1$	$m + [hp] - [dm]$
$F_1 \times P_1$	$m + \frac{1}{2}[dp] + \frac{1}{2}[hp] + [hm]$
$F_1 \times P_2$	$m - \frac{1}{2}[dp] + \frac{1}{2}[hp] + [hm]$
$F_1 \times F_1$	$m + \frac{1}{2}[hp] + \frac{1}{2}[hm]$
$F_2 \times P_1$	$m + \frac{1}{2}[dp] + \frac{1}{2}[hp] + \frac{1}{2}[hm]$
$F_2 + \times P_2$	$m - \frac{1}{2}[dp] + \frac{1}{2}[hp] + \frac{1}{2}[hm]$
$F_2 \times F_1$	$m + \frac{1}{2}[hp] + \frac{1}{2}[hm]$
$F_2 \times F_2$	$m + \frac{1}{2}[hp] + \frac{1}{2}[hm]$
$BC_1 \times P_1$	$m + \frac{3}{4}[dp] + \frac{1}{4}[hp] + \frac{1}{2}[dm] + \frac{1}{2}[hm]$
$BC_2 \times P_2$	$m - \frac{3}{4}[dp] + \frac{1}{4}[hp] - \frac{1}{2}[dm] + \frac{1}{2}[hm]$

where $[dp]$ = additive component of progeny's own genotype,

$[hp]$ = dominance component of progeny's own genotype,

$[dm]$ = additive component of maternal genotype.

and $[hm]$ = dominance component of maternal genotype.

The use of these models in explaining the hybrid vigor will be discussed in the chapter on heterosis and inbreeding depression.

9.2.1.2 Generation variances

As far as variances of different generations are concerned, the maternal effect is constant in all the members within families as they come from the same mother and so they do not contribute to within families variances. But such is not the situation when F_3 and other advancing generations are concerned as can be seen in Table 9.1

The F_3 families will have three kinds (AA, Aa, aa) of F_2 mother. Assuming additive dominance model

$$\sigma_w^2 F_3 = \frac{1}{4} d^2 + \frac{1}{8} h^2$$

$$\sigma_b^2 F_3 = \frac{1}{4}(d + dm)^2 + \frac{1}{2}\left(\frac{1}{2} h + hm\right)^2$$

9.4 Biometrical Genetics–Analysis of Quantitative Variation

Table 9.1 Means and variances of F_3 families in case of maternal effects

F_2 mother		AA	Aa	aa	Mean
		$+ dm$	$+ hm$	$- dm$	$\frac{1}{2}(h + hm)$
F_3 family		AA	AA Aa aa	aa	
	Mean	$+ d + dm$	$\frac{1}{2}h + hm$	$- d - dm$	$\frac{1}{2}\left(\frac{1}{2}h + hm\right)$
	Variance	0	$\frac{1}{2}d^2 + \frac{1}{4}h^2$	0	

$$+ \frac{1}{4}(-d - dm)^2 - \left(\frac{1}{4}h + \frac{1}{2}hm\right)^2$$

$$= \frac{1}{2}(d + dm)^2 + \frac{1}{4}\left(\frac{1}{2}h + hm\right)^2$$

Thus we see that additive genetic and dominance genetic variance are inflated in the presence of maternal effects.

When the simple additive dominance model fails and there is progeny genotype × maternal effect interaction then the within family variances of P_1, P_2 and F_1 generations are not affected but within F_2 family variances and both between and within family variances of F_3 and other advancing generations are affected as show in Table 9.2.

The within F_2 family variance ($\sigma^2_w F_2$) is obtained as

$$\sigma^2_w F_2 = \frac{1}{4}(d + d \cdot hm)^2 + \frac{1}{2}(h + h \cdot m)^2$$

$$+ \frac{1}{4}(-d - dhm)^2$$

$$= \frac{1}{2}(d + dm)^2 + \frac{1}{4}(h + h \cdot hm)^2$$

Further it can be shown that both within F_3 family variance ($\sigma^2_w F_3$) and between F_3 family variance ($\sigma^2_w F_3$) are affected in the presence of progeny genotype × maternal effect interaction.

9.2.2 Multiple mating designs

Considering one locus with 2 alleles (A, a) and with u and v being the frequencies of alleles A and a, respectively, the three genotypes (AA, Aa and aa) in a random mating population, their frequencies and mean values will be as follows:

Genotype	AA	Aa	aa
Frequency	u^2	$2uv$	v^2
Mean value	$+ dm$	$+ hm$	$- dm$

The total maternal variation in a diallel set of crosses in a random mating population (Mather and Jinks, (1971) is

$$\frac{1}{2}D_{Rm} + \frac{1}{4}H_{Rm}$$

Table 9.2 Means and variances of F_2 and F_3 families in case of progeny genotype × maternal effect interaction

	AA	Aa	aa	Mean
F_2 genotype	AA	Aa	aa	
(F_1 mother)	$d + d \cdot hm$	$h + h \cdot hm$	$- d - d, hm$	$\frac{1}{2}(h + h \cdot hm)$
F_3 family	AA	AA Aa aa	aa	
	$d + d, \frac{1}{2}hm$	$\frac{1}{2}h + d \cdot \frac{1}{2}hm$	$- d - d \cdot \frac{1}{2}hm$	

where $D_{Rm} = \Sigma\, 4_{uv}\, [dm + (u-v)\, hm]^2$

$H_{Rm} = \Sigma\, 16u^2v^2\, hm^2$

The variance between maternal arrays = $\frac{1}{2} D_{Rm} + \frac{1}{4} H_{Rm}$

The variance between parental arrays = 0.

The difference between the variances of maternal array means and paternal array means = $\frac{1}{2} D_{Rm} + \frac{1}{4} H_{Rm}$ which can be attributed to maternal effects. Thus there is no interaction between maternal and paternal array. The maternal effect estimated as the difference between the male and the female arrays raises a problem (Topham, 1966) in that it is an interaction term.

Although $W_r - V_r$ analysis is developed for the analysis of summed reciprocals, reciprocal differences can be examined. If there is a pronounced difference between the reciprocal crosses of two parents, examination of $W_r - V_r$ graph of the reciprocal means and substitution of each reciprocal in turn in both reciprocal cells will provide information on gene/cytoplasm interaction.

In the *NCII* design, the maternal effect, $\frac{1}{2} D_{Rm} + \frac{1}{4} H_{Rm}$ is confounded with mean squares due to dam and thus if this mean squares is greater than mean squares due to sire, it shows the presence of maternal effects. Thus additive variance (D), dominance variance (H) and maternal effects can be estimated. In the presence of maternal effect by progeny genotype interaction, $\sigma^2_{m \times f}$ will contain this interaction in addition to dominance (or σ^2_{sca} = variance due to *sca*) and thus will inflate *HR*. The mean squares due to maternal as well as paternal parents will also contain maternal effects × progeny genotype interaction effects. Thus it is not possible to test the adequacy of the additive-dominance model in this type of experiment. The genetic variance is tested by paternal mean squares/maternal × parental mean squares and the genetic interaction by maternal paternal mean squares/remainder error. In NCI the mean squares due to dam within sire is confounded (with maternal effect, see chapter 6) and thus dominance variance cannot be separated and is thus inflated whereas mean squares due to sire will provide estimate of additive genetic variance. The maternal effects in case of BIPs is confounded with between families variance.

9.3 Detection of Epistasis and Linkage of Epistatic Genes in the Presence of Reciprocal Differences

In the presence of reciprocal differences the following scaling tests, a modified form of Mather and Jinks (1971, 1982), can be used to detect epistasis.

Scaling test	Comparison
A''	$2\overline{BC}_1\,(P_1 \times F_1) - \overline{F}_1\,(P_1 \times P_2) - \overline{P}_1 = 0$
B''	$2\overline{BC}_2\,(P_2 \times F_1) - R\overline{F}_1\,(P_2 \times P_1) - \overline{P}_2 = 0$
C''	$2\overline{F}_2\,(F_1 \times F_1) - R\overline{BC}_1\,(F_1 \times P_1)$ $- 2\overline{BC}_2\,(F_1 \times P_2) = 0$

The A'' scaling test is based upon generation involving P_1 as mother whereas scaling tests B'' and C'' are based upon generation involving P_2 and F_1 mother, respectively. The D scaling test can be applied as such it is not influenced by the reciprocal differences.

In case of triple test cross, the test of epistasis is affected by the presence of reciprocal differences. In the standard *TTC* where F_2 population is used as male or *TTC* where F_2 population is replaced by a sample of inbred/pure breeding lines, $L_1 + L_2 - 2L_3$ will not provide a test for epistasis until reciprocal difference is assumed absent but it does provide estimate of additive and dominance components of variation free from maternal effects. But when F_2 population is used as mother for each of the three generations (P_1, P_2, F_1), $L_1 + L_2 - 2L_3$ comparison does provide a test for epistasis.

As far as the detection of linkage of interacting genes in concerned, the standard tests of Jinks (1978) cannot be applied as they will be biased in the presence of reciprocal differences but they have been modified (Pooni et al. 1987) as follows:

Test	Comparison
1	$\frac{1}{2}(\bar{B}_1 + R\bar{B}_1) - \bar{L}_1 - \frac{1}{4}(\bar{F}_1 - R\bar{F}_1)$
2	$\frac{1}{2}(\bar{B}_2 + R\bar{B}_2) - \bar{L}_2 - \frac{1}{4}(\bar{F}_1 - R\bar{F}_1)$
3	$\frac{1}{2}(\bar{B}_1 + R\bar{B}_1) + \frac{1}{2}(\bar{B}_2 + R\bar{B}_2) - \bar{L}_1 - \bar{L}_2$

9.4 Reciprocal Effects Model

Various models for reciprocal effects were presented by different workers (Yates 1947; Jinks 1954; Henderson 1948, 1952; Griffing 1956). The models of Henderson and Yates and Hayman are as follows:

$$Y_{ij} = \mu + g_i + g_j + m_j + s_{ij} + r_{ij} + e_{ij}$$

Model A (Henderson)

$$Y_{ij} = \mu + g_i + g_j - m_j + m_j^* + s_{ij} + r_{ij} + e_{ij}$$

Model B (Yates and Haymen)

m_j^* is added in model B so that it agrees with m_j in model A.

A comparison of parameters of model A and model B given below shows that the model A is superior to model B in that its parameters exhibit the least confounding of biologically determined meaningful parameters (Cockerham 1983).

Model A	Model B
μ	μ
$g_j^* = g_i - m_j$	$g_j = g_j^* + \frac{1}{2} m_j$
$m_j^* = 2 m_j$	$m_j = \frac{1}{2} m_j$
s_{ij}	$=$	s_{ij}
r_{ij}	$=$	r_{ij}

The maternal effect m_j is never a part of g in the model A. Griffing's (1956) diallel model has no term for maternal or general reciprocal effects but only a term for A's reciprocal effects. Neither g_i is general combining ability nor s_{ij} is specific combining ability of Sprague and Tatum (1942). Although the difference between g_i and general combining ability is trivial, the s_{ij} in addition to general combining ability contains a component for the average deviations of the crosses from their respective mid-parents. This component was termed mean dominance deviation by Hayman (1954) who tested this fraction of the S_{ij} variation.

Wearden (1964) worked out the expectations of mean squares for maternal effects and reciprocal effects for fixed as well as random models. Jinks and Broadhurst (1963), Griffing (1956), Henderson (1952) and Kempthorne 1952) chose to treat all reciprocal differences in one category $d_i - d_j + r_{ij}$ with a composite component of variance of $\sigma_r^2 + 2\sigma_d^2$. Wearden's model takes the following form:

$$Y_{ij} = \mu + g_i + g_j + m_j + s_{ij} + b_k + e_{ijk} \text{ (maternal effect)}$$

$$Y_{ij} = \mu + g_i + g_j + s_{ij} + r_{ij} + b_k + e_{ijk} \text{ (reciprocal effect).}$$

The expectations of mean squares for Hayman's factorial and Henderson's analysis are given in Table 9.3 under fixed and random model, respectively.

In case of fixed model Hayman's analysis yields the most powerful test for σ_g^2 under reciprocal effects but the relative power of Hayman's σ_g^2 will be probably less than in factorial and Henderson analysis for maternal effect, the reason being the appearance of the term σ_m^2, the variance due to maternal effect in the expectation of a mean squares which makes it essential to be tested against c for the presence of genic effects. If σ_m^2 is large then it adversely affects the power of the test of genic effect. By the same token it can be shown that Hayman's analysis will probably yield a more powerful test for maternal effect. Under reciprocal effect, Hayman's analysis yields the most powerful test of σ_r^2 when c, d mean squares are pooled and tested against random variation.

In the factorial analysis maternal mean squares/paternal mean squares tests the significance of maternal effects in maternal effects model in both fixed and random sample. Under reciprocal effects in both fixed and random models factorial analysis is inappropriate and provides neither test of significance nor estimates of different components of variation.

Table 9.3 Expectation of mean squares for a single replicate of a p^2 diallel cross in random as well as fixed model

Item	d.f.	Fixed model Maternal effect	Fixed model Reciprocal effect	Random model Maternal effect	Random model Reciprocal effect
Hayman analysis (a) Parental lines	$(p-1)$	$\sigma^2 + 1/2\sigma_m^2$	$\sigma^2 + 2p\sigma_g^2 + 2p\sigma_g^2$	$\sigma^2 + p/2\sigma_m^2$	$\sigma^2 + 2\sigma_s^2 + 2p\sigma_g^2 + 2\sigma_s^2 + 2p\sigma_g^2$
(b) Genetic interaction	$\frac{1}{2}(p-1)$	$\sigma^2 + 2p/(p-1)\sigma_s^2$	$\sigma^2 + 2p/(p-1)\sigma_s^2$	$\sigma^2 + 2\frac{(p-1)}{p}\sigma_s^2$	$\sigma^2 + 2\frac{(p-1)}{p}\sigma_s^2$
(c) Av. maternal effect	$(p-1)$	$\sigma^2 + p/2\,\sigma_m^2$	$\sigma^2 + 2\sigma_r^2$	$\sigma^2 + p/2\,\sigma_m^2$	$\sigma^2 + 2\sigma_r^2$
(d) Reciprocal effect	$\frac{1}{2}(p-1)(p-2)$	σ^2	$\sigma_2^2 + 2\sigma_r^2$	σ^2	$\sigma^2 + 2\sigma_r^2$
Factorial analysis Maternal lines	$(p-1)$	$\sigma^2 + p\,\sigma_g^2 + p\,\sigma_m^2$	$\sigma^2 + \sigma_r^2 + p\,\sigma_g^2$	$\sigma^2 + \sigma_s^2 + p\,\sigma_g^2 + p\,\sigma_m^2$	$\sigma^2 + \sigma_r^2 + \sigma_s^2 + p\,\sigma_g^2$
Paternal lines	$(p-1)$	$\sigma^2 + p\,\sigma_g^2$	$\sigma^2 + \sigma_r^2 + p\,\sigma_g^2$	$\sigma^2 + \sigma_s^2\,p\,\sigma_g^2$	$\sigma^2 + \sigma_r^2 + \sigma_s^2 + p_g^2$
M × P	$(p-1)(p-1)$	$\sigma^2 + \sigma_s^2$	$\sigma^2 + (p-2)/(p-1)\sigma_e^2 + \sigma_s^2$	$\sigma^2 + \sigma_s^2$	$\sigma^2 + \frac{(p-2)}{p-1}\sigma_r^2 + \sigma_s^2$
Henderson analysis Dams	$(p-1)$	$\sigma^2 + p\,\sigma_g^2 + p\,\sigma_m^2$	$\sigma^2 + \sigma_r^2 + p\,\sigma_g^2$	$\sigma^2 + \sigma_s^2 + p\,\sigma_g^2 + p\,\sigma_m^2$	$\sigma^2 + \sigma_r^2 + \sigma_s^2 + p\,\sigma_g^2$
Sires	$(p-1)$	$\sigma^2 + p\,\sigma_g^2$	$\sigma^2 + \sigma_r^2 + p\,\sigma_g^2$	$\sigma^2 + \sigma_s^2\,p\,\sigma_g^2$	$\sigma^2 + \sigma_r^2 + \sigma_s^2 + p\,\sigma_g^2$
Cross	$\frac{1}{2}(p)\times(p-1)$	$\sigma^2 + 2p(p-1)\sigma_s^2$	$\sigma^2 + 2p(p-1)\sigma_s^2$	$\sigma^2 + 2\frac{(p-1)\sigma_s^2}{p}$	$\sigma^2 + 2\frac{(p-1)\sigma_s^2}{p}$
Remainder	$\frac{1}{2}(p-1)\times(p-2)$	σ^2	$\sigma^2 + 2\sigma_r^2$	σ^2	$\sigma^2 + 2\sigma_r^2$

Table 9.4. Expectation of mean squares in case of single replicate

Item	D.F.	S.S.	E.M.S.
a	$n-1$	$\Sigma (y_i. + y._i)^2/2n - 2y^2../n^2$	$\sigma^2 + \frac{n}{2}\sigma^{*2}_m + 2n\sigma^{*2}_g$
b	$n(n-1)/2$	$\underset{i>j}{\Sigma}(y_{ij}+y_{ji})^2/2 + \Sigma y_{ii}^2$ $-\Sigma(y_i. + y._i)^2/2n + y^2.../n^2$	$\sigma^2 + \frac{2n}{(n-1)}\sigma^{*2}_s$
c	$n-1$	$\Sigma(y_i. - y.i)^2/2n$	$\sigma^2 + \frac{n}{2}\sigma^{*2}_m$
d	$(n-1)(n-2)/2$	$\underset{i>j}{\Sigma}(y_{ij}-y_{ji})^2/2 - (y_i. - y.i)^2/2n$	$\sigma^2 + 2\sigma^{*2}_n$

*σ^2 indicates an average of squared effects and not a true variance.

In Henderson analysis dam mean squares/sire mean squares tests the significance of maternal effects in maternal effects model in fixed as well as random sample. Under reciprocal effect it provides test for σ^2_r, mean squares due to reciprocal and its estimates can be obtained.

Topham (1964) extended the concept of combining ability of Sprague and Tatum (1942) to estimate maternal effects and maternal × paternal interaction effects. The model takes the form in case of a full set of diallel crosses as

$$Y_{ij} = \mu + g_i + g_j + m_i + s_{ij} + n_{ij} + e_{ij}$$

where Y_{ij} is the total for the cross between the ith female and the jth male,
μ is the mean,
g_i or g_j is the ith or jth parental effects,
m_i is the ith maternal effect,
s_{ij} is the genic interaction between the ith and the jth parental effects,
n_{ij} is the interaction between the ith maternal effect, and the jth parental effect,
and e_{ij} is the error,
with $\sum_i^n g_i = 0, \sum_i^n m_i = 0, \sum_{i=1}^n s_{ij} = \sum_{j=1}^p s_{ij} = 0$
$= \sum_{l=1}^n n_{ij} = \sum_{j=1}^n n_{ij}, n_{ij} + n_{ji} = 0$

Using these restraints, the expectations of the mean squares in Hayman's analysis take the form as given in Table 9.4

When the reciprocal differences are more general, the c item measures maternal effect whereas the d item measures maternal × paternal interaction effects. As the c mean squares does not designate whether a significant effect is maternal or paternal, factorial or Henderson's analysis can be used to designate the sex of the parents to which these effects are due. b mean squares/d mean squares provides a test for significance of variance arising from genetic interaction. There is no test for genic variance under this model. Further he pointed out that replacing the maternal effects by paternal effects in Jinks' model (where the maternal effect is estimated from the average difference between the maternal and paternal arrays) will not lead to any alteration in the analysis. What proportion of the differential effect is maternal can also be estimated. In case the differential effect is wholly maternal, the expectations of row and column mean squares will be $\sigma^2 + n\sigma^2_g + n\sigma^2_m$ and $\sigma^2 + n\sigma^2_g$, respectively. The expectations are reversed if the differential effect is entirely paternal. σ^2_m is thus estimated from both the row and the column mean squares using estimates of σ^2 and σ^2_g from Hayman's analysis. The relative importance of maternal influence in differential effect is estimated as

$$\text{Maternal influence} = \frac{\sigma^2_m(\text{row})}{\sigma^2_m(\text{row}) + \sigma^2_m(\text{column})}$$

The different parameters of the model are calculated as follows:

$$\mu = y..\,/n^2$$
$$m_i = (y_i \cdot - y \cdot _i)/n$$
$$g_i = (y_i \cdot + y \cdot _i)/2n - m_i/2 - \mu$$
$$s_{ij} = (y_{ij} + y_{ji})/2 - g_i - g_j\, m_i/2 - m_i/2 - m_j/2 - \mu$$
$$s_{ij} = y_{ii} - 2g_i - m_i - \mu$$
$$n_{ij} = Y_{ij} - s_{ij} - g_i - g_j - m_i - \mu$$

The variances of the above parameters are calculated as

$$\text{Var}(\mu) = \frac{1}{n^2}$$
$$\text{Var}(m_i) = 2(n-1)/n^2$$
$$\text{Var}(g_i) = (n-1)/n^2$$
$$\text{Var}(s_{ii}) = (n-1)^2/n^2$$
$$\text{Var}(s_{ij}) = [(n-1)(n-2) + n]/2n^2$$
$$\text{Var}(n_{ij}) = n - 2/2n$$

Durrant (1965) noted that these above methods are insufficient for recognition of some pattern of reciprocal differences and there is no one universally recognised mechanism determining reciprocal differences. He described the nature of gene action underlying the maternal effect. The reciprocal difference was separated into Alpha (α) inheritance where the contributions of homozygote and heterozygote are increased or decreased to the same extent on the male side, or on the female side or by unequal amounts on the male and female side and Beta (β) inheritance where the change in dominance is due to different male and female gametic environmental differences. The former has constant effect on all the members of an array while the later leads to a change in dominance. The two forms of inheritance can be detected from the W female/W male (covariance female/covariance male). The test of significance of item α inheritance (see Table 9.5) is the same as comparing Hayman's c and d items. A comparison of c and d items leads to a test for both α and β. Parental contributions in the β inheritance are estimated from a set of simultaneous equations and its significance is tested against residuals. Alpha and Beta inheritance are not completely separatable and where they are both present, their combined estimate can be partitioned into three items, namely

Table 9.5 ANOVA in the analysis of maternal effects

Item	d.f.
Total	$\frac{n}{2}(n-1)$
(a) Alpha inheritance	
Alpha (c)	$n-1$
Residual (d)	$\frac{1}{2}(n-1)(n-2)$
(b) Beta inheritance	
Beta	n
Residual	$2n-3$
(c) Alpha and beta inheritance	
Alpha plus beta	
a'	$2n-3$
b'	$n-2$
$\bar{b}$	$n-2$
Residual	$\frac{1}{2}(n-3)(n-2)$
b'/a'	
$b'/\bar{b}$	

a', b' and $\bar{b}$ where $\bar{b}$ is a measure of overall maternal or paternal inheritance; a' value measures paternal contribution to α inheritance (maternal, paternal or cytoplasmic effects) and b' value measures parental contributions to β inheritance (change in dominance). a' (G/n) values are the relative amounts by which α inheritance of the respective parents adds to each member of their female arrays or subtracts from each member of their male arrays. β inheritance of any one parent is measured as the difference between the regression coefficients of its male and female arrays, respectively on to the parents. β inheritance occurs when each of any number of genes has different amount of dominance in the reciprocal crosses. If a/c is not significant, it shows maternal effects and if b/d is not significant then it shows no specific combining ability, SCA or dominance/epistasis. A comparison of c item against d item yields evidence on these points. The significant c and d items of Hayman's analysis show average maternal effects and specific reciprocal interaction, respectively. Thus if c item is significant, α inheritance is mainly responsible for this effect and when d item is significant it show β inheritance to be involved. When c mean squares is no greater than d mean squares and both are significant it shows that both types of inheritance are there.

The populations thus can be grouped into two, male dominance or female dominance. Durrant noted that the reciprocal differences are more pronounced when the experiment is conducted in stress environment without irrigation, fertilization, etc. Cockerham (1963) worked out the expectation of mean squares for maternal and reciprocal items appeared in the model of Yates (1947) and translated the components of variances and covariances into genetical components of variation. The expected mean squares in case of diallel with F_1's and RF_1's are given in Tables 9.6 and 9.7 where both these items are separate and combined, respectively.

Table 9.6 Expected mean squares when maternal and reciprocal effects are separate

Item	d.f.	E.M.S.
General combining ability	$n-1$	$\sigma^2 + r\sigma_r^2 + 2r\sigma_s^2 + r(n-2)$ $\sigma_m^2 + 2r(n-2)\sigma_g^2$
Specific combining ability	$n(n-3)/2$	$\sigma^2 + r\sigma_r^2 + 2r\sigma_s^2$
Maternal	$n-1$	$\sigma^2 + r\sigma_r^2 + rn\sigma_m^2$
Reciprocal	$(n-1)(n-2)/2$	$\sigma^2 + r\sigma_r^2$
Error	$(r-1)(n^2-n-1)$	σ^2

If reciprocal mean squares is significant, σ_r^2 is not zero and indicates maternal × paternal interaction

Table 9.7 Expected mean squares when both effects are combined

Item	d.f.	E.M.S.
General combining ability	$n-1$	$\sigma^2 + 2r\sigma_{s'}^2 + 2n(n-2)\sigma_{g'}^2$
Specific combining stability	$n(n-3)/2$	$\sigma^2 + 2r\sigma_{s'}^2$
Reciprocal	$n(n-1)/2$	$\sigma^2 + r\sigma_{r'}^2$
Error	$(r-1)(n^2-n-1)$	σ^2

r is the number of replications

(interaction of reciprocal effects). If σ_r^2 is zero then σ_m^2 is most likely due to maternal effects in addition to paternal ones. If the reciprocal effects are entirely due to maternal effects (i.e. $\sigma_r^2 = 0$ and are additive to the paternal effects then

$$\sigma_m^2 = \frac{1}{2}(\text{Cov }(HS)_m - \text{Cov }(HS)_p)$$

$$\sigma_g^2 = \text{Cov}(HS)_p$$

$$\sigma_s^2 = 0 \text{ Cov }(FS) - \text{Cov }(FS)_m - \text{Cov }(HS)_p$$

where Cov $(HS)_m$ is the covariance maternal half-sib and Cov $(HS)_p$ is the covariance paternal half-sib.

Cockerham and Weir (1977) attempted to estimate separately both maternal and paternal variances from reciprocal crosses. Earlier models of Yates, Henderson and Griffing of diallel crosses with reciprocals were largely concerned with estimation of effects. Their biomodel (model I), diallel model (model II) factorial model (model III) are given below:

$$G_{ij} = n_i + n_j + t_{ij} + m_i + p_j + k_{ij} \text{ Model I}$$

where n_i and n_j represent the nuclear contribution the parents, t_{ij} is the interaction effect, m_i is maternal extra-nuclear effect, p_j is the paternal extranuclear effect and k_{ij} involves interaction between nuclear and extra nuclear effects.

$$G_{ij} = g_i + g_j + s_{ij} + d_i - d_j + r_{ij} \qquad \text{Model II}$$

$$G_{ij} = M_i + P_j + (MP)_{ij} \qquad \text{Model III}$$

where M_i is the maternal effect, P_j is the paternal effect and $(MP)_{ij}$ is the interaction effect and

$$g_i = (M_i + P_j)/2$$

$$d_i = (M_i - P_j)/2$$

$$s_{ij} = [(MP)_{ij} + (MP)_{ji}]/2$$

$$r_{ij} = [(MP)_{ij} - (MP)_{ji}]/2$$

The biomodel is related to diallel model and factorial model in the manner given on next page:

Biomodel	Diallel model	Factorial model
σ_n^2	$\sigma_g^2 - \sigma_d^2$	Cov MP
σ_p^2	$4\sigma_d^2$	$\sigma_p^2 -$ Cov MP
σ_m^2		$\sigma_M^2 -$ Cov MP
σ_t^2	$\sigma_s^2 - \sigma_r^2$	Cov. $M \times P$
σ_k^2	$2\sigma_r^2$	$\sigma^2 MP -$ Cov. MP

where $\sigma^2 MP$ is the variance of interaction effect, Cov MP is the covariance of M_i and P_i and Cov $M \times P$ is the covariance of $(MP)_{ij}$ and $(MP)_{ji}$. The five components to be estimated in the biomodel are σ_n^2, σ_t^2, σ_m^2, σ_p^2 and σ_k^2 from the analysis of reciprocal crosses. However, the variances arising from extra-nuclear effects are σ_m^2, σ_p^2 and σ_k^2 which reflect a mixture of interaction effects involving both nuclear and extra nuclear effects. For estimation of components of variances at least some parts of the analysis corresponding to the factorial model are required to separate the maternal and paternal variances. A least square partitioning of the sum of squares according to the diallel model but with expectation expressed in terms of the biomodel provides most of the tests of hypotheses of interest. They further noted that the diallel model of effects is most useful in terms of linear estimation and separation of sum of squares in the analysis of variance and it should be carried over into variance components estimation.

9.5 Covariances between Relatives

Wilham (1963) derived a general formula for the genetic covariances between relatives x and y which receive a maternal effect from relatives w and z, respectively. The problem with this approach is that $\sigma^2 dm$ (maternal dominance) is confounded with maternal additive effects. Eisen (1967) proposed three designs which are suitable for estimating the causal components of variances as designed by Wilham and found that diallel is not the best design for estimating $\sigma^2 d.dm$. Further stability of maternal variances and covariances over environments is consistent with the maternal effects which derives from the environment provided to the developing seed but is not universal and there are examples of cytoplasmic × environment interaction. Thus we have seen that if the reciprocal difference is expressed as maternal effect, it will inflate the estimate of additive genetic variance, total genetic variance and which in turn will inflate the estimate of heritability (h_n^2) and if the reciprocal difference is expressed as maternal interactions, it will inflate dominance, total genotypic variance and the heritability (h_b^2). Thus it shows how important it is to detect and estimate these reciprocal differences. We can also visualize that the actual response based on the parameters will be less than predicted. When reciprocal difference is transient, selection should be practised on mature plants with the hope that the maternal effect declines as the plants mature.

9.6 Sex Linkage

Most flowering plants are monoecious and therefore do not have chromosomes grouped as autosomes and sex chromosomes as we see in animal kingdom. With the exception of Melandrium where there is chromosomal sexuality in which pistillate plants possess XX and staminate plants possess XY chromosomes and these two are two distinct types of saprophyte, dioecism in crop plants is under control of a single gene. Thus the phenomenon of sex linkage where genes for a quantitative character reside on sex chromosomes particularly X-chromosome (Y chromosomes might carry either no gene or few genes) and which occurs in animal species is an important source of genetic variation. Sex linkage involves sexually dimorphic species. We will also consider complication arising in case of sex limited traits—the genes which are present in both sexes but the trait appears in only one sex and the genes for such traits reside on autosomes.

As usual the analysis of sex linked traits involves analysis of generations derived from the cross between two inbred lines or analysis of one of the several multiple mating designs when the interest lies in the population rather than the fixed set of just two lines.

There are two kinds of sex-homogametic (XX) and heterogametic (XY). Considering one locus with two alleles (A, a), there will be three possible genotypes namely $X_A X_A$, $X_A X_a$ and $X_a X_a$ in homogametic sex and two genotypes, $X_A Y$ and $X_a Y$ in heterogametic sex. The gene in homogametic sex can be either in homozygous or heterozygous condition but it is in hemizygous condition in heterogametic sex. Like autosomal gene effects there will be for sex linked genes additive sex linked effect (dx), dominance sex linked effect (hx) and interaction effect for homogametic sex whereas in case of heterogametic sex the effect of a gene in hemizygous conditions can be represented by $d'x$. The genotypic values of the above genotypes will then be $+ dx$ ($X_A X_A$), hx ($X_A X_a$), $- dx$ ($X_a X_a$), $+ d'x$ ($X_A Y$) and $- d'x$ ($X_a Y$), respectively. Thus the equivalent parameter of $[d]$ and $[h]$ of autosomal genes will be $[dx] = \Sigma\, dx$, $[d'x] = \Sigma\, d'x$ and $[hx] = \Sigma\, hx$ for sex linked genes in case of polygenes. The expectations of means of P_1, P_2, F_1, F_2 and backcross generations (Mather and Jinks, 1971, 1982) show that in the presence of sex linkage

1. $\overline{F}_1 (+d'x) \neq \overline{RF}_1 (-d'x)$ in case of heterogametic sex.

2. $F_{2F_1 \times F_1} \left(\frac{1}{2} dx + \frac{1}{2} hx \right)$

   $\neq F_{2RF_1 \times RF_1} \left(-\frac{1}{2} dx + \frac{1}{2} hx \right)$

   in case of homogametic sex.

3. $\overline{BC}_1 \neq \overline{RBC}_1$ and $\overline{BC}_2 = \overline{RBC}_2$ in both sexes.

Within a generation there are males and females and the expectation will differ for the two sexes for the same generation. So analyses on males and females are carried out separately. In case of maternal effect the phenotypic difference appears because of difference in cytoplasm rather than in sex chromosome as in sex linkage and it usually affects both male and female offspring and thus studied together. If the analysis is carried out on the averages over reciprocal crosses, the effect of sex linkage disappears which shows adequacy in this situation. Weighted least squares estimates of the parameters can be obtained and the adequacy of the model can be tested. Also, r_d, the coefficient of dispersion of alleles can be incorporated in the model and its value can be determined.

While fitting model it would be proper to fit the usual autosomal parameters first and then to add the sex linkage parameter and finally to substitute the sex limitation model for the sex linkage model. The relative sizes of the χ^2 will provide a guide as to which types of gene action are important (Perkins and Jinks, 1970). Thus we can separately estimate the autosomal and sex linked parameters. The additive-dominance model can be extended to include non-allelic interaction with the difference in that in homogametic sex, the interaction parameters are similar to that of the autosomal interaction parameters described earlier but for the heterogametic sex the interaction parameters are similar to that for the haploid system which we will see later.

While studying the generation variances, a simple difference between the averages of F_2 homogametic individuals $\left(\frac{1}{2} D + \frac{1}{4} H + E \right)$ and F_2 heterogametic individuals $\left(\frac{1}{2} D + \frac{1}{4} H + \frac{1}{4} Dx + \frac{1}{4} Hx + E \right)$ shows the presence of sex linkage. Unlike generation mean analysis, the parameters for autosomes and sex linked genes cannot be estimated separately as the parameters Dx and Hx appear in the expectations with the same coefficient and sign. Eberhart and Gardner's (1966) model can be applied for estimating sex linked genes. Like all the parameters in case of autosomal genes we can have parameters like additive sex linked effect a_i^s (male or female), dominance heterotic effect $h_{ii'}^s$ and additive x additive gene interaction. The item a_i^s, additive sex linked effect (male) is estimated from the male population separately as well as from both sexes combined by using a single population plus the diallel set of two-way crosses. At least four populations are required for the estimation of specific effects. The additive sex effect due to female is confounded with a_i^s and non-allelic interaction in the corresponding populations. However, inclusion of either backcross or three way cross would permit one to obtain a_i^s separately.

The dominance heterotic effects for sex-linked genes $h_{ii'}^s$ can be expressed only in female sub-population and can be partitioned into components similar to that for autosomal genes as follows.

$$h_{ii'}^s = h_i^s + h_{i'}^s + hh_{ii'}^s$$

When additive × additive gene effects exist, the additive × additive heterotic effect $aa_{ii'}$ can be partitioned in a similar way to that for dominance.

$$aa_{ii'} = \overline{a}\,\overline{a}^h + aa_i^h + aa_{i'}^h + aa_{ii'}^h$$

The combined model for the unsexed offspring has disadvantages. First, the coefficients of parameters $a_i^s\, \sigma$, $a_i^s\, ♀$, $d_{i'}^s$, and $h_{ii'}^s$ are halved and thus results in larger standard error for the estimates. Secondly, as the number of individuals decreases, the sampling variation in sex ratio tends to inflate the error. Finally, for all those parameters that are common to both sexes, one is forced to estimate only the average of the parameters for the two sexes.

9.7 Sex Limitation

Killick and Kearsey devised a model for traits showing sex-limitation. The distinction between dx and $d'x$ allows for sex-limitation on the sex chromosome. As the effect of a gene (e.g. milk yield in cattle, egg production in fowl) is determined by the sex of the individual, in this phenomenon each of the autosomal parameters, m, d and h is increased by some function fm, fd and fh, respectively in the female and decreased by the same amount in the male. The model for sex-limitation takes the form as:

	Parameters					
	m	fm	d	fd	h	fh
AA	1	1	1	1	0	0
aa	1	1	−1	−1	0	0
Aa	1	1	0	0	1	1
AA	1	−1	1	−1	0	0
aa	1	−1	−1	1	0	0
Aa	1	−1	0	0	1	−1

The parameters are orthogonal and are defined as

$$m = 1/4\,(AA + aa), ♀ + ♂$$

$$d = 1/4\,(AA - aa), ♀ + ♂$$

$$h = 1/4\,[2Aa - (AA + aa)], ♀ + ♂$$

$$fm = 1/4\,(AA + aa), ♀ - ♂$$

$$fd = 1/4\,(AA - aa), ♀ - ♂$$

$$*fh = 1/4\,[2Aa - (AA + aa)], ♀ - ♂$$

Multiple mating designs:

Considering one locus with alleles (A, a) and u and v being the frequencies of allele A and a in a random mating population, the genotypes, their frequencies and mean values can be written as follows:

	Sex				
	Homogametic			Heterogametic	
Genotype	AA	Aa	aa	A	a
Frequency	u^2	$2uv$	v^2	u	v
Mean value	$+dx$	$+hx$	$-dx$	$+d'x$	$d'x$

The total variation arising from sex linked genes in a random mating population in case of homogametic sex is

$$\frac{1}{2} D_{Rx} + \frac{1}{4} H_{Rx} + E_w + E_b$$

and for heterogametic sex is

$$D'_x + E_w + E_b$$

where $D_{Rx} = \Sigma\, 4uv\,[dx + (u - v)\,hx]^2$

$H_{Rx} = \Sigma\, 16u^2v^2hx^2$

$D'_x = \Sigma\, 4uv\, d'_x{}^2$

The important covariances are the covariances parent/offspring and covariances full-sib. Irrespective of the sex of the parents and of the offspring, the covariances parent/offspring are expected to be similar and equal to $\frac{1}{4} DR$ in the absence of sex linked variation. But in its presence covariance among heterogametic parent and heterogametic offspring is expected to be smaller then covariance homogametic parent or homogametic offspring or both. The covariances between parents/offspring in a random mating population are as follows:

Mother − daughter $\frac{1}{4} DRx$

Mother – son $\quad \frac{1}{2} D'_{xx} - \frac{1}{4} F'_{xx}$

Father – daughter $\quad \frac{1}{2} D'_{xx} - \frac{1}{4} F'_{xx}$

Father – son $\quad 0$

where $D'_{xx} = \Sigma\, 4uv\, dx\, d'_x$

and $F'_{xx} = \Sigma\, 8uv\, d'x\, (u-v)hx$

The value for all the above four parent-offspring covariances is $\frac{1}{4} DR$ for autosomal genes. The values of covariances mother-son and father-daughter are equal and highest and is followed by co variances mother-daughter and father-son which reflects the presence of linkage.

The different types of covariances full-sib are as follows:

Covariance sister-sister $= \frac{3}{8} DRx + \frac{1}{8} HRx + Eb$

Covariance brother-brother $= \frac{1}{2} D'_x + Eb$

Covariance brother-sister $= D'_{xx} - \frac{1}{8} F'_{xx}$

The covariance full-sib equals $\frac{1}{4} DR + \frac{1}{6} HR$ in case of autosomal genes and thus is different than the covariance sister-sister.

Intra-class correlation coefficients calculated using the above covariances will be as follows:

Full-sib correlation (assuming no sex linkage)

$$= \frac{(1/4)\, DR + (1/16)\, HR + Eb}{(1/2)\, DR + (1/4)\, HR + Ew + Eb}$$

Full sib correlation (sister - sister)

$$= \frac{(3/8)\, DRx + (1/8)\, HRx + Eb}{(1/2)DRx + (1/4)\, HRx + Ew + Eb}$$

Assuming $HR = 0$ and $Ew = Eb = 0$, the maximum value of full-sib correlation will be 0.5 in case of autosomal genes whereas it will be greater than 0.5 in case of correlation between homogametic sib in the presence of linkage. The expectations of these above covariances assume no sex limitation and there is no test available for its detection and estimation.

That is why any estimates of sex-linkage parameters derived from the covariances between relatives will have large standard error.

Killick (1970) compared the relative efficiency of BIPs, NCI, NC 2 and NC 3 for separation of sex linked and autosomal gene effects. The mating designs were compared under conditions, namely, when the sexes were separate and when the sexes were pooled. The expectations of statistical variance components of the different design are given in Table 9.8.

When the sexes are pooled, the expectation becomes more complicated as besides sex linkage parameters (D_{Rx}, H_{Rx} and D'_x) and autosomal parameters (D_R and H_R) there will appear two more sex-linkage parameters, D'_{xx} and F'_{xx} and two sex-limitation parameters, D_{Rs} and H_{Rs} where

$$D_{Rs} = \Sigma 4uv\, [fd + (v-u)\, fh]^2$$

and $\quad H_{Rs} = \Sigma 16 u^2 v^2 f h^2$

Comparative study showed that $NC2$ is probably the most satisfactory mating design. This design allows estimation of D_R in both sexes and if the two values are similar sex-limitation may be taken to be unimportant. The only restriction here being that the dominance effect of the sex-chromosome cannot be separated from that of the autosomes. Further as the sex-linkage is not the only possible cause of reciprocal differences, cytoplasmic and maternal effects will tend to inflate σ'^2_f which will lead to over estimation of D'_x, in male progeny and under estimation of D_{Rx} in female progeny. As there is no test for the detection of these effects in the presence of sex-linkage they must be assumed absent though they will distort the estimates of the components of variation. But if we have a diallel made of inbred lines, sex-linkage can be detected by comparing reciprocal crosses assuming no maternal inheritance. As mentioned in the test for sex linkage, the individuals in the F_1 for heterogametic sex will not show differences between reciprocals whereas they do in case of homogametic sex. Thus if the analysis of variance is carried out separately for both sexes, interaction between sex and reciprocal difference will indicate the presence of sex-linkage.

Table 9.8 Genetic components of variation when the sexes are separate

| | Sex linkage parameters | | | Autosomal parameters | |
| | Females | | Males | Either sex | |
	DRx	HRx	D'_x	DR	HR
I. Biparental progenies					
σ_b^2	$\frac{3}{8}$	$\frac{1}{8}$	$\frac{1}{2}$	$\frac{1}{4}$	$\frac{1}{16}$
σ_w^2	$\frac{1}{8}$	$\frac{1}{8}$	$\frac{1}{2}$	$\frac{1}{4}$	$\frac{3}{16}$
II. North Carolina Mating I					
σ_m^2	$\frac{1}{4}$	–	–	1/8	–
$\sigma_{f/m}^2$	1/8	1/8	1/2	1/8	1/16
σ_w^2	1/8	1/8	1/2	1/4	3/16
III. North Carolina Mating II					
σ_m^2	1/4	–	–	1/8	–
σ_f^2	1/8	–	1/2	1/8	–
$\sigma_{m \times f}^2$	–	1/8	–	–	1/16
σ_w^2	1/8	1/8	1/2	1/4	3/16
IV. North Carolina Mating III					
	$D_1 x$	$H_1 x$		D	H
σ_m^2	1/4	–	–	1/8	–
$\sigma_{t \times m}^2$	–	1/4	–	–	1/8
σ_w^2	–	–	–	1/8	1/8

The triple test cross provides a test for epistasis which should be made prior to the estimation of additive and dominance genetic variance components since these are liable to be distorted by non-allelic interaction. In case of sex-linked gene it can be shown that the value of $\bar{L}_{1i} + \bar{L}_{2i} - 2\bar{L}_{3i}$ is zero for male progeny, female progeny and for the two sexes pooled irrespective of which allele the male parent possesses and hence the variances of these values must be zero and thus the test is not impaired by sex-linkage. In case where the sexes are pooled it can be shown that the sex-limitation has no effect on the test. But another test for epistasis, $L_{1i} + L_{2i} - P$ works only for females and has a value of $m + hx$, a constant for all i and thus has zero variance. For females

$$\bar{L}_{1i} + \bar{L}_{2i} - \bar{P}_i = m + d'_x$$
if male has sex-linked allele 'A'

$$\bar{L}_{1i} + \bar{L}_{2i} - \bar{P}_i = m - d'_x$$
if male has sex-linked allele 'a'

The variance of these values comes to D'_x and the test thus fails. In the presence of sex-linkage the estimates of additive and dominance variance components are confounded with sex-linked gene effects. The genetical variance components describing the sex-linked variation analogous to these for the autosomal genes are defined as

$$D_{1x} = \Sigma 4uvd^2 x$$

$$H_{1x} = \Sigma 4uvh^2x$$

$$F_{1x} = \Sigma 4uvdxhx$$

The analysis of sex-linked trait is complicated because of the simultaneous existence of autosomal variation and with the possibility that the trait in question may be sex-limited.

References

Barnes, B.W 1968. Maternal control of heterosis for yield in *Drosophila melanogaster*. Heredity. **23**: 563–572.

Carbonell, E.A., Nyguist, WE. and Bell, A.E. 1983. Sex linked and maternal effects in Eber hart-Gardner general genetics model. Biometrics, **39**: 607–619.

Cockerham, C.C. 1963. Estimation of genetic variances, In: Statistical Genetics and Plant Breeding, WD. Hanson and H.F. Robinson (eds.) Pub. 982, Washington, NAS-NRC.

Cockerham, C.C. and Weir, B.S. 1977. Quadratic analysis of reciprocal crosses. Biometrics, **33**: 187–203.

Durrant, A. 1965. Analysis of reciprocal differences in diallel crosses. Heredity, **20**: 573–607.

Eberhart, S.A. and Gardner, C.O. 1966. A general model for genetic effects. Biometrics, **22**: 864–881.

Eisen, E.J., Bohren, B.B. and McKean, H.E. 1966. Sex-linked and maternal effect in the dial lei cross. Aust. J. Biol. Sci., **19**: 1061–1071.

Griffing, B. 1956. Concept of general and specific combining ability in relation to diallel crossing system. Aust. J. Biol. Sci., **9**: 463–493.

Henderson, C.R. 1948. Estimation of general, specific and maternal combining abilities in crosses among inbred lines of swine. Ph. D. the-sis, Iowa State Univ. Library, Ames, Iowa.

Henderson, C.R. 1952. Specific and general combining ability. In: Heterosis, J.W. Gowen (ed.) 352–370, Ames, Iowa State Univ. Press.

Jinks, J.L. 1983. Biometrical genetics of heterosis. In: Franklin, R. (ed.) Heterosis-Reappraisal of the theory and practice, Springer-Verlag, pp. 1–46.

Jinks, J.L. and Broadhurst, P.L. 1963. The detection and estimation of heritable differences in behaviour among individuals. Heredity, **20**: 97–115.

Kearsey M.J. and Pooni, H.S. 1996. The general analysis of Quantitative traits. Chapman and Hall, London.

Kempthorne, O. 1952. The design and analysis of experiments. Wiley Inc., New York.

Killick, R.J. 1970. Sex linkage and Sex-limitation in quantitative inheritance. I. Random mating population. Heredity, **25**: 175–188.

Mather, K. and Jinks, J.L. 1982. Biometrical Genetics. 3rd end. Chapman and Hall, London.

Perkins, J.M. and Jinks, J.L. 1970. Detection and estimation of genotype × environment, linkage and epistatic components of variation for a metrical trait. Heredity, **25**: 157–177.

Pooni, H.S., Coombs, D.T., Virk, P.S. and Jinks, J.L. 1987. Detection of epistasis and linkage of interacting genes in the presence of reciprocal differences. Heredity, **58**: 257–266.

Topham, P.B. 1966. Diallel analysis involving maternal and maternal interaction effects. Heredity, **21**: 665–674.

Wearden, S. 1964. Alternative analysis of the diallel cross, Heredity, **19**: 669–680.

Yates, F. 1947. The analysis of dates from all possible reciprocal crosses between a set of parental lines. Heredity, **1**: 287–301.

10

Linkage and Epistasis

10.1 Linkage

Linkage is the linear sequence of a group of non-allelic genes along the chromosome which forms a physical distinct group from other linked group of genes and thus deals with the physical (or more especially distributional) aspect of 'non-alleles whereas epistasis deals with the functional interdependence or interaction of non-alleles. In the simplest case of 2 loci, each with 2 alleles (A, a and B, b)

$f(AB)$	=	$f(A) + f(B)$
Frequency of gametes containing A and B alleles		Product of the frequencies of gametes contain only A allele and B allele

This relationship holds only when the association is independent i.e. the two loci are segregating independently. A measure of lack of independence which will be measure of linkage disequilibrium will be equal to the observed frequency of AB gametes minus the expected frequency calculated assuming independent assortment. Linkage equilibrium refers to independence of genes at different loci.

Considering the same above two loci with two alleles each, there will be four possible types of gametes, namely, AB, aB, Ab and ab. Suppose that they occur with the frequencies $C_1(AB)$, $C_2(aB)$, $C_3(Ab)$ and $C_4(ab)$, respectively in such a way that

$$C_1 + C_2 + C_3 + C_4 = 1$$

The frequencies of the alleles A, a, B and b will be as follows:

$$f(A) = ua = C_1 + C_3$$
$$f(a) = va = C_2 + C_4$$
$$f(B) = ub = C_1 + C_2$$
$$f(b) = vb = C_3 + C_4$$

Now assuming random mating, the mean of the F_2 population can be shown to be following Mather and Jinks (1971) as:

$$\overline{F}_2 = m + (ua - va)\, da + (ub - vb)\, db + 2\, ua\, va\, ha + 2ub\, vb\, hb$$

As no C term appears in the expectation of the F_2 mean so linkage disequilibrium does not affect the generation mean although we will see later that linkage disequilibrium does affect the generation means in the presence of epistasis. Assuming $ua = va = ub = vb = \frac{1}{2}$ the above expectation of F_2 mean changes to

$$\overline{F}_2 = m + \frac{1}{2} ha + \frac{1}{2} hb$$

The variance of the F_2 population takes the form as

$$VF_2 = [2\, ua\, va\, da^2 + 2\, ubvb\, db^2 \\
+ 4\, (C_1 C_4 - C_2 C_3)\, da\, db] \\
+ [2\, uava\, (va - ua)^2\, ha^2 + 4ua^2\, va^2\, ha^2 \\
+ 2ub\, vb\, (vb - ub)^2\, hb^2 \\
+ 4ub^2\, vb^2\, hb^2 + [(C_1 - C_4 - C_3 + C_4) \\
- 2\, (C_1 C_4 - C_2 C_3)]\, (C_1 C_4 - C_2 C_3)\, hahb]$$

10.2 Biometrical Genetics–Analysis of Quantitative Variation

$[4uava (ua - va) daha + 4ub\ vb\ (ub - vb)\ dahb]$
$+ [4(C_1C_3 - C_2C_4) - ua\ vb\ (va - vb)]\ dahb$
$[4(C_1C_2 - C_3C_4) - ua\ va\ (ub - vb)]\ dbha$

This shows that the variance is affected by linkage disequilibrium. The term $(C_1C_4 - C_2C_3)$ determines the coefficient of linkage. If $(C_1C_4 - C_2C_3)$ is positive, it shows that there is an excess of coupling or association phase linkage whereas if it is negative, it indicates an excess of repulsion or dispersion phase linkage.

Assuming coupling phase linkage the frequencies C_1, C_2, C_3, and C_4 of gametes AB, aB, Ab and ab, respectively can be replaced by $\frac{1}{2}(1-p)$, $\frac{1}{2}p$, $\frac{1}{2}p$ and $\frac{1}{2}(1-p)$, respectively, where p is the recombination frequency and assuming $ua = va = ub = vb = \frac{1}{2}$, the variance of F_2 can be obtained as

$$VF_2 = \frac{1}{2}da^2 + \frac{1}{2}db^2 + \frac{1}{4}ha^2 + \frac{1}{4}hb^2$$
$$+ (1-2p)\ dadb + \frac{1}{2}(1-2p)^2\ hahb.$$

With an excess of coupling phase linkage, covariances da, db and ha, hb will be positive and thus the F_2 variance is inflated whereas with repulsion phase covariances da, db and ha, hb are negative and thus the variance is deflated.

Linkage disequilibrium in population can develop due to (i) chance/sampling small population, genetic drift (ii) selection, (iii) linkage *per se* in the parents from which the population has been derived. Inbreeding also causes correlation of allelic genes. (iv) Mutation and mixture (Crossbreeding).

10.2 Effect of Random Mating on Linkage Disequilibrium

Considering the gametic frequencies described above, the linkage disequilibrium after one generation of random mating will be

$$(C'_1C'_4 - C'_3C'_4) = (1-p)(C_1C_4 - C_2C_3)$$

where C'_1 $C'2$, C'_3 and C'_4 are the frequencies of gametes AB, aB, Ab and ab, respectively produced as a result of one round of random mating. After n generations of random mating the linkage disequilibrium will be $(1-p)^n (C_1C_4 - C_2C_3)$. This shows that if $p = 0.5$ i.e. when there is no linkage at all, the linkage equilibrium will be reached faster. Further in case of polygenes the attainment of linkage disequilibrium will be slower.

10.3 Effect of Linkage on Variances of Advancing Generations Derived by Selfing Cross Between Pairs of Purebreeding Lines

We have seen above that under the condition $ua = va = ub = vb = \frac{1}{2}$ and p being the recombination frequency, the variance of F_2 in case of coupling linkage becomes

$$V_1F_2 = \frac{1}{2}[da^2 + db^2 + 2(1-2p)\ dadb]$$
$$+ \frac{1}{4}[ha^2 + hb^2 + (1-2p)^2 - hahb]$$

Thus it differs from what we see in case of no linkage as

$$\frac{1}{2}[da^2 + db^2] + \frac{1}{4}[ha^2 + hb^2]$$

The terms $2(1-2p)\ dadb$ and $2(1-2p)^2\ hahb$ are the linkage disequilibrium components. The variance of $F_2(V_1F_2)$ in case of repulsion phase linkage becomes

$$V_1F_2 = \frac{1}{2}[da^2 + db^2 - 2(1-2p)\ dadb]$$
$$+ \frac{1}{4}[ha^2 + hb^2 + 2(1-2p)^2\ hahb]$$

When the results from the two loci case is extended to polygenic case then

$$V_1F_2 = \frac{1}{2}D + \frac{1}{4}H + E$$

where $\quad D = \Sigma d^2 \pm_R^C \Sigma 2(1-2p)\, dadb$

$$H = \Sigma h^2 + \Sigma 2(1-2p)^2\, hahb$$

The changes in estimates of the genetical components, D and H depend on (1) phase of linkage, (2) sign of dominance effect, h, (3) equality in effects of d and h or h/d and (4) recombination frequency. The estimate of D is inflated or deflated depending upon whether coupling is in excess or repulsion is in excess whereas dominance component (H) is inflated irrespective of the phase of linkage provided that the dominance is unidirectional i.e. dominance effects at all loci are all positive or negative. If some loci show dominance effects positive whereas some other loci negative then dominance component (H) is deflated.

In polygenic case, linkage will show its maximum effect when all pairs of genes are in association/coupling phase and all h's are reinforcing. The maximum effect of repulsion and opposition is expected to be maximum when the adjacent loci are repulsed and h's opposed and the effects lessen as the neighbouring loci tend to become associated. For example, with five loci (3 increasing loci and 2 decreasing loci) the maximum D will be when the genes are distributed in the fashion $+-+-+$ followed by $++--+$, $-+++-$ and $+++--$. And, of course, D, is maximum when all five pairs are increasing loci and are in association phase, $+++++$. As with more than two loci all pairs of loci can never be in repulsion phase which thus limits the deflation that could have been expected otherwise. The effects of linkage on variances and covariances would be zero if all the parents are associated and dissociated for similar number of loci and equal number of loci shows positive and negative dominance effects respectively.

By selfing we can obtain F_2, F_4, F_5, etc. generations. The total variances of the different advancing generations will be

$$VF_3 = \frac{3}{4}D + \frac{3}{16}H$$
$$VF_4 = \frac{7}{8}D + \frac{7}{16}H$$
$$VF_5 = \frac{15}{16}D + \frac{15}{256}H$$

$$\cdot \quad \cdot \quad \cdot$$
$$\cdot \quad \cdot \quad \cdot$$
$$\cdot \quad \cdot \quad \cdot$$

$$VF_\infty = D + 0$$

Thus we see that in the F_∞ generation all the variances among pure breeding lines will be made up of additive genetic variance. In the above expectations D's and, H's are summed over all ranks. In the presence of linkage rank $1D$ and H, rank $2D$ and H and rank $3D$ and H are as follows:

$$D_1 = \Sigma d^2 \pm_R^C \Sigma 2(1-2p)\, dadb$$
$$H_1 = \Sigma h^2 + 2(1-2p)^2\, hahb$$
$$D_2 = \Sigma d^2 \pm_R^C \Sigma 2(1-2p)^2\, dadb$$
$$H_2 = \Sigma h^2 + \Sigma(1-2p)^2\, hahb + \Sigma(1-2p)^4\, hahb$$
$$D_3 = \Sigma d^2 \pm_R^C \Sigma(1-2p)^3\, dadb$$
$$H_3 = \Sigma h^2 + \Sigma 2(1-2p)^2\, hahb$$
$$\quad + \Sigma 2(1-2p)^4\, hahb + \Sigma 2(1-2p)^6\, hahb$$

The general formula for calculating D of any rank is

$$D_r = \sum_{i=1}^{k} di^2 \pm_C^R \sum_{i=1}^{k} \sum_{j=i+1}^{k} 2(1-2p)^r\, dadb$$

where r is the round of recombination which produces that particular statistics. A look at the different rank H's shows that by 4 or 5th generation no h's will be left. The additive genetic variances of the above mentioned generations in terms of ranks of D's would look like as

$$DVF_2 = \frac{1}{2}D_1 + \frac{1}{4}H_1$$
$$DVF_3 = \frac{1}{2}D_1 + \frac{1}{4}D_2$$
$$DVF_4 = \frac{1}{2}D_1 + \frac{1}{4}D_2 + \frac{1}{8}D_3$$

$$\cdot$$
$$\cdot$$

$$DVF_\infty = \frac{1}{2}D_1 + \frac{1}{4}D_2 + \frac{1}{8}D_3 + \frac{1}{16}D_4$$
$$+ \ldots + \left(\frac{1}{2}\right)^\infty D_\infty$$

$$= \sum_{i=1}^{\infty} \left(\frac{1}{2}\right)^2 D_r$$

Thus the total additive genetic variance in the nth generations of selfing (DVF_∞) is

$$\sum_{r=1}^{\infty} \left(\frac{1}{2}\right)^2 D_r$$

When n is large, the total additive genetic variance among pure breeding lines (DVF_∞) is found as the sum of Ds of $r = 1$ to ∞. At equilibrium

$$DF_\infty = \Sigma d^2 \pm_R^C \Sigma \frac{2(1-2p)}{1-2p} dadb$$

$$HF_\infty = \Sigma h^2 + \frac{2(1-2p)}{1+2p} hahb$$

Thus linkage disequilibrium in selfing series does not reduce to zero whereas in random mating population linkage disequilibrium reduces to zero and here

$$DF_\infty = \Sigma d^2$$

When $p = 0.0$ or 0.5, all the D's and H's from different generations are equal assuming $h/d = 1.0$ at all loci. The difference between D's and between H's from different ranks is greatest at intermediate recombination frequencies ($p = 0.15$ and $p = 0.35$)

Free and potential forms of variation:

As we know that linkage affects the proportion of different genotypes in the F_2 so it controls the form of variation. The relative magnitude of free and potential form depends on the recombination frequency. Parents in the coupling or repulsion phase would produce variation in the proportion given below (Mather, 1973).

Cross	Linkage Phase	Free	Homozygous Potenital
AABB × aabb	Coupling	$2(1-p)d^2$	$2pd^2$
AAbb × aaBB	Repulsion	$2pd^2$	$2(1-p)d^2$

In the outbreeding population (the random mating population) at equilibrium, the distribution of variability between free and homozygotic potential is independent of linkage. The tighter the linkage the less readily is homozygous potential variability freed and the longer the time necessary for it to become available for utilization by selection.

10.4 Detection of Linkage Disequilibrium

Linkage disequilibrium is detected by changes in D and H with ranks. For example, the total F_3 families variance will contain D's and H's of rank 1 and rank 2, respectively and in the absence of linkage

rank $1D$ = rank $2D$

rank $1H$ = rank $2H$

Thus the test of linkage is a test of homogeneity of D's and H's over ranks within generation. The coefficients of the dominance components are too small to give worthwhile estimates of H_1 and H_2 and hence the contribution of dominance to the variances can be safely ignored and thus whether H_1 is greater than or smaller than H_2 can be glossed over. In the absence of non-allelic interaction and genotype × environment interaction, the test for linkage can be done by fitting additive-dominance model (D_1, H_1, E_1 and E_2) to the statistics V_1F_2, V_1F_3, W_1F_3 and V_2F_3 in addition to the variances of P_1, P_2 and F_1, generations. The additive-dominance model will fail because of heterogeneity of D and H over variances of different ranks. Assuming no linkage if a model with parameters D_1, H_1, $(D_2 + 1/2\ H_2)$, E_1, and E_2 which take care of linkage bias is fitted, the model should be adequate. If the estimate of D_1 is greater than D_2, the predominant linkage phase is coupling but if D_1 is smaller than D_2, it shows the predominance of repulsion phase linkage. As each breeding system, i.e. selfing, backcrossing, sib mating, random mating biases the estimates of D and H in its own characteristic way, the variances used for testing the homogeneity of different ranks must have been obtained from the same breeding system. For example, D, H and E are estimated from the variances of P_1, P_2, F_1, BC_1 and BC_2 families using the following equations:

$$\hat{E} = VP_1, = VP_2 = VF_1$$

$$\hat{D} = 4\ VF_2 - VBC_1 - 2VBC_2$$

where
$$\hat{H} = 4VBC_1 + 4VBC_2 - 4VF_2 - 4E$$
$$\hat{D} = D \pm {}_R^C \Sigma\, 2(1-2p)\, dadb$$
$$- \Sigma\, 4p(1-2p)\, hahb$$
$$\hat{H} = H + \Sigma\, 2(1-4p)^2\, hahb$$

Thus these estimates of D and H are seriously biased in the presence of linkage.

Finally, it can be seen that the magnitude of the linkage bias on D and H decreases with the increase of ranks. Thus in coupling phase linkage

$$D_1 > D_2 > D_3 > D_4 > D$$

but in case of repulsion phase linkage

$$D_1 < D_2 < D_3 < D_4 < D$$

and irrespective of the linkage phase, $H_1 > H_2 > H_3 > H_4 > H$ when the dominance is reinforcing whereas $H_1 < H_2 < H_3 < H_4 < H$ when there is ambidirectional dominance, Comparisons of means and variances of L_1 ($F_2 \times P_1$) with BC_1, L_2 ($F_2 \times P_2$) with BC_2 and L_3 ($F_2 \times F_1$) with F_2 will provide test for linkage (Van der Veen, 1959; Perkins and Jinks, 1970; Jinks 1978). In the absence of linkage the following relationships hold within their sampling errors between the mean of F_2, BC_1 and BC_2 families and those of the L_1, L_2 and L_3 families of an F_2 triple test cross under all circumstances i.e, irrespective of the presence of non-allelic interaction or of genotype-environment interaction.

$$\overline{F_2} - \overline{L_3} = 0$$
$$\overline{BC_1} - \overline{L_1} = 0$$
$$\overline{BC_2} - \overline{L_2} = 0$$

and
$$V_1F_2 = VL_3$$
$$VBC_1 = VL_1$$
$$VBC_2 = VL_2$$

The above relationships between means of different families hold even in the presence of linkage unless linkage is between pairs of genes which also show non-allelic interaction but the variances of the families are no longer identical in the presence of linkage i.e,

$$VBC_1 \neq \sum_{r=1}^{2} VL_2$$
$$VBC_2 \neq \sum_{r=1}^{2} VL_2$$
$$V_1F_2 \neq \sum_{r=1}^{2} VL_3$$

The linkage is in coupling phase and dominance reinforcing when

$$VBC_1 > VL_1$$
$$VBC_2 > VL_2$$
$$V_1F_2 > VL_3$$

The above order is reversed when repulsion phase linkage is in excess with opposing dominance effects.

There is another method of detection of linkage based on comparison of the variances for single character and for phenotypic correlations between pairs of characters of dihaploids derived from the F_1 and F_2 generations and therefore, of different ranks (Snape and Simpson, 1981) and is discussed in Chapter 12.

10.5 Generation that Yields an Approximate Estimate of DVF_∞

The coefficients of D_r follow the geometric series $\left(\frac{1}{2}\right)^r$, the successive terms make a rapidly decreasing contribution to the total variance and also are the coefficients of the linkage components of D_r, $2(1-2p)^r\, dadb$, the successive D_r's show rapidly falling differences. Thus

$$|D_1 - D_2| > |D_2 - D_3| > |D_3 - D_4| > |D_4 - D_5|$$

... Therefore, $\frac{1}{2}D_1 + \frac{1}{4}D_2 + \frac{1}{4}D_3$ assuming that $D_3 = D_4 = D_5$ ... which means $\frac{1}{8}D_3 = \frac{1}{16}D_4 = \frac{1}{32}D_5$ = ... is a close approximation of DF_∞ for all values of the recombination frequency, p. Further, it can be seen that $\frac{1}{2}D_1 + \frac{1}{2}D_2$ assuming $D_2 = D_3 = D_4$

again which means $\frac{1}{4}D_2 = \frac{1}{8}D_3 = \frac{1}{16}D_4 + \frac{1}{32}D_5$ + ... will be an acceptable approximation to DVF_∞ (Jinks and Pooni, 1982). They further showed from an experiment with values of D_r for $r = 1$ to 5 for each pair of gene linked in coupling and in repulsion each with an additive effect $d = 1$ and recombination frequency of $p = 0.1$, 0.25 and 0.4 that irrespective of the linkage phase or recombination frequency $\frac{1}{2}D_1 + \frac{1}{2}D_2 + \frac{1}{4}D_3$ and even $\frac{1}{2}D_1 + \frac{1}{2}D_2$ gave close approximation to the total genetic variance among the means of the pure breeding families of F_∞ generation (DVF_∞). Thus any breeding programme yielding estimates of D_1, D_2 and possibly D_3 will allow to predict the estimate of total genetic variance that the means of pure breeding lines derived by selfing from a cross of two pure breeding lines will show. As in a selfing series the contribution of non-additive sources of variance (dominance) to the variation is non significant, D_1, D_2 can be estimated from F_3 or S_3 families and D_3 also by F_4 or S_4 families. But in the presence of dominance D_1 and D_2 can be obtained from F_3 and S_3 families and D_3 also by F_4 and S_4 families. A randomised and replicated experiment with P_1, P_2, F_1 F_3 and S_3 families will yield estimates of the following statistics from which the value of $\frac{1}{2}D_1 + \frac{1}{2}D_2$ can be obtained.

F_3 families	S_3 families
Between families means-	
$V_1 F_3 = \frac{1}{2}D_1 + \frac{1}{16}H_1$	$V_1 S_3 = \frac{1}{4}D_1 + \frac{1}{16}H_1$
Within families	
$V_2 F_3 = \frac{1}{4}D_2 + \frac{1}{8}H_2 + E$	$V_2 S_3 = \frac{1}{4}D_2 + \frac{3}{16}H_2 + E$

From the statistics V_1F_3, V_2F_3, V_1S_3 and V_2S_3, D_1, D_2, H_1, and H_2 are obtained as

$$D_1 = 4(V_1F_3 - V_1S_3)$$
$$D_2 = 12\,V_2F_3 - 8V_2S_3 - 4E$$
$$H_1 = 32\,V_1S_3 - 16\,V_1F_3$$
$$H_2 = 16(V_2S_3 - V_2F_3)$$
$$DVF_\infty = \frac{1}{2}D_1 + \frac{1}{2}D_2 = 2V_1F_3 + 6V_2F_3$$
$$- 2V_1S_3 - 4V_2S_3 - 2E$$

The estimate of E, the environmental component of variation is obtained from P_1, P_2 and F_1 generations.

The best estimate of D normally available in the early generation of a cross is that from an F_2 triple test cross design which of course, yields estimate of D_1. Although D_1 will underestimate D, this underestimation is small and rather negligible for tight linkages (p approaching zero) and loose linkage (p approaching 0.5). Thus D_1 can reasonably be the total genetic variance among the population of pure breeding lines derived from the cross between a pair of pure breeding lines.

10.6 Effect of Linkage on Covariances

The effect of linkage on covariances between relatives has been studied by Cockerham (1956), Schnell (1961, 1963) and Van Arde (1975). They pointed out that linkage affects the epistatic component of the covariances between full-sibs and half-sibs but not the covariances of offspring and parents. Linkage has no effect on the expectation of covariances when the parents are homozygous, since crossingover will form no new gametes but when parents are non-inbred, the effect of linkage is to increase the coefficients of the epistatic terms. As the linkage increases, the effect will be to reduce the magnitude of epistatic terms. Cockerham (1963) pointed out that linkage presents two troublesome features on the estimation of variances, (i) the base or reference population may not be in linkage equilibrium and (ii) the recombination value less than 0.5 affects some of the covariances between relatives. The linkage bias would not be of much importance unless the average recombination frequency is 0.1 or less.

The effect of linkage disequilibrium on additive and dominance components of BIP's is given by Comstock and Robinson (1948) and for the self progenies by Gates et al. (1957). The dominance components in BIP's is increased by linkage disequilibrium. The results are otherwise too complex for any trend to be predicted and the same is true for $\sigma^2_{A \times A}$. Lack of free recombination affects only and will increase the $\sigma^2_{A \times A}$ terms in BIP's.

10.7 Linkage vs. Pleiotropy

Association between two characters say X and Y is expected when either they share genes in common or they share common environment. That is, correlation between characters can be due to genetic or environmental causes. If the two traits have no genes in common we would expect them to be uncorrelated. Further, if the genetic systems seem to be correlated or have a common genetic basis then the genes might act in different ways. When a gene has manifold effects, its action is called pleiotropic. There are numerous examples of such genes in classical genetics. Such genes can also be thought of operating for characters showing quantitative variation. Falconor (1960) advocates that the association between characters is more due to pleiotropy of genes than linkage. However, Mather and Jinks (1971) hold a different view and they saw linkage, in general, the cause of correlation. The three parameters namely, additive genetic correlation, genotypic correlation and phenotypic correlation can throw light on the nature of the underlying genetic mechanism of association between characters. The presence of epistatic variance always inflates the estimates of additive and dominance variance. The estimate of additive genetic covariance, used for measuring additive genetic correlation coefficient, will be inflated or deflated depending upon the signs and the relative magnitudes of the epistatic component of covariance. In the event the signs and magnitudes of the components of covariances are such that all other genetic covariances except $\sigma_{A \times A}$ vanish, the additive genetic correlation coefficient will still be under-estimated since the denominator is always inflated in the presence of epistasis.

We can have a whole range of pleiotropy, ranging from partial to complete pleiotropy. When all genes affect both characters X and Y, the genes may increase or decrease both equally or genes may increase X but decrease Y equally and vice-versa. Some genes affect both X and Y while others affect only X or Y or some genes affect only X while others affect only Y. Further, the same gene might affect the two traits unequally. The correlation arising from pleiotropy is as a result of the overall effects of all the segregating genes that affect both traits. Thus pleiotropy does not necessarily cause detectable correlation. That is correlation coefficient could be zero. Selection can distinguish between linkage of genes and pleiotropy. If the genotypic correlation is due to pleiotropic genes, selection for the two traits X and Y in the same direction will cause a positive change in the genotypic correlation. Simultaneously selection for X and Y in the opposite direction will cause a positive change in genotypic correlation. If the negative genotypic correlation is due to pleiotropy, it is unlikely that a new combination of characters can be obtained. But the situation is different when the correlation is due to linkage of genes. Here the linkage can be broken by intercrossing (more especially recombination) in the segregating population. Thus individuals showing all the four extreme combinations of X and Y such as high X and high Y, low X and low Y, Low X and High Y and high X and Low Y can be selected. In other words, individuals showing positive and negative correlation can be isolated but if linkage is complete ($p = 0$) then we will fail in isolating these combinations. The distinction between linkage disequilibrium and pleiotropy as a cause of correlation can be made following the study of magnitude and sign of correlation coefficient between characters in inbreds/pure breeding developed through single seed descent and dihaploidy (Jinks et al. 1985) and is discussed in Chapter 12.

10.8 Epistasis

We have seen in Chapter 5, the different types of digenic interaction parameters, their detection and estimation. The corresponding parameters in variances to digenic interaction parameters of means are

$$I = \Sigma\, i^2$$

$$J = \Sigma\, j^2$$

$$L = \Sigma\, l^2$$

which are homozygous × homozygous, homozygous × heterozygous and heterozygous × heterozygous interactions, respectively. In the presence of non-allelic interaction the variance of F_2 (VF_2) derived from a cross between pairs of pure breeding lines becomes

$$VF_2 = \left[\frac{1}{16}(da + db + iab)^2 + \ldots\right.$$
$$\left. + \frac{1}{16}(-da - db + iab)^2\right]$$
$$- \left(\frac{1}{2}ha + \frac{1}{2}lab + \frac{1}{4}lab\right)^2$$
$$\vdots$$
$$= \frac{1}{2}\left(da + \frac{1}{2}jab\right)^2 + \frac{1}{2}\left(db + \frac{1}{2}jba\right)^2$$
$$+ \frac{1}{4}\left(ha + \frac{1}{2}lab\right)^2 + \frac{1}{4}\left(hb + \frac{1}{2}lab\right)^2$$
$$+ \frac{1}{4}iab^2 + \frac{1}{8}(jab + jba)^2 + \frac{1}{16}lab^2$$
$$= \frac{1}{2}D + \frac{1}{4}H + \frac{1}{4}I + \frac{1}{8}J + \frac{1}{16}L$$

whereas the variance of $F_2(VF_2)$ in the absence of non-allelic interaction is

$$VF_2 = \frac{1}{2}(da^2 + db^2) + \frac{1}{4}(ha^2 + hb^2)$$
$$= \frac{1}{2}D + \frac{1}{4}H$$

Thus it differs when there is non-allelic interaction. In the presence of non-allelic interaction D and H are no longer $= \Sigma d^2$ and $= \Sigma h^2$, respectively and they take the following forms:

$$D = \Sigma\left(da + \frac{1}{2}j\right)^2$$
$$H = \Sigma\left(ha + \frac{1}{2}l\right)^2$$

Thus in the presence of non-allelic interaction d is confounded with j and h is confounded with l.

The F_3 family variance (V_1F_3) takes the form as

$$V_1F_3 = \frac{1}{2}\left(da + \frac{1}{4}jab\right)^2 + \frac{1}{2}\left(db + \frac{1}{4}jba\right)^2$$
$$+ \frac{1}{16}\left(ha + \frac{1}{4}lab\right)^2 + \frac{1}{16}\left(hb + \frac{1}{4}lab\right)^2$$
$$+ \frac{1}{4}iab^2 + \frac{1}{32}(jab + jba)^2 + \frac{1}{25}lab^2$$

$$= \frac{1}{2}D + \frac{1}{16}H + \frac{1}{4}I + \frac{1}{32}J + \frac{1}{256}L$$

The variances V_2F_3 and covariances F_2F_3 become

$$V_2F_3 = \frac{1}{4}D + \frac{1}{8}H + \frac{5}{16}I + \frac{7}{16}J + \frac{1}{32}L$$

$$\text{Cov } F_2F_3 \, (W_1F_{32}) = \frac{1}{2}D + \frac{1}{8}H + \frac{1}{4}I$$
$$+ \frac{1}{16}J + \frac{1}{64}L$$

In these statistics D and H take the following form:

$$D = \Sigma\left(d + \frac{1}{4}j\right)^2$$
$$H = \Sigma\left(h + \frac{1}{4}l\right)^2$$

Thus the coefficients of j and l are now reduced to $\frac{1}{4}$. This shows that the definition of D changes from F_2 to F_3 generation. Also it can been seen that in the first rank variances and covariances, the coefficients of I, J and L are the products of the coefficients of D and H but this relationship does not hold with variances and covariances of higher rank. Further it can also be observed that bias due to epistasis is different in different statistics within the same rank (V_1F_2, V_1F_3) and also bias is different in different statistics of different ranks. V_1F_2 and V_1F_3 on one hand and V_2F_3 on the other. Thus additive-dominance model will fail when fitted to the similar rank statistics like V_1F_2, V_1F_3, V_1F_4 and so on from different generations and it will fail also when fitted to statistics of different ranks. We have seen that in the presence of linkage. the additive-dominance model fits when applied to the statistics of similar rank although obtained from different generations but fails when applied to statistics of different ranks and so we can not distinguish between whether the failure is due to epistasis or linkage alone or both.

Hayman and Mather (1955) showed that the i type interaction which does not affect the definition of D and H, may resemble linkage (particularly repulsion) effect. Opsahl (1956) showed that under certain conditions upto approximately 40% of the variation legitimately attributable to interaction can be taken up by allowing for linkage and vice-versa. He showed that linkage phase is correlated with the type of non-allelic interaction. Association or coupling phase linkage is positively correlated with duplicate gene effects whereas dispersion phase effect is positively correlated with complementary gene effects. The model fitting with parameters D, H, E_1 and E_2 will yield a sum of squares for deviation depending upon the effects of linkage and non-allelic interaction. After fitting extra parameters to allow explicitly for either linkage or interaction any significant reduction in the residual sum of squares must indicate the presence of linkage or interaction effects provided there is no correlation between linkage and non-allelic interaction. The sum of squares remaining after allowing for linkage or interaction can be attributed unambiguously to interaction or linkage effects, respectively. But then a simple scaling test is more sensitive and more specific than sequential model fitting procedures for detecting complex effects such as linkage, non-allelic interaction and genotype × environment interaction and correlations.

10.9 Effect of Non-allelic Interaction on Estimation of Variance Components

With the introduction of interaction variance there will be seven parameters namely D, H, E, F, I, J and L to be estimated from six basic generations (P_2, P_2, F_2, F_2, BC_1 and BC_2) and consequently it does not appear feasible to estimate genetic variances and covariances from a joint analysis of generations in the presence of non-allelic interaction. In fact there is no biometrical genetical analysis which will allow simultaneous estimation of D, H, I, J and L when these sources of variation are present. In general, the presence of interaction components of variation biases the estimates of D and H although these biases vary with the breeding design from which they are estimated. Pooni and Jinks (1979, 1982) showed that these biases are complex but their effects can be shown by examining the estimates of D and H obtained in case of two loci with two alleles each (A, a and B, b) assuming non-allelic interaction.

In the selfing series when genes are associated with the parents, the estimates of D and H from P_1, P_2, F_1, F_2, BC_1 and BC_2 generations become

$$D' = D + \frac{1}{4}(iab^2 + jab^2 + jba^2) + dajab + dbjba$$
$$+ \frac{1}{2} jab\,jba* + dajba* + dbjab*$$
$$+ iab\,(ha + hb)* + \frac{1}{2} iab\,lab*$$
$$H' = H + \frac{1}{2}\left(iab^2 + jab^2 + jba^2 + \frac{3}{2} lab^2\right)$$
$$+ lab\,(ha + hb) - 2\,dajba* - 2\,dbjab*$$
$$- 2\,lab\,(ha + hb)* - iab\,lab^2$$

The sign of the terms marked with an asterisk reverses if the genes in the parents are dispersed. Also, the cross product terms with the exception of $jab\,jba$ and $iab\,lab$ may take positive or negative sign depending upon the kind of interaction present. In case of complementary interaction dj, hi and hl terms will be positive whereas they will be negative for duplicate interaction. In addition the majority of the terms appear with opposite signs in the interaction biases of D' and H'. Thus we see that the magnitudes and signs of the biases are functions of the type of interaction and the distribution of the interacting genes and that if one of the components (D, H) is inflated then the other is deflated. It can be seen that with associated complementary interaction or dispersed duplicate interaction D is inflated and H is deflated whereas with associated duplicate it is just the reverse.

10.9.1 Biases in D and H components obtained from T.T.C.

Again when the genes in the parents are in association phase we have

$$D'' = D + \frac{1}{4} iab^2 + \frac{13}{16}(jab^2 + jba^2) + \frac{1}{9} lab^2$$
$$+ dajab + dbjba + \frac{2}{3} jabjba* - 2\,dajba*$$
$$- \frac{2}{3} dbjab* + \frac{1}{3} iab\,lab*$$

$$H'' = H + \frac{1}{2} iab^2 + \frac{1}{8}(jab^2 + jba^2) + \frac{1}{2} lab^2 + \frac{1}{4} jab\, jba$$

As before the terms marked with an asterisk reverse their sign if the genes in the parents are dispersed and the cross product of terms d and j may be positive or negative depending upon whether the interaction is complementary or duplicate, respectively. However, the biases of D and H are not negatively correlated as we found in case of D and H obtained from basic generations. H'' is always inflated by interaction irrespective of the kind of interaction or distribution of alleles in the parents. However, D may be inflated or deflated by interaction biases. With dispersed duplicate genes, it may be deflated but in all other cases it is more likely to be inflated and more often than H'' (Jinks, 1983). Thus in most circumstances both D'' and H'' are inflated and are inflated by similar amounts. The bias on dominance ratio based on the estimates of D'' and H'' obtained from TTC will therefore, be relatively small in comparison to those based on estimates obtained from basic generations.

The triple test cross not only detects non-allelic interaction but also partitions the interaction into $[i]^2$ and $\left(j + \frac{1}{2}l\right)^2$ and provides their estimates. But here $[i]^2 = r_i^2 (\Sigma i)^2$ which is different than I which equals Σi^2.

$$[i]^2 = \Sigma i^2 \text{ only when } r_i \underline{\Omega} \pm \sqrt{\frac{1}{k}}$$

where k is the number of genes involved in the determination of a character. If $r_i > +\sqrt{\frac{1}{k}}$ or $< -\sqrt{\frac{1}{k}}$ which signifies an excess of coupling or repulsion among the interacting pairs of genes, respectively then $[i]^2$ will be greater than 1.

10.10 Comparison of Tests of Epistasis

If we compare the different available tests for epistasis we find that under the assumptions $iab = lab$ and $jab = jba$, the test, $\overline{L}_1 + \overline{L}_2 - \overline{P}_i$ will fail to detect epistasis but the test, $L_1 + L_2 - 2L_3$ will not as in case of later the variance of $(\overline{L}_1 + \overline{L}_2 - 2\overline{L}_3)$ equals $[i]^2$ $+ \Sigma l^2 + \Sigma j^2$ and not the i, j and l i.e. it is the squared value and that it why it will not fail. In case of scaling test the individual scaling tests will all fail in the presence of above mentioned assumptions but the joint scaling test will not fail. The Wr, Vr graph will often fail under such assumptions but certainly fail under some other assumptions.

10.11 Epistasis in Random Mating Population

Mather (1974) considered the consequences of digenic interactions on the variances and covariances estimated from randomly mating population. In the presence of non-allelic interaction, expectations of covariance offspring-parent (Cov OP), covariance full-sib (Cov FS) and covariance half-sib (Cov HS) are as follows:

$$\text{Cov } OP = \frac{1}{4} DR + \frac{1}{16} IR$$

$$\text{Cov } FS = \frac{1}{4} DR + \frac{1}{16} HR + \frac{1}{16} IR + \frac{1}{16} JR + \frac{1}{256} JR$$

$$\text{Cov } HS = \frac{1}{8} DR + \frac{1}{64} IR$$

The total variance of random mating population equals

$$\frac{1}{2} DR + \frac{1}{4} HR + \frac{1}{4} IR + \frac{1}{8} JR + \frac{1}{16} LR.$$

The components of variation DR, HR, IR, JR and LR take the following form.

$$DR = 4uava\,\{[d_a + 2ubva\,jab + (ub - vb)\,iab]$$
$$- (ua - va)\,[ha + (ub - vb)\,jba$$
$$+ 2ubvblab]\}^2 + 4ubvb\,\{[db + 2\,uavajba$$
$$+ (ua - va)\,iab] - (ub - vb)$$
$$[hb + (ua - va)\,jab + 2\,uavaLab]\}^2$$

$$HR = 16\,ua^2\,va^2\,[ha + (ub - vb)\,jba$$
$$+ 2ubvblab]^2$$

$$+ 16 ub^2vb^2 [hb + (ua - va) jab + 2 uavalab]^2$$

$$IR = 16 uavaubvb [iab - (ub - vb) jab$$
$$- (ua - va) jba + (ua - va)(ub - vb) lab]^2$$

$$JR = 64 uavaub^2vb^2 [jab - (ua - va) lab]^2$$
$$+ 64 ua^2va^2 ubvb [jba - (ub - vb) lab]^2$$

$$LR = 256 ua^2va^2ub^2vb^2lab^2$$

The expectation of DR, HR, IR, JR and LR reduces to D, H, I, J and L, respectively of the F_2 of a cross from two pure breeding lines where $ua = va = ub = ub = \frac{1}{2}$. The expectations of the various statistics show that IR appears with DR and JR and LR appear with HR. Thus it is not easy to separate IR from DR and JR from HR and the estimates of DR and HR are biased in the presence of interaction variances. The estimates of DR and HR not only depend on the gene frequencies but also on the magnitudes and signs of i, j and l. The interaction variance, IR can be estimated as the difference between the parent-offspring covariance and the covariance half-sib as

$$\text{Cov } OP - 2 \text{ Cov } HS = \left(\frac{1}{4} DR + \frac{1}{16} IR\right)$$
$$- 2\left(\frac{1}{8} DR + \frac{1}{64} IR\right) = \frac{1}{32} IR$$

The problem with this estimate of interaction variance is that the estimates of covariances are subject to large standard error.

Estimates of epistatic genetic variance can be obtained from genetic variances and covariances from a combination of different designs such as NCM I and NCM II using parents with different levels of inbreeding (Cockerham 1952, 1956). Combining different mating designs with materials of different levels of inbreeding provides additional covariances of relations and differences in coefficients of genetic variance components for covariances of relatives thereby permitting detection and estimation of epistatic variance. Using NCM I with random mating individuals ($F = 0$) covariance full-sib and covariance half-sib can be obtained. From the same random mating population inbred lines can be derived without selection and using NCM II covariance HS and covariance FS can be computed. Cockerham (1956) showed that the epectation of Cov FS and Cov HS under different levels of inbreeding will take the following forms:

$$\text{Cov } FS = \sum_{i,j} \left(\frac{1+F}{2}\right)^{i+2j} \sigma^2 ij$$

$$\text{Cov } HS = \sum_{i} \left(\frac{1+F}{4}\right)^{i} \sigma^2 ij$$

where i and j indicate the number of times additive and dominance appear in the nomenclature, respectively, and F is the level of inbreeding of parents Thus,

$$\text{Cov } FS (F = 0) = \frac{1}{2} \sigma_A^2 + \frac{1}{4} \sigma_D^2 + \frac{1}{4} \sigma_{AA}^2$$
$$+ \frac{1}{8} \sigma_{AD}^2 + \frac{1}{16} \sigma_{DD}^2 + \frac{1}{8} \sigma_{AAA}^2$$

$$\text{Cov } FS (F = 1.0) = \sigma_A^2 + \sigma_D^2 + \sigma_{AA}^2 + \frac{1}{8} \sigma_{AD}^2$$
$$+ \sigma_{DD}^2 + \sigma_{AAA}^2$$

$$\text{Cov } HS (F = 0) = \frac{1}{4} \sigma_A^2 + \frac{1}{16} \sigma_{AA}^2 + \frac{1}{64} \sigma_{AAA}^2$$

$$\text{Cov } HS (F = 1.0) = \frac{1}{2} \sigma_A^2 + \frac{1}{4} \sigma_{AA}^2 + \frac{1}{8} \sigma_{AAA}^2$$

The total genetic variance, σ_G^2 becomes

$$\sigma_G^2 (F = 1.0) = 4 \sigma_{f/m}^4$$
(NCMI)
$$= \sigma_A^2 + \sigma_D^2 + \frac{3}{4} \sigma_{AA}^2 + \frac{1}{2} \sigma_{AD}^2$$
$$+ \frac{1}{4} \sigma_{DD}^2$$

Here $\sigma_{f/m}^2 = \text{Cov } FS (F = 0) - \text{Cov } HS (F = 0)$
$$\sigma_m^2 = \text{Cov } HS (F = 0)$$
$$\sigma_G^2 (F = 1.0) = \sigma_m^2 + \sigma_f^2 + \sigma_{f\times m}^2$$
(NCMII)
$$= \sigma_A^2 + \sigma_D^2 + \sigma_{AA}^2 + \sigma_{AD}^2 + \sigma_{DD}^2$$

Here $\sigma_m^2 = \sigma_f^2 = \text{Cov } HS (F = 1.0)$

10.12 Biometrical Genetics–Analysis of Quantitative Variation

$$\sigma^2_{m \times f} = \text{Cov } FS\,(F=1.0) - 2\,\text{Cov } HS\,(F=1.0)$$

The interaction variance (σ^2_{Ig}) is calculated as

$$\sigma^2_{Ig} = \sigma^2_G\,(F=1.0) - \sigma^2_G\,(F=0.0)$$

$$= \frac{1}{4}\sigma^2_{AA} + \frac{1}{2}\sigma^2_{AD} + \frac{3}{4}\sigma^2_{DD}$$

As interaction variance (σ^2_{Ig}) can be partitioned into additive and dominance type interaction

$$\sigma^2_{Ig} = \sigma^2_{I_a} + \sigma^2_{I_d}$$

the estimate of σ^2_{Ia} can be obtained as

$$\sigma^2_A = \sigma^2_A\,(F=1.0) - \sigma^2_A\,(F=0.0)$$

$$= \frac{1}{4}\sigma^2_{AA} + \frac{3}{16}\sigma^2_{AAA} + \frac{7}{16}\sigma^2_{AAAA}$$

where

$$\sigma^2_A\,(F=1.0) = \sigma^2_m\,(F=1.0) + \sigma^2_f\,(F=1.0)$$

$$\sigma^2_A\,(F=0.0) = 4\,\sigma^2_m\,(F=0.0)$$

Similarly, σ^2_{Id} can be estimated as

$$\sigma^2_{Id} = \sigma^2_D\,(F=1.0) - \sigma^2_D\,(F=0.0)$$

$$= \frac{1}{2}\sigma^2_{AD} + \frac{3}{4}\sigma^2_{DD} + \frac{3}{8}\sigma^2_{AAA}$$

The less biased function of additive and dominance variances, in the presence of epistasis, are

$$\sigma^2_{A'} = 8\,\sigma^2_m\,(F=0.0) - (\sigma^2_m\,(F=1.0)$$
$$+ \sigma^2_f\,(F=1.0)$$

$$= \sigma^2_A - \left(\frac{1}{8}\sigma^2_{AA} + \frac{3}{32}\sigma^2_{AAA} + \frac{7}{128}\sigma^2_{AAA} + ...\right)$$

$$\sigma^2_{D'} = 4\,(\sigma^2_{flm}\,(F=0.0) + \sigma^2_m\,(F=0.0)$$

$$- 2\,(\sigma^2_m + \sigma^2_f)\,F=1.0$$

$$= \sigma^2_D - \left(\frac{1}{2}\sigma^2_{AD} + \frac{1}{4}\sigma^2_{DD} + \sigma^2_{AAD}\right)$$

The expected variances of the variance estimates are computed as follows:

$$\text{Var}\,(\hat{\sigma}^2) = 2\,\Sigma\,ai^2\,Mi^2\,(fi+2)$$

Where ai's are the linear coefficients used in computing $\hat{\sigma}^2$, Mi's are the mean squares and fi's are the degrees of freedom.

The problem with this method of estimation of epistatic components is that the test of epistasis is associated with large standard error. Standard errors are quadratic forms and are also comprised of linear functions of several mean squares as a result the SE's tend to be large in comparison with the variance components estimates (Stuber et al. 1966). Stuber and Moll (1969) obtained estimates of epistatic variance by comparison of mean squares obtained from the study of F_1 crosses in NCM II and derived S_1 progenies from the cross. They noted that the difference between results, whether present or absent, of epistasis from two groups of material depends on: (i) experimental approach and the material used and (ii) the past selection history of the material used. They further suggested that for detecting and estimating epistasis, the better population is the selfing generations derived from the cross of two pure breeding lines as selfing enhances loci, which in homozygous state interact with other loci by increasing their frequency. This then enhances the probability of detecting epistasis whereas in random mating population, the combination of loci giving epistatic effects may occur too infrequently or with limited effects and thus they are not detectable. Results from a large number of studies (designed to estimate epistatic variance) have shown that they all failed to obtain realistic estimates of epistatic variance components (Hallauer and Miranda, 1981). Thus it seems that it may be either impossible to estimate epistatic variance component with the present genetic models or the relative portion of total genetic variance as the epistatic variation is quite small. Chi et al. (1969) indicated that the higher correlation between coefficients of digenic and trigenic epistatic components of variance and the additive and dominance variance components results into greater standard errors of the estimates of genetic components of variance and thereby reduces the sensitivity of model for detecting and estimating epistasis. Smaller coefficients of epistatic terms relative to the coefficients of additive and dominance terms are also an important factor in the detection and estimation of epistatic component of variance. No statistical design has been developed for separation of effects of epistasis and linkage from the genetic variance or for estimating the mean and variance of the recombination value and there is a problem of interpreting the genetic variances if epistasis is present.

10.12 Detection of Linked Epistasis

Jinks and Perkins (1969) proposed a model for detecting and estimating linked digenic interaction parameters. Their model included a large number of new parameters in addition to m, d, h, i, j and l which depended on the nature of generation included in the analysis. To specify the 21 generation means, they needed fourteen parameters, namely, m, d, h, i, j, l, pi, p^2i, pj, p^2j, pl, p^2l, p^3l and p^2l. Because of the complex correlation between these parameters m, d, h, i and l were estimated as $m + h + l$ and $m + i$ as m, h, l, were completely correlated and so were the m and i. This problem was overcome when Jinks (1978) described an unambiguous test for linked digenic interaction. In the presence of linkage, the expected variances of a BC_1 family and a L_1 generation, a BC_2 family and a L_2 generation and an F_2 family and a L_3 generation are no longer the same but the expected means are not affected by linkage unless epistasis is present i.e.

$$\overline{BC}_1 = \overline{L}_1, \overline{BC}_2 = \overline{L}_2 \text{ and } \overline{F}_2 = \overline{L}_3 = \overline{F}_2 \text{ bips.}$$

In the presence of linked epistasis the expectations of means of BC_1, L_1, BC_2, L_2, F_2 and L_3 generations are as follows:

$$\overline{BC}_1 = m + \frac{1}{2}[d] + \frac{1}{2}[h] + \frac{1}{2}[i] + \frac{1}{2}[l]$$
$$- \frac{1}{2}[pi] + \frac{1}{2}[pj] - \frac{1}{2}[pl]$$

$$\overline{L}_1 = m + \frac{1}{2}[d] + \frac{1}{2}[h] + \frac{1}{2}[i] + \frac{1}{2}[l]$$
$$- \frac{3}{4}[pi] + \frac{1}{2}[p_i^2] + \frac{3}{4}[p_j] - \frac{1}{2}[p_j^2]$$
$$- \frac{3}{4}[pl] + \frac{1}{2}[p_l^2]$$

$$\overline{BC}_2 = m - \frac{1}{2}[d] + \frac{1}{2}[h] + \frac{1}{2}[i] - \frac{1}{2}[l]$$

$$\frac{1}{2}[pi] - \frac{1}{2} - [pj] - \frac{1}{2}[pl]$$

$$\overline{L}_2 = m - \frac{1}{2}[d] + \frac{1}{2}[h] + \frac{1}{2}[i] + \frac{1}{2}[l]$$
$$- \frac{3}{4}[pi] + \frac{1}{2}[p_i^2] - [pj] + \frac{1}{2}[p_j^2]$$
$$\frac{3}{4}[pl] + \frac{1}{2}[p_l^2]$$

$$\overline{F}_2 = m + \frac{1}{2}[h] + \frac{1}{2}[i] + \frac{1}{2}[l] - [pi] - [pl]$$
$$+ [p_l^2]$$

$$\overline{L}_3 = m + \frac{1}{2}[h] + \frac{1}{2}[i] + \frac{1}{2}[l] - \frac{5}{4}[pi]$$
$$+ \frac{1}{2}[p_i^2] - \frac{5}{4}[pl] + [p_i^2] - [p_i^3]$$

Thus,

$$\overline{BC}_1 - \overline{L}_1 = \frac{1}{4}([p(1-2p)i] - [p(1-2p)j]$$
$$+ [p(1-2p)l])$$

$$\overline{BC}_2 - \overline{L}_2 = \frac{1}{4}([p(1-2p)i] + [p(1-2p)j]$$
$$+ [p(1-2p)l])$$

$$\overline{F}_2 - \overline{L}_3 = \frac{1}{4}([p(1-2p)i] + [p(1-2p)]^2 l])$$

The expectations of $(\overline{BC}_1 - \overline{L}_1)$ $(\overline{BC}_2 - \overline{L}_2)$ and $(\overline{F}_2 - \overline{L}_3)$ are zero when either there is no epistasis i.e. $i = j = l = 0$ or there is no linkage i.e. $p = \frac{1}{2}$. Thus in the presence of linked epistasis

$$\overline{BC}_1 \neq \overline{L}_1, \overline{BC}_2 \neq \overline{L}_2 \text{ and } \overline{F}_2 \neq \overline{L}_3$$

Let us now see how the estimates of parameters of digenic interaction m, d, h, i, j and l are biased when estimated assuming no linkage when linkage is present. The estimates of above parameters under the assumptions, namely, no linkage and linkage are given in Table. 10.1

The estimates of the parameters in Table 10.1 show that if we estimate m, d, h, i, j and l parameters assuming no linked epistasis when it is there then h is over-estimated whereas i, j and l, in general, are under-estimated and the tighter the linkage, the greater the biases become. We will see in Chapter 11

Table 10.1 Estimates of digenic interaction parameters

No linkage	Linkage
m	$m + i - 2\,[pi] - 2\,[pl] + 4\,[p_l^2]$
$[d]$	d
$[h]$	$[h] - [i] + 2\,[pi] + [l] + 2\,[pl] - 8\,[p_l^2]$
$[i]$	$2\,[pi] + 2\,[pl] - 4\,[p_l^2]$
$[j]$	$2\,[pj]$
$[l]$	$4\,[p_l^2]$

that h, and l are components of heterosis so then the contribution of dominance to heterosis is overestimated whereas the contribution of non-allelic interaction to heterosis is underestimated.

In normally self-fertilizing species all the evidence are that heterozygote × heterozygote (l) or homozygote × heterozygote (j) types of interactions are less important than a homozygote × homozygote (i) type so it can be safe to assume only additive type of epistasis (Matzinger and Cockerham, 1963).

The dominance and epistasis arise not so much as a means of obtaining a particular genetic value, but of securing a high frequency of these values in the sexual offspring of outbreeders. As the material with breeder becomes highly selected, the importance of dominance and the interaction becomes greater. Dominance and epistasis are likely to be ambidirectional in natural population subjected to stabilizing selection while directional selection will tend to produce unidirectional dominance and duplicate type interaction (Mather, 1973).

Information on amount of additive × additive variance is of importance for developing effective breeding strategies and for predicting response due to selection. If epistatic variance is high then one should go for selection between families and line breeding. If additive epistasis is present then in the conventional breeding method, selection should not be too severe in early stages of breeding for allowing desirable epistatic combination to come together. Where a large amount of epistasis is present more inbreeding prior to final selection and recombination will doubtless be worthwhile.

References

Chi, R.K., Eberhart, S.A. and Penny, L.H. 1969. Covariance among relatives in a maize variety (*Zea mays* L.). Genetics, **63**: 511–520.

Cockerham, C.C. 1954. An extension of the concept of partitioning hereditary variance for analysis of covariances among relatives when epistasis is present. Genetics, **39**: 859–882.

Cockerham, C.C. 1956. Effects of linkage on the covariances between relatives, Genetics, **41**: 138–141.

Cockerham, C.C. 1963. Estimation of genetic variances. Symposium on Statistical Genetics and Plant Breeding, NAS, NRC 982, 53–94.

Comstock, R.E. and Robinson, H.P. 1948. The components of genetic variance in populations of biparental progenies and their use in estimating the average degree of dominance. Biometrics, **4**: 254–266.

Gates, C.E., Comstock, R.E. and Robinson, H.P. 1957. Generalized genetic and covariance formulae for self fertilized crops assuming linkage. Genetics, **42**: 749–763.

Hallauer, A.R. and Miranda Fo, J.B. 1988. Quantitative Genetics in Maize Breeding. Iowa State Univ. Press, Ames, Iowa, USA.

Jinks, J.L. 1978. Unambiguous test for linkage of genes displaying non-allelic interactions for a metric trait. Heredity, **40**: 171–173.

Jinks, J.L. 1981. The genetic frame work of plant breeding. Phil. Trans. Res. Soc. Lond. B. **292**: 407–419.

Jinks, J.L. and Perkins, J.M. 1969. The detection of linked epistatic genes for a metrical trait. Heredity, **24**: 465–475.

Jinks, J.L. and Pooni, H.S. 1982. Predicting the properties of pure breeding lines extractable from a cross in the presence of linkage. Heredity. **49**: 265–270.

Jinks, J.L., Pooni, H.S. and Chowdhury, M.K.U. 1985. Detection of linkage and pleiotropy between characters of *Nicotiana tabacum* using inbred lines produced by dihaploidy and single seed descent. Heredity, **55**: 327–333.

Kearsey, M.J. 1985. The effect of linkage on additive genetic variance with inbreeding an F_2. Heredity, **55**: 139–143.

Kearsey, M.J. and Pooni, H.S. 1996. The Genetical Analysis of Quantitative Traits. Chapman and Hall, London.

Mather, K. 1938. The Measurement of Linkage in Heredity. Methuen, London.

Mather, K. and Jinks, J.L. 1982. Biometrical Genetics. 3rd edn. Chapman and Hall, London.

Moreno–Gonzalez, J. and Dudley, J.W. 1981. Epistasis in related and unrelated maize hybrids is determined by three methods. Crop Sci., **21**: 644–651.

Perkins, J.M. and Jinks, J.L. 1970. Detection and estimation of genotype × environment, linkage and epistatic components of variation for a metrical trait. Heredity, **25**: 157–177.

Pooni, H.S. and Jinks, J.L. 1986. Estimation of true additive genetic variance in the presence of linkage disequlibrium, Heredity, **57**: 341-344.

Pooni, H.S. and Jinks, J.L. 1981. The true nature of non-allelic interaction in *Nicotiana rustica* revealed by association crosses. Heredity, **47**: 253-258.

Pooni, H.S. and Jinks, J.L. 1982. Comparative analyses of dispersion and association crosses to detect linkage and epistatic components of variations. Heredity, **49**: 211–220.

Schnell, F.W. 1961. Some general formulations of linkage effects in inbreeding. Genetics, **46**: 947–958.

Schnell, F.W. 1963. The covariance between relatives in the presence of linkage, In: Statistical Genetics and Plant Breeding, W.D. Hanson and Robinson, H.F., eds, pp. 468–483, NAS, NRC Pub. 982.

Stuber, C.W. and Moll, R.H. 1969. Epistasis in maize (*Zea mays* L.) 1. FI hybrids and their S_1 progeny, Crop Sci., **9**: 124–127.

Van der Veen, J.H. 1959. Tests for non-allelic interaction and linkage for quantitative characters in generations derived from two diploid pure lines. Genetica XXX: 201–232.

Van, Ooijen, J.W. 1989. Estimation of additive genotypic variance with the F_3 of autogamous crops. Heredity, **63**: 73–81.

11

Heterosis and Inbreeding Depression

11.1 Heterosis

When two inbred lines of the cross-fertilizing crop like maize are crossed, the, F_1 always outperform its parents in vigor and fertility. This phenomenon is called heterosis or hybrid vigor (Shull, 1909). Likewise, the F_1 of the two pure breeding lines from self-fertilizing crop such as cereals, may show some heterosis but the magnitude of heterosis is comparatively very low. In cross-pollinated crop, heterosis can be over 200% as in corn in comparison to about 10% in self-pollinated crop like wheat. In some cases the heterosis may not occur at all and in some other cases the F_1's may be inferior to its parents. Thus overall we have the following two situations.

1. When F_1 means exceeds the mean of the better parent—a case of positive heterosis.
2. When F_1 mean is less than the mean of the low scoring parent—a case of negative heterosis

11.2 Theories of Heterosis

As with any other phenomena we would first discuss the biometrical genetics of heterosis. In other words, we would like to know the types of gene action involved in this phenomenon of heterosis. The three theories put forward relate to the type of gene action involved in heterosis.

These theories are:
1. Dominance theory (Davenport, 1908, Bruce 1910; Keeble and Pellow, 1910 and Jones 1917)
2. Overdominance theory or Single gene heterosis (East, 1936 and Hull, 1952)
3. Epistasis theory (Power, 1944 and Jinks, 1955)

The proponents of the dominance theory suggested that the F_1 is heterotic because the recessive genes which cause deleterious effects when present in homozygous condition as in parents are masked by dominant genes in F_1. Also, the number of loci where the recessive genes are in homozygous condition is reduced in F_1 in comparison to parents. Thus, F_1 shows superiority in performance over better parent. This theory leads one to believe that the superiority of performance is due to outcrossing advantage which is caused by dominance. But then this outcrossing advantage might be caused by the avoidance of sib competetion, i.e. the outcrossed progenies interfere less with each other or by some other frequency dependent selection mechanism (e.g. minority advantages caused by predators or pathogens).

The overdominance as a cause of heterosis relates hybrid vigour with the heterozygosity *per se*. In other words there is superiority of heterozygous genotypes in relation to homozygous genotypes. Lerner (1954) while giving his concept of homeostasis stated that heterozygotes display a greater developmental homeostasis than do homozygotes whose phenotypes are more variable due to increased susceptibility to environmental stresses, a theory later disproved by Mather and Jinks (1955) and others.

Crow (1948, 1954) demonstrated that with the three population parameters, n (number of loci), μ (mutation rate) and s (selection coefficient of a recessive phenotype) the average gain in yield of a population in which all the deleterious genes have been replaced by dominant genes would not be expected to greatly exceed 5% for a reasonable values of n, μ and s. But studies by Gardner (1961, 1969) and Marck and Gardner (1979) showed that the mass selection in maize resulted in an increase in yield which was reasonably greater than the 5% maximum and thus supporting the overdominance theory. Also, double cross hybrids from parents derived by selfing from the same open pollinated variety out-yielded the parental performance by over the 5% maximum. Hull (1945, 1952) while arguing in favor of the overdominance theory stated that the dominance theory is inadequate as it does not explain:

I. Why hybrids often exceed the sum of the parents?
II. Why mass selection does not result in increase in yield?
III. Why hybrids from inbreds of second or third cycle yield little than those involving original parental lines?

The II and III reasons were later found to be incorrect (Gardner 1961, 1969). Mather and Jinks (1971) and Jinks and Perkins (1976) demonstrated that how a pure breeding line derived from the F_1 of the two parental lines can yield equal to or better than the F_1, which is discussed in Plant Breeding (Roy 2012).

Studies by Bliss and Gates (1968), Bailey (1976) and Bailey and Comstock (1976) also confirmed that the average genetic value of the pure breeding lines derived from a single cross would be expected to approximate the mean of the parents and that the probability of isolating a line superior to the parent is very low but this probability of isolation of superior line is increased by the presence of coupling linkages in parents and further improved by allowing intermating among selected lines at any cycle. When heterosis, depending upon over-dominance, is due to relatively smaller number of loci at which the recessive (inactive) loci are closely linked with dominance (active) loci then strains assembling dominant (favourable) alleles at all loci are almost impossible to arrive at (Jones, 1917). It is possible that additivity of loci and partial dominance is the rule for most of the loci that are concerned and deviations from these depend on only a few major loci. Sprague and Miller (1950) suggested a recurrent selection scheme to provide information on the relative importance of dominance and over-dominance in yield heterosis wherein the two populations are crossed with a common tester parent and selection is practiced in each population on the half-sib (test cross) performance. Successive cycles of selection will bring about change in gene frequencies. If overdominance were involved in the heterosis of yield then these two advancing cycle populations should become more genetically similar than in comparison to the original starting population and thus the population cross of the two advanced populations should yield less than the population cross involving original population. But if dominance were involved in the heterosis then the results would be just the reverse. Experimental results by Russell et at. (1971) and Russell (1977) in maize showed that $C_5 \times C_5$ population cross yielded significantly higher than $C_0 \times C_0$ cross which demonstrated that dominance rather than over-dominance was the kind of gene action involved in yield heterosis. Further when these populations $C_0 - C_5$ in both studies were evaluated with a different tester, the test cross results indicated that the gene frequency changes brought about by the common tester remain maintained which thus indicates that the additive gene effects with complete dominance were primarily responsible for the yield heterosis. Moll et at. (1980) and Moll and Hanson (1984) showed that in full-sib recurrent selection with varieties Jarvis and Indian chef the improvement of yield of the population *per se* increased the yield of the cross between the populations but it was associated with a reduction of the relative heterosis. Thus heterosis seems to be fixable for a great part. Further large improvement in two populations and the relatively small increase in heterosis suggests that many loci may be showing complete or nearly complete dominance although over-dominance may also be involved. At molecular level also over-dominance is clearly demonstrated. There are few examples of over-dominance for major Mendelian loci. There are some cases of over-dominance with enzymatic system but no correlation between heterosis and the degree of heterozygosity for different systems has been observed and thus enzymatic system seems to be neutral for heterosis.

Jinks (1955) showed how interaction is involved in the expression of heterosis. That epistasis does play a role in heterosis was demonstrated from the experiments of Bauman (1959), Gorsline (1961) and Sprague et al. (1992) in maize when they found that the means of the single crosses differ significantly from the means of the three way crosses all involving three parents. The estimate of epistatic component did not differ appreciably between the selected and unselected group of lines which should if the selection identifies favourable epistatic combinations which does. Wright demonstrated that the mean performance is linearly related to the level of the inbreeding irrespective of the degree of dominance, unless linkage or epistasis is involved. There are reports of both linear and non-linear responses in maize but in majority of cases the inbreeding depression can be accounted for by an increase in the frequency of recessive deleterious loci and thus again heterosis seems to be a result of genes showing additive effects with complete dominance and the bias introduced as a result of epistasis is relatively small and can safely be ignored. Results indicate epistasis to be involved in heterosis in a classical genetical sense. Although Stuber and Moll (1970) reported that less than 10% of the total variability is attributed to epistasis, Otsuka et al. 1972; Stuber et al. 1973; Stuber and Moll 1971 concluded that either epistasis or genotype × environment interaction may introduce significant bias in the prediction of hybrids from inbred lines.

11.3 Expressing Heterosis in Terms of Biometrical Parameters

11.3.1 Generation Means

Let us now see how these parameters contribute to heterosis. Considering additive-dominance model the expected means of P_1, P_2 and F_1 will take the form as

$$\overline{P}_1 = m + [d]$$
$$\overline{P}_2 = m - [d]$$
$$\overline{F}_1 = m - [h]$$

As mentioned earlier heterosis is + ve when $\overline{F}_1 > \overline{P}_1$ the better parent, i.e. $\overline{F}_1 - \overline{P}_1 = +$ ve and $\overline{F}_1 < \overline{P}_1$ in case of negative heterosis i.e. $\overline{F}_1 - \overline{P}_1 = -$ve. In terms of additive [d] and dominance [h] parameter, the positive and negative heterosis becomes

$$\overline{F}_1 - \overline{P}_1 = [h] - [d]$$
$$\overline{F}_1 - \overline{P}_2 = [h] + [d]$$

It shows that irrespective of the kind of heterosis, the estimate of the dominance component, $|[h]|$ has to be greater than $|[d]|$, the additive genetic component and atleast estimate of [h] should not be equal to zero. To meet this requirement

1. the estimate of [h] should not be equal to zero ($[h] \neq 0$) i.e. there should be atleast net directional dominance but the heterosis would be maximum when there is a uni-directional dominance.

2. there is either overdominance i.e.

$$|[h]| > |[d]| \text{ or } \left|\sum_{i=1}^{k} hi\right| > r_d \sum_{i=1}^{k} di$$

or

there is dispersion of completely or incompletely dominant alleles ($r_d < 1$) i.e.

$$\left|\sum_{i=1}^{k} hi\right| > r_d \sum_{i=1}^{k} di \text{ where } \left|\sum_{i=1}^{k} hi\right| = \sum_{i=1}^{k} di$$

i.e. $r_d < 1$ and there is complete dominance

or $\left|\sum_{i=1}^{k} hi\right| > r_d \sum_{i=1}^{k} di \text{ where } \left|\sum_{i=1}^{k} hi\right| < \sum_{i=1}^{k} di$

i.e. $r_d < 1$ and there is incomplete dominance.

The reduction in heterosis shown by F_2 equals

$$\overline{F}_2 - \overline{P}_1 = \frac{1}{2}[h] - [d]$$

i.e. half of the mean heterosis shown by F_1. The reduction in heterosis shown by F_n generation equals

$$\overline{F}_n - \overline{P}_1 = \left(\frac{1}{2}\right)^{n-1}[h] - [d]$$

Thus we can see that mean of any generation and heterozygosity is linearly related in the absence of epistasis.

In case of non-allelic interaction, the positive and negative heterosis becomes

$$\overline{F}_1 - \overline{P}_1 = [h] + [l] - [d] - [i]$$
$$\overline{F}_1 - \overline{P}_2 = [h] + [l] + [d] - [i]$$

The magnitude of heterosis will be maximum when (i) $[h]$ and $[l]$ and $[i]$ have similar sign i.e., when there is complementary gene action and (ii) the estimate of $\pm [d] + [i]$ is minimum and this is obtained when $r_d = 0$ and $r_i = -1$ i.e. when there is complete dispersion of domanent alleles. Although value of r_i never goes beyond -0.6, $[i]$ takes an opposite sign to that of $[h]$ and $[l]$ and $[d]$ will be zero. In this situation the expected heterosis will be equal to $[h] + [l] + [i]$. The magnitude of heterosis can be smaller in cases when there is less than complete dispersion ($0 < r_d < 1$; $-1 < r_i < 1$) and in the presence of duplicate interactions. But as we are not estimating r_d and r_i so we do not know its effect on the estimates of $[d]$ and $[i]$ and thus in this situation only $[h]$ and $[l]$ reflect the net direction of the underlying gene action and from these can be deduced the net direction of dominance and the classification of the types of non-allelic interaction. The experiments have been designed (see chaper on analysis of mean) to estimate r_d and r_i so that contribution of dispersion to heterosis can be obtained. We have seen that the presence of dispersion of complete or incomplete dominant allele leads to underestimation of additive genetic effects (d) and thereby leading to dominance ratio showing over-dominance when it does not exist.

The amount of heterosis shown by F_2 becomes

$$\overline{F}_2 - \overline{P}_1 = \frac{1}{2}[h] + \frac{1}{4}[l] - [d] - [i]$$
$$\overline{F}_2 - \overline{P}_2 = \frac{1}{2}[h] + \frac{1}{4}[l] - [d] - [i]$$

Thus in polygenic case with non-allelic interaction the heterosis in F_2 is not reduced by half as we could have expected.

11.3.2 Variances

Let us now see how we can distinguish between the overdominance ($\Sigma hi < \Sigma di$) and the dispersion of completely or incompletely dominant allele ($rd < 1$). With two parents P_1 and P_2 and their F_1, we can compare the estimate of dominance $[h]$ with additive $[d]$ only when $r_d < 1$, i.e. the genes are associated in the parents and if it is not the case then we will have to resort to the estimation of variance components and the dominance ratio, $\sqrt{H/D}$ will provide information on whether over-dominance($\sqrt{H/D} > 1$) or complete dominance ($\sqrt{H/D} = 1.0$) or partial dominance ($\sqrt{H/D} < 1$) is involved in the determination of a character showing heterosis. The problem with this parameter is that this ratio equals $\Sigma hi/\Sigma di$ only when the ratio $hi/di = f$ constant at all loci i.e. dominance ratio is equal at all loci. But as this ratio is a weighted estimate of dominance ratio, the weighting favours loci with the above average contribution to D and H. We can have estimate of the dominance ratio from any biometrical design discussed in Chapter 6. Biparental, Augmented biparental, NC I, NC II and *NC* III provide estimate of dominance ratio either in a sample of inbred lines with equal gene frequency or F_2 of a cross between two pure breeding lines. The estimates of D and H based on basic generations analysis are highly negatively correlated and hence the dominance ratio has a large sampling variance. In order to obtain reliable estimates of the variance components, large samples of F_2, BC_1 and BC_2 populations should be included in the experiment, NC III and TTC provide unbiased estimates of dominance ratio even when gene frequencies are not equal provided the correct genotypes are used as testers and only diallel and TTC provide test for detecting biases in the estimates arising from non-allelic interaction and linkage disequilibrium.

11.4 Effect of Linkage Disequilibrium on Dominance Ratio

In practice the additive (D) and dominance (H) components are biased because of linkage and non-allelic interaction and as discussed in Chapter 10, the estimates of D and H will change as a consequence of the above phenomena and consequently distort the dominance ratio and thus wrongly interpret the cause of heterosis. We have seen that the nature and extent of biases in the estimates of D and H differs from one breeding design to another but in a characteristic way. The dominance ratio, $\sqrt{H/D}$ will therefore no longer unambiguously diagnose the cause of heterosis. In particular where heterosis is due to dispersed directionally dominance genes with complete or incomplete dominance ($\sqrt{H/D} \leq 1$) the linkages are likely to be mainly in the repulsion phase and the dominance deviation of the same sign i.e., when linked genes display net directional dominance. Comstock and Robinson (1948) and Robinson et al. (1949) pointed out that the dominance ratio exceeds 1.0 because of upward bias due to linkage being maximum in F_2 population and minimum in the F_n generation obtained after n generations of random mating. The effect of linkage on the dominance ratio can be demonstrated using suitable mating designs like NC III and TTC which provide estimates of D and H which are independent and reliable. The dominance ratio in the F_2 population assuming linkage disequilibrium will take the form (Jinks, 1981) as

$$\sqrt{\frac{\sum_{i=1}^{k} hi^2 + \sum^{1/2\,k(k-1)} 2(1-2p_{ij})\,h_{ihj}}{\sum_{i=1}^{k} di^2 \pm {}_{R}^{C}\sum^{1/2\,k(k-1)} 2(1-2p_{ij})\,d_{idj}}}$$

and in the F_n advancing generation as

$$\sqrt{\frac{\sum_{i=1}^{k} hi^2 + \sum^{1/2\,k(k-1)} \frac{2(1-2p_{ij})}{(1+2p_{ij})}\,h_{ihj}}{\sum_{i=1}^{k} di^2 \pm {}_{C}^{R}\sum^{1/2\,k(k-1)} \frac{2(1-2p_{ij})}{(1+2p_{ij})}\,d_{idj}}}$$

When n is large the linkage disequilibrium will be zero and thus the H and D and consequently the dominance ratio will be free from any linkage disequilibrium component. Gardner (1963) reviewed the estimates from different populations which showed that the dominance ratio obtained from F_2 showed over-dominance (dominance ratio greater than 1.0) whereas the dominance ratio estimates obtained from generations after a large number of generations of random mating were either below 1.0 indicating partial dominance or approximately 1.0 indicating complete dominance. If the linkage disequilibrium is the real cause of bias, it will decrease rapidly in the beginning with the number of generations of random mating but after a number of generations, the decrease will be slow until it stabilizes at its true value where the population would have approached equilibrium which is a result of breaking of linkage groups through generations of random mating which provided opportunity for recombination. But if the dominance ratio greater than in F_2 persists over many generations of random mating it is true over-dominance and is the real cause of heterosis. In this situation estimate of D from the usual source will be deflated while the estimate of H inflated and thus the dominance ratio will indicate over-dominance.

11.5 Effect of Interaction on Dominance Ratio

The non-allelic interaction like linkage also biases the estimates of D and H and these biases vary considerably with the breeding design from which they are estimated. Study of relation between different types of non-allelic interaction, i.e. combination of complementary and duplicate with associated or dispersed gene and the dominance ratio in each condition will provide information on whether or not over-dominance plays a role in heterosis. When estimates are obtained from TTC, the dominance ratio because of the reason discussed in chapter 10, is relatively less biased in comparison to the dominance ratio based on estimate of D and H obtained from basic generation analysis and thus it is less likely to lead to misinterpretation of the cause of heterosis. We can avoid misinterpretation if we have prior knowledge of the type of non-allelic interaction present from the generation mean analysis. Heterosis due to dispersed complementary genes which results in inflation of D while deflate H, could give overdominance (dominance ratio greater than 1.0) when the dominance is complete or partial. In the presence of dispersed complementary genes or associated duplicate genes, if we get dominance showing partial or complete dominance then it confirms that overdominance is playing no role in heterosis. On the other hand, if the true cause of heterosis is overdominance, associated complementary or dispersed duplicate genes will give dominance ratio showing partial or complete dominance but again if we get a ratio showing overdominance it is a confirmation that the overdominance is the real cause of heterosis.

Jinks (1954, 1955, 1956) developed a method for testing the presence of non-allelic interaction and calculated dominance ratio in the absence of it when the genes frequencies are not equal. He showed that how complementary genes, dispersion and repulsion linkage all lead to inflation of H relative to D and thus lead to a high dominance ratio (dominance ratio greater than 10) which can be misinterpreted as the presence of over-dominance and thus provided the evidence that over-dominance which Hull (1946) proposed as the basis of hybrid vigour could be attributed to the biases arising from non-allelic interaction. The duplicate genes, association and coupling linkage all lead to deflation of H relative to D and thus in turn lead to a deflation of dominance ratio.

Although linkage alone does not affect the mean of any generation and thus has no effect on the magnitude of heterosis, it does affect the mean of the populations in the presence of epistasis and hence biases the estimates of components of heterosis (d, h, i, l) and thereby the relative contribution of the different kinds of gene action and interaction to heterosis is altered. It has been shown (Mather and Jinks 1971) that in case of epistasis but no linkage, alongwith $[h]$, $[i]$ and $[l]$ are contributing to the heterosis but in the presence of linked interaction, the contribution of dominance component $[h]$ to heterosis is over estimated and the contribution of non allelic interaction components (i, l) are underestimated and thus dominance component (h) appears to be the sole cause of heterosis.

11.6 Environmental Heterosis

Role of genotype × environment interaction

Various hypotheses discussed above have been developed on purely genetic grounds but the phenomenon of heterosis can be viewed as a result of the interaction between genetic and environmental stimulation. Two of the most important factors of physical environment which may influence heterosis are temperature and nutritional status of the soil. In the biotic environment competition may be a factor. Thus heterosis can be nutrient dependent, temperature dependent, development dependent and group dependent forms of heterosis. The heterotic expression will be different in different stages of development.

We have seen in the chapter on $G \times E$ interaction that homozygotes and heterozygotes differ in response to changes in environment. Further we have seen that in case of either positive or negative heterosis, the heterozygote mean (F_1) is compared with homozygote (parental) mean and so the magnitude of heterosis will vary with the environment. Accepting this then like heterosis in yield, we can have heterosis in environmental sensitivity.

Considering the expectations of means of P_1, P_2 and F_1 assuming genotype × environment interaction the positive and negative heterosis in the jth environment takes the form as

$$\overline{F}_{1j} - \overline{P}_{1j} = [h] - [d] + g_{hj} - g_{dj}$$
$$\overline{F}_{1j} - \overline{P}_{2j} = [h] + [d] + g_{hj} - g_{dj}$$

The above expectations for positive and negative heterosis work only when we assume that the jth is the environment in which $\overline{P}_{1j} > \overline{P}_{2j}$. If in an environment $\overline{P}_{2j} > \overline{P}_{1j}$, $\overline{F}_{1j} > \overline{P}_{1j}$ gives the expectation for negative heterosis and $\overline{F}_{1j} > \overline{P}_{2j}$ for positive heterosis. Another point that we must remember is that although on average over all environments, $\overline{P}_1 > \overline{P}_2$ but we may have many environments in which $\overline{P}_1 = \overline{P}_2$ or even $\overline{P}_1 < \overline{P}_2$. In the former case $\overline{P}_1 > \overline{P}_2$ by 2 $[d]$ while the later will occur when $-2 [d] = 2g_{dj}$ and $-2 [d] > 2g_{dj}$, respectively. Now considering $g_{dj} = b_d e_j$ and $g_{hj} = b_h e_j$ the positive and negative heterosis becomes

$$\overline{F}_{1j} - \overline{P}_{1j} = [h] - [d] + (b_h - b_d)e_j$$
$$\overline{F}_{1j} - \overline{P}_{2j} = [h] - [d] + (b_h - b_d)e_j$$

Thus we see that the heterosis will change with the change in environment if $g_d \neq g_h$ or $\beta_d \neq \beta_h$. If $\beta_d \neq \beta_h$ heterosis will be constant but if $\beta_d > \beta_h$ then in this condition, the magnitude of heterosis will fall off linearly as the environment improves at a rate equal to $\beta_d = \beta_h$. Hence the greater stability of the heterozygote to changes in the environment leads to a lower response to an improving environment than is shown by its better parent. But then equally the heterozygote is less affected by poor environment. Thus in such condition heterozygote (F_1) is preferred over homozygotes (parents) because it combines high mean yield with low sensitivity to environment.

The heterozygosity for regulatory system must lead to greater homoestasis in a variable environment (internal or external) and so heterosis will be a result of genotype × environment interaction and such a mechanism will be a fundamental property of the diploid level heterosis. In autogamous crops such a mechanism could have been fixed by the allopolyploidisation or by duplication. This could be another reason to have less heterosis in self pollinated crops then in cross-pollinated crops. Positive heterosis reported seems to be due to:

(i) spaced planting of F_1 plants, (ii) too small population and (iii) growing condition (nursery bed, green house different from those in the field). Further hybrid superiority can be due to density dependent or having higher competitive ability. If the environment is poor (nutritional status of the soil is maintained at a low level so that nutrients severely limit growth) then hybrids cannot be expected to out yield highly selected parental strains. Such a situation is likely to occur with density planted cereals where extreme competition between plants aggravates the nutritional stress on an individual plant basis. All these thus indicate that heterosis is environ-mentally dependent.

11.7 Dominance Ratio

We can see that the potence ratio will change with the change in environment provided $g_{dj} \neq g_{hj}$ or $\beta_d \neq \beta_h$ and so will be the dominance ratio based on the variance components as additive and dominance component will change with the change in environment. Thus the cause of heterosis will vary with the environments and we will have to be very cautious while interpreting the result. Like epistasis and linkage, genotype × environment interactions are potential sources of bias in the dominance ratio but it appears most likely that all would operate to increase the estimate of dominance variance (H) proportionately more than that of additive genetic variance (D). Hence the estimate of H/D is biased upward rather than downward.

11.8 Maternal Effects in Heterosis

We have seen in Chapter 9 that in the presence of maternal effect, the maternal parent affects the mean of the progeny and thus there arises a difference in the means of F_1's and RF_1's. It is then easy to see that the maternal effect will also affect the heterosis. Assuming additive-dominance model with no epistasis and no interaction between progeny genotype and maternal effect, the positive heterosis will become

$$\overline{F}_{1j} - \overline{P}_1 = [h] + [dm] - [d] - [dm] = [h] - [d]$$

$$\overline{RF}_1 - \overline{P}_1 = [h] - [dm] - [dm] - [dm]$$
$$= [h] - [d] - 2[dm]$$

The average estimate of heterosis will then become

$$[h] - [d] - [dm]$$

When F_1 and RF_1 are allowed to produce F_2 and RF_2 families, the heterosis shown by these families would be equal and become

$$\overline{RF}_2 - \overline{P}_1 = \overline{F}_2 - \overline{P}_1 = \frac{1}{2}[h] + [hm] - [d] - [dm]$$

This is because both F_1 and RF_1 mothers are genotypically similar and as the maternal effect is genotypically determined they are contributing equally to their progenies. If the maternal effects are determined by the phenotype of F_1 and RF_1 then we should expect differences between F_2 and RF_2. It also demonstrates that the magnitude of heterosis depends on whether homozygote (P_1, P_2) or heterozygote ($\overline{F}_1$) is used as mother. The $\overline{F}_2 - \overline{P}_2$ or $\overline{RF}_2 - \overline{P}_1$ will be greater than the average of [($\overline{F}_1 - \overline{P}_1$) + ($\overline{RF}_1 - \overline{P}_1$)] when $[hm] - 1/2 [h]$ is positive, i.e. F_2 and RF_2 will be more heterotic than the average F_1 and RF_1 when $[hm]$ is greater than 2 $[h]$. This can happen only if we assume that F_1 as mother is superior to either P_1 or P_2 as mother because of its genotype and thus its progeny, the F_2 will be more heterotic than the F_1.

11.8 Biometrical Genetics–Analysis of Quantitative Variation

In an experiment, if one finds $\overline{P}_1$ vs. $\overline{P}_2$ being not significant, $\overline{P}$ vs. $\overline{F}_1$ significant and $\overline{F}_2$ vs. $\overline{F}_1$. significant with $\overline{F}_2 > \overline{F}_1$, it shows the manifestation of maternal effects. Further if BC_1 $(P_1 \times F_1)$ vs. RBC_1 $(F_1 \times P_1)$ or BC_2 $(P_2 \times F_1)$ vs. RBC_2 $(F_2 \times P_2)$ is significant with $\overline{F_1 \times P_1}$ or $\overline{F_1 \times P_2} > \overline{P}_1$ or $\overline{P_2 \times F_1}$ it shows the superiority of F_1 as mother. From the above results it can be inferred that the progeny's own genotype and the maternal genotype are important in determining a trait. There are thus two systems of heterosis, one system is dependent on the individual's genotype and the other is dependent upon the maternal genotype. The F_1 heterosis is associated with the genotype of the progeny whereas the superiority of F_2 progeny is a property of the progeny's own genotype and the genotype of the F_1, mother and the two effects can be separated. Which of the two components is important in determining heterosis can be known by raising P_1, P_2, F_1, F_2, BC_1 and BC_2 generations and fitting a model with four parameters namely m, d, h and f which is due to genotype of F_1, mother and this appears in F_2, BC_1 and BC_2 generations expectation with a coefficient of + 1as explained in Chapter 9. If such a model is adequate and the magnitude of $[f]$ is greater than $[h]$, it can be concluded that progeny's own genotype is of minor importance in comparison to F_1 maternal genotype. To separate out the effects of progeny's own genotype from those of maternal genotype, Barnes (1968) proposed a test: A comparison of the magnitude of the components m, dp, hp, dm, and hm, will distinguish between the two types of effects. The dp and hp are the additive and dominance components of progeny's own genotype, respectively. The model is presented in Chapter 9. If $[hp] > [dp]$, then the high level of heterosis is determined by the progeny's own genotype. High heterosis is also by maternal genotype, if $[hm] > [dm]$. If $[l]$ is of opposite sign to $[hp]$ it suggests that any heterosis due to progeny's own genotype is of minor importance since it is determined as

$$[hp] + [l] > [dp] + [i]$$

Heterosis when $[l]$ is of opposite sign to $[hm]$ and $[hp]$ could most probably due to overdominance or dispersion of dominant increasing alleles in the parental lines rather than to non allelic interaction. Extra vigor of F_1 hybrid may be a manifestation of superior gene array reflecting increased heterozygosity following dispersion of increasing alleles between the inbred parents.

Estimates of additive (D) and dominance (H) components unbiased by the maternal effects can be obtained from the basic generations and from an F_2 triple test cross where P_1, P_2 and F_1 are used as the maternal parent. The dominance ratio then calculated provides evidence of the genetical basis of heterosis.

So far our conclusion about the role of dominance or over-dominance in heterosis is based on the dominance ratio but this dominance ratio suffers from the problem that it is the weighted average dominance ratio and thus complete or incomplete dominance inferred from the dominance ratio estimate can in reality mask a whole range of dominance ratio at individual loci including over-dominance at some. The dominance theory thus does not rule out the possibility of over-dominance playing a role although in limited sense in heterosis. Jinks (1983) suggested that in such situation pure breeding lines superior to the heterotic F_1 can be extracted (see Roy (2012) for detail) but even the best line in cross with other can produce F_1 superior to either parent and this heterosis in F_1 is now not fixable. This also implies that inbred lines can be improved by cyclic hybridization and re-extraction. The F_1 produced by the original inbred lines should show complete or incomplete dominance as basis of heterosis while the $[F_1]$ from the improved lines will show over-dominance as the basis of heterosis. The limited existence of over-dominance in heterosis can also be explained through dispersion of genes or repulsion linkage. Suppose genes at all loci are showing incomplete dominance but some loci are unlinked and in dispersion while others are linked and some others are lightly linked and in dispersion. In this case we will find incomplete or complete dominance first and superior inbred lines can be extracted but again the superior lines in cross will show heterosis but it can be now said to be due to over-dominance although often tight linkage can be broken by recombination. Thus there is a parallelism between variable linkages between loci with tight linkage at some loci and the variable dominance ratio with over-dominance at some loci.

The causes of heterosis will decide which breeding strategy should be followed. If over-dominance is high, inbreeding with development of hybrid can be employed whereas if non-allelic interaction is the cause then one can obtain purebreeding or inbred lines as yielding as the F_1 (hybrid).

11.9 Heterosis in Population Cross

Suppose there are two populations, Pop 1 and Pop 2 with gene frequencies p, q and r, s, respectively. Following Falconer (1960) the means of the two populations would be

$$\overline{Pop}_1 = d(p - q) + 2pqh$$
$$\overline{Pop}_2 = d(r - s) + 2rsh$$

The mid-parental value (MP) becomes

$$MP = \frac{Pop_1 + Pop_2}{2} = \frac{1}{2}[d(p - q + r - s) + 2(pq + rs)h]$$

The $\overline{F}_1$ (Pop 1 × Pop 2) mean becomes

$$\overline{F}_1 = (pr - qs)d + (ps + qr)h$$

The heterosis defined as the superiority of F_1 over mid-parent becomes

$$\overline{F}_1 - MP = d\left[(pr - qs) - \frac{1}{2}(p - q + r - s)\right] + [(ps + qr) - (pq - rs)]h$$

$$= (p - r)^2 h$$

The heterosis therefore depends on the difference in the gene frequencies between the two populations and the level of dominance. Further it can be shown that the heterosis shown by F_2 population then becomes

$$\overline{F}_2 - MP = \frac{1}{2}(p - r)^2 h$$

which is half the heterosis shown by F_1 population.

In case of cross between two inbreds or pure breeding lines for P_1, $p = 0$ or $p = 1$ depending upon whether it is homozygous recessive or homozygous dominant and correspondingly for P_2, $r = 0$ or $r = 1$ and further considering the two parents together then $p = 0$ and $r = 1$ or vice versa it can be shown that the heterotic response depends on the number of contrasting loci besides the level of dominance.

Gardner and Eberhart (1966) partitioned the total heterosis ($h_{kk'}$) into three components when 4 or more varieties are included in the diallel.

$$h_{kk'} = \overline{h} + h_k + h{k'} + S_{kk'}$$

where $h_{kk'}$ measures the heterosis and this parameter is due to the difference in gene frequencies in k and k' varieties and to dominance. $\overline{h}$ is the average heterosis due to the particular set of varieties in the diallel, h_k is the average heterosis due to the variety k and measured as deviation from average heterosis ($\overline{h}$) and $S_{kk'}$ is the specific heterosis in the cross of varieties k and k'. The average heterosis ($\overline{h}$) will be zero if there is no difference in gene frequency among varieties. The h_k will be negative if the gene frequency of variety k equals the average gene frequency of all parents and h_k will be positive if the kth variety is having may loci at a high gene frequency or at a low gene frequency or the variety shows a dispersion of gene frequency (high and low) in comparison to the average gene frequency at each locus. In other words $\overline{h}_k$ will be positive if the gene frequency of kth variety differs from the average gene frequency of all the parents. The specific heterosis (or specific combining ability) depends on the size n of the diallel set, average heterosis ($\overline{h}$), heterotic component of general combining ability and the difference in gene frequency between k and k' varieties. The general combining ability (g_k) is calculated as

$$g_k = \frac{1}{2}v_k + h_k$$

where $v_k = a_k + d_k$

This shows that the general combining ability depends on both varietal effect (v_k) and dominance effects (h_k). a_k and d_k the additive and dominane effects and are the contribution of homozygous and of heterozygous loci, respectively to the varietal mean. The problem with this type of partitioning of heterosis is that the heterosis effects are not wholly contained in the specific effect ($S_{kk'}$). Besides, the average heterosis ($\bar{h}$) is confounded with the mean mean and the variety heterosis (h_k) is confounded with the general combining ability effect (g_k). The average variety heterosis (h_k) in the above equation becomes

$$h_k = g_k - \frac{1}{2} v_k$$

It shows that there is negative correlation between average heterosis and line performance (or variety effect) and it is possible to increase the value of inbred lines without the loss on the F_1 value.

11.10 Misinterpretation of Heterosis

Some workers have defined heterosis as F_1 performance above the mid-parental value and hybrid performance better than the best available commercial pure line cultivar. Heterosis had been claimed to be observed because a hybrid cultivar out-performed the best cultivar in the area where the trial was conducted with the definition in one case. These definitions thus differ from the real definition of heterosis explained above.

Testing the significance of heterosis

The critical differences (CD) for testing the significance of heterosis are as follows:

1. Critical difference for heterosis over mid-parent (MP)

$$CD\ (MP) = \sqrt{\frac{3\ Me}{2r}} \times t$$

2. Critical difference for heterosis over better parent

(BP) or standard check variety

CD (BP or standard check variety)

$$= \sqrt{\frac{2\ Me}{r}} \times t$$

where Me is the error mean squares, r is the number of replication and 't' is the table value of t at 5 or 1 per cent level of significance.

11.11 Genetic Distance/Divergence

We have seen above that the magnitude of heterosis depends on the differences in the gene frequencies in the parents and thus is a measure of the genetic divergence of the parental stocks but then lack of heterosis cannot be used to infer a lack of genetic divergence. This is because the magnitude of heterosis also depends on the dominance and if the dominance effects at some loci are positive and at others negative then the net dominance effect would be little or no and thus we would expect little or no heterosis even though there is a diference in gene frequency. Thus the validity of evaluating the degree of genetic divergence based on the heterotic effects is subject to question. There are numerous examples of unexpectedly poor hybrid performance in spite of superior parental stock and genetic diversity. Linkage and epistasis have been proposed as the causes of this poor performance. Even in absence of epistasis and linkage, Cress (1966) showed how the effect of multiple al1eles can account for this variable performance. Moll et al. (1965) in their study on the relationship between the magnitude of heterosis and the degree of divergence concluded that although the amount of heterosis is a linear function of differences in al1ele frequency for loci having dominance or over-dominance effects, the linearity disappears when highly differentiated populations or inbreds or pure breeding lines are crossed. Thus there is an optimum degree of genetic divergence for a maximum expression of heterosis and this optimum occurs within a range of divergence that is narrow enough so that incompatibility barriers are not apparent. Within this range the amount of heterosis is linear function of the difference in allele frequency, i.e. the amount of heterosis increases with the increase of divergence.

There are different indices of genetic distance (i) Morphological quantitative distance, (ii) Biochemical distance. The genetic distance from morphological data is particularly useful for inter group classification below the species level. As natural selection affects morphological traits linked to adaptive characters, genetic distances allow inferences about adaptation and co-adaptation pattern of populations. The divergence of quantitative traits on the basis of

$$Aa = 2uv(1 - F_n)$$
$$aa = v^2 + 2uv\, F_n$$

Assigning these genotypes the values $+d, h$ and $-d$ respectively, it can be shown that the new population mean differs from the original population by $-2\,uvh\,F_n$.

Thus,
$$H_n = 2uv(1-F_n)$$

Therefore, if we can find a recurrence relation for F we can also find the recurrence relation for H.

The population mean changes as a result of inbreeding in the following way. The original population mean is

$$\Sigma(u-v)d + 2\Sigma uvh = \mu_o$$
$$\mu_1 = (u-v)d + uvh$$
$$\mu_2 = (uv)d + \frac{1}{2}uvh$$
.
.
.
$$\mu_\infty = (u-v)d$$

Thus it can be shown easily that ($F = 1.0$) will result in loss of heterozygosity and that the amount of inbreeding depends on the level of dominance, inbreeding coefficient and over and above the gene frequencies. We can also visualize that as a result of selfing, the recessive genes which were hidden in the form of heterozygotes are now in homozygous state and thus show their effects and thus lowers the population mean. In general the frequency of harmful recessive allele is low. The frequency of such genes exceeds 0.15 in several species (Kornaki, 1982). The frequency of harmful alleles is determined by mutation, selection against them when in homozygous form and random genetic drift which is discussed in Chapter 19.

Extending the model to a polygenic case as in case of quantitative traits we find that the loss of heterozygosity is not as rapid as for one factor where the heterozygosity is reduced by 1/2 every generation. In other words, the more the number of factors involved in the determination of a character the less rapid will be the loss of heterozygosity. In terms of inbreeding depression the rate will be slow in case of polygenes. Besides the number of factors determining a character the rate of depression will also depend on the linkage and epistasis. The inbreeding depression is thus not only as a result of homozygosity *per se* but also the deleterious recessive genes are uncovered by homozygosity due to break down of balanced linkages. Heterosis and inbreeding depression are interpretable in terms of dominance and lack of linear relationship between means and inbreeding in terms of epistasis. However, lack of inbreeding depression or heterosis, is not positive indication of lack of dominance or epistasis because positive and negative gene effects can cancel. There is existence in several cases of strong additive variance and of a strong inbreeding depression. This can be explained by the role of natural selection that has built up segregating units integrating dominance and epistasis and thus a particular genetic organization has developed. The rate of inbreeding depression will also depend on the method of inbreeding which ranges from most intense form, selfing to milder form like full-sib and half-sib mating. To obtain similar level of homozygosity, say 0.5, one needs one generation of selfing, 3 generations of full-sib mating or 6 generations of half-sib mating (Figure 11.2). To obtain the same degree of homozygosity, selfing in autotetraploid takes 8.8 times as long with no double reduction (dr) and 2.9 times as long with $\alpha = 0.14$. In diploid 7 – 8 generations of selfing is required to obtain 99% homozygosity whereas polyploids need 27-28 generations (at least 20 generation is needed). Further sib mating is 3.3 times slower than selfing in diploids (Figure 11.3).

The rate of reduction in heterozgosity per due to inbreeding is ½ with self-fretilization. ¼ with full-sib matings, 1/8 with half-sib matings and 1/16 with cousin matings.

So far we have seen that the magnitude of inbreeding depression depends on genetic variables, that is, gene action and interactions, number of genes controlling a trait. It also depends on the environmental variation as we have seen that homozygotes and heterozygotes respond differently to environmental change. In other words the viability differential between out-cross

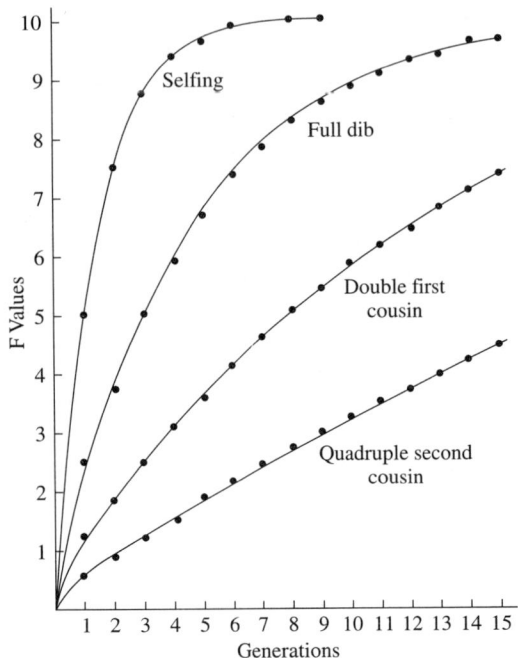

Fig 11.2 Showing inbreeding coefficients at senerations 1 to 15 in different systems of inbreeding (Adapted from Strickber)

progeny (the heterozygote) and inbred progeny (the homozygote) is influenced by environmental quantity. Under condition of drought, diseases and other stress conditions, the relative variability of inbreds is much less than under favourable condition (Allard and Hansche; Pawsey, 1964; Koski, 1973; Libby et al. 1981). In addition to these factors, the level of inbreeding depression is also governed by the tolerance to higher level of homozygosity (Mayo, 1980). Inbreeders whose genetic system are adapted to relatively higher levels of homozygosity are likely to show the least depression with inbreeding whereas most of the outbreeders will show considerable depression. Then within normally cross-fertilizing crops there is crop like alfalfa which cannot tolerate more than 2–3 generations of inbreeding. Inbreeding like heterosis is not a universal phenomenon in normally self-fertilizing species.

Single population characterization of inbreeding depression may not be especially representative because of the important roles which genetic drift and hitch-hiking may play for recessive lethal genes. Since the genetic structure of population varies from population to population in space, the relationship between inter parental distance and inbreeding depression for a population may not generally hold for other population. Inbreeding may be substantial even when the genetic systems of species are adapted for outcrossing.

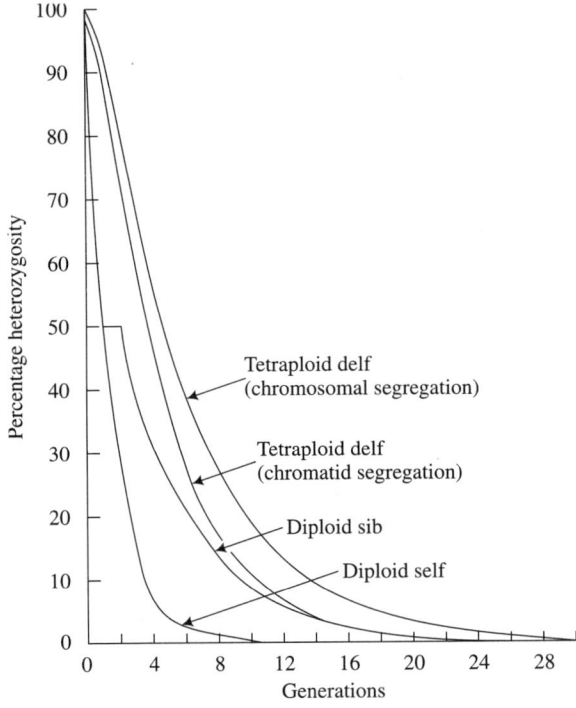

Fig 11.3 Showing decrease in heterozygosily on inbreeding (Adapated from Bradshaw, 1994)

11.14 Estimation of Inbreeding Coefficient

Inbreeding coefficient of an individual and inbreeding coefficient of a population

Coefficient of coancestry – The mating $A \times B$ can be written as $A_1A_2 \times B_1B_2$. The coancestry (ϕ_{AB}) of individuals A and B (Coancestry = Coefficient of kinship = Coefficient de parte of parentage) can be defined as the probability (P) that a pair of alleles taken at random from each of two individuals, A

and B are identical by descent. Thus, Coefficient of coancestry

$$\Phi_{AB} = \frac{1}{4}[P(A_1 \equiv B_1) + P(A_1 \equiv B_2) + P(A_2 \equiv B_1) + P(A_2 \equiv B_2)]$$

Now the progeny in the cross $A \times B$ will be $A_1 B_1$, $A_1 B_2$ or $A_2 B_2$, each with probability ¼. Therefore, the coefficient of inbreeding of the progeny will be

$$\frac{1}{4}[P(A_1 \equiv B_1) + P(A_1 \equiv B_2) + P(A_2 \equiv B_1) + P(A_2 \equiv B_2)] \quad ...(i)$$

as they are mutually exclusive events and so their probabilities have been added. Thus for a diploid, the inbreeding of coefficient of an individual (progeny) (probability that the two alleles at a locus in that individual are identical by descent) equals the coancestry of its parents.

In case of the following mating

$$\begin{array}{cc} A \times B & C \times D \\ \downarrow & \downarrow \\ E & G \end{array}$$

The above two crosses can be written as $A_1 A_2 \times B_1 B_2$ and $C_1 C_2 \times D_1 D_2$, the coancestry ϕ_{EG} can be written as $F_{A \times B, C \times D}$

$$\Phi_{EG} = 1/16 \times [P(A_1 = C_1) + P(A_1 = D_1) + P(B_1 = C_1) + P(B_1 = D_1) + P(A_1 = C_2) + P(A_1 = D_1) + P(B_1 = C_2) + P(B_1 = D_1) + \text{etc.}]$$

$$= \frac{1}{4}(\Phi_{AC} + \Phi_{AD} + \Phi_{BC} + \Phi_{BD})$$

Thus we see that

$$\Phi_{A \times B, C \times D} = \frac{1}{4}(\Phi_{AC} + \Phi_{AD} + \Phi_{BC} + \Phi_{BD}) \quad ...(ii)$$

By the same token

$$\Phi_{A, B \times C} = \frac{1}{2}(\Phi_{EG} + \Phi_{AC}) \quad ...(iii)$$

And further,

$$\Phi_{A \times B, A \times B} = \frac{1}{4}(\Phi_{AA} + 2\Phi_{AB} + \Phi_{BB}) \quad ...(iv)$$

Provided Φ_{AA} is defined as coancestry of A with itself.
Considering the mating $A_1 A_2 \times A_1 A_2$, Φ_{AA} equals

$$\Phi_{AA} = \frac{1}{4}[P(A_1 = A_1) + P(A_1 = A_2) + P(A_2 = A_1) + P(A_2 = A_2)]$$

.
.
.

$$= \frac{1}{2}[1 + F_A] \quad ...(v)$$

where F_A is the coefficient of inbreeding. In other words, the coefficient of parentage of A with itself equals one plus coefficient of inbreeding of A divided by 2.

We can apply all rules (i to v) to cases of individuals obtained through different types of mating. Let us consider its application to full- sib mating (brother-sister mating).

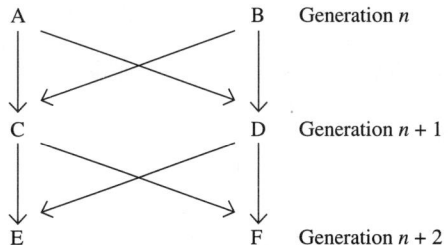

Now $F_{n+2} = F_E = F_G = \Phi_{CD} = \Phi_{A \times B, A \times B} = \frac{1}{4}(\Phi_{AA} + 2\Phi_{AB} + \Phi_{BB})$

$F_{n+1} = F_C = F_D = \Phi_{A \times B}$

Here,

$\Phi_{A \times A} = \frac{1}{2}(1 + F_A) = \frac{1}{2}(1 + F_n)$
$\Phi_{B \times B} = \frac{1}{2}(1 + F_B) = \frac{1}{2}(1 + F_n)$

So, $4 F_{n+2} = \frac{1}{2} + \frac{1}{2} F_n + 2 F_{n+1} + \frac{1}{2} + \frac{1}{2} F_n$

Therefore, $4 F_{n+2} - 2 F_{n+1} - F_{n-1} = 0$
Assuming $(1 - F_n) = P_n$ and so $F_n = 1 - P_n$, the above equation can take the form as

$$4 P_{n+2} - 2 P_{n+1} - P_n = 0$$

Now putting $H_n = 2uv(1 - F_n) = 2uv P_n$ in the above equation we find

$$4 H_{n+2} - 2 H_{n+1} - H_n = 0$$

Assuming $H_n = A\lambda^n$ where A and λ are constants, the λ will take the following values:

$$\lambda = 0.809 = \text{say}, \lambda_1$$

Or

$$\lambda = -.309 = \text{say}, \lambda_2$$

There are thus two solutions

$$H_n = A\lambda_1^n$$

$$H_n = B\lambda_2^n$$

but then the one general solution will be as

$$H_n = A\lambda_1^n + B\lambda_2^n$$

Where A and B depend on the foundation mating, i.e. the mating with which one starts. But if n is large then λ_2^n will be negligible and we can use $H_n = A\lambda_1^n$

The recurrence relationship described above for sib mating differs from other systems of inbreeding but for all systems the following relationship holds when n is large.

$$H_n = A\lambda_D^n$$

Where λ_D depends on the system of inbreeding, A depends on the system of inbreeding and the foundation mating. In fact, A differs little from one case to other and thus the critical thing is the value of λ_D which is called 'dominant latent root (leading eigen value).

Coefficient of coancestry between inbreds can be obtained by tabular analysis (Emik and Terril, 1949). The inbreeding coefficient of an individual X (F_X) takes the following from (Wright, 1922; Lush, 1945).

$$F_X = \tfrac{1}{2} \Sigma[(1/2)^n (1 + F_A)] \ldots\ldots (i)$$

Where n is the number of generations between individuals in a line through which the parents are related and F_A is the inbreeding coefficient of the common ancestor A from which the line of descent arose. Σ indicates that each such path of relation between male and female is to be evaluated separately and then all results are to be added together.

Measuring relationship between inbreds to be used in the hybridization program for development of hybrids is essential. Relationship between two inbreds is simply probability that as they are related by descent, they will be similar in more of their genes than unrelated inbreds. Thus measuring relationship means just evaluating the probability that two related inbreds will have duplicate genes. Here the first thing would be to examine the pedigrees and find all paths or line of descent by which the two inbreds are related. Each path or line is evaluated separately and then results from all paths are added to obtain the total probability of likeness in their genes. The coefficient of inbreeding shows how much decrease in heterozygosity is to be expected from a particular type of inbreeding. Although the inbreeding coefficient says little about one pair of genes in one individual but it tells much about the average condition of one pair of genes in one population and also tells much about the average heterozygosity of the whole group of genes in an individual (Lush, 1945). If we say that the inbreeding coefficient of an individual, X is ¼, it means that individual X is probably homozygous for ¼ of the genes which were heterozygous in the ancestors at foundation or base date to which the pedigree of X was traced. For example, if the ancestor was heterozygous for 500 pairs of genes then the calculated inbreeding coefficient of ¼ relative to the base date indicates that the individual will be heterozygous for about 375 pairs of genes.

Inbreeding in polyploids

In diploid, the inbreeding coefficient (F) is the probability that the two alleles at a locus are identical by descent. The F depends on the mating system with self-pollination being the strongest from of inbreeding. The inbreeding coefficient in tetraploid is defined as the probability that a randomly chosen pair of alleles at a locus are identical by descent. Thus inbreeding coefficient for an $X_iX_jY_kY_l$ individual formed form X_iX_j and Y_kY_l gametes takes the following form:

$$F = 1/6 \, P(X_i \equiv X_j) + (X_i \equiv Y_k) + (X_i \equiv Y_l) + (X_j \equiv Y_k)$$
$$+ (X_j \equiv Y_k) + (X_j \equiv Y_l) + (Y_k \equiv Y_l)$$
$$= 1/6 \, (F_X + 4\Phi_{XY} + F_Y)$$

This is because of the reason that a gamete can contain two alleles identical by descent, the probability of which is by definition the inbreeding coefficient of the parent from which they came.

In diploids, the number of generations required to reach 99% homozygosity is 7 whereas in case of tetraploid it is 27 (considering chromosomal segregation and 20 (considering chromatid segregation). With n loci there are $2n$ factors (diallelic interactions) in diploids whereas in autotetraploid there are $4n$ factors. The levels of each factor are different alleles-tri allelic and tetra allelic interactions. There is importance of triallelic and tetra-allelic interactions in autotetraploid. The frequency of tetra-allelic genotypes declines rapidly under selfing being multiplied by 1/6 in each generation. The effect of one generation of selfing on F, the coefficient of inbreeding assuming double reduction is

$$F_1 = 1/6 (1 + 2\alpha) + 1/6(5-2\alpha) F_0$$ (Kempthorne, 1957)

When both F_0 and α equal zero. F_1 equals 1/6 compared with ½ for diploids.

Approach to equilibrium in case of tetraploid—In case of diploid species, the effect of inbreeding can be removed by one generation of random mating since the offspring of a cross between unrelated parents are not inbred regardless of the inbreeding coefficients of parents. In other words, equilibrium between gene and genotypic frequencies can be established at a single by just one generation of random mating in diploid. In case of tetraploid the gametes can contain two identical alleles by descent. Thus the inbreeding coefficient of the offspring depends on the inbreeding coefficients of the parents and the probability that two pairs of alleles are identical by descent is greater then zero if both parents are inbreds. When two inbreds and non-related autotetraploid genotypes are crossed, the hybrid has an inbreeding coefficient of $F = 0.33$. The effect of parental inbreeding partially remains in the progenies and does not disappear as in a diploid where the inbreeding coefficient in the progenies will be zero. This is the reason that one should go for exploiting hybrid vigor through production of double cross hybrids or synthetic varieties in autotraploid crops. In case of tetraploid the single locuse equilibrium is not attained in just one generation of random mating as the genotype frequencies at equilibrium are not the product of gene frequencies unless $\alpha = 0$. In case of tertasomic inheritance with $\alpha = 0$ (no double reduction) at least three generations of random mating would seem desirable before carrying out at genetical analysis (Crow and Kimura, 1970) where it is essential to work with equilibrium population. The three generations of random mating would reduce the difference to 3.7% of the initial value.

The equilibrium frequency of a deleterious recessive gene arising by mutation is higher with tetrasomic than with disomic inheritance and could explain the greater inbreeding depression which is sometimes seen in established tetraploids when compared with newly synthesized.

11.15 Other System of Inbreeding

Bartlett, Haldane and Fisher used 'generation matrices' for calculation of inbreeding coefficient. In spite of many advantages the methods offered like it, can take account of differences between genotypes in viability and the method can be extended to cover all loci simultaneously in case where all genotypes are equally viable and thus can calculate the probability of an individual in homozygous at every locus simultaneously using theory of junctions, the method did not become popular as it involved heavy algebra. Then came the concept of coefficient of inbreeding by Wright (1921) and Malecot (1948). Wright considers gametes produced by a given generation at time at which these gametes are uniting to form a zygote. He defines F, the coefficient of inbreeding or inbreeding coefficient as equal to the 'correlation between the uniting gametes'. Malecot considers a given locus in a given individual. The two alleles at this locus in an individual could be either identical by descent i.e. both alleles of the same ancestral allele which appeared earlier in the pedigree or independent in origin. Malecot defines F as the probability that the two alleles are identical by descent. Wright and Malecot's definitions are equivalent and give similar results but Wright's method is preferred because it is comprehensible and the analysis is much simpler.

Like for any regular system of inbreeding our aim was to find a formula, similar to our $H_{n+1} = 1/2\ H_n$ for selfing which connects the probability that an individual is heterozygous at a given locus, in a given generation with the corresponding probability in earlier generation. Such a formula is called recurrence relationship for probability that an individual is heterozygous. When we come to use such relationship we need to take account of the starting population for which we assume that all alleles are independent in origin i.e. in generation O, the coefficient of inbreeding is zero. Now we shall see how the small population sampling which will result in fixation of genes i.e. some genes will be randomly brought into homozygous state (a phenomenon called random drift) can be determined by measuring the coefficient of inbreeding of the population. Now assuming diploid inheritance suppose N individuals were randomly selected to form the generation 1 population. The N individuals will produce $2N$ gametes and thus the coefficient of inbreeding of this newly constituted generation population (F_1) would be the probability that the pairs of uniting gametes carry identical alleles which would be $1/2\ N$. If this new population is allowed to breed, the inbreeding coefficient from the following two sources instead of one in the F_1 in the successive generation populations will contribute to the total inbreeding coefficient of the population.

I. Inbreeding coefficient will arise afresh in generation which will be equal to $\frac{1}{2N}$.

II. Inbreeding coefficient will arise from the already fixed loci (homozygous loci) of the previous generation.

$$\left[1 - \left(\frac{1}{2N}\right)\right] F_1$$

Thus the total coefficient of inbreeding in generation 2 population will be

$$F_2 = \frac{1}{2N} + \left(1 - \frac{1}{2N}\right) F_1$$

By the same token the inbreeding coefficient in generation n will be

$$F_n = \frac{1}{2N} + \left(1 - \frac{1}{2N}\right) F_{n-1}$$

Falconer (1960) called $\frac{1}{2N}$ portion in the above generation equation as the 'increment' which measures the rate of inbreeding. Substituting $\frac{1}{2N}$ by ΔF, F_n becomes

$$F_n = \Delta F + (1 - \Delta F)\ F_n$$

Therefore,

$$\Delta F = \frac{F_n - F_{n-1}}{1 - F_{n-1}}$$

With this parameter ΔF, the proportionate increase in the rate of inbreeding, the effects of various methods of inbreeding can be compared. Further it can be shown that the change in the variance of gene frequency which will occur as a result of random genetic drift can be expressed in terms of ΔF, the rate of inbreeding.

$$\sigma^2_{\Delta v} = u_O v_O\ F$$

where u_O, v_O are the gene frequencies, respectively in the generation O population, the starting or base population. The estimation of inbreeding coefficient will have applicability in practical plant breeding programme especially recurrent selection breeding which we shall see in Plant Breeding (Roy, 2012).

The inbreeding affects the estimation of genetic parameters and the selection study. As the coefficient of inbreeding, F increases the coefficients of components of genetic variance increase. With the increase in inbreeding, the amount of all types of genetic variances increases. The relative increase in the genetic variances varies with the types of genetic variances and is generally larger in polygenic case and the more times dominance is present. In the selection programme it affects the consequences of selection within crosses. Gain from selection will be markedly affected by the inbreeding of the lines if much of the genetic variance is dominant and epistatic. The inbreeding experiments and selection experiments provide information on the nature of genetic differences present in the population.

References

Bauman, L.F. 1959. Evidence of non-allelic interaction in determining yield, ear height, and kernel row number in corn. Agron. *J.*, **51**: 531–534.

Bailey, T.B.Jr. 1976. Selection limits in self-fertilizing populations following the cross of homozygous lines. In: Pollak, E., Kempthorne, O., Bailey, T.B. Jr. (eds.). Proc. Int. Conf. Quant. Genet. Iowa State University Press. Ames. Iowa, pp. 399–412.

Bailey, T.B.Jr. and Comstock, R.E. 1976. Linkage and the synthesis of better genotypes in self-fertilizing crop species. Crop. Sci., **16**: 363–370.

Bennett, J.H. 1976. Expectations for inbreeding depression on self-dertilization of tetraploids. Biometrics, 32: 449-452.

Bliss, FA. and Gates, C.E. 1968. Directional selection in simulated populations of self-pollinated plants. Aust. J. Biol. Sci., **21**: 705–719.

Bradshaw, J.E. 1994. Quantitative Genetic Theory for Tetrasomic Inheritance. In "Potato Genetics.J.E. Bradshaw G.R. Mackay (eds). CABI, U.K., pp. 71-99

Bruce, A.B. 1910. The Mendelian theory of heredity and the augmentation of vigor. Science, **32**: 627–628.

Camussi, A., Ottaviano, E., Calinski, T. and Kaczamarer, Z. 1985. Genetic distance based on quantitative traits. Genetics. **111**: 945–962.

Comstock, R.E. and Robinson, H.E 1948. The components of genetic variance in populations of biparental progenies and their use in estimating the average degree of dominance. Biometrics. **4**: 254–266.

Cress, C.E. 1966. Heterosis of hybrid related to gene frequencies between the populations. Genetics. **53**: 269–174.

Crow, J.E 1948. Alternative hypothesis of hybrid vigor. Genetics, **33**: 477–487.

East, E.M. 1936. Heterosis. Genetics, **21**: 375–397.

Ellis, J.R.S., Brunion, C.J. and Palmer, J.M. 1973. Can mitochondrial complementation be used. as a tool for breeding hybrid wheat cereals? Nature, **241**: 45–47.

Falconer, D.S. 1989. Introduction to Quantitative Genetics. Longman, Burnt Mill.

Fisher, R.A. 1965. The theory of Inbreeding. 2ed Edn. Oliver and Boyd, London. Crow and Kimura, 1970. Population Genetics.

Gale, J.S. 1980. A statistical study at Chinese head measurement. J. Asiatic Soc. Bengal. 25: 301-377.

Gardner, C.O. 1963. Estimates of genetic parameters in cross-fertilizing pants and their implications in plant breeding. In: Hanson, W.D. and Robinson, H.F (eds.) Statistical Genetics and Plant Breeding. Natl. Acad. Sci., Washington, D.C, pp. 225–252.

Gardner, C.O. 1969. Genetic variation in irradiated and control population of corn after 10 cycles of mass selection for high grain yield. In: Induced mutations in Plants. IAEA. Vienna, pp. 469–477.

Gardner, C.O. and Eberhart, S.A. 1966. Analysis and interpretation of the variety cross diallel and related populations, Biometrics, **22**: 439–452.

Giles, B.E. 1984. A comparison between quantitative and biochemical variation in the wild barley *Hordeum murenum*. Evolution, 38: 34–41.

Griffing, B. 1990. Use of controlled nutrient experiment to test heterosis hypotheses. Genetics, **126**: 753–767.

Hallauer, A.R. and Miranda Fo, J.B. 1981. Quantitative Genetics in Maize Breeding. Iowa State Univ. Press. Ames, Iowa, USA.

Hill, W.G. 1982. Dominance and epistasis as components of heterosis. Z. Tierz Züchungsbiol., **99**: 161–168.

Hull, F.H. 1952. Overdominance and recurrent selection. In: Gowen, J. W. (ed) Heterosis, Chapter 28, Iowa state University Press, Ames, Iowa, pp. 451–473.

Hull, EH. 1945. Recurrent selection for specific combining ability in com. Jour. Amer. Soc. Agron., **37**: 134–137.

Hull, EH. 1946. Regression analysis of com yield data. Genetics, **31**: 219.

Jinks, J.L. 1955. A survey of the genetical basis of heterosis in a variety of diallel crosses. Heredity. **9**: 223–238.

Jinks, J.L. 1981. The genetic frame work of plant breeding. Phil. Trans. R. Soc. Lond. B. **292**: 407–419.

Jinks, J.L. 1983. Biometrical genetics of heterosis. In: Frankel, R. (ed) Heterosis-Reappraisal of the theory and practice. Springer-Verlag, pp. 1–46.

Jinks, J.L. and Jones, R.M. 1958. Estimation of components of heterosis. Genetics, Princeton **43**: 223–234.

Jones, D.E 1917. Dominance of linked factors as a means of accounting for heterosis. Genetics, 2: 466–479.

Kearsey, M.J. and Pooni, H.S. 1996. The Genetical Analysis of Quantitative Traits. Chapman and Hall, London.

Keeble, J. and Pellow, C. 1910. The mode of inheritance of stature and of timing of flowering in peas. (*Pisum sativum*) J. Genet., **1**: 47–58.

Kempthorne, O, 1969, An Introduction to Genetic Statistics. Iowa State University Press. Ames, Iowa.

Leforte-Buson, M. and Vienne D. De. 1985. Les distances genetiques. Estimates et Applications. INRA, Paris, pp. 181.

Lerner, I.M. 1954. Genetic Homoestasis. John Wiley, New York.

Malecot, G. 1948. Les mathematiques de 1 Here dete, Masson et cie. Paris.

Mareck, J.H. and Gardner, C.O. 1979. Response to mass selection in maize and stability of resulting populations. Crop Sci., **19**: 770–783.

Mather, K. and Jinks, J.L. 1982. Biometrical Genetics. 3rd edn. Chapman and Hall. London.

Minvielle, F. 1987. Dominance is not necessary for heterosis: a two locus model. Genet. Res., **49**: 245–247.

Moll, R.H. and Hanson, W.D. 1984. Comparisons of effects of intrapopulations versus interpopulation selection in maize. Crop Sci., **24**: 104–152.

Moll, R.H., Lonnquist, J.H., Fortuno. J. and Johnson, E.C. 1965. The relationships of heterosis and genetic divergence in maize. Genetics, **52**: 139–144.

Mungoma, C. and Pollak, L.M. 1988. Heterotic patterns among ten cornbelt and exotic maize populations. Crop Sci., **23**: 500–504.

Price, S.C., Shumaker, K.M., Kahaler, A.L., Allard, R.W., and Hill, J.E. 1984. Estimates of population differentiation obtained from enzyme polymorphism and quantitative characters. J. Heredity., **75**: 141–142.

Robinson, H.F, Comstock, R.E., and Harvey, P.H. 1949. Estimation of heritability and the degree of dominance in com. Agron. J., **41**: 354–359.

Sarkissian, I. V. and Srivastava, H.K. 1967. Mitochondrial polymorphism in maize. 2. Further evidence of a correlation of mitochondrial complementation and heterosis. Genetics **57**: 843–863.

Sarkissian, I.V. and Srivastava, H.K; 1971. Mitochondrial polymorphism 3. Heterosis, complementation and special properties of purified cytochrome oxidase of wheat. Biochem. Genet., **5**: 57–63.

Sarkissian, I.V. and Srivastava, H.K. 1973. Some molecular aspects of mitochondrial complementation and heterosis. In: Srb. A.M. (ed). Genes, enzymes and populations. Plenum Press, New York.

Shull, G.H. 1909. A pure line method of corn breeding. Amer. Breed. Assoc. Rept., **5**: 51–59.

Sinha, S.K. and R. Khanna, 1975. Physiological, biochemical and genetical basis of heterosis. Adv. Agron., **27**: 123–174.

Snell, F.W. and Cockerham, C.C. 1992. Multiplicative vs. arbitrary gene action in heterosis. Genetics, **131**: 461–469.

Sprague, G.F. 1983. Heterosis in maize. In: R. Frankel (ed). Heterosis-Reappraisal of the theory and practices. Springer-Verlag. pp. 47–67.

Stuber, C.W. and Moll, R.H. 1971. Epistasis in maize 3. comparison of selected with non-selected populations. Genetics. **67**: 137–149.

Willham, R.L. and Pollak, E. 1985. Theory of heterosis. J. Dairy Sci., **68**: 2411–2417.

Wright, S. 1921. Systems of mating. Genetics, **6**: 111–178.

Wright, S. 1922. Coefficient of inbreeding and relationship. Amer. Nat., **56**: 330–338.

Zoble, R.C., Fishback, F.N. and Laizko, E. 1972. Complementation of isolated rnitochodria from several wheat varieties. Plant Physiol., **50**: 790–791.

12

Polyploids and Haploids

12.1 Definition and Classification

The basic chromosome set designated as x of a species is called genome. The saprophytic or diplophase of a plant is designated as $2n$ whereas the gametophytic phase is designated as n. In $2n$ stage if the plants are having only one x it is called haploid whereas if $2x$ i.e. 2 chromosome sets are there, the plant is diploid. The variation in chromosome number can be of two types: (1) Euploid variation where the variation in chromosome number is in the multiple of basic chromosome set (x). Thus $2n = 3x$ is called triploid; $2n = 4x$ is called tetraploid, etc. Triploids ($3x$), tetraploids ($4x$), pentaploids ($5x$), hexaploids ($6x$), septaploids ($7x$), octaploids ($8x$) and so on are called polyploids. (2) Aneuploid variation where variation is in the multiple of a particular chromosome of chromosome set. Polyploids can further be classified into (1) autopolyploids where the same genome is multiplied, for example, in case of triploid it is AAA or in tetraploids it is $AAAA$, (2) allopolyploid which is the multiple of genomes which are not identical. For example, $AABB$ where A and B genomes are not identical (Kihara and Ono, 1926). Allopolyploids can be further partitioned into (1) classical allos (or genome allos or amphidiploids) where A and B genomes are very dissimilar, there is homologous pairing during meiosis and thus give rise to disomic inheritance and are functional diploids, (2) segmental allos where the two genomes A and B are having some similarities which permit considerable pairing of chromosomes or segment of chromosomes (Stebbins, 1947). This represents an intermediate situation between autopolyploids and amphidiploids. The individual shows homologous as well as homeologous pairing during meiosis. In wheat (*Triticum aestivum*) homeologous pairing is suppressed and homologous pairing takes place and thus it gives rise to disomic inheritance. In aneuploidy we can have variation of single or a few chromosomes. As $2n = 2x$ is called the disomic, $2x - 2$, $2x - 1$ and $2x - 1 - 1$ are called nullisomic, monosomic and double monosomic, respectively whereas $2x + 1$ and $2x + 1 + 1$ or $2x + 2$ are called trisomic and tetrasomic, respectively. All these hypoloids (nullisomics, monosomics, double monosomics) and hyperploids (trisomics, tetrasomics) come under primary aneuploids. We can have secondary trisomics when the extra chromosome is an isochromosome but when the extra chromosome is a translocated chromosome it is called tertiary trisomic.

Polyploidy results in increase in cell and nuclear size. In tomato it results in size of guard cells. It shows gigas effects i.e. leaves and petals are larger. There is a change in proportion of different parts of plants. The gigas effect is confined to newly produced autotetraploids. Plants show less branching and have larger fruits. Plants are late in flowering and fruiting. Tetraploid rye and red clover, however, show less frost resistance as compared to diploids. Triploids sugarbeets ($3x$) have higher sugar percentage in comparison to diploid. In vegetatively propagated fruits, forage and root crops polyploidy has been a success. Stability of triploids can be exploited to produce seedless fruits such as seedless banana and seedless watermelon. In ornamental crops larger and different flowers have got much value. Polyploidy is not successful in cereals crops. In most cases polyploids result by failure of nuclear division during mitosis which results in somatic doubling or by failure of cell division. During meiosis unreduced gametes may be produced which upon fertilization may either produce $4x$ zygote or $3x$ zygote. Such a situation can arise after temperature shock or artificial wounding of a plant followed by callus production and regeneration. Many plants are natural polyploids such as wheat, tobacco, etc. Polyploidy can be induced with high frequency by treatment with colchicine, an alkaloid derived from *Colchicicum autumnale* (Blakeslee and Avery, 1937) and used in plant breeding. The amphidiploid (*AABB*) is formed by doubling the chromosome set of the hybrid (*AB*) between two species. The hybrid (*AB*) is sterile which upon doubling may lead to complete normal meiotic pairing and fertility. Thus interspecific hybridization followed by chromosome doubling leads to the formation of a new species. These occur frequently among wild species. Autopolyploid does not produce true to its type in most cases because of irregular chromosome segregation and thus the behaviour is quite unpredictable. Aneuploids are formed by non-disjunction. In $2x$ plants loss of a chromosome is usually lethal. Polyploids can tolerate aneuploidy better than others. Hyperploidy is much more tolerated than hypoploidy. A classical example of practical use of aneuploidy has been the transfer of genes for leaf rust resistance from *Aegilops umbelluta* to wheat through induced translocation.

Breaking and rejoining of chromosomes through radiation and chemical treatment produces changes in chromosome structure. Such structural arrangements could be stable if only a small segment of chromosome is involved and unstable if it involved larger segment of a chromosome. The causes of unstability are (1) the mechanical inefficiency and (2) genetic loss (lethal). The structural rearrangement can be of three types: (1) Duplication, deletion, (2) Inversion and (3) Reciprocal translocation: Duplication may give rise to entirely new phenotype as in case of Bar eye in Drosophila. The barley production of α – amylase has improved by duplication of this locus. Also duplicated region through mutation can give rise to new gene. Deletion has largely mortality effects. The extent of effect depends on the size of segment deleted. Inversion, translocation and transposition, however, may be associated with position effects because of disruption of physiological relations between genes at opposite sides of a rearrangement. Incorporation of one or more chromosomes or chromosome segments into a population from other population may occur through hybridization followed by backcrossing — a process known as introgression (Anderson, 1949).

There are several crop plants such as potato, coffee, lucerne and several forage grasses which are autotetraploids and in which the pattern of inheritance is tetrasomic, not disomic. Thus like diploid, it is essential to know the genetical architecture of the metrical traits so that a suitable method of breeding and selection can be devised in order to obtain improved cultivars. Polyploids are frequently found to be heterozygous.

12.2 Biometrical Genetics of Autotetraploids

With two alleles A and a there will be five possible genotypes in tetraploids. These are:

$AAAA$ or $A4$ quadruplex

$AAAa$ or $A3a$ triplex

$AAaa$ or A_2a_2 duplex

$Aaaa$ or Aa_3 simplex

$aaaa$ or a_4 nulliplex

Following Mather and Jinks (1971) the genetical values of these above genotypes will be

$$A_4 = m + d_a$$
$$A_3 = m + ha_3$$
$$A_2a_2 = m + ha_2$$
$$Aa_3 = m + ha_1$$
$$a_4 = m - d_a$$

where triplex, duplex and simplex genotypes have been given the unique values of ha_3, ha_2 and ha_1, respectively and this approach thus avoids making any assumptions about the relative genetical values of the genotypes which other approaches do.

The another extreme way of assigning genetical-values to these group of genotypes would be to just follow the expectation as in diploids (Haldane, 1930; Mather 1949) i.e. to assume that $ha_3 = ha_2 = ha_1, = ha$.

Then the expectations of the above five genotypes would be

$$A_4 = m + da$$
$$A_3a = A_2a_2 = Aa_3 = m + ha$$
$$a_4 = m - da$$

The problem with this model is that the genetic effect is not proportional to the dosage of alleles which it should be and is in the earlier model. This problem can be overcome by assigning duplex, the value ha ($ha_2 = ha$) and expressing the values of triplex and simplex as the mean of duplex and the homozygous genotypes.

$$\overline{A}a_3 = \frac{\overline{A_2a_2} + \overline{a}_4}{2} = -\frac{1}{2}da + \frac{1}{2}ha$$

$$\overline{A}_3a = \frac{\overline{A}_4 + \overline{A_2a_2}}{2} = +\frac{1}{2}da + \frac{1}{2}ha$$

The shortcoming with this type of model is that with complete dominance ($h = d$), the simplex takes the value of m when it should take $m + d(= m + h)$. But again this situation can be avoided if the coefficient of h in the expectation of triplex and simplex is changed to 4. Dessureaux (1959) gave the expectations of A_3a and Aa_3 as

$$A_3a = m + \lambda d$$
$$Aa_3 = m + \lambda d$$

Where $\quad \lambda = \frac{1}{2} - \frac{1}{2} h/d \text{ if } h \le d$

or $\quad \lambda = \frac{1}{2} + \frac{1}{2} h/d \text{ if } h \le -d$

With this the expectations of A_3a and Aa_3 came out to be

$$A_3a = m + \frac{1}{2}d + \frac{1}{2}h$$
$$Aa_3 = m - \frac{1}{2}d + \frac{3}{2}h$$

The difficulty with this model is that first, the parameters are not independent and secondly, the contribution of h to simplex is three times greater than that of triplex which is unrealistic.

Two parents, P_1, P_2 (AAAA and aaaa) are selected and crossed to derive F_1, F_2, BC_1, BC_2 and other generations in order to estimate parameters like in diploids. But in tetraploids unlike diploids, there is occurrence of double reduction (Mather, 1936) wherein during meiosis sister chromatids may enter the same gemete which results from the recombination between the centromere and distal genes followed by a particular orientation of the chromosome on the second metaphase spindle. The coefficient of double reduction (a) represents the portion of gametes in which sister genes occur. It has a theoretical maximum value of 1/7 (0.1428) and a minimum value of zero (when segregation is chromosomal, i.e. the genes are transmitted as if they were completely linked to the centromere). The five genotypes produce gamete in the frequencies as given in Table 12.1 (Fisher, 1949).

Table 12.1 Gametic output in an autotetraploid

Parental Genotype	Gametes			Divisor
	AA	Aa	aa	
A_4	1	–	–	1
A_3a	$2 + \alpha$	$2(1 - \alpha)$	α	4
A_2a_2	$1 + 2\alpha$	$4(1 - \alpha)$	$1 + 2\alpha$	6
Aa_3	a	$2(1 - \alpha)$	$2 + \alpha$	4
a_4	–	–	1	1

The frequencies of five genotypes (quadruplex, triplex, duplex, simplex and nulliplex) in the generations derived from the cross of two inbred lines are given in Table 12.2 assuming $\alpha = 0$ and $\alpha = 0.14$ from which the expected generation mean of any generation can be written (Killick, 1971).

For other generations Killick's (1971) paper can be seen. The double reduction causes changes in the proportion of the genotypes which would have a marked effect on the estimation of parameters from generation means. Double reduction causes inbreeding by allowing sister chromatids to enter the same gamete and thus causes the proportions of quadruplex and/or nulliplex to increase and consequently the overall proportion of heterozygous genotypes (simplex, duplex and triplex) to decline and further it increases the proportion of the rarest genotype and decreases the proportion of the

12.4 Biometrical Genetics–Analysis of Quantitative Variation

Table 12.2 Frequency of different genotypes at $\alpha = 0$ and $\alpha = 0.14$

Generation	Quadruplex	Triplex	Duplex	Simplex	Nulliplex
P_1	1.000	–	–	–	–
	1.000	–	–	–	–
P_2	–	–	–	–	1.000
	–	–	–	–	1.000
F_1	–	–	1.000	–	–
	–	–	1.000	–	–
F_2	0.028	0.222	0.500	0.222	0.028
	0.046	0.245	0.420	0.245	0.046
BC_1	0.167	0.667	0.167	–	–
	0.213	0.573	0.213	–	–
BC_2	–	–	0.167	0.667	0.167
	–	–	0.213	0.573	0.213

commonest genotype in any generation except the first back-cross selfed and the second back-crosses. Thus assuming ha_3, ha_2 and ha_1 being the genetical values of triplex, duplex and simplex, respectively and no double reduction, i.e. $\alpha = 0$ the expected means of basic generations will be as follows:

Generation	Expected mean
P_1	$m + da$
P_2	$m - da$
F_1	$m + ha_2$
F_2	$m + 2/9 ha_1 + 1/2 ha_2 + 2/9 ha_3$
BC_1	$m + 1/6 da + 1/6 ha_2 + 2/3 ha_3$
BC_2	$m - 1/6 da + 2/3 ha_1 + 1/6 ha_2$

The parameters can be estimated following weighted least square procedure.

12.3 Scaling Tests

A, B and C scaling tests when applied to the general model using ha_1, ha_2 and ha_3 will all fail as with only additive and dominance gene actions, the expected totals are not zero. Even when the three heterozygous genotypes are pooled, i.e. $ha_1 + ha_2 + ha_3 = h$, these tests do not hold. But assuming $A_3a = m + 1/2d + 1/2h$ and $Aa_3 = m - 1/2d + 1/2h$ the following six tests are satisfactory even with double reduction.

$$A = \overline{P}_1 + \overline{F}_1 - 2\overline{BC}_1$$

and $$B = \overline{P}_2 + \overline{F}_1 - 2\overline{BC}_2 \text{ (Mather, 1949)}$$

$$BC_{11} = 3\overline{P}_1 + \overline{F}_1 - 4\overline{BC}_{11}$$

$$BC_{12} = \overline{P}_2 + 3\overline{F}_1 - 4\overline{BC}_{12}$$

$$BC_{22} = 3\overline{P}_2 + \overline{F}_1 - 4\overline{BC}_{22}$$

and $$BC_{21} = \overline{P}_1 + 3\overline{F}_1 - 4\overline{BC}_{21} \text{ of Hill (1966)}$$

The two scaling tests which are satisfactory using the general definition are

$$\overline{P}_1 - \overline{F}_1 + 2\overline{BC}_{21} - 2\overline{BC}_{11} = 0$$

$$\overline{P}_2 - \overline{F}_1 + 2\overline{BC}_{12} - 2\overline{BC}_{22} = 0$$

which also hold for disomic inheritance. These two scaling tests have been derived from the above four relationships which hold for any relative values of h_1 h_2 and h_3.

12.4 Variances and Covariances

Killick (1971) worked out the coefficients of the parameters Σd^2, Σha_3^2, Σha_2^2, Σha_1^2, $\Sigma ha_3 ha_1$ $\Sigma ha_3 ha_2$, $\Sigma ha_2 ha_1$, $\Sigma daha_3$, $\Sigma daha_2$, $\Sigma daha_1$, assuming chromosomal segregation for variances and covariances of different generations but given below in Table 12.3 are expectations in case of few generations.

It is evident from the table that the additive and dominance variance components cannot be separated

Table 12.3 Expectation of variances assuming $\alpha = 0$

Variances	d_a^2	ha_3^2	ha_2^2	ha_1^2	ha_3ha_2	ha_3ha_1	ha_2ha_1	$daha_3$	$daha_2$	$daha_1$
F_2	1/18	14/81	1/4	14/81	−2/9	−8/81	−2/9	–	–	–
BC_1	5/36	2/9	5/36	–	−2/9	–	–	−2/9	−1/18	–
BC_2	5/36	–	5/36	2/9	–	–	2/9	–	1/18	2/9

Covariance

F_2/F_3 1/18 5/18 5/72 5/81 −2/81 −8/81 −2/81 1/18 −1/18

from the estimates of F_2, back-cross and other generations but assuming $ha_1 = -\frac{1}{2}d + \frac{1}{2}h$, $ha_2 = ha$ and $ha_3 = \frac{1}{2}da + \frac{1}{2}ha$, the expectations of F_2 and backcross generation variances reduce to

$$F_2 = \frac{1}{6}da^2 + \frac{1}{12}ha^2$$

$$BC_1 = \frac{1}{12}da^2 + \frac{1}{12}ha^2 - \frac{1}{6}daha$$

$$BC_2 = \frac{1}{12}da^2 + \frac{1}{12}ha^2 + \frac{1}{6}daha \text{ and}$$

thus it is possible to separate additive and dominance variance. The problem with estimation of genetical components of variation from estimate of variances and covariance is that the coefficients of the components of variation are small and thus their estimation is subject to a large error. Thus in tetraploids first degree statistics seem to be more useful source of information than second degree statistics.

12.5 Random Mating Population

Kempthorne (1955) provided a model for the analysis of quantitative traits in autotetraploid populations. He partitioned the total genetic variance (σ_G^2) into components as

$$\sigma_G^2 = \sigma_A^2 + \sigma_D^2 + \sigma_T^2 + \sigma_F^2 + \sigma_{AA}^2 + \sigma_{A \times D}^2 + \sigma_{A \times T}^2 + \sigma_{A \times F}^2 + \sigma_{D \times D}^2 + \dots \text{ etc.}$$

Where σ_A^2 is the additive genetic variance,

σ_D^2 is the variance due to digenic effect (refers to interaction between 2 alleles),

σ_T^2 is the variance due to trigenic effects,

and σ_F^2 is the variance due to quadrigenic effects:

σ_A^2 and σ_D^2 are analogous to the additive and dominance variance components of diploid. At autotetraploid level, the term dominance includes digenic, trigenic and quadrigenic effects. The general method of subdivision of the genetic variance in random mating population developed by Kempthorne (1957) and others depends on fitting the effects of single alleles and their successively higher orders of interaction by least square. For the case of an autotetraploid locus with two alleles, Li (1957) proposed a different method by which the components are found as the variance associated with successive terms in a polynomial regression of genotypic value on to allele frequency. Changes in allele frequency can be affected by the genetic background and in particular by selection at closely linked loci.

Mating designs used in the analysis in diploids can also be used for autotetraploids and the components of variance arising from a given mating design can be equated to the same covariances between relatives regardless of the ploidy level. The difference between analyses of diploid and autotetraploid populations lies in the interpretation of covariances between relatives in terms of genetic variance. Kempthorne worked out the expectation of covariances in terms of genetic components of variation in autotetraploid population. A comparison of the expectations of covariances in diploids and autotetraploids is shown in Table 12.4

With these expectations of Cov (HS) and Cov (FS) in random mating autotetraploid population, the statistical variances obtained from any mating design can be translated in terms of genetical variance components as can be seen in Table 12.5 in case of NCMI.

Table 12.4 Comparison of coefficients of genetic variance of covariances in diploids and autotetraploids

Covariance	Ploidy level	Coefficients of components of genetic variance						
		σ_A^2	σ_D^2	σ_T^2	σ_F^2	σ_{AA}^2	$\sigma_{A\times D}^2$	$\sigma_{D\times D}^2$
Cov (HS)	2n	1/4	0	–	–	1/16	0	0
	4n	1/4	1/36	0	0	1/16	1/144	1/1296
Cov (FS)	2n	1/2	1/4	0	0	1/4	1/8	1/16
	4n	1/2	2/9	1/12	1/36	1/4	1/9	4/81
Cov (OP)	2n	1/2	0	–	–	1/4	0	0
	4n	1/2	1/6	0	0	1/4	1/12	1/36

Table 12.5 Expectation of statistical variances in NCMI in autotetraploid population

Components of variance	Coefficients of genetic variance						
	σ_A^2	σ_D^2	σ_T^2	σ_F^2	σ_{AA}^2	$\sigma_{A\times D}^2$	$\sigma_{D\times D}^2$
σ_m^2	1/4	1/36	0	0	1/16	1/144	1/1296
$\sigma_{f|m}^2$	1/4	7/36	1/12	1/36	3/16	15/144	63/1296

Table 12.6 Comparison of expectations of GCA and SCA variances in diploid and autotetraploid

Components or variance	Ploidy level	Coefficients of genetic variance						
		σ_A^2	σ_D^2	σ_T^2	σ_F^2	$\sigma_{A\times A}^2$	$\sigma_{A\times D}^2$	$\sigma_{D\times D}^2$
σ_g^2	2n	1/4	0	–	–	1/16	0	0
	4n	1/4	1/36	0	0	1/16	1/144	1/1296
σ_s^2	2n	0	1/4	–	–	1/8	1/8	1/16
	4n	0	1/6	1/12	1/36	1/8	7/72	31/648

The general combining ability (σ_g^2) and specific combining ability (σ_s^2) variances can be expressed as functions of the components of genetic variance assuming both diploid and autotetraploid inheritance (Levings and Dudley, 1963) as given in Table 12.6.

Table 12.6 shows that it is not possible to have a clean estimate of average degree of dominance using GCA and SCA variances in autotetraploids as in case of diploid species. The best estimate of narrow sense heritability (h_n^2) in autotetraploids as in diploids, probably results from doubling the regression of offspring on parent and becomes

$$h_n^2 = (\sigma_A^2 + 1/3\,\sigma_D^2 + 1/2\,\sigma_{AA}^2 + 1/18\,\sigma_{DD}^2 + 1/6\,\sigma_{AD}^2 + \text{etc})/\sigma_P^2$$

In case of triallel analysis in autotetraploids, the expectations of statistical variances take the form as given in Table 12.7

12.6 Triploids

In diploid species, the endosperm of plant is triploid. The endosperm in cereals is rich in starch whereas it is rich in oil in castor. The models of analysis of quantitatively inherited characters expressed in

Table 12.7 Expectation of statistical variances in triallel analysis

Statistical variance	Coefficients of components of genetic variance						
	σ_A^2	σ_D^2	σ_T^2	σ_F^2	$\sigma_{A\times A}^2$	$\sigma_{A\times D}^2$	$\sigma_{D\times D}^2$
σ_g^2	1/9	4/729	0	0	1/64	49/46,656	361/3,779,136
σ_{S2}^2	0	1/162	1/243	$\frac{1}{2916}$	$\frac{25}{1152}$	$\frac{567}{93,312}$	$\frac{11293}{7558,272}$
σ_{S3}^2	0	0	0	0	0	1/288	29/293,328
σ_{01}^2	1/16	25/1296	0	0	9/256	12/20.736	1225/1.679,616
σ_{02a}^2	0	1/24	1/108	1/1296	1/128	29/2804	1403/419,904
σ_{02b}^2	0	0	1/432	0	0	35/20,726	75/93.312
σ_{03}^2	0	0	0	0	0	1/384	1485/419,984

triploid endosperm would be different than in case of diploid. We have seen above the models of analysing the polyploid and more specifically the autotetraploid assuming no non-allelic interaction, no linkage or no double reduction. Although Gate (1976) suggested a model for analysing the triploid endosperm assuming no non-allelic interaction, Bogyo et al. (1988) proposed a model considering non-allelic interaction but uses a scale different than that of Gate's. With one locus 2 alleles (B, b) the endosperm can have four types of genotypic constitution, namely, BBB, BBb, Bbb, and bbb with the genotypic values given below:

Genotype	Genotypic value	
	No dominance	Dominance
BBB	$m + \frac{3d}{2}$	
BBb	$m + \frac{d}{2}$	$m + \frac{a}{2} + h_1$
Bbb	$m - \frac{d}{2}$	$m - \frac{a}{2} + h_2$
bbb	$m - \frac{3d}{2}$	

Where m is the mid-parent between the parents BBB and bbb; d is the additive genetic effect and h is the dominance effect. The genotypic value of an individual (G) can thus be expressed in terms of (d), h and the interaction terms as

$$G = m + [d] + [h_1] + [h_2] + [dd] + [dh] + [hh]$$

This model differs from Mather and Jinks' model of diploid in that the additive value is not scaled from the mid-parent. Also, the dominance effects are deviations from the additive effects (under the assumption of no dominance) rather than deviation from the mid-parent as shown in Figure 12.1.

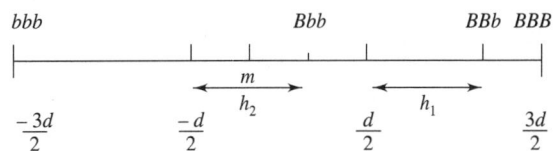

Fig 12.1 Additive (d) and dominance genetic effects (h_1, h_2) in case of triploids (adapted from Bogyo et al. 1988).

In this model interaction parameters can also be included but the interaction components cannot be derived simply by multiplying the coefficients of main effect parameter. With two parents (BB and bb), the genotypic constitution of endosperm of F_1, RF_1, F_2, BC_1 and BC_2 generations can be written and their expectations of means can be worked out as given in Table 12.8.

Model fitting and estimates of parameters can be obtained by weighted least square technique as described in case of diploids. Like the expectations of means, the expectations of variances of basic

12.8 Biometrical Genetics–Analysis of Quantitative Variation

Table 12.8 Coefficients of components of means of basic generations

Generation	Parameters						
	m	$[d]$	$[h_1]$	$[h_2]$	$[dd]$	$[dh]$	$[hh]$
P_1	1	3/2	0	0	1	0	0
P_2	1	−3/2	0	0	1	0	0
F_1	1	1/2	1	0	0	0	1
F_2	1	0	1/4	1/4	0	0	1/4
BC_1	1	1/2	0	1/2	1/4	1/2	1/4
BC_2	1	−1/2	1/2	0	1/4	1/2	1/4

Table 12.9 Coefficients of components of variances of basic generations

Generation	Parameters					
	D	H_1	H_2	DH_1	DH_2	H_1H_2
F_2	5/4	3/16	3/16	1/4	−1/4	−1/8
BC_1	1	0	1/4	0	−1	0
BC_2	1	1/4	0	1	0	0

generations can be worked out in terms of additive (D), dominance (H) and interaction variances as given in Table 12.9 and estimates of these can be obtained following unweighted least square procedure (Hayman, 1960). However, this procedure is correct when the observed variances and covariances are uncorrelated and have equal errors but these conditions are seldom met.

Further expectations of different covariances can also be worked out in terms of genetical components of variation. Thus we find that by studying the means, variances and covariances one can determine the genetical architecture of endosperm characters.

12.7 Haploids

Plants having gametic or haploid number of chromosomes are called haploids. In other words, haploid is a sporophyte with gametic chromosome number. In most plants loss of single chromosome results in inviability but can tolerate whole set and survive. Thus there operates some kind of genetic balance and this has led to the interest for developing and studying haploids. Haploids in comparison to their diploids are not very vigorous. Also deleterious gene is expressed in haploids which is more frequent and common in haploids of outbreeding species. Further haploids are weak and there is general physiological imbalance which can lead to sterility. In haploid rye no viable gametes are produced and thus there is complete sterility.

12.7.1 Production of haploids

Haploids arise spontaneously in many species at lower rate through parthenogenesis. The seed develops from an unfertilized egg cell and in this process pollination is essential but no fertilization. In maize the rate of production of parthenogenetic haploids ranges from 0 to 1.8% whereas in *Brassica napus* it ranges from 0 to 6.8%. The phenomenon of androgenesis in which male gamete nucleus develops in egg cytoplasm (embryosac) is rare. Twin seedlings (poly embryony) although not important commercially is a source of haploid as a very small percentage of population produces twin seedlings and there is a higher frequency of haploids in seedlings rather than in normal population. For example, in rye out of 300,000 seeds 82 twins were observed and further examination showed only one haploid whereas in asparagus out of 1,50,000 seeds 325 were twins and only in 27 plants haploidy was confirmed. In capsicum haploidy is the most frequent mutation occurring with a frequency of one out of 1000 to one per 10,000 plants and most capsicum haploids have occured naturally from $n-2n$ twin seedlings of polyembryony. Cytological investigation has shown that the $n-2n$ twins originate from a synergid and a fertilized egg nucleus whereas $n-n$ type of twins originates from a synergid and unfertilized egg nucleus or from a synergid nucleus by division. The other synergid disintegrates after having passed the pollen tube. The antipodal cells in peppers disintegrate before fertilization. Besides $n-2n$ and $n-n$ twins which occured in high frequency relative to twins of other genetic constitution such as $4n-2n$, $2n-2n$, conjoined $2n-2n$ unattached, $2n-2n-2n$ (triplets) $2n-2n-2n-2n$ (quadruplets). $n-n-2n$ triplets with haploid conjoined occured rarely. $n-n-2n$ triplets arise from a haploid proembryo developed from a synergid plus a zygote nucleus. As the frequency of polyembryony is often under genetic control, its frequency can be increased through selection. The different varieties differ in the frequency of polyembryony.

Haploids can be produced by interspecific hybridization and by anther culture and pollen culture techniques (*in vitro* pollen and another culture). In the first method there is chromosome elimination following wide cross hybridization as in case of barley in which the cross (*Hordeum vulgare* × *H. bulbosum*) gives rise to haploids ($2n = x = 7(v)$ of *H. vulgare*). The chromosome elimination takes place during first cell division after zygote formation. Haploids can be developed directly from pollen grain. It was first reported in Datura (Guha and Maheshwari, 1964). Haploids can now be developed from anther as well as pollen culture. There are different pathways for developing haploids from pollen-grains. Either plantlet can be obtained directly from the pollen as in case of *Nicotiana, Petunia, Oryza* or from pollen callus which gives rise to plantlet as in *Brassica, Hordeum* and *Solanum*. There are two things that are important when the development of haploid from pollen is considered. The first is the stage of pollen chosen for culturing. The pollen-grain or microspores are uninucleate formed as a result of meiosis and thus one pollen mother cell (PMC) gives rise to four uninucleate microspores. The ideal stage is the pollen grain just before first mitosis and the second important thing is the culturing medium. The anther culture method gives higher yield in terms of number of haploid than the pollen grain culture because of the nutritive tissue (tapetum) in the anther.

The foreign pollen, active or inactive (dead) from a different species stimulates the egg to develop into a haploid. For example, in potato, haploid is produced when *S. tuberosum* ($2x = 48$) is crossed to a related wild species like *S. phureja* ($2x = 24$). Haploids can also be produced by the inactivated 'self pollen' of the same species. Pollen can be inactivated by irradiation (temperature, X-ray, U.V. light) or by chemical treatment. The treatment of female parent by chemical shock (toluidine blue, laughing gas, formaldehyde) and by physical method can lead to production of haploid without pollination. Finally, haploids can also be produced by somatic reduction using various treatments such as colchicine, fluorophenylalanine, chloramphanicol.

12.7.2 Identification of haploids

Haploids can be recognised by morphology. Haploids are generally smaller, weak, less vigorous as compared to their diploids. Haploidy can be ascertained by counting the chromosome number. It can also be recognized by counting the nucleoli. In the interphase, the diploid ($2x$) will have two nucleoli, whereas the haploid (x) will have only one. Marker gene can also be used to identify the haploids. In maize, purple aleurone is used to detect haploid.

12.7.3 Advantages and disadvantages

The most practical use of haploids will be to go for production of dihaploids. Haploids produced can be treated with colchicine and the resultant is homodiploid which is homozygous at every locus irrespective of the genetic constitution earlier. Thus pure breeding homozygous lines can be produced spontaneously which normally takes 6 – 7 generations of selfing following hybridization. Thus there is a great saving of time and is a shortcut method of obtaining homozygous diploid in one step. The genetical investigations of the effectiveness of pure breeding lines developed from haploids have been carried out by Nei (1963), Walsh (1974), Demarly (1975), Griffing (1975) and Jinks and Pooni (1981). Walsh (1974) concluded that in comparison to pedigree method, the line developed through dihaploidy showed lower performance and this was attributed to the repeated opportunity of recombination in pedigree method but then inadequate recombination in the F_1 derived dihaploids can be compensated by deriving dihaploid lines from F_2 or backcross generations. Although haploids are inferior in fertility and stability which has been attributed to attainment of absolute homozygosity which is considered undesirable in many crops. However, Thevenin produced superior F_1 asparagus hybrids of greater uniformity through use of haploids. In case of self sterile allogamous crops and even heterogametic dioecious plants complete homozygosity can be obtained. *Asparagus officinale* is a dioecious plant; female plants are *XX* and male plants are *XY*. Through anther culture haploids for *X* and *Y* can be obtained which upon chromosome doubling yields *XX* and *YY* (super male) plants which are completely homozygous and the F_1 progenies would be very uniform (*XX XXY – XY*). We can go for dihaploid breeding. In this method from autotetraploid ($4x$), haploid ($2X$) can be produced and breeding can be started at the haploid level and improved through selection and finally $4X$ plants would be restored. Whereever cytoplasm determines a particular desirable trait androgenic haploid can be used as a means of transferring cytoplasm between varieties. One can start mutation breeding at haploid level as all the genes are expressed in haploid but not at diploid level, so useful gene mutation can be immediately isolated and fixed by doubling the chromosomes. One of the main stumbling blocks in mutation breeding of higher plants is the formation of chimera's following the treatment of multicellular organism. The mutation will not be stable if seed propagated plants are used but if plants originate from one mutant cell, it will be genetically uniform. Cell suspensions, protoplasts, microspores and tissues of haploid plants can be exposed to various physical and chemical mutagens (Nitzsche and Wenzel, 1977). Screening of mutants showing resistance to fungal and bacterial toxins and to other environmental inorganic chemicals such as NaCl, SO_2, Pb salt, herbicides, insecticides can be done but isolation of mutants with undesirable characters or lacking other desirable character and the character showing deviation from Mendelian inheritance and the differences in results between calluses and the regenerated plants show that this method has yet to show a promise. Besides there is problem of the regeneration of plants from a single cell and even within a species, different genotypes differ in their capacity to generate plants and thus this method of generating useful mutants has not become a reality.

When plant is pollinated with irradiated pollen, the progeny showed the maternal morphology (matromorphy). There are examples of gene transfer through irradiated pollen (Pandey, 1975, 1976; Jinks et al. 1981). Pandey observed that when *N. forgetiana* was pollinated with irradiated pollen from *N. alata*, the progeny, pathogenic diploid, showed maternal morphology as well as colored flower and/or an incompatibility allele characters transferred from the male parent. The explanation proposed was that the irradiation led to pulverised generative nucleus which during pollination led to the transfer of only few fragments of the male genome.

But there is question of cost, economy involved in the production of haploid. Haploid production needs facilities like good laboratory and technical staff. Besides one will have to decide which cross will give rise to superior recombinant inbred/pure breeding lines. In case of the study of genetical architecture of disease resistance traits, the breeder is not only working with plants but also with plant pathogens such as bacteria, fungi, viruses, etc. In most fungi there exists a dominant haploid phase. An independent sexual diploid phase is restricted to some yeasts, Myxomycetes and Oomycetes. Basidiomycetes, however, possess an independent dikaryotic phase which resembles the diploid state (Simchen and Jinks, 1964). Thus we see that a breeder besides encountering the haploid crop plants, has to work with haploid pathogen when studying the host-pathogen interaction.

Therefore, it is essential to have knowledge of the analyses of metric traits in haploid plants as well as haploid plant pathogens so that improvement can be brought about in haploid plants as well as in haploid microorganism particularly industrial micro organism besides providing an insight into the host-pathogen interaction and to formulate a methodology for breeding for resistance in crop plants.

12.7.4 Biometrical genetics of haploids

12.7.4.1 Generation means

Considering one locus with two alles A, a, the two parents P_1 and P_2 will be of genotypic constitution A and a, respectively. Following Mather and Jinks approach the expectations of P_1 and P_2 means assuming additive-dominance model will be as follows in the polygenic case.

$$\overline{P}_1 = m + [d]$$
$$\overline{P}_2 = m - [d]$$

The F_1 from the cross ($P_1 \times P_2$) will have genotypic constitution Aa, a diploid genotype, which is transitory and ultimately results in individuals of A and a genotypes. Thus the expectation of F_1 generation mean would become

$$\overline{F}_1 = m$$

By selfing F_2, F_3, F_4, etc. generations will be produced and the expectations of these generation means can be shown to be

$$\overline{F}_2 = \overline{F}_3 = \overline{F}_4 \ldots = \overline{F}_n = m$$

As in case of diploid back-cross generations can be generated and the expectations of these would become

$$\overline{BC}_1 = m + 1/2 \, [d]$$
$$\overline{BC}_2 = m - 1/2 \, [d]$$

From the above generation means, the estimates of m and $[d]$ parameters are obtained as

$$m = 1/2 \, (\overline{P}_1 - \overline{P}_2)$$
or
$$m = \overline{F}_1$$
And
$$[d] = 1/2 \, (\overline{P}_1 - \overline{P}_2)$$

or
$$[d] = (\overline{BC}_1 - \overline{BC}_2)$$

Thus if additive model is adequate

$$[d] = 1/2 \, (\overline{P}_1 - \overline{P}_2) = (\overline{BC}_1 - \overline{BC}_2)$$

The variances of the parameters m and $[d]$ are estimated as

$$V_{(m)} = V\overline{F}_1$$
$$V[D] = 1/4 \, V\overline{P}_1 + 1/4 \, V\overline{P}_2$$

In case of non-allelic interaction, the expectations of P_1, P_2, F_1, F_2, BC_1 and BC_2 generations become

$$\overline{P}_1 = m + [d] + [i]$$
$$\overline{P}_2 = m - [d] + [i]$$
$$\overline{F}_1 = \overline{F}_2 = m$$
$$\overline{BC}_1 = m + 1/2 \, [d] + 1/4 \, [i]$$
$$\overline{BC}_2 = m - 1/2 \, [d] + 1/4 \, [i]$$

The estimates of m, $[d]$ and $[i]$ parameters are obtained as

$$m = \overline{F}_1 = \overline{F}_2$$
$$[d] = 1/2 \, (\overline{P}_1 - \overline{P}_2)$$
$$[i] = 1/2 \, (\overline{P}_1 + \overline{P}_2) - \overline{F}_1$$

Here the variance of $[i]$ is estimated as

$$V[i] = V\overline{F}_1 + 1/4 \, V\overline{P}_1 + 1/4 \, V\overline{P}_2$$

12.7.4.2 Generation variances

The variances within parents (P_1, P_2) will be all environmental (E) whereas the variance of the F_1 will become

$$VF_1 + D + E$$

The estimate of D, the additive genetic variance is estimated from the analysis of variance of F_1 individuals using clonal replicate as shown in Table 12.10.

12.12 Biometrical Genetics–Analysis of Quantitative Variation

The environmental variance component (E) is estimated as

$$\sigma_w^2 = E$$

and the additive genetic variance (D) becomes

$$\sigma_b^2 = D$$

Table 12.10 Analysis of variance of F_1 individuals derived from the cross of two haploid parents

Item	D.F.	E.M.S.
Between individuals	$(S-1)$	$\sigma_w^2 + r\sigma_b^2$
Within individuals	$S(r-1)$	σ_w^2

S is the number of F_1 individuals; r the number of replicate clones.

In the presence of non-allelic interaction, the variances of F_1, F_2 and back-cross generations will be as follows:

$$VF_1 = VF_2 = D + I + E$$

$$VBC_1 + VBC_2 = 1/2 D + 3/8 I + E$$

Thus from the study of above generations variances, we can estimate additive genetic (D), interaction (I) and environmental variance (E) components of variation.

12.8 Doubled Haploids

Haploids as we have seen can be produced by means of various production techniques such as anther culture, interspecific hybridization (Balbosum method) and so on (Kasha, 1974), which upon chromosomes doubling by colchicine form doubled haploids. Steps involved in the production of doubled haploids (DH) lines can be seen in Figure 12.2. Two inbreds/pure breeding lines parents are crossed to produce F_1. The male gametes (pollens) from the F_1 plants are cultured to generate haploid plants which are treated with colchicine to produce DH progeny. DH lines can also be produced from individuals from F_2 or F_3 generation segregating population. All lines produced by Bulbosum technique are a random sample of gametes from F_1 plants (John, 1974).

Parents $AABB \times aabb$
F_1 $AaBb \rightarrow F_2$
Gametes $\dfrac{1-p}{2} AB + \dfrac{1-p}{2} ab + \dfrac{p}{2} Ab + \dfrac{p}{2} aB$
$\quad\quad\quad\quad\downarrow\quad\quad\quad\downarrow\quad\quad\quad\downarrow\quad\quad\quad\downarrow$
Generation of Haploids
$\quad\quad\quad\quad AB \quad\quad ab \quad\quad Ab \quad\quad aB$
Colchicine treatment
DH progeny $\downarrow\quad\quad\downarrow\quad\quad\downarrow\quad\quad\downarrow$
$(1-p)/2\ AABB + \dfrac{1-p}{2} aabb + \dfrac{p}{2} AAbb + \dfrac{p}{2} aaBB$

Fig. 12.2 Production of doubled haploid lines, p is the frequency of recombination.

The doubled haploids production technique thus offers the opportunity of production of homozygous lines immediately from the gametes of F_1 hybrids and thereby the time required to develop new variety through pedigree, bulk or single seed descent method may be shortened by 2 to 3 generations and thus has the breeding application in that it is the quickest method of fixing the genes in homozygous state. Besides this, it offers the potential for the study of quantitative inheritance. The analyses of DH lines as we will see in this section provide the following information:

1. The partitioning of the genetical variation into components of variation.
2. The detection of linkage between genes controlling a quantitative character.
3. The estimation of the number of effective factors.

For obtaining information on the above, we will have to study the first degree statistics, second degree statistics, third degree statistics and fourth degree statistics in the doubled haploid lines.

12.8.1 Analysis of means

Following Mather and Jinks (1971), the expectations of means of the parents and DH progeny under various models will be as follows:

1. No epistasis and Linkage equilibrium

$$\overline{P}_1 = m + [d]$$
$$\overline{P}_2 = m - [d]$$
$$1/2\,(\overline{P}_1 + \overline{P}_2) = m$$
$$F_1(DH) = F_2(DH) = F_3(DH) = m$$

2. Epistasis but no linkage disequilibrium

$$\overline{P}_1 = m + [d] + [i]$$
$$\overline{P}_2 = m - [d] + [i]$$
$$F_1(DH) = F_2(DH) = F_3(DH) = m$$

Thus significant difference between means of parental lines and a population of DH lines will provide a test for epistasis, i.e.

$$1/2\,(\overline{P}_1 + \overline{P}_2) - \overline{F}_1(DH) \text{ or } \overline{F}_2(DH) \text{ or } \overline{F}_3(DH) \neq 0$$

In the absence of linkage disequilibrium, the frequency of any gamete in F_1, F_2, F_3 and S_3 generations is the same, namely, $(1/2)^k$ is where k is the number of segregating loci. Thus the frequency of homozygous genotype in DH population will be $(1/2)^k$ and consequently the expected means of DH lines derived from F_1, F_2, F_3 and S_n generations will all have the same mean m.

3. No epistasis and linkage disequilibrium

In the presence of linkage, the genotypic frequency in the gametes of different population (F_1, F_2, F_3 and S_3) will differ as a result of different rounds of recombination and so will differ the genotypic frequencies of these generations. However, linkage will cause difference in means only in the presence of non-allelic interaction. Thus in case of no epistasis and linkage disequilibrium, the mean will remain the same and equals m which is also the mid parental value, $1/2\,(\overline{P}_1 + \overline{P}_2)$.

4. Epistasis and Linkage Disequilibrium

In the presence of non-allelic interaction linkage well affect the means of DH populations derived from different generations and the expectations of means of DH lines derived from F_1, F_2, F_3 and S_3 generations will become

$$\overline{F}_1(DH) = m \pm \tfrac{C}{R} \Sigma\,(1 - 2p)_i$$
$$\overline{F}_2(DH) = m \pm \tfrac{C}{R} \Sigma\,(1 - p)(1 - 2p)_i$$
$$\overline{F}_3(DH) = m \pm \tfrac{C}{R} \Sigma\,1/2\,(1 - 2p)(2 - 3p + 2p^2)_i$$
$$\overline{S}_3(DH) = \pm \tfrac{C}{R} \Sigma\,(1 - p)^2 (1 - 2p)_i$$

where i is homozygous × homozygous interaction.

Thus the means of DH populations will depend on the degree of linkage disequilibrium, recombination frequency and epistasis. Further in the presence of coupling linkage with complementary epistasis or repulsion linkage with duplicate epistasis, the means of different generations will be in the following order.

$$\overline{F}_1(DH) > \overline{F}_2(DH) > \overline{F}_3(DH) > \overline{S}_3(DH)$$

This order is reversed when there is coupling linkage with duplicate epistasis or repulsion linkage with complementary epistasis. Thus a comparison of the means of $F_1(DH)$ and $S_3(DH)$ will provide the most sensitive test for linked epistasis whilst a comparison of the means of $F_2(DH)$ and $F_3(DH)$ the least sensitive (Snape and Simpson, 1981).

12.8.2 Analysis of variances

In the absence of non-allelic interaction the expectations of the variances of different generations will be as follows:

$$VP_1 = E$$
$$VP_2 = E$$
$$VF_1(DH) = VF_2(DH) = VF_3(DH) = D + E$$

And under condition of epistasis but no linkage disequilibrium, the expectations of variances of parental populations and different DH populations would become

$$VP_1 = E$$
$$VP_2 = E$$
$$VF_1(DH) = VF_2(DH) = VF_3(DH) = D + I + E$$

Table 12.11 Partitioning of variances in double haploids

Statistics	Generation		
	F_2	F_3	S_3
Average variance between DH lines within parental individuals	$\frac{1}{2}D + \frac{3}{4}I$	$\frac{1}{4}D + \frac{7}{16}I$	$\frac{1}{2}D + \frac{3}{4}I$
Average variance between means of parental individuals within families	–	$\frac{1}{4}D + \frac{5}{16}I$	$\frac{1}{4}D + \frac{11}{64}I$
Parental individuals within families Variance of means of families	$\frac{1}{2}D + \frac{1}{4}I$	$\frac{1}{2}D + \frac{1}{4}I$	$\frac{1}{4}D + \frac{5}{64}I$
Total variance between DH lines	$D + I$	$D + I$	$D + I$

Table 12.12 ANOVA of F_2 derived DH lines

Item	D.F.	M.S.	E.M.S.
Replication	$r - 1$		
Double haploids	$nm - 1$		
Between families	$m - 1$	MS_b	$\sigma_e^2 + r\sigma_w^2 + rn\sigma_b^2$
Within families	$n(m - 1)$	MS_w	$\sigma_e^2 + r\sigma_w^2$
Error	$(r - 1)(mn - 1)$	MS_e	σ_e^2

σ_b^2 and σ_w^2 are calculated as: $\sigma_b^2 = \frac{1}{rn}(MS_b - MS_w)$; $\sigma_w^2 = \frac{1}{r}(MS_w - MS_e)$.

In the F_2, F_3 and S_3 derived DH populations, a hierarchial family structure can be obtained where more than one DH line can be derived from each parental individual and these parental individuals in case of F_3 and S_3 generations, can be assigned to families. Thus the total variance between F_2 derived DH lines can be partitioned into (1) variance of means of families (σ_b^2) and (2) average variance between DH lines within families (σ_w^2) whereas the total variance between F_3 or S_3 derived DH lines can be partitioned into (1) variance of means of families, (2) average variance between DH lines within families and (3) average variance between means of parental individuals within families. The expectations of these statistics are given in Table 12.11.

12.8.3 Estimation of additive and additive × additive interaction variances

If n haploids are produced from each of the mF_2 plants and thus $n \times m$ plants are evaluated in a randomized complete block design with r replications. The ANOVA will take the form as given in Table 12.12.

From the estimates of σ_b^2 and σ_w^2, σ_A^2, the additive genetic variance and σ_{AA}^2, the additive × additive epistasis are calculated as

$$\sigma_A^2 = 3\sigma_b^2 - \sigma_w^2$$

$$\sigma_{AA}^2 = 2(\sigma_w^2 - \sigma_b^2)$$

where
$$\sigma_b^2 = \frac{1}{2}\sigma_A^2 + 1/4\sigma_{AA}^2$$

$$\sigma_w^2 = 1/2\,\sigma_A^2 + 3/4\sigma_{AA}^2$$

The sampling variance of the estimate of σ_A^2 or σ_{AA}^2 can be obtained by finding the variance of the linear fraction of the appropriate mean squares (Crump, 1946). Choo (1981) suggested that there is an optimum combination of n and m for the smallest sampling error in a fixed number (fixed *nmr*) of *DH* lines. With a sample size as large as 300 *DH* lines only the larger variance, i.e. either σ_A^2 or σ_{AA}^2 is greater than twice its sampling error. Therefore, a large sample size is required in order to detect additive and additive × additive genetic variances by the analysis of variance technique.

12.8.4 Estimation of dominance variance and the degree of dominance

If the F_2 population is raised alongwith F_2 derived *DH* lines in a replicated experiment, the estimates of dominance genetic component can be obtained. In the absence of non-allelic interaction

$$\sigma_b^2 = \frac{1}{2} \sigma_A^2$$

$$\sigma_{F_2}^2 = \frac{1}{2} \sigma_A^2 + \frac{1}{4} \sigma_d^2$$

Therefore,

$$\sigma_{F_2}^2 - \sigma_b^2 = \frac{1}{4} d^2 = \frac{1}{4} VD$$

In the ANOVA the significant between populations mean squares with 1 d.f. clearly demonstrates the presence of dominance effects. The degree of dominance is calculated as

$$\sqrt{\frac{H}{D}} = \sqrt{\frac{4(\sigma_{F_2}^2 - \sigma_b^2)}{3\sigma_b^2 - \sigma_w^2}}$$

12.8.5 Linkage disequilibrium

As linkage disequilibrium differentially affects the variances of the *DH* populations irrespective of the contribution of epistasis because of the covariance between individual additive effects caused by the different degrees of linkage disequilibrium, the two *DH* populations derived from plants of different generations are needed in order to provide a test of linkage disequilibrium. The expectations of the total variances of *DH* populations derived from different generations are

$$VF_1(DH) = D + \Sigma \{4p(1-p)^2\} i^2$$

$$\pm \frac{C}{R} \Sigma \{2(1-2p)\} \, dadb$$

$$VF_2(DH) = D + \Sigma \{(1-2p)^2 (1-p)^2\} i^2$$

$$\pm \frac{C}{R} \Sigma \{2(1-2p)(1-p)\} \, dadb$$

$$VF_3(DH) = D + \Sigma \{(1 - 1/4(1-2p)^2$$

$$\times (2 - 3p + 2p^2)\} i^2 \pm \frac{C}{R} \Sigma$$

$$\times \{(1-2p)(2 - 3p + 2p^2)\} \, dadb$$

$$VS_3(DH) = D + \Sigma \{1 - (1-2p)^2 (1-p)^4\} i^2$$

$$\pm \frac{C}{R} \Sigma \{2(1-2p)(1-p^2)\} \, dadb$$

In case of no epistasis and linkage disequilibrium coupling phase linkage will result in the following order of variances of different *DH* populations.

$$VF_1(DH) > VF_2(DH) > VF_3(DH) > VS_3(DH)$$

This order is reversed in case of repulsion phase linkage. In case of epistasis and linkage disequilibrium for both linkage phases, the order of variances of different *DH* populations will be

$$VF_1(DH) < VF_2(DH) < VF_3(DH) < VS_3(DH)$$

This shows that a combination of epistatic and non-epistatic linkages will reinforce one another and increase the differences between generations when linkage is in repulsion phase and act in opposition and decrease the differences when linkage is in coupling phase.

The total variance of different *DH* populations is estimated from the analysis of variance. When only one *DH* line from each parental family is grown, the between family mean squares is a direct estimate of the total generation variance. In case of F_1, and F_2 derived *DH* lines, significant F_1 vs. F_2-derived lines mean squares shows that the F_1-derived *DH* line mean differs significantly from F_2-derived line mean. Similarly the testing of homogeneity of mean squares of $F_1(DH)$ and $F_2(DH)$ is done by F test. The inequality suggests the presence of linkage disequilibrium. The biggest differences are expected between $F_1(DH)$ and $S_3(DH)$ populations and a comparison of these provides the most sensitive test of linkage disequilibrium whereas a comparison of F_2 and S_3-derived populations the least sensitive. The magnitude of differences varies with the recombination frequency and it is maximum at intermediate rather than extreme frequencies. The differences in variances are caused by changes in the relative proportions of extreme to intermediate genotypes in the DH population. Thus we see that linkage disequilibrium can be detected by comparing parental means, cross means, variance of parents and DH lines and variances among and within crosses provided that the number of DH lines produced from each F_1 is considerably large.

Like $F_2(DH)$ population, the back-cross-derived DH population, namely, $BC_1(DH)$ and $BC_2(DH)$ will yield two statistics: (1) Variance between families (σ_b^2) and (2) variance within families (σ_w^2) each. In the absence of linkage, the DH line obtained from these two backcrosses does provide estimate of genetic variances but because of large sampling error associated with the estimates of σ_A^2 and σ_{AA}^2, the genetic variances should be estimated from F_2-derived DH population. The test of linkage can be made by comparing the statistics of the two backcross derived DH populations:

$$\sigma_b^2(BC_1DH) \neq \sigma_b^2(BC_2DH)$$

or

$$\sigma_w^2(BC_1DH) \neq \sigma_w^2(BC_2DH)$$

The above comparison confirms the presence of linkage and in the absence of epistasis

$$2\sigma_b^2(BC_1DH) - \sigma_w^2(BC_1DH) = 2\sigma_b^2(BC_2DH) - \sigma_w^2(BC_2DH) = \pm p(1-2p)\,dadb$$

and thus it can provide the indication of the two parental configuration and of the presence or absence of epistasis.

12.8.6 Estimation of recombination value

In the absence of i type epistasis, the variances among and within F_2-derived DH families become

$$\sigma_{bi}^2 = \frac{1}{2}da^2 + \frac{1}{2}db^2 \pm (1-2p)\,dadb$$

$$\sigma_{wi}^2 = \frac{1}{2}da^2 + \frac{1}{2}db^2 \pm (1-2p)^2\,dadb$$

A comparison of these two variances suggests that when the genes in the two parents are associated $\sigma_b^2 > \sigma_w^2$ whereas when the genes are dispersed in the parents, $\sigma_b^2 < \sigma_w^2$. The two variances are equal at $p = 0.5$ or $p = 0.0$. Further when the two parents which supply the estimate of variance of the parents which is

$$\sigma_p^2 = da^2 + db^2 \pm 2dadb$$

are raised with F_2-derived DH population, it will yield the weighted mean of recombination value ($\bar{p}$) which is as follows:

$$\left| \frac{\sigma_{wi}^2 - 2\sigma_{bi}^2 + 1/2\,\sigma_p^2}{2\sigma_{bi}^2 - \hat{\sigma}_p^2} \right| = \frac{p^2 dadb}{p\,dadb} = \bar{p}$$

The back-cross-derived DH analysis can also provide estimate of the weighted mean of recombination value in the presence of epistasis only and it includes the recombination values of interacting genes but not those of non-interacting genes.

12.8.7 Skewness and Kurtosis

The skewness of the frequency distribution of a double haploid population derived from F_2 plants in a two locus system can be shown as

$$K_3 F_2 DH = 6\,idadb - 6^2\,q^2\,idadb \pm 2q(1-q^2)\,i^3$$

$$= 6\,idadb \text{ if } p = 0.5$$

where $q = (1 - 3p + 2p^2)$

When $i = 0$, $K_3 F_2 DH = 0$ but it is greater or smaller than zero in the presence of complementary or duplicate interaction, respectively. Thus the presence or absence of i, the additive × additive type interaction can be determined by studying the skewness of the frequency distribution of DH lines. The finding based on the study of skewness may be further substantiated by studying the kurtosis of the frequency of distribution of DH lines. The distribution of DH line is leptokurtic only in the presence of 'i' type epistasis and it is always platykurtic or mesokurtic when 'i' epistasis is absent.

12.8.8 Estimation of number of effective factors

Through *DH* method information on the number of effective factors determining a quantitative character can be obtained which is discussed in Chapter 16.

12.8.9 Covariances/correlation

Genetic correlation between characters can be due either to pleiotropy or linkage and if linkage is the cause then the correlation between characters in F_1, F_2, F_3, and S_n-derived *DH* populations may differ as a result of difference in the frequencies of recombinant genotypes because of different rounds of recombination. The total covariance between two characters due to linkage of genes in different populations can be shown to be

$F_1 \, DH \pm \frac{C}{R} \Sigma \, (1 - 2p) \, dadb$

$F_2 \, DH \pm \frac{C}{R} \Sigma \, \{(1 - p)(1 - 2p)\} \, dadb$

$F_3 \, DH \pm \frac{C}{R} \Sigma \, \{1/2 \, (1 - 2p)(2 - 3p + 2p^2)\} \, dadb$

$S_3 \, DH \pm \frac{C}{R} \Sigma \, \{(1 - p)^2 (1 - 2p)\} \, dadb$

This shows that coupling phase linkage produces a positive covariance whereas repulsion phase linkage a negative covariance. Further, difference in rate of recombination will result in differences in covariances over different *DH* populations. The direction of the difference will depend on the relative contribution of linkage and pleiotropy to the total covariance although the covariances will always rank in the following order.

Cov F_1DH > Cov F_2DH, Cov F_3DH, Cov $S_3 \, DH$

And if a positive correlation between characters is due entirely to coupling linkages then the recombination will result in

Cov F_1DH > Cov F_2DH > Cov $F_3 \, H$ > Cov S_3DH

which in turn results in

$\bar{r}F_1DH > \bar{r} \, F_2DH > \bar{r}F_3DH > \bar{r}S_3DH$

Thus the test for the role of linkage in correlation between characters can be carried out by comparing the total covariances and correlations in different *DH* populations. But as correlation coefficients are estimated as ratios of covariance to variances, the correlation coefficients between different *DH* populations could change due to changes in the variances rather than the covariance and so when there is linkage of gene for individual characters, total covariances rather than correlation coefficients should be used.

The distinction between linkage disequilibrium and pleiotropy as a cause of correlation can be made following the study of magnitude and sign of correlation coefficient between characters in inbred/pure lines developed through single seed descent and dihaploidy (Jinks et al. 1985). Supposing two characters A and B where A is determined by k_A loci and B is determined by k_B loci, the genetical correlation between two characters becomes

$$r_g AB = \frac{D_{AB}}{(DA \cdot DB)^{1/2}}$$

where D_{AB} is the additive genetic covariance between characters A, B and D_A and D_B are their additive genetic variances. When we talk of the linkage disequilibrium, it is between k_A and k_B loci and it does not necessarily mean that there will be a similar linkage disequilibrium within the two sets of loci. In the absence of pleiotropy and linkage disequilibrium there will be no correlation between the random samples derived either through *SSD* or *DH* whereas in the presence of linkage disequilibrium the correlation between characters in the sample of inbreds will depend on their method of extraction in so far as this affects the opportunity for recombination. The values of D_{AB}, D_A and D_B in respect of *DH* (F_1-derived) and SSD will be as follows:

Statistics	DH
D_{AB}	$\pm \frac{C}{R} \sum_{i=1}^{k_A} \sum_{s=1}^{k_B} (1 - 2 \, pis) \, dids$
D_A	$\sum_{i=1}^{k_A} di^2 \pm \frac{C}{R} \sum_{i=1}^{k_A-1} \sum_{j=i+1}^{k_A} 2(1 - 2\, p_{ij}) \, didj$
D_B	$\sum_{s=1}^{k_B} ds^2 \pm \frac{C}{R} \sum_{s=1}^{k_B-1} \sum_{i=s+1}^{k_B} 2(1 - 2\, psi) \, dsdi$

In case of SSD, the linkage coefficients $(1 - 2pis)$, $(1 - 2pij)$ and $(1 - 2psi)$ are replaced by $(1 - 2pis)/(1 + 2pis)$, $(1 - 2pij)/(1 + 2pij)$ and $(1 - 2psi)/(1 + 2psi)$, respectively and thus are smaller if p's are greater than zero, i.e. when there is recombination. Thus, in the presence of linkage disequilibrium, the absolute value of D_{AB} will be larger in DH than in SSD. Furthermore, D_{AB} is positive if coupling is in excess and negative if repulsion is in excess. The values of D_A and D_B in DH and SSD in case of linkage disequilibrium will be as given below:

Linkage phase	DH	SSD
Coupling In excess	D_A or D_B larger	D_A or D_B smaller
No linkage ($C = R$ or $C = R = 0$)	,,	,,
Repulsion in excess	Smaller	Larger

Thus although the higher value of D_{AB} for DH leads to expectation that the genetical correlation coefficient will be higher but infact it will be lower if the fall in the value of D_{AB} between the DH and SSD is exceeded by a fall in the value of $\sqrt{D_A D_B}$, in case of excess of coupling phase linkage. The contribution of pleiotropy to correlation will not be affected by the recombination between genes controlling them. Therefore, the contribution of pleiotropy to the genetic correlation will be same for the SSD and DH samples of inbreds. The effect of pleiotropic action of genes is thus indistinguishable from complete linkage. Considering the significance and the relative magnitudes of correlation coefficients in DH and SSD samples which can be explained using the above theoretical expectations, distinction between linkage of genes and pleiotropic action of genes as cause of genetic correlation can be made as given in Table 12.13.

Multiple mating design:

Multiple mating designs discussed in Chapter 6 can be used with doubled haploid method for estimation of genetic components of variance. In this method the progenies are developed using the conventional design but instead of evaluating the cross progeny these progenies are used as parents to produce DH lines which are then evaluated for performance. Thus DH method of analysis differs from conventional analyses in that it is the homozygous pure lines and not the cross progenies which are studied and further this DH population differs in the gene frequencies. So the additive genetic variance estimated from DH method may not relate directly to those obtained in conventional analysis using random inbred or out

Table 12.13 Comparison of correlation coefficients and inferences

Correlation	DH	SSD	Conclusion
r_{gAB}	NS	NS	(i) No linkage or no linkage disequilibrium ($C = R$) (ii) No pleiotropy or positive pleiotropy cancels out negative pleiotropy
r_{gAB}	SIG (Similar in magnitude)	SIG	(i) Pleiotropy or tight linkage (ii) Linkage disequilibrium (Coupling linkage among genes controlling one or both characters)
r_{gAB}	SIG	NS	(i) Linkage disequilibrium but no pleiotropy (ii) Repulsion phase linkage among genes controlling one or both characters.
r_{gAB}	SIG larger	SIG Smaller	(i) Pleiotropy (ii) Linkage disequilibrium
r_{gAB}	SIG Smaller Or NS Smaller	SIG Larger SIG larger	(i) Linkage disequilibrium (coupling in excess or repulsion in excess) oppose each other or pleiotropy and linkage disequilibrium.

NS = Not significant; SIG = significant.

bred population. The additive genetic variance component in the *DH* population has the expectation

$$\sigma^2_{DH} = \sigma^2_p = 2\sigma^2_A = 2uvd^2$$

which equals $\frac{1}{2}D_1$ where D_1 is the diallel additive genetic' variance parameter of Mather and Jinks. This expectation is different than $\frac{1}{2}DR$ (where *DR* is the additive genetic variance parameter in random mating population defined by Mather and Jinks) or *VA* of Falconer (1960). But then in the analysis of single crosses this problem does not arise and the D_1, the additive genetic variance, estimated in both analyses has similar expectation. There are two problems in case of estimation of genetic components of variance from the F_1 diallel derived *DH* lines. First, the number of parents cannot be increased and secondly the genes in the parents may not be in equilibrium. Both will introduce sampling errors which will affect the accuracy of the estimates. But considering the problems associated with the diallel, the use of *DH* lines to estimate genetic variances in self-pollinated crops appears to give better estimate of genetic variances and understanding about the parents and inheritance of quantitative traits than using F_1 or F_2 diallel (Choo et al. 1979).

Thus we see that from the analysis of *DH* lines, we can have information on additive genetic variance, additive × additive interaction variance, linkage and genotype × environment interaction.

From the study of *DH* lines Choo (1981) showed that linkage affects Cov *HS* the most followed by Cov grand-parent/grand-offspring and covariance full- sib which is affected the least as shown below:

Cov *FS* is increased by $\sum_{i<j}(1-2p_{ij})^2 \sigma^2_{ij}$

Cov grand-offspring/grand-parent increased by

$$\sum_{i<j}(1-2p_{ij})\sigma^2_{ij}$$

Cov *HS* increases by $\frac{1}{2}\sum_{i<j}(1-2pij)^2 \sigma^2_{ij}$

$$+\frac{1}{4}\sum_{i<j}(1-2p_{ij})^2 \sigma^2_{ij}$$

where p_{ij} is the recombination value between *i*th and *j*th locus and $\sum_{i<j}\sigma^2_{ij}$ is the additive × additive variant. The covariance offspring/parent is not affected linkage.

12.8.10 Generation of deriving DH lines

A breeder is hybridising two parental lines in the hope of isolating superior recombinant line (*s*). Considering two loci the two parents would be *AAbb* and *aaBB*, the frequencies of superior recombinant (*AABB*) in the F_1-, F_2- and F_3 derived *DH* populations can be shown to be $p/2$, $1/4(3p-2p^2)$ and $1/8(7p-8p^2+4p^3)$, respectively. The differences in the frequencies of *AABB* in the three populations at various recombination values can be calculated At $p=0$ and $r=0.5$, the three populations will have the best recombinant in equal frequency. Thus in this situation one can go for producing *DH* lines from F_1 population itself. The F_3-derived *DH* population has only a slightly higher frequency of superior recombinant than the F_2-derived *DH* population. Thus considering the small advantage of F_3 population but loss of time one can go for producing *DH* lines from the F_2 population with the hope of extracting superior recombinant. It is generally. proposed that doubling of haploids can be better applied to F_2 plants after selection rather than to F_1 plants.

12.8.11 Efficiency of DH method

Linkage, selection and the ease with which *DH* lines can be derived are the decisive factors of efficiency of *DH* method. Desirable genotypes for practical breeding aims are expected to occur very infrequently in the early stage of selection and might therefore be missed by sampling error. The expected segregation frequency of the desirable genotypes and the size of the population determine the probability of loss. Any increase in recombination with repulsion linkages will result in greater proportion of extreme genotypes and vice versa for coupling linkages. Thus in case of coupling linkages, the superiority of conventional method is reduced but the linkage of desirable alleles in coupling phase will strengthen the superiority of the *DH* method. In *DH* breeding about 1/5 or hopefully 1/2 as many test plants are needed to be raised as in conventional breeding to achieve the same level of success. Using Monte Carlo computer simulation Yonezawa et al. (1987) found that *DH* breeding method can efficiently be used when one or more of the following conditions are met:

(i) A relatively small number of loci, presumably 10 or less is involved in the breeding objective concerned.

(ii) Desirable alleles are recessive to undesirable alleles at most, if not all of the segregating loci and

(iii) the genes are not strongly linked.

Significant differences between *DH* lines and the parental inbreds or conventionally derived fixed lines have rarely been observed (Jinks et al. 1985; Yonezawa et al. 1987). Nei (1963) mathematically showed that the *DH* method may be much more efficient than the conventional breeding methods in manipulation of quantitative traits under control of large number of loci. But a number of explanations have been put forward by various workers for the comparable differences between *DH* lines and lines produced by conventional breeding.

(i) Greater inbreeding depression in the dihaploids because they lack the residual heterozygosity present in the conventionally bred inbred lines.

(ii) Random or systematic mutation during anther culture; Chromosomal and plasmic mutation may be easily identified and eliminated but gene mutations may be transmitted to the next generation causing genetic segregation in the progeny.

(iii) A cytoplasmic deficiency due to conventional source of cytoplasm in dihaploids obtained from anther culture.

(iv) Preferential selection among microspores during anther culture.

(v) Reduced vigor and/or poor adaptibility to different environments. Poor seed quality in the first generation of dihaploids as a consequence of colchicine treatment.

The double haploid breeding method has not yet come to be routinely applied in practical plant breedng inspite of improvement in its efficiency. The main reason for this seems to be the rate at which it can be produced. Although with bulbosum technique of *DH* production in wheat and barley, high rate of *DH* production is possible but this technique suffers from the problem that the rate of production of *DH* varies in different crosses.

References

Allard, R.W. 1960. Principles of plant Breeding. Wiley and Toppan.

Anderson, E. 1949. Introgressive Hybridization. John Wiley, New York.

Baker, R.J. 1984. Quantitative genetic principles in plant breeding. In: Gene Manipulation in Plant Improvement (ed.) Gustafson, J.P. Plenum Press, New York. pp. 147–176.

Blakeslee, A.F. and Avery, A.G. 1937. Methods of inducing doubling of chromosomes in plants by treatment with colchicine. J. Heredity, **28:** 392–411.

Bogyo, T.P., Lance, R.C.M., Chevalier, P. and Nilan, R.A. 1988. Genetic models for quantitatively inherited endosperm characters. Heredity, **60:** 61–67.

Caten, C.E. 1979. Quantitative genetic variation in fungi. Quantitative Genetic Variations. Academic Press. New York, pp. 35–57.

Choo, T.M. 1981. Doubled haploids for studying the inheritance of quantitative characters, Genetics, **99:** 525–540.

Choo, T.M. 1988. Cross prediction in barley using doubled haploid lines. Genome, **30:** 366–371.

Choo, T.M., Christie, B.R. and Reinbergs, E. 1979. Doubled haploids for estimating genetic variance and scheme for population improvement in self pollinating crops, Theo. Appl. Genet., **54:** 267–271.

Dessureaux, 1. 1963, Polyploids and polygenes. In: Hanson W.D. and Robinson, H.F. (eds.) Statistical Genetics and Plant Breeding. NAS, NRC, Washington DE. pp. 522.

Foroughi-Wehr, B. and Wenzel, G. 1993. Andro-and partheno-genesis. In: Plant Breeding: Principles and Prospects. Hayward, M.O.; Bose-mark, **N.O.** and Romagosa, 1. (eds). Chapman & Hall, pp. 262–277.

Foroughi-Wehr, B. and Wenzel, G. 1990. Recurrent selection alternating with haploid steps in a rapid breeding procedure for combining agronomic traits in inbreeders. Theor. Appl. Genet., **80:** 564–568.

Gallais, A. 1981. Quantitative genetics and Breeding theory of autotetraploids. In: Gallais, A. (ed) Quantitative Genetics and Breeding Methods. INRA, Lusignan, pp. 189–216.

Guha, S. and Maheshwari, S.C. 1964. *In vitro* production of embryos from anthers of Datura. Nature, 207.

Haldane, J.B.S. 1930. The theoretical genetics of autopolyploids. J. Genet. **22**: 359–372.

Hermsen, J.G.T. 1984. Nature, Evolution and Breeding of Polyploids, Iowa State J. Res. **58**: 411-412.

Hoffmann, F., Thomas, E. and Wenzel, G. 1982. Anther culture as a breeding tool in rape. II. Progeny analysis of androgenetic lines and of induced mutants from haploid cultures. Theor. Appl. Genet., **61**: 225–232.

Jinks, J.L., Caligari, P.D.S. and Ingram, N.R. 1981. Gene transfer in *Nicotiana rustica* by means of irradiated pollen followed by selection. Nature. vol. 291, No. 5816, 586–588.

Jinks, J.L., Pooni, H.S. and Chowdhury, M.K.U. 1985. Detection of linkage and pleiotropy between characters of *Nicotiana tabacum* using inbred lines produced by dihaploidy and single seed descent.Heredity, **55**: 327–333.

Johns, W.A. 1974. A preliminary evaluation of haploidy as a breeding technique in barley (H. vulgare L.). Ph. D. thesis. Univ. of Guelph.

Kasha, KJ. (ed.). 1974. Haploids in higher plants. Univ. of Guelph, Guelph, Canada.

Kearsey, M.J. and Pooni, H.S. 1996. The Genetical Analysis of Quantitative Traits. Chapman & Hall, London.

Keller, W.A. and Stringam, G.R. 1978. Production and cutilizations of microspore derived haploid plants. In: Frontiers of Plant Tissue culture, Thorpe, T. (ed.). Univ. of Calgary Press, Calgary, pp. 113–122.

Kempthome, O. 1969. An Introduction to Genetic Statistics. Iowa State Univ. Press, Ames, Iowa.

Kihara, H. and Ono, T. 1926. Chromosomenzahlen and systematische gruppieruna der Rumex-arten (in German). Zeitzchr, Zellforch. : 475–481.

Killick, R. J. 1971. The biometrical genetics of autotetraploids, I. Generations derived from a cross between two pure lines. Heredity, 27: 331–346.

Laurie, D.A. and Snape, J.W. 1990. The agronomic performance of wheat doubled haploid lines derived from wheat × maize crosses Theor. Appl. Genet., **79**: 813–816.

Levin, D.A. 1983. Polyploidy and novelty in flowering plants. Amer. Nat., **122**: 1–25.

Levings, C.S. and Dudley, J. W. 1963. Evaluation of certain mating designs for estimation of genetic variance in autotetraploid alfalfa. Crop Sci., **3**: 532–535.

Lewis, W.H. 1980. Polyploidy: Biological relevance. New (ed.) York, Plenum Press.

Mac Key, J. 1987. Implications of polyploidy breeding. BioI. Zent bl., **106**: 257–266.

Mather, K. 1936. Segregation in autotestraploids. J. Genet., **32**: 298–314.

Mather, K. and Jinks, J.L. 1982. Biometrical Genetics. 3rd edn. Chapman and Hall, London.

Mendiburu, A.O. and Peloquim, S.l. 1977. The significance of $2n$ gametes in potato breeding. Theor. Appl. Genet., **49**: 51–63.

Moore, P.H. Nagai, C. and Fitch, M.M.M. 1989. Production and evaluation of sugarcane haploids, Proc. Int. Soc. Sugarcane Technol., **20**: 599–607.

Nitzche, W. and Wenzel, G. 1977. Haploids in plant breeding. Adv. in Plant Breeding (Suppl, J. Plant Breeding), Vol. 8.

Pandey, K.K. 1975. Sexual transfer of specific genes without gametic function. Nature, **256**: 310–313.

Peloquin, S.J and Ortiz, R. 1992. Techniques for introgressing unadapted germplasm to breeding populations. In: Stalker, H.T. and Murphy, J.P. (eds.). Plant Breeding in the 1990s. Red Wood Press, Melksham, U.K. pp. 485–507.

Simchen, G. and Jinks, J.L. 1964. The determination of dikaryotic growth rate in the Basidiomycetes, Schizophyllum commune, a biometrical analysis. Heredity, **19**: 629–649.

Snape, J.w. 1988. The detection and estimation of linkage using doubled haloids of SSD population. Theor. Appl. Genet., 76; 125–128.

Snape, J.W. and Simpson, E. 1981. The genetical expectations of doubled haploids lines derived from different generations. Theor. Appl. genet., **60**: 123–128.

Snape, J.W. and Simpson, E. 1986. The utilization of doubled haploid lines in quantitative genetics. Bull. Soc. bot. Fr., 133, Actualites bot., **4**: 59–66.

Snape, J.W., Stuh, L.A., Simpson, E. and Parker, B.B. 1988. Test for the presence of gene to clonal variation in barley and wheat doubled haploids produced using *Hordeum bulbosum* system. Theor. Appl. Genet., **75**: 509–513.

Stebbins, G.L. 1947. Types of polyploids: Their classification and significance. Adv. Genet. **1**: 403–429.

Stebbins, G.L. 1980. Polyploidy in plants: unresolved problems and prospects. In: Polyploidy: Biological Relevance, Lewis W.H. (ed). Plenum. Press, New York.

Tsay, H.S., Miao, S.H. and Widholm, 1.M. 1986. Factors affecting haploid plant regeneration from maize anther culture. 1. Plant Physiol., **126**: 33–40.

Uhrig, H. and Salamini, F. 1987. Dihaploid plant production from 4 × genotypes of potato by the use of efficient anther plant producing tetraploid strains. (4 × EAPP–clones)-a proposal of a breeding methodology. Plant Breeding, **98**: 228–235.

Vasil, I.K. 1980. Androgenetic haploids. Int. Rev. Cytol., suppl. 11B: 195–217.

Welch, 1.E. (1962). Linkage in autotetraploid maize. Genetics, **47**: 367–396.

Wricke, G. and Weber, W. E. 1986. Quantitative genetics and selection in plant breeding. Walterde Gruyter, Berlin and New York.

13

Competition

13.1 Definition and Types

Competition has been defined and studied in many ways. It arises when two or more organisms seek a common resource such as light, temperature, water and/or nutrients whose supply falls below their combined demands (Donald, 1963). In Mather's term a competitor does no more than set a specific limit on the environmental resources of its associates.

Competition is also a spatial process in which the performance of any individual depends on its height relative to its neighbour and its distance from them. Thus competition is a stochastic process and the success depends on the probability of having smaller neighbours. Thus to have competition there should be at least two individuals. The competition between these two individuals can be of the following types depending upon the genotype of the individuals.

1. Intragenotypic competition—when the two individuals are of the same genotypic constitution.

2. Intergenotypic competition—when the two individuals differ in their genotypic constitution. The two individuals considering one locus with two alleles system could be either two different homozygotes (AA or aa) or one homozygote (AA or aa) and the other heterozygote (Aa). Further the two individuals can be either from the same species or from different species. For example, the intergenotypic competition occurs in multiline and mixture varieties or in multiple cropping system where cereal is intercropped with cereal or cereal is intercropped with legume or legume is intercropped with legume or non legumes. In multiple cropping systems, the objective is to find such combination of crops where the inter crops yield greater than yield of the most productive component in monoculture.

In practical plant breeding experiments both types of competition occur. Intragenotypic competition occurs when plants are sown at wide spacing than the normal commercial practice or in other words at a different density than the usual commercial practice or at commercial density but in a smaller plot so that border effects mask a significant contribution to plot yield. When we are selecting crosses for further inbreeding as in case of pedigree or bulk method on the basis F_1's performance which are being tested under wider spacing because of limited supply of seeds, the intragenotypic competition takes place. The inter-plot competition effects are more pronounced in early stage selection in yield trial where the entries are single row than in trials using the 3, 4, 5 or 6 rows plot. We encounter intergenotypic when selection is practiced in inbreeding generation of pedigree or bulk method of advancing generation on the performance of plants of different genotypes shown at different than the normal commercial planting density. Intergenotypic (or inter varietal) competition is important particularly when the plot size is small and is without guard rows where the yield of a variety may be depressed by more aggressive genotypes. The interplot competition can cause adjacent plot yields to be negatively correlated. The effect of this interplot competition on the difference between varieties and a method which provides more accurate estimates of the expected difference between varieties have been discussed by Kempton (1982).

13.2 Yield-Density Relationship

The study of competition has evolved the establishment of a number of experimental designs. In the additive experiment, the density of one component is held constant while that of other is varied and thus both the total density and the density of the competitor vary and in the substitution or replacement series, the densities of both competitors are varied simultaneously so as to hold the total plant density constant. The densities of both competitors thus vary inversely. The competition (intra-genotypic) among plants in monoculture is studied by means of systematic variation in plant density. Models of yield/density relationship in mixtures have generally been based on addition or replacement series. Yield per unit area depends on the number of plants per unit area and the spatial arrangement of the plants. Variation in yield of individual plants is important where the size of individual plant is an attribute of yield.

Various statistical models have been presented to describe the relationship between yield and density in monoculture.

(i) Asymptotic
(ii) Parabolic
(iii) Polynomial
(iv) Reciprocal equations.

In asymptotic model with increase in density, yield increases to a maximum and is then relatively constant at higher densities whereas in parabolic model yield rises to a maximum but then declines at higher densities. These two models are not entirely valid. The polynomial equation describing yield/density relationship is restricted to a relatively narrow range of density around the point of maximum yield. The resulting curves from the above mentioned yield density relationships will take the form as described in Chapter 4. The reciprocal equation (Shinozaki and Kira, 1956) describing yield/density relationship accurately and meaningful is

$$\frac{1}{W} = a + bp$$

or

$$y = \frac{p}{a + bp}$$

where W is mean weight per plant; y is yield per unit area; p is plant density; a is the intercept; and b is the slope.

y approaches the value of $1/b$ i.e. asymptote with increasing density while the expected plant weight at low densities similarly approaches an asymptote of $1/a$. This equation does not represent a very realistic situation for it ignores the fact that yield per plant levels off at densities too low for competition to occur. Assuming that $1/a$ gives some indication of yield/plant in a competition free environment then a can perhaps be regarded as a meaningful factor indicative of genetic potential. Likewise assuming that asymptote of yield/unit area gives a measure of the potential of a given environment then b is a meaningful factor indicative of the environmental potential. The yield of plant at a given density is thus a product of the potential of the plant and the forces of competition that are acting on it. The ratio of b/a is an intraspecific competitive stress. Because of heterogeneity of variances around the line, weighted regression analysis should be used to estimate these parameters.

This model of yield/density in monoculture can be extended to allow for the parabolic yield-density relationship using either as a power index on $W(1/W^\theta = a + bp)$ or as quadratic term cp^2 i.e. $1/W = a + bp + cp^2$ (Bleasdale and Nelder, 1960). Taken together the two parameters in the above equations define the extent to which plant yield is decreased as density is increased. Holiday (1960) referred the term $(1 + bp/a)^{-1}$ as the competition function of the particular monoculture at density p. This inverse polynomial relationship has been found to describe the yield/density relationship in a number of crops. This equation of yield/density in monoculture can be extended to binary mixture (Spitters, 1984) which takes the form as

$$Y_{i\,mix} = bi\, Y_{i\,mono}$$

where $Y_{i\,mix}$ is the yield of ith genotype in mixture; b_i is the competitive ability of the ith genotype; and $Y_{i\,mono}$ is the yield of ith genotype in monoculture. Thus the effects of intraspecific competition and interspecific competition can be estimated.

De Wit's (1960) competition model defines relative yield, selective value and competition ability in terms of parameters. He derived a single species yield-density equation based on linear relationship between the reciprocal of yield per unit area and the space per plant. Further he along with other workers used graphical analysis to demonstrate the effects of competition in binary mixtures. In this technique the performances of binary mixtures and its individual components are plotted against the different mixture proportions and the shape of the curve reveals the effect of competition. Through the use of such models, we can compare the yield ability of different genotypes in monoculture at different densities and also a comparison of yielding ability of a genotype in a mixture of genotypes in which there is a possibility of inter-genotypic competition, can be made.

13.3 Mixture Diallels

Breese and Hill (1973) introduced the concept of general competitive ability and specific competitive ability which is analogous to general and specific combining ability of the diallel mating design. They showed that the performance of the species in competition could largely be specified by three parameters, namely, the species mean (v), the regression coefficient (b) and the mean effects of the associates (a) which respectively measures the general vigor of the species, its sensitivity to competition and its aggressiveness. The general competitive ability can be measured for each species by its mean and regression coefficient and specific combining ability which is a feature of the individual mixtures or monocultures is detected by deviation from regression line which represents co-adaptational including co-operational effects and mutual antagonism. The three parameters provide descriptive and predictive measurement of the competitive advantages of species in particular combinations and of the mixture performances relative to the performance of other mixtures or monocultures The analysis is done following Perkins and Jinks (1968). Positive complementation occurs when b and a are negatively correlated whereas negative complementation occurs when b and a are positively correlated. The magnitude of complementation in mixture relative to the yields of monocultures depends on the correlation on v with a and hence with b. Over-complementation occurs when v and a are positively correlated. This may be positive with mixtures out-yielding the best monoculture or negative with mixture yielding less than either monoculture depending upon the relationship between b and a. Where competition is defined in Donald's or De Wit's way v will be largely negatively correlated with a. This correlation between v and a will be weakened where one or more species show a general tolerance of crowding effects such as shade or plant chemical exudates, etc. or if one or more species have a general antagonistic effects on neighbours. Such phenomena will affect b and which in turn will determine the relative competitive abilities and the degree and direction of complementation in mixture.

Regression analysis of competition suffers from the same drawback as regression analysis of genotype × environment interaction. Regression on environmental means (= associate means) introduces bias to the estimates of the coefficients and invalidates the test of significance in the analysis of variance. Also, deviations from regression are not independent of the slopes and that this arises inescapably from fitting lines to data which can only be properly represented in several dimensions.

Norrington–Davies (1967) partitioned competition effects into three components. These are alpha (α) competition measuring a constant increase or decrease in one species or more species when grown with others and thus the value of α could be positive or negative; Beta (β) competition measures the extent to which the effect of one species upon another is a function of the species grown by themselves (as pureline cultures) and is measured as the difference between the regression coefficients of its row and column, respectively on to the unmixed species and $\overline{\beta}\left(\sum_{i=1}^{n} \beta\right)$ is the average β overall species and is a measure of the extent to which large species suppresses small species and/or are themselves increased in size when grown with small species. α and $\overline{\beta}$ are so some extent confounded (Breese and Hill, 1973).

13.4 Intergenotypic Competition

The inter-genotypic competition between genotypes, A and B can be studied by adding individual of B genotype to A (addition design) or substituting individual of A for individual of B in substitution or replacement design. In addition experiment starting with N individuals of A, the primary genotype, we can add X, 2X, 3X, 4X, etc. individuals of B, the secondary genotype and observe the character shown by N individuals. When the performance of primary genotype is regressed onto the number of secondary genotype, the resulting regression coefficient, b_d, then measures intergenotypic competition between individuals of A and B genotypes. Now if $b_d = 0$, i.e. when there is no intergenotypic competition, regression line will be parallel to the abscissa. In the above experiment if the added individuals X, 2X, 3X and 4X are from the genotype A itself we will have estimate of b_m from this experiment. b_m measures the slope of the line and it estimates the amount by which the addition to the culture further reduces the expression of a character or to put it in the other way round, the amount by which the subtraction of an individual raises the expression of that character and thus measures the intragenotypic competition. If $b_d = b_m$ then the intensity of intergenotypic competition equals that of intragenotypic competition among individuals of A genotype. Comparison of the relative magnitudes of b_d and b_m will provide the following information.

(i) Nearer the value of b_d to 0, the weaker the intergenotypic competition

(ii) Nearer the b_d value approaches b_m, the stronger the intergenotypic competition.

(iii) Greater negative value of b_d in comparison to b_m indicates intergenotypic competition stronger than intragenotypic competition effect and the genotype is a strong competitor whereas a lower value (< 0) of b_d shows intergenotypic competition weaker than intragenotypic competition and the genotype is called a weak competitor.

(iv) Positive value of b_d indicates that the secondary genotype is a facilitator rather than a competitor of primary genotype.

In the substitution experiment, the significances of b_m and b_d are the reverse of that in addition experiment. When $b_d = b_m$, it shows an absence of intergenotypic competition and when $b_d = 0$, it shows that intergenotypic competition effect equals intragenotypic effect of the primary genotype. A genotype is a weak competitor when the value of b_d falls in between b_m and 0, whereas a genotype is a strong competitor when $b_d < 0$ and it is a facilitator when $b_d > b_m$. Facilitators, and competitors are shown in Figures 13.1 a and b in substitution and additive experiments, respectively.

The estimates for intra- and inter-genotypic competition effects are generally referred to as competition or C-values. It is highly desirable to use a character which shows a linear relation between its performance and densities as in case of characters showing curvilinear relation, the number of regression coefficient to be calculated increases which in turn will make comparisons between b_m's and b_d's difficult and less meaningful.

Mather and Caligari (1981) described a general method for obtaining least square estimates of a, b_m and b_d (or b_d's if more than one series of duo-culture is raised each with a different genotype). We have seen that both addition and substitution designs estimate a and b's but because of the reversal in the relation of b_d to b_m and O, the value of b_d is these two experiments will be different except in the special case of $b_d = \frac{1}{2} b_m$ and thus the experiments can be combined to have a combined estimate of b_d as estimates of a and b_m are expected to be the same in the two designs. In the substitution design, the independent variable (x) is negative and the useful feature of this design is that with the inclusion of monoculture density series for the secondary genotype, the roles of indicator/primary and associate/secondary genotypes can be changed without having to raise any further duo cultures.

13.4.1 Estimation of a, b_m and b_d

If y be the character showing competition effect, the expected value of y in monoculture of A genotype at any density can be written as

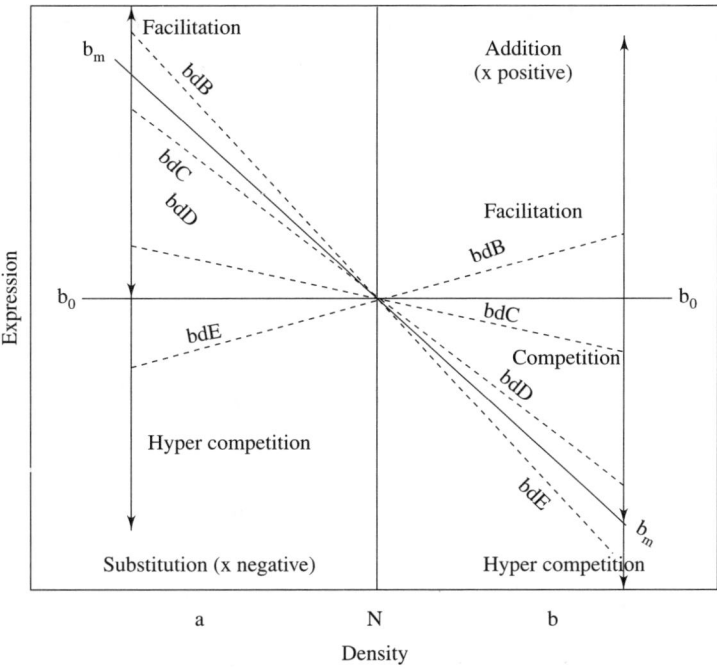

Fig. 13.1(a-b) The regression lines show the expression of the character in the indicator genotype A when raised in monocultures (bm) and in duo cultures (d_{bB}) when the added genotype, B, is a facilitator, b_{dC} when C is a weak competitor, b_{dD} when D is a strong competitor and d_{bE} when E is hyper competitor (Adapted from Mather and Caligari, 1981).

$$y_m = a + b_m x_m$$

Similarly, the expected value of y of A genotype in duo-culture with B or C genotype can be written as

$$Y_{dB} = a + b_{dB}\, x_{dB}$$
$$Y_{dC} = a + b_{dC}\, x_{dC}$$

where a is the mean value of the character at the reference density, N. n_m, n_{dB}, n_{dC} etc. are, respectively the numbers of different values of x used in the series of mono-cultures, in the series of duo-cultures with B, duo-cultures series with C, etc. and $n = n_m + n_{dB} + n_{dC}, \ldots$. With observed values of $y_{m1}, y_{m2}, \ldots y_{mn}$ of Y in monocultures series of A genotype, $y_{dB1}, y_{dB2}, \ldots, y_{dBn}$ in duo-cultures series (A with B) and $y_{dC1}, y_{dC2}, \ldots, y_{dCn}$ in duo-cultures of A with C, the least squares equations of estimation then become

$$an + b_m S(x_m) + b_{dB} S(x_{dB}) + b_{dc} S(x_{dC})$$
$$= S(y_m) + S(y_{dB}) + S(y_{dC})$$
$$= S(y)$$

$$aS(x_m) + b_m S(x_m^2) = S(y_m x_m)$$
$$aS(x_{dB}) + b_{dB} S(x_{dB}^2) = S(y_{dB} x_{dB})$$
$$aS(x_{dC}) + b_{dC} S(x_{dC}^2) = S(y_{dC} x_{dC})$$

which in matrix notation takes the form as

$$\begin{bmatrix} n & S(x_m) & S(x_{dB}) & S(x_{dC}) \\ S(x_m) & S(x_m^2) & 0 & 0 \\ S(x_{dB}) & 0 & S(x_{dB}^2) & 0 \\ S(x_{dC}) & 0 & 0 & S(x_{dC}^2) \end{bmatrix} \begin{bmatrix} a \\ b_m \\ d_{dB} \\ d_{dC} \end{bmatrix} = \begin{bmatrix} Sy \\ S(y_m x_m) \\ S(y_{db}\, x_{dB}) \\ S(y_{dC}\, x_{dC}) \end{bmatrix}$$

$$J \qquad \hat{M} \qquad S$$

The estimates of the parameters a, b_m, b_{dB} and b_{dC} are obtained as

$$\hat{M} = J^{-1}S$$

The general inverted matrix J^{-1} takes the form as

$$\begin{bmatrix} 1 & r_m & r_{dB} & r_{dC} & \cdots \\ r_m & r_m^2 + (D/Sx_m^2) & r_m r_{dB} & r_m r_{dC} & \cdots \\ r_{dB} & r_{dB} r_m & r_{dB}^2 + (D/Sx_{dB}^2) & r_{dB} r_{dC} & \cdots \\ r_{dC} & r_{dC} r_m & r_{dC} r_{dB} & r_{dC}^2 + (D/Sx_{dC}^2) & \cdots \\ \vdots & \vdots & \vdots & & \end{bmatrix}$$

All elements are divided by $D (= n - r_m Sx_m - r_{dB} Sx_{dB} - r_{dC} Sx_{dC} \ldots$ where $r_m = S_{xm}/Sx_m^2$ etc.).

Using the estimates of $\hat{a}$, $\hat{b}_m$, $\hat{b}_{dB}$ and $\hat{b}_{dC}$, the expected values of Y in various cultures can be obtained and the adequacy of the model can be tested by comparing the observed values y with expected values Y. The null hypothesis which is that all regression lines pass through the reference point, is accepted when $S(y-Y)^2/f$ will not significantly exceed the error variance of individual observations.

Competition *per se* cannot be assumed to differ between the two designs. De Miranda and Eggleston (1987) following Mather and Caligari's (1981) analysis showed that the competition values for intra and intergenotypic competition obtained from substitution and addition design experiments were equivalent. They used different reference density for addition and substitution design in comparison to a single reference density proposed for both designs by Mather and Caligari (1981). Competitive ability depends on the choice of character. Spitters (1983 b) also showed in plant crop yield-density experiments that the total biomas or any component of the biomas can be used to measure competitive success but then the competitive values will differ from character to character.

13.4.2 Partitioning of C-values

Mather and Caligari (1983) partitioned the competition values into components of aggression and response. When two organisms are competing with one another, one individual exerts a competitive pressure on the other fellow individual and simultaneously it will be responding to the total pressure exerted on it by its fellow individuals. This distinction between pressure or aggressiveness and response or sensitivity was recognised by Breese and Hill (1973). Competitive interaction among like genotypes in a monoculture can be measured as the rate of change of a character under observation on the density of the culture. The rate of change is measured by slope of the regression line (b_m) relating expression to density and represents the effects of reducing the competitive interaction by eliminating the contribution of one individual to the overall competition. If each individual of the primary (or indicator) genotype so omitted had been replaced by another of a different (secondary or associate) genotype, which however made an exactly equivalent contribution to the competitive pressure on the primary genotype, the regression of the expression of the character shown by this primary genotype on its density in the culture would have been zero and the regression that it would have yielded is denoted by b_o. Thus the intragenotypic competitive interaction is measured by $b_o - b_m = -b_m$. It is the change in the intensity of intragenotypic competition due to changes in the density of the individuals. When the omitted individuals of primary genotype are replaced by equal number of secondary genotype having a different competitive effect, the regression of the character shown by the primary genotype on their density will not be zero and is estimated by b_d. The competitive value of the associate is thus measured by $b - b_m$. For a pair of genotypes say A and B there will be four C values for each character.

$$\begin{array}{cc} A & B \\ b_{mA} & b_{mB} \\ b_{dA} & b_{dB} \end{array}$$

$$C_{AA} = -b_{mA} \quad C_{BB} = -b_{mB}$$
$$C_{AB} = b_{dA} - b_{mA} \quad C_{BA} = -b_{dB} - b_{mB}$$

C_{AA} denotes the response shown by A to its own competition whereas C_{AB} denotes its response to competition from B. Similarly, C_{BB} denotes the response displayed by B to its own competition and C_{BA} denotes its response to competition from A. The four C values for each character have 3 degrees of freedom among them and the overall sum of squares can be partitioned into following three items, each for 1 df using an orthogonal analysis.

$$R = \frac{1}{4}(C_{AA} + C_{AB} - C_{BA} - C_{BB})^2$$

$$A = \frac{1}{4}(C_{AA} - C_{AB} + C_{BA} - C_{BB})^2$$

$$I = \frac{1}{4}(C_{AA} - C_{AB} - C_{BA} + C_{BB})^2$$

where R tests the significance of the difference between A and B in their overall response to competition pressure exerted on them by the two of them as associate; A tests the significance of the difference between A and B in the competitive pressures that they exert as associates; and I tests the interaction ($R \times A$) between competitive pressure and response. Assuming additivity of the effects of pressure and response on the expression of the relevant character C_{AA}, C_{AB}, C_{BA} and C_{BB} can be written as

$$C_{AA} = \bar{c} + r + a$$
$$C_{AB} = \bar{c} + r - a$$
$$C_{BA} = \bar{c} - r + a$$
$$C_{BB} = \bar{c} - r - a$$

where r and a are the deviations from mean ($\bar{c}$) ascribable to the effect of response and pressure. Estimates of r and a are obtained as

$$r = \frac{1}{4}(C_{AA} + C_{AB} - C_{BA} - C_{BB})$$

$$a = \frac{1}{4}(C_{AA} - C_{AB} + C_{AA} - C_{BB})$$

The differences between two lines in response and pressure are $2r$ and $2a$, respectively. Thus differences in aggressiveness and in response can be measured in terms of parameters, a and r, respectively and a value of each parameter, expressed as a deviation from the mean of the experiment, can be assigned to each genotype. The inter-relations of a and r can be measured by the interaction parameters. Thus we see that the values of a, r and i can be derived from the competition values (c) of the various pairs of genotypes which in turn are derived from the regression of character expression on the density of the indicator (or responder) genotype in mono- and duo-cultures. There are thus four parameters which between them describe the competitive interactions in mixed cultures.

1. The absolute performance of a genotype at standard reference density (e-value).
2. The effect of monoculture density on performance (intragenotypic competition).
3. The influence of a genotype on the performance of other genotypes (inter-genotypic pressure).
4. The response of a genotype to the pressure exerted by associated genotypes (inter-genotypic sensitivity).

Of these four parameters 3rd and 4th parameters determine performance in duo-culture. Of the four competition parameters, the role of e value in the process of competition is not clear. Caligari and Mather (1988) have questioned whether the e value can be considered as a competitive parameter. It is based on static rather than a dynamic measurement, taken at a single density (the reference density). All other competitive parameters are measured as a rate of change in some characters over a series of densities and there dynamic nature is quite clear. For these reasons, the e value presumably reflects agents which have an effect on the fitness of a genotype but which are not necessarily dependent on population density. To estimate these parameters there is a need for control over the environmental conditions necessary for the measurement of competition such as density and the amount of food.

13.5 Nature of the Underlying Genetical Control of Competition and Competitive Ability

In a chromosome assay experiment using *D. melanogaster* substitution lines, Caligari and Mather

(1988) found that all three major chromosomes (I, II, III) were involved in the determination of competitive ability. However, the relative importance of these chromosomes varied among the competitive parameters. They showed that the genes carried by chromosome III affected both aggression and response whereas those on chromosome II primarily affected aggression only. The involvement of chromosome I appeared to be limited to interaction with chromosome I, with respect to the control of response. Hemmat and Eggleston (1990) conducted a biometrical analysis of four continuously varying traits, namely, e value, intragenotypic competition, intergenotypic pressure and intergenotypic sensitivity using two lines, their F_1, F_2 and backcross generations. The above traits were measured using yield-density regression analysis of each genotype in monoculture and in duoculture with a phenotypically distinguished tester. Model fitting was done following Mather and Jinks (1973) to define the genetical architecture of competition. The e value showed additive genetic effects whereas intragenotypic competition showed dominance effects. Intergenotypic sensitivity showed not only additive and dominance effects but also dominance maternal effect. The mechanism of competition will be very different in green plants and flies. Model fitting must be carried out for competitive parameters in other organisms in order to confirm this type of genetical architecture.

Both competitive sensitivity and pressure are determined partly by the environment created by competitors themselves, through the process of interference and detoxification. Determination of genetic effects is dependent upon the amount of competitive stress or selective pressure. This is itself determined by a wide and complex range of parameters. Thus any alternation to the competitive environment may have a profound effect on the estimation of the genetic parameters.

13.6 Competition in Natural Population

So far we have seen that competition effects can be measured using pairs of pure breeding lines, If we are working with a natural population where inbreeding as well as outbreeding occurs, the population would comprise homozygotes as well as heterozygotes. If we start with a synthetic population comprising a number of homozygous pure breeding lines, then after a number of generations, the population would reach an equilibrium where the frequency of homozygotes and heterozygotes would be determined by the rate of outcrossing. But as the population contains homozygotes and heterozygotes, competition can occur between homozygotes and heterozygotes and this may augment or oppose this equilibrium. Whether homozygotes or heterozygotes are the better competitor can be known by comparing the observed frequencies of homozygotes or heterozygotes with the expected.

13.7 Estimation of Outcrossing Rate

As heterozygosity is associated with hybrid vigor and competitive ability, it would be worthwhile to estimate the amount of heterozygosity and in turn the rate of outcrossing which gives rise to the heterozygosity. When only random or cross fertilization occurs, the proportion of heterozygosity at equilibrium is $2uv$ with $u = v = 0.5$, where u is the frequency of gene A and v is the frequency of gene a but in the case of only self fertilization, the heterozygosity approaches to zero as a limit and the resulting population is having $u = 1.0$ and $v = 0.0$ or vice versa. However, in case of partial self and cross fertilzation, the heterozygosity approaches some limit other than zero depending upon the level of self fertilization and selection.

There are two ways of estimating the heterozygosity, one using the qualitative trait and the other using the quantitative trait.

(i) Use of qualitative trait

In this approach a qualitative trait such as ovary colour, as in case of *Nicotiana rustica* which is under control of one major gene with two alleles (A, a) with black ovary dominant over green (Pooni and Jinks 1981) is selected. The population could be a synthetic population in any generation ($S_1, S_2 \ldots S_n$) except S_o, the starting population where the expected frequency of heterozygote is zero. A large number of plants are randomly selected and scored for ovary colour-black or green and self-pollinated. As the black ovary is dominant over green ovary, the plants with black ovaries may be homozygous dominant (AA), or heterozygous (Aa). 20 progenies of each black ovaried plant are grown in a tray in the glass house and at least 16 plants of each progeny are scored for ovary colour in order to keep the probability of mis-classification of a parent below one per cent. If all the 16 plus plants have black ovaries, the parent is classified as homozygous black (AA) and if one or more of 16 sibs have a green ovary, the parent is classified as heterozygous black (Aa). Then the rate of outcrossing is estimated by a method of Boughey et al. (1981).

The frequency of heterozygotes in advancing generations of synthetic population is calculated as

$$\beta_n = \frac{AA + Aa}{AA + Aa + aa} \times \frac{(Aa)'}{(AA + Aa)'}$$

where β_n is the frequency of heterozygotes at the A, a locus in the nth generation of the synthetic population, ($AA + Aa + aa$) is the total number of plants in the S_n generation raised in the assessment experiment, ($AA + Aa$) is the number of these plants with a black ovary, ($AA + Aa$)' is the total number of black ovary plants in the sub sample of the S_n generation which were chosen for progeny testing and (Aa)' is the number of black ovary plants classified as heterozygotes on the basis of the progeny test.

The frequencies of the dominant black ovary colour, u and the recessive green ovary colour, v alleles are calculated after classifying the sampled plants into three genotypic classes AA, Aa and aa for ovary colour. In the present case

$$AA = \frac{((AA + Aa)' - (Aa)') \times (AA + Aa)}{(AA + Aa)'}$$

$$Aa = \frac{(Aa)' \times (AA + Aa)}{(AA + Aa)'}$$

$$u = \frac{AA + AA/2}{(AA + Aa + aa)}$$

$$v = \frac{aa + 1/2\, Aa}{(AA + Aa + aa)}$$

The outcrossing rate $(1 - f_n)$ in any generation is calculated by the general formula,

$$\beta_n = 2uv \sum_{i=1}^{n} (f/2)^{i-1} (1 - f)$$

where n is the number of generations and f is the rate of selfing (inbreeding coefficient). This formula can be used to calculate the expected frequency of heterozygotes in any generation (β_n) provided the gene frequencies and the outcrossing rate in the starting population are known as given below:

Generation	Expected frequency of heterozygotes
S_0	$\beta_0 = 0$
S_1	$\beta_1 = 2uv(1 - f)$
S_2	$\beta_2 = wuv - uvf(1 + f)$
S_3	$\beta_3 = 2uv - uvf - 1/2\, uvf^2(1 + f)$
S_4	$\beta_4 = 2uv - uvf - 1/2\, uvf - 1/4\, uvf^3(1 + f)$

Assuming no selection and u, v and f constant over generations, the frequency of heterozygote will increase with $n = 1, 2, 3$ and 4 and so on until, when n is large, it reaches the equilibrium where

$$\beta_\infty = 2uv \frac{2(1 - f)}{(2 - f)}$$

a conclusion proposed earlier (Ali and Hadley, 1955.

(ii) *Use of quantitative trait*

A method of determining the level and distribution of heterozygosity in a synthetic population by studying quantitative variation was proposed by Roy and Jinks, 1983. Suppose that two varieties, namely, A and B are crossed and F_1 and F_2 are obtained. A random sample of plants, say n, are selected in F_2 population and selfed and thus nF_3 families are produced. If r sibs/n families are grown individually randomised in the field and the different quantitative characters are scored, then the analysis of variance will take the following form as shown in Table 13.1.

13.10 Biometrical Genetics–Analysis of Quantitative Variation

Table 13.1 ANOVA in case of F_3 families

Item	D.F.	M.S.	E.M.S.
Between F_3 families	$(n-1)$	MS_1	$\sigma^2_{WF_3} + r\sigma^2_{bF_3}$
Within F_3 families	$n(r-1)$	MS_2	$\sigma^2_{WF_3}$

The expectations of $\sigma^2_{bF_3}$ and $\sigma^2_{WF_3}$ are

$$\sigma^2_{bF_3} = \frac{MS_1 - MS_2}{r} = 1/2\, D + \frac{1}{16} H$$

$$\sigma^2_{WF_3}\, V_2 F_3 = 1/4\, D + 1/8\, H + E$$

where
$$D = \sum^{k_1} di^2 = D_1$$
$$H = \sum^{k_1} hi^2 = H_1$$

Assuming $\frac{1}{16} H$ negligible

$$2\sigma^2_{bF_3} = 2V_1 F_3 = D = D_1$$

This estimate of D or $D + 1/8\, H$ is the summed effect of segregation at all of the say k_1 loci at which A and B differ. Now if the inbred lines constituting the S_0 population were derived from the F_2 of $A \times B$ cross, then the plants in the S_0 generation would also differ at these same k_1 loci. Since some outcrossing occurs in the population, any one plant in the S_n generation population may be heterozygotes for up to all of these k_1 loci but will probably be homozygotes at some or most of them. Any plant chosen from the S_n generation will be like an F_1 in respect of the $k_2 \leq k_1$ loci at which it is heterozygous. On selfing it will produce the equivalent of an F_2 segregation in respect of these k_2 loci. If a random sample of n plants of this k_2 ($F_2 S_n$) are self pollinated, it will produce n families equivalent to an F_3. If r replicates of each of these families ($F_3 S_n$) are raised, we will have for each plant sampled from the population, the following analysis of variance table.

Table 13.2 ANOVA in heterozygosity experiment

Item	D.F.	M.S.	E.M.S.
Between n selfs	$(n-1)$	MS_{11}	$\sigma^2_{WF_3} S_n + r\sigma^2_{bF_3} S_n$
Within selfs	$n(r-1)$	MS_{22}	$\sigma^2_{WF_3} S_n$

Here
$$\sigma^2_{WF_3} S_n = \frac{MS_{11} - MS_{22}}{2}$$
$$= \frac{1}{2} D + \frac{1}{16} H$$

where
$$D = D_2 = \sum^{k_2} di^2$$
$$H = H_2 = \sum^{k_2} hi^2 \text{ where } k_2 \leq k_1$$

The ratio of $\dfrac{D_2}{D_1} = \dfrac{\sum^{k_2} di^2}{\sum^{k_1} di^2}$ will be a measure of relative heterozygosity, $\dfrac{k_2}{k_1}$ weighted for the additive effects of the loci of the plants sampled from the S_n generation population. If we measure k_2 on a scale where the F_1 level $k_1 = 1$ then $D_2/D_1 = k_2$. If a random sample of m plants are sampled from the S_n generation, the average value of D_2/D_1 over m will measure the level of heterozygosity ($\beta_n = \bar{k}_2$) of the S_n population. Two samples can be taken from the population-random and stratified. The random sample is to ensure an unbiased estimate of the β_n, the stratified sample to ensure that the distribution of heterozygosity across the whole phenotypic distribution of the S_n generation is measured. If dominance is negligible, $D_2/D_1 = k_2$ may be estimated as $\sigma^2_{bF_3,Sn}/\sigma^2_{bF_3}$. If, however, dominance is not negligible, one can still estimate k_2 in this way provided that the d's and h's are either constant over loci or the ratio h/d = dominance ratio, is constant over all loci. For example, if the d's and h's are constant

$$\sum^{k_1} di = dk_1 \bar{d}^2 \text{ and } \sum^{k_1} hi^2 = k_1 \bar{h}^2$$

Similarly $\sum^{k_2} di^2 = k_2 \bar{d}^2$ and $\sum^{k_2} hi^2 = k\bar{h}^2$

Therefore, $\dfrac{1}{2} D_1 + \dfrac{1}{16} H_1 = k_1 \left(\dfrac{1}{2} \bar{d}^2 + \dfrac{1}{16} \bar{h}^2 \right)$

$\dfrac{1}{2} D_2 + \dfrac{1}{16} H_2 = k_2 \left(\dfrac{1}{2} \bar{d}^2 + \dfrac{1}{16} \bar{h}^2 \right)$

Therefore, $\dfrac{1/2\, D_2 + 1/16\, H_2}{1/2\, D_1 + 1/16\, H_1} = \dfrac{k_2}{k_1} = k_2$

when $\dfrac{h}{d} = $ constant $= f$, $h^2 = d^2$

$$\dfrac{1}{2} D_2 + \dfrac{1}{16} H_2 = \dfrac{1}{2} k_2 d^2 + \dfrac{1}{16} f \sum^{k_2} d^2$$

$$= \left(\frac{1}{2} + \frac{1}{16}f\right) \sum^{k_2} d^2$$

$$\frac{1}{2}D_1 + \frac{1}{16}H_2 = \frac{1}{2}\sum^{k_1} d^2 + \frac{1}{16}f\sum^{k_1} d^2$$

$$= \left(\frac{1}{2} + \frac{1}{16}f\right)\sum^{k_1} d^2$$

Therefore, $\quad \dfrac{\frac{1}{2}D_2 + \frac{1}{16}H_2}{\frac{1}{2}D_1 + \frac{1}{16}H_1} = \dfrac{k_2}{k_1} = k_2$

To test whether an individual in the sample of m plants can be said to be heterozygous at the same number of loci as the F_1 of $A \times B$, i.e. $k_2 = k_1$ it is necessary to test whether the set of F_3 S_n families is identical with the normal set of F_3 families. This can be carried out by fitting the following model to variances of these families. If the model fits the F_3S_n and F_3 do not differ significantly.

		Model	
	Variance	σ_b^2	σ_w^2
F_3	Between families	1	1
	Within families	0	1
F_3S_n	Between families	1	1
	Within families	0	1

In the absence of selection, the estimates of heterozygosity obtained from the study of qualitative variation and the quantitative variation should be the same.

13.8 Detection of Competition

Competition is detected in advancing synthetic populations by comparison of observed means, variances, skewnesses and kurtosis of generations with the expected statistics based on the estimates of the rate of outcrossing.

For two alleles system assuming no selection and u, v and f constant the coefficients of $[h]$ and $[l]$ in the advancing generations of synthetic population are calculated in the following way:

The expected means and variances of S_n generation synthetic population would become

$$S_n = m + \beta_h [h]$$
$$VS_n = (1 - \beta_n)D + \beta_n(1 - \beta_n)H + E_W$$

References

Alii M. and Hadley, H.H. 1955. Theoretical proportion of heterozygosity in populations with various proportions of self and cross fertilization. Agron. Journal, **47**: 589–590.

Benjamin, L.R. 1982. A model to predict the effect of complex row spacings on the yields of root crops. J. Agric. Sci., Camb., **98**: 113–139.

Bleasdale, J.K.A. and Nelder, J.A. 1960. Plant population and crop yield. Nature, 188, 342.

Breese, E.L. and Hill, J. 1973. Regression analysis of interactions between competing species. Heredity, **31**: 181–200.

Boughey, H.J., Doughlas, C., Al-Banna, M.K.S. and Jinks, J.L. 1981. The frequency of heterozygotes maintained in synthetic populations of *Nicotiana rustica*. Heredity, **47**: 257–278.

Caligari, P.D.S. and Mather, K. 1988. Competitive interactions in *Drosophila melanogaster*. 4. Chromosome assay. Heredity, **60**: 355–366.

De Miranda, R. and Eggleston, P. 1987. A comparison of substitution and addition designs for the analysis of competitive interactions in *Drosophila melanogaster*. Heredity, **58**: 279–288.

De, Miranda, J.R. and Egzleston, P. 1988c. Genetic analysis of larval competition in *Drosophila melanogaster*. Heredity, **61**: 339–346.

De Wit, C.T. 1960. On Competition. Versl. Landbouwk, Onderz. Ned. **66**: 1–82.

Donald, C.M. 1963. Competition among crop and pasture plants. Advances in Agron., **15**: 1–118.

Eggleston, P. 1985. Variation for aggression and response in the competitive interactions of *Drosophila melanogaster*. Heredity, **54**: 43–51.

Hemmat, M. and Eggleston, P. 1990. The biometrical genetics of competitive parameters in *Drosophila melanogaster*. Heredity, **64**: 223–231.

Holiday, R. 1960. Plant population and crop yield. Nature, **186**: 22–24.

Kempton, R.A. 1982. Adjustment for competition between varieties in plant breeding trials. J. Agric. Sci., Camb., **98**: 599–611.

Mather, K. 1983. Response to selection. In: Ashburner, M., Carson, H.L. and Thompson, J.N. Jnr (eds.). The Genetics and Biology of *Drosophila* Vol. 3c. Academic Press, pp. 152–215.

Mather, K. and Caligari, P.D.S. 1981. Competitive interactions in *Drosophila melanogaster*. 2. Measurement of competition. Heredity, **46:** 239–254.

Mather, K. and Caligari, P.D.S. 1983 Pressure and response in competitive interaction. Heredity, **51:** 435–454.

Mather, K., Hill, J. and Caligari, P.D.S. 1982. Analysis of competitive ability among genotypes of perennial rye grass. Heredity, **48:** 421–434.

Mead, R. and Riley, J. 1981. A review of statistical ideas relevant to intercropping research. J. Royal Stat. Soc. A., **144:** 462–509.

Minjas, A.N. and Runeckles, V.C. 1984. Application of monoculture yield/density relationships of plant competition in binary additive series: Annals of Botany, **53:** 599–606.

Norrington-Davies, J. 1968. Diallel analysis of competition between grass species. J. Agric. Sci., Camb., **71:** 223–231.

Pooni, H.S. and Jinks, J.L. 1981. Colour of floral parts in *Nicotiana rustica*. Heredity, **46:** 273–275.

Roy, D. 1983. Application of biometrical genetics to synthetic populations of *Nicotiana rustica*. Ph.D. Thesis. The University of Birmingham.

Spitters, C.J.T. 1983. An alternative approach to the analysis of mixed cropping experiments 1. Estimation of competition effects. Neth, J. Agric. Sci., **31:** 1–11.

Spitters, C.J.T., Kropf, M.J. and De Groot, W. 1986. Use of hyperbolic yield density equation to describe crop weed competition. Annals of Appl. Biol.

Wright, A.J. 1981. The analysis of yield density relationship in binary mixtures using inverse polynomials. J. Agric. Sci., Camb., **96:** 561–567.

14

Environmental Variation

14.1 Definition and Types

The non-heritable variation (or the error variation) must be reduced in order to get precise estimate of different statistics through which we get information on the types of gene action and interaction. The different sources of non-heritable variation are as follows:

1. E_1 (or E_w): It is the environmental variation among individuals within families. It arises from sampling variation, error of measurement and development variation. The developmental variation refers to the variation in the character appearing at different times on an individual and this can be estimated by taking the repeated measurement of the character. Most of the environmental variation within families is attributed to this type of variation. It also includes the plant to plant variation because of microenvironmental effects. The error of measurement will be larger with quasi-quantitative trait in comparison to the quantitative trait.

2. E_2: It is the environmental variation of the means of families. Development of families from parents bred in different environments gives rise to such variation. In case of animals the pre-and post-natal maternal environment may contribute to E_2. E_2 is same as VEC, the variance due to common environment and E_1 is similar to VE_1, the environmental variation of Falconer (1960). In plants E_2 is mostly zero.

3. Eb: It is the environmental variance due to plot effects. Microenvironmental variation within a plot will be small but becomes larger between plots and thus attributes to the Eb. It is the additional source of non-heritable variation between families. Total random error due to plant-plant variation can be reduced by increasing the size of the plot.

The variance between plot means (Vm) depends on the individual variance $V_{(i)}$ and the number of the plants in the plot (Smith, 1938).

$$Vm = \frac{V(i)}{m^b}$$

Here b accounts for all ways in which the members of the group interacted. With self-fertilizing crops, b is a function of soil heterogeneity whereas with cross fertilizing crops, plant to plant genetic variation can contribute to the value of b. Estimates of b ranged from 0.16 to 0.80 with an average of about 0.5

$$Vm = \frac{VG}{m} + \frac{VE}{m^b}$$

The genetical variance of the mean depends inversely on the sample size while the environmental variance would change according to Smiths rule. In the notations used here

$$E_2 = \frac{VE}{m^b} = \frac{1}{m}\sigma_w^2 + \sigma_p^2 = \frac{1}{m}E_1 + Eb$$

(Mather and Jinks, 1971)

14.2 Estimation of Environmental Variation

Let us see how these various components appear in the analysis of variance table and estimated. Suppose there are n families with k number of sibs per family. Now if each family is raised in p plots of size m ($k = p \times m$) the analysis of variance takes the forms as given in Table 14.1.

14.2 Biometrical Genetics–Analysis of Quantitative Variation

Table 14.1 Estimation of environmental variation

Item	d.f.	E.M.S.
Between families	$n-1$	$\sigma_w^2 + m\sigma_p^2 + mp\sigma_B^2$
Between plots with families	$n(p-1)$	$\sigma_w^2 + m\sigma_p^2$
Within plots	$np(m-1)$	σ_w^2

Now if the families are inbred then σ_w^2 measures the environmental variation within plots and equals E_1. σ_p^2 which measures the environmental variation between plots equals Eb. But when the families considered are segregating families, then σ_w^2 also includes genetic differences among individuals within plot. Further, σ_B^2 not only measures the genetic differences between families but also includes E_2.

14.3 Individual Plant Randomization

The ANOVA given above takes the form as given in Table 14.2 when $m = 1$, i.e. number of plants per plot is 1. There is, thus, complete individual plant randomization.

Table 14.2 ANOVA is case of individual plant randomization

Item	d.f.	M.S.	E.M.S.
Between families	$(n-1)$		$\sigma_w^2 + \sigma_p^2 + p\sigma_B^2$
Between plots within families	$n(p-1)$		$\sigma_w^2 + \sigma_p^2$

In this design then E_1 and E_b are confounded and other source of variation does not exist.

When all the sibs are raised together in a plot i.e. when there is only one plot per family then the ANOVA takes the form as given in Table 14.3.

Table 14.3 ANOVA is case of randomized complete block design

Item	d.f.	M.S.	E.M.S.
Between families	$(n-1)$		$\sigma_w^2 + m(\sigma_p^2 + \sigma_B^2)$
Within plot	$n(m-1)$		σ_w^2

Here we can see that σ_B^2 is confounded with σ_p^2, the block-effects and thus is not the appropriate design for estimating variances. A design is most appropriate if it is able to pick up even the smaller differences between families and in this sense individual plant randomization design is the most appropriate. It can be shown that as m becomes smaller with fixed nk where $k = p \times m$, i.e. with a given experimental size, the number of replications increases which in turn increases the number of degrees of freedom for error item in the ANOVA and thus the error is more precisely estimated. Further F becomes larger and as a result even a smaller genetic difference between families is detected and estimated.

When the aim is to estimate variances of different ranks working with plot totals instead of individual plant means will result in loss of information on genotypic variance within line variance. As the plot size increases there will hardly be any informations left on genotypic variance within line variance. For example, there are variances of two ranks, namely, V_1F_3 and V_2F_3 in F_3 family. V_2F_3 will be difficult to estimate; its estimator will have a larger standard error and standard error of D_1, the estimator of additive genetic component will increase but the precision of the estimator of V_1F_3 can be raised by increasing the plot size and thus the standard error of D_2, the estimator of additive genetic component will decrease. Thus there should be an optimum allocation of experimental size regarding plot size, number of plots and number of lines.

As the heritability of a character decreases, the experimental size required for obtaining precise estimation of genetic components of variation increases and where it will be difficult to score every individual plant and in that case one can go for experimental design based on plot total (or plot means).

The individual plant randomization design can be used without any problem in vegetable and fruit crops and crops like tobacco, cotton, etc. where one can work with individual plants comfortably as plant to plant and row to row distances are larger but the adoption of such experimental design is not feasible in cereals, legumes and other densely planted crops. And if there are more individuals in a plot i.e. when m is larger we can detect more plot effects. Where information on means is more important as in case of screening a large number of selections, there it is essential to go for raising different entries in plots which represents the ideal situation as a variety will be finally grown on a larger plot in commercial planting in pure stand.

14.4 Sampling Variation

Sampling error results from error in the selection of sampling units, method of selection of sampling units and the sample size. The variance of family means will contain sampling variation arising from the variation within family. Thus V_1F_3 will contain sampling variation equal to $(1/n)$ V_2F_3 where each family contains n individuals but if n varies from family to family then n is the harmonic mean of these numbers. Similarly, V_1F_4 will contain sampling variation equal to $(1/n)$, V_2F_4, where n is the number of families in each group of F_4 families. V_2F_4 will contain sampling variation equal to $(1/n)$ V_3F_4 where n is the number of individuals in each family. The sampling variation thus equals $1/n$ of the within family variance.

Plants within a plot should be randomly selected for taking observation otherwise it will result in biased estimate of statistics. For example, if F_2 population made up of *AA*, *Aa,* and *aa* individuals is raised in a plot and if the sampling is not random then only *AA* and *Aa* plants may be selected and then estimate of variation obtained will not represent the F_2 population. Further as the variation among plants could be due to microenvironmental variation so if only good looking plants within a plot are selected it will lead to serious error.

14.5 Other Experimental Designs

The experimental designs reduce the residual errors (E_1, E_b). The E_2 is zero with the single plant randomization. There are complete block designs such as randomised complete block design, latin square and incomplete block design such as lattice. The use of a particular design depends on (1) the magnitude of soil heterogeneity, (2) type and the number of families to be tested and (3) the degree of precision desired. An appropriate design is one which includes replication, randomization and local control. Replication and local control help in reducing the experimental error whereas randomization and replication provide the unbiased estimate of error. One of the simplest means of increasing precision is to increase the number of replications but as the number of replications increases, the plot size increases given an experimental area.

These different designs differ in their efficacy in reducing the experimental error and thus the statistical precision of the estimates varies with the design. Randomized block designs have been found to be 67 per cent as efficient as latin squares and 50 per cent as efficient as lattice. The latin square is more efficient a design than the *RBD* because of the reason that in the latin square, the variances due to rows, columns and entries are subtracted from the total variance and thus the residual error variance is smaller in comparison to the *RBD* for testing the differences between treatment means. But then the latin square can accomodate up to 8 entries. In the incomplete block design, the experimental error is reduced through the principle of local control by grouping the plots into blocks of moderate size with each block containing only a portion of the entries to be compared.

If our main objective is to have precise estimate of means as would be in the case of selection of a particular entry from among a number of entries, lattice designs are more useful in comparison to other experimental designs. As in incomplete block designs, genetic and environmental differences among means are confounded, these designs are not satisfactory for estimating variances. Since variety and block effects are confounded, the variety mean must be adjusted. In case the soil heterogeneity is little, the data generated using this design can be analysed in a *RBD* fashion. In balanced incomplete blocks when large number of cultivars are to be studied, the cultivars are associated with one another in groups of sets and thus results in large experimental error. To reduce this error all possible groups of sets are made and every cultivar replicated the same number of times. As specific cultivars are paired, direct comparison can be made while other cultivars can be compared indirectly using the average of associated cultivars. Balanced incomplete block requires a large number of replications in order to attain the balanced layout.

Lattice design allows indirect comparison through intermediate cultivars and so reduces the need for large replication.

When blocking of genetic material is necessary to control the environmental variation, replication in blocks and blocks in replication designs (Robinson and Comstock, 1952) can be used. A large number of progenies are required to obtain reasonable estimates of genetic components of variance. Individual plant randomization is most useful when the objective is to estimate variances of the families.

Complete block designs do provide unbiased estimates of family means and variances. *RBD* with 8–10 replications can be used to obtain estimates of means with standard error of 10% or less for characters having low heritability. This design is used when comparisons among a smaller number of lines are required. The number of replications chosen should be such that the degree of freedom for error is not less than 10.

If the objective is to assess cultivars under large range of conditions, one should go for split plot design as it would probably extract enough information to allow recommendation of cultivars for well defined cultural conditions. Through the use of this design, comparison of yield is more informative if half of each plot is heavily fertilized and while the remaining half receives the normal dose of fertilizer. Here one can select another factors such as irrigation and no irrigation or date of planting or different dates of harvesting or cutting as in case of fodder crops, etc. Depending upon the size it could be worthwhile to split each plot into even more parts so long as each part still remained large enough to give meaningful results.

14.6 Estimation of Experimental Error

The magnitude of experimental error in an experiment can be expressed in terms of coefficient of variation which is estimated as:

$$\text{C.V.} = \frac{\text{Standard deviation}}{\text{mean}} = \sqrt{\frac{\sigma_w^2}{\overline{X}}}$$

where $\overline{X}$ is the mean of treatments. The size of coefficient of variation indicates the influence of soil heterogeneity. The size, shape and orientation of a plot affects the magnitude of the experimental error. As the C.V. decreases with the improvement of the fertility of the experimental sites, the C.V. can be used to verify a poor site. The C.V. of an experiment varies from one station to another and from one type of experiment to another.

14.7 Augmented Design

In the early stages of selection programme, the breeder has a large number of new selections but the seed of the new selections is so little that he cannot conduct a replicated yield trial. As he would like to compare his new selections with one another and with the existing checks for yield he can then adopt Augmented Design developed by Federer (1956, 1961) and described by Federer and Ragavarao (1975).

In this design, the checks are replicated while the selections are unreplicated. If C be the number of checks, then the number of blocks should be such that the $(C - 1)(b - 1)$, the degrees of freedom for the error in the ANOVA for the checks should be at least 10. The checks should be randomly assigned to plots within each block. The number of selections per block would be S/b where S is the total number of selections. The trial is most efficient if the number of plots per block is the same in all blocks. The assigning of selections to blocks and to plots within a block should be random. The analysis can proceed even if the blocks may not contain the same number of plots.

The plot yields of new selections and the checks are recorded and the ANOVA for the checks is constructed which is same as that of ANOVA of a *RBD* experiment. An adjustment factor (r_j) is computed for each block which is as follows:

$$r_j = \frac{1}{C}(B_j - GM) \text{ where } \sum_j r_j = 0$$

where r_j is the adjustment factor for jth block,

C is the number of checks included in the design,

B_j is the total of all checks in the jth block,

and GM is the grand mean of all the checks.

If Y_{ij} is the yield of ith selection in the jth block then the adjusted mean of the ith selection, Y_i equals

$$\hat{Y}_i = Y_{ij} - r_j$$

Thus the yield of the ith selection is adjusted for the block effect.

Now if MSE be the error mean squares in the ANOVA for checks, the variances for making different kinds of comparisons are the following.

(i) The variance of the difference between adjusted yields of two new selections in the same block is 2MSE.

(ii) The variance of the difference between adjusted yields of the two new selections in different blocks is $2MSE(1 + C)/C$.

(iii) The variance of the difference between the adjusted yields of a new selection and a check mean is $MSE(1 + b)(1 + C)/bC$.

(iv) The average variance of the difference between adjusted yields of the two new selection is $MSE(2C + 1)/C$.

For most cases of comparison between two new selections the *LSD* based on the average variance is satisfactory.

14.8 Environmental Covariance

The covariance does not contain environmental component of variation as the environmental variance which arises as a result of correlation between non-heritable components of offspring and parents can be eliminated by way of randomization of offspring and parents in the experiment, which all the plant breeding experiments do, but the covariances in the animal breeding experiments do contain the environmental variance E, as the mother provides the pre-and post-natal environment to offspring which results into environmental correlation.

References

Cochran, W.G. and Cox, G.M. 1950. Experimental Designs, Wiley, New York.

Falconer, D.S. 1988. Introduction to Quantitative Genetics, Longman, London.

Federer, W.T; 1974. Experimental Design. Oxford & IBH Publishing Co., Delhi.

Fisher, R.A. 1949. The Design of Experiment. 5th edn. Oliver and Boyd, London.

Gauch, H.G. Jr. and Zobel, R.W. 1996. Optimal replications in selection experiment. Crop Sci., **36**: 838-843.

Gomez, K.A. and Gomez, A.A. 1976. Statistical procedures for agricultural research. The Int. Rice Res. Inst. Los Banos, Philippines.

Kempthorne, O. 1952. The Design and Analysis of Experiments. Wiley, New York.

Mather, K. and Jinks, J.L. 1982. Biometrical Genetics. Chapman and Hall. London.

Nishida, A. and Abe, T. 1974. The distributions of genetic and environmental effects and the linearity of heritability. Can. J. Genet. Cytol, **16**: 3-10.

Panse, V.G. and Sukhatme, P.Y. 1978. Statistical methods for Agricultural Workers, ICAR, New Delhi.

Sokal, R. and Rohlf, F. 1969. Biometry. W.H. Freeman and Company, San Francisco.

Robertson, A. 1977. The non linearity of offspring-parent regression. In: Proc. of the Intl. Conf. on Quant. Genet., Ames, Iowa, pp. 297-304.

15

Heritability

15.1 Definition and Types

We have seen that a trait manifested in an individual of a population is a result of genetic and environmental causes, i.e. $P = G + E$ at mean level and $VP = VG + VE$ at variance level, assuming no $G \times E$ interaction. It is the genetic variance of trait in the population which can be transmitted from parents to offspring. How much of the phenotypic variability will be transmitted is measured by a parameter called heritability. Heritability thus provides a measure of the overall importance of heriditary determination of a trait. In other words, heritability for a character is a measure of the relative importance of heredity and environment. There are two types of heritability.

1. Heritability in the broad-sense (h_b^2)
2. Heritability in the narrow-sense (h_n^2)

The broad-sense heritability is measured as

$$h^2(b) = \frac{VG}{VP} = \frac{1/2DR + 1/4HR}{1/2DR + 1/4HR + E}$$

$$= \frac{1/2D + 1/2H_1 - 1/4H_2 - 1/2F}{1/2D + 1/2H_1 - 1/4H_2 - 1/2F + E}$$

where DR and HR are random mating population parameters and D_1, H_1, H_2 and F are diallel parameters and E is environmental component of variation. And thus h_b^2 is a measure of the relative proportion of genetic variance in the total variance. This can be called as genetic heritability.

We have further seen that the genetic value (G) can be further partitioned into additive, dominance and interaction genetic values and thus at variance level, the genetic variance can be partitioned into additive genetic (V_A), dominance (V_D) and interaction (V_I) variance and it is the additive genetic variance portion of the total genetic variance which is transmitted to the offspring and thus is of more concern to us in the selection programme. Thus the relative proportion of the additive genetic variance in the total variance gives an estimate of heritability which is called narrow-sense heritability and is measured as

$$h^2_{(n)} = \frac{VA}{VP} = \frac{1/2DR}{1/2DR + 1/4HR + E}$$

$$= \frac{1/2D + 1/2H_1 - 1/2H_2 - 1/2F}{1/2D + 1/2H_1 - 1/4H_2 - 1/2F + E}$$

This additive genetic heritability thus measures the speed of progress of a trait under selection. The heritability estimate tells about the extent of transmission of values of a trait from parent to offspring. If the parent and offspring completely resemble each other for a trait we say that the transmission of the trait has been complete and the heritability is 100% but then this is wrong as the resemblance could be for any reason, genetical or environmental.

There is another way of looking at heritability. Resemblance between the values for a trait among offspring and parents defines heritability. The resemblance between individuals is measured by correlation/regression and so this correlation/regression can be expressed as a function of heritability narrow-sense. The resemblance between individuals is influenced by the kinship relation (such as offspring-parent, half-sib, full-sib, etc.). As correlation can be due to genetic or environmental reasons so we do not have explanation what so ever for the observed resemblance. The regression coefficient (b_{op}) which is obtained by regressing the offspring values on parental values in a population equals $\frac{1}{2} h_n^2$ whereas the regression of offspring values on mid parental values ($b_{\overline{op}}$) equals h_n^2 which can be shown to be derived as follows:

$$b_{op} = \frac{Cov(OP)}{VP} = \frac{(1/2)VA}{VP} = \frac{1}{2}\frac{VA}{VP} = \frac{1}{2} h_n^2$$

$$b_{o\overline{p}} = \frac{Cov(O\overline{P})}{VP} = \frac{(1/2)VA}{(1/2)VP} = \frac{VA}{VP} = h_n^2$$

Further, intra-class correlation (t) which is the ratio of between families variance to total variance, yields estimate of heritability. As we can have full-sib or half-sib families so we can estimate heritability from the estimate of intra-class correlation in full-sib or half-sib families as

$$t \text{ (half-sib)} = \frac{(1/8)\,DR}{(1/2)DR + (1/4)HR + E_1} \approx \frac{1}{4} h_n^2$$

$$t \text{ (half-sib)} = \frac{(1/4)\,DR + (1/6)HR}{(1/2)DR + (1/4)\,HR + E_1} \approx \frac{1}{2} h_n^2$$

15.2 Heritability a Characteristic of Population

Heritability characterizes not only the character itself but the population and the environment in which the character is studied. We have seen that the estimates of variances or the covariances of relatives in a population depend on the genetic structure of the population. The genetic components of variation such as DR and HR are a function of gene frequencies which shows that the same character may well have totally different heritability in two populations. Further, it can be shown that although the heritability broad-sense will be 1.0 for trait, the heritability narrow-sense will be vary small. Supposing that a character is determined by one locus with 2 alleles as can be in case of recessive genetic disease with no environmental effects so h_b^2 will be 1.0. Following Kempthorne (1953) it can be shown that

$$h_{(n)}^2 = \frac{VA}{VA + VD} = 2v/1 + v$$

This clearly shows that although the h_b^2 can be 1.0 but if the frequency of recessive allele is very less in a population, the h_n^2 will be almost zero.

15.3 Estimates of Heritability

The heritability can be estimated, (1) indirectly from the estimates of components of variation and (2) directly from the estimates of regression coefficients (b's) and intra class correlations (t's) and finally, (3) we can have estimate of heritability from the selection experiment.

The estimates of components of variation can be obtained through the study of basic generations (P_1, P_2, F_1, F_2 and backcross generations) and various mating designs discussed earlier. The h_b^2 and h_n^2 from the estimates of variances of basic generations can be calculated as

$$h_b^2 = \frac{VF_2 - 1/3(VP_1 + PV_2 + VF_1)}{VF_2}$$

and

$$h_n^2 = \frac{2VF_2 - (VB_1 + VB_2)}{VF_2}$$

The above heritability estimates are based on the assumption that epistasis is absent.

In self-pollinated crops following pedigree method the regression approach i.e. the regression of F_3 progeny values on F_2 parental values can be used to calculate the heritability as:

$$b_{F_2F_3} = \frac{Cov\,F_2\,F_3}{VF_2}$$

$$= \frac{VA + (1/4)D}{VA + VD + VE}$$

$$= h_n^2 \text{ assuming no dominance}$$

In a sexually propagated crop as VG is completely transmitted from parent to offspring, we can have only estimate of heritability broad-sense, h_b^2.

When heritability is based on estimates of components of variation, if narrow-sense heritability is low, we know that it can be due to presence of dominance (HR) or due to higher E_1 and E_2 components but the problem with this method of estimation of heritability is that it is difficult to estimate the standard error of heritability as the various components of variation are correlated. When heritability estimate is based on covariances such as Cov OP or Cov $O\overline{P}$ if the heritability is high we do not know the reason as high estimate of h_n^2 could be due to genetic or environmental causes but the advantage with this method of estimation is that the standard error of h_n^2 can be estimated. Between the two methods, regression of offspring on parent and regression of offspring on mid-parent, lower standard error is associated with the later approach as less environmental variation is found around mid-parental value. Estimate of heritability using covariance full-sib is least reliable amongst the different direct methods of estimation because of presence of large amount of E_2, the common environmental effect and the estimate obtained using Cov FS sets an upper limit to the heritability. Parent-offspring regression is sometimes preferred over other methods because heritability estimates based on full-sib correlations, contain components of dominance and environmental covariance. Estimates of heritability from regression analysis are more conservatives whereas heritabilities from variance-covariance analysis are over estimates.

We can have estimate of heritability, the realized heritability from the selection experiments as

$$h_n^2 = R/S$$

where R is the response due to selection and S is the selection differential. The h_n^2 is calculated using the prediction formula, $R = h_n^2 \times S$. Heritability is here defined as the proportion of the selection differential which is transmitted to the progeny.

15.4 Effect of $G \times E$ Interaction

In the presence of $G \times E$ interaction the phenotypic variance for a character X, $VP(X)$ becomes

$$VP(X) = VG(X) + VE(X) + 2 \,\text{Cov}\,(G, E)$$

Under normal circumstances $G \times E$ interaction variance $(VG \times E)$ can not be separated and thus can not be estimated and is best considered a part of the environment variance but if the environmental differences are large then $VG \times E$ can be estimated.

And when a genotype is raised over a number of locations (L) and over a number of years (Y), the phenotypic variance becomes

$$\sigma_P^2 = \sigma_G^2 + \sigma_E^2 + \sigma_{GLY}^2 + \sigma_{GL}^2 + \sigma_{GY}^2$$

This shows that unless $G \times E$ interaction variances are separated from the genetic variance we will not have valid estimate of heritability. Multisites and multiyears testing will permit separation of the $g \times e$ interaction variances but in case of one location, one year testing the interaction variances cannot be separated. The heritability in these situations will be

$$h_b^2 = \frac{\sigma_G^2}{\sigma_G^2 + \sigma_{GLY}^2 + \sigma_{GL}^2 + \sigma_{GY}^2 + \sigma_E^2}$$

$$h_b^2 = \frac{\sigma_G^2 + \sigma_{GLY}^2 + \sigma_{GY}^2 + \sigma_{GL}^2}{\sigma_G^2 + \sigma_{GLY}^2 + \sigma_{GY}^2 + \sigma_{GL}^2 + \sigma_E^2}$$

Thus the heritability estimate obtained from one location, one year data is inflated in comparison to the estimate obtained after separation of $g \times e$ interaction variances from multisites and multiyears data. This further shows that the heritability estimate will vary from environment to environment as the genotype (s) will respond differently in different environments. Thus heritability for a character also characterizes the environment in which the population is raised. It further shows that $G \times E$ interaction causes the heritability estimate to be greater than 1.0. Heritability on a single plot basis has limited utility and it should be estimated while having two replications within two environments.

15.5 Uses of Estimates of Heritability

The estimates of heritability can be used in the following way:
1. To know which character is expected to respond more to selection pressure.
2. Prediction of response to selection.
3. Determination of the size of population required to be maintained under selection.
4. Determining the various alternatives of selection scheme.

Keeping other factors constant, the speed of selection is proportional to the magnitude of heritability. A relatively higher population size is required in case of selection for characters, of lower heritability in comparison to highly heritable characters. If heritability is high for a character, we can go for individual (or mass) selection whereas in case of characters of lower heritability we can employ pedigree, sib or progeny test. If heritability is low because of higher E_1, the environmental variance, the family selection should be used whereas if low heritability of a character is due to larger E_2, the common environmental effect, one should go for within family selection.

15.6 Changing Estimates of Heritability

During selection programme, the estimate of heritability is not constant as selection changes the gene frequencies in the population. So, theoretically the estimate of heritability should change every generation but in practice its predictive value may hold for several generations. Further, as during selection, the additive genetic variance is used up over generations, its estimate will be higher in the early generation and little in the advanced generations. Similarly, heritability estimate in early generation segregating population derived from a cross of widely diverged materials will be relatively higher than in the advanced generation when the variation is fixed.

Although the original concept of heritability was developed for random mating population i.e. heritability is a population parameter and generalization cannot be applied to individual plants, in practice we are working with selected group of plants which represents the extreme in the population and here in fact, the mating is assortative. The regression coefficient in case of assortative mating will be larger and with less standard error in comparison to random mating.

15.7 Problems in Use of Estimate of Heritability

The problem in using estimate of heritability in practical plant breeding is that the various field designs and experimental procedures differ in their efficiency in reduction of environmental component of variation and thus although the additive genetic variance is the same, the different experiments will yield different estimates of heritability which thus reduces its use. Another point that must be kept in mind is that the h^2 estimate should be computed on the basis of the type of selection units (including number of replications, years, locations, types of progenies such as full-sib, half-sib, etc. and plot total or plot means) which will be used in the selection programme. For example, heritability narrow-sense computed on plot means will be

$$h_n^2 = \frac{(1/2)DR}{(1/2)DR + (1/4)HR + E_1 + mE_2}$$

where m is plot size and h^2 (full-sib family) would be equal to

$$\frac{(1/4)DR(1 + 1/m)}{(1/m)E_1 + E_b + (1/2)DR + (1/16)HR}$$

where m is the number of individuals in the family.

References

Cavalli's froza, L.L. and Bodmer, W. 1971. The Genetics of Human Populations, San Francisco, Freeman.

Dudley, J.W. and Moll, R.H. 1969. Interpretation and use of estimates of heritability and genetic variances in plant breeding. Crop Sci., **9**: 257–261.

Falconer, D.S. 1988. Introduction to Quantitative Genetics, Longman, London.

Fisher, R.A. 1918. On the correlation between relatives on the supposition of mendelian inheritance. Trans, R. Soc. Edinburgh, **52**: 399–433.

Hanson, W.D. 1963. Heritability. In: Statistical genetics and Plant Breeding. Hanson, W.D. and Robinson, H.F. (edited). National Academy of Science, Washington. pp. 125–139.

Jacquard, A. 1983. Heritability: One Word, Three Concepts. Biometrics, **39**: 465–477.

Kempthorne, O. 1957. An Introduction to Genetic Statistics, New York, Wiley.

Kempthorne, O. 1978. Logical epistemological and statistical aspects of nature-nurture data interpretation. Biometrics, **34**: 1–23.

Mather, K. and Jinks, J.L. 1982. Biometrical Genetics, Chapman and Hall, London.

Pollack, E. and Kempthorne, O. 1977. In: Proceedings of the International Conference on Quantitative Genetics. Ames, Iowa, Iowa State University Press.

Robinson. H.P., Comstock, R.E. and Harvey, P.H. 1949. Estimates of heritability and the degree of dominance in corn. Agron. J. **41**: 435–441.

Smith, J.D. and Kinman, M.L. 1965. The use of parent offspring regression as an estimator of heritability. Crop Sci., **5**: 595–596.

16

Estimation of the Number of Effective Factors

In the expectations of mean, variances and covariances of a continuously varying character we have assumed that polygenes are involved in the determination of a trait. Estimates of the number of genes contributing to the variance of quantitative characters within and between populations are fundamental for the study of mechanism of heredity and evolution. Heritable variation in quantitative traits within a population is caused by segregation of polygenes and that large evolutionary changes in quantitative traits generally occur through Darwinian process of accumulation of numerous genetic factors with individually small effects. Evidences supporting the above hypothesis are from (i) artificial selection experiment where the populations have evolved far beyond the limit of variation in the original base population (Falconer, 1960), (ii) correlation of quantitative traits with multiple genetic markers introduced in crosses between divergent lines (Dobzhansky 1936; 1957 and Smith 1937) and (iii) pattern of segregation following hybridization between widely different lines, races or species (Wright, 1968).

As in a polygenic system, many loci with small effects are involved, some loci may be tightly linked with no chance of recombination among them whereas some other loci may be loosely linked so in fact what can be estimated is the number of independently segregating block of genes, i.e. of effective factors. In case there is no linkage, i.e. one gene per block, the number of effective factors and genes are the same but if there is linkage, then the number of genes will be greater than the effective factor estimated. The effective factor is merely a segment of chromosome acting as a unit of inheritance and separated from other units by an average recombination frequency of 50%. Since chiasma may vary in position, a further breakdown of effective factor must occur in later generations by recombination and thus the number of effective factors estimate in successive generations must increase.

The selective potentiality of a population for a given character depends on magnitude and number of genes besides other factors. Number of genes influences the limits of progress from recurrent selection. A given amount of genetic variation can be generated either by a small number of genes with large effects or by a large number of genes with small effects. With a given amount of genetic variation, a small number of genes will produce less response than a large number of genes and if a given amount of variation is produced by few genes, the effects are greater than if many genes are involved. Further, character under control of few genes can be fixed with relatively small population size and early in the advancing segregating generation.

16.1 Approaches to Estimation of Number of Genes

As genes/effective factors affecting a quantitative trait cannot be observed directly, the number of genes/effective factors must be obtained indirectly. There are three approaches to the estimation of number of genes

1. Chromosome assay (Mather and Harrison 1949; Breese and Mather 1957; Thoday 1961; Law 1967).

2. Statistical properties of distribution (Wright 1934; Panse 1940; Mather 1949).

3. Genotype as say (Jinks and Towey 1976).

16.2 Chromosome Assay

In chromosome assay technique identification of genetic effects related to the specific segment of the chromosome is made. Analysis of intrachromosomal recombinants also provides information about the locations of genes. This method uses the genetic markers and is never likely to be more generally applicable.

16.3 Statistical Properties of Distribution

There are two ways in which information on the number of effective factors is obtained.

1. Comparison of theoretical and observed distributions when simultaneously considering different generations, e.g. pure line parents, F_1, F_2 and backcross generations (powers 1934, 1963; Gates 1963).

2. Inferences from phenotypic means and genetic variances of the character in parental, F_1, F_2 and backcross generations (Wright 1934; Panse 1940; Mather, 1949). The problem with the first method of estimation of number of effective factors is that although the observation of a continuous distribution of phenotypes in a segregating population is frequently interpreted as the involvement of a large number of genes or effective factors, there are cases in diploids where two or three loci are sufficient to produce distributions which are in practice indistinguishable from normal distributions (Thoday and Thompson, 1976). Also P_1, P_2, F_1, BC_1 and BC_2 generations will respond differently to environmental variation and this will result in untrue value of K, the number of effective factors.

In the second method there are two ways in which the estimate of number of effective factors can be obtained.

1. Use of parental range and additive genetic variance.

2. Use of variances of an F_2 derived population. In the first method the two parents, say P_1 and P_2, are supposed to represent the two genetic extremes. P_1, the high scoring parent contains all increasing alleles at all relevant loci whereas P_2 contains all decreasing alleles. Assuming no epistasis, the expectations of means of P_1 and P_2 become

$$\overline{P}_1 = m + [d] = m + kd$$

$$\overline{P}_2 = m - [d] = m - kd$$

Therefore, $\overline{P}_1 - \overline{P}_2 = 2kd$ or $1/2(\overline{P}_1 - \overline{P}_2) = kd$.

The estimate of D, the additive genetic variance ($= kd^2$) can be obtained from basic generations or from other mating designs. The ratio of the square of half the parental difference and the additive genetic variance provides the estimate of the number of effective factors k.

$$\frac{[(1/2(P_1 - P_2)]^2}{D} = \frac{(kd)^2}{kd^2} = \frac{k^2 d^2}{kd^2} = k$$

Wright (1934 b) used the genetic variance of F_2 as an estimate of D but as this F_2 genetic variance contains dominance variance, the estimate of D is inflated and thus the estimate of effective factors is biased downward.

This method is based on a number of assumptions, namely, (i) segregating genes are in one parent only. i.e. P_1 contains all increasing alleles and P_2 contains all decreasing alleles, (ii) no linkage, (iii) genes have equal effects, (iv) genes have equal degree of dominance and act in the same direction and (v) no epistasis. If these assumptions are not met then the estimate of number of effective factors can be heavily biased but in the right conditions it can give reliable estimates. Thus we see that all those factors which affect d and D, will also affect the precision of the estimate of the number of effective factors. Let us now examine the effect of inequality of gene effect. Considering $da = d(1 + \alpha a)$, $db = (1 + \alpha b)$, ..., $dk = (1 + \alpha k)$, it can be shown that the estimate of number of effective factors say k_1, becomes

$$k_1 = \frac{k}{1 + V_\alpha}$$

where V_α is the variance of α. In case gene effects are equal, $V_\alpha = 0$. Since V_α will have (except under certain conditions) positive value, the estimate of the number of genes k_1 will be less than k and is thus underestimated.

When parental lines contain increasing as well as decreasing alleles, i.e. when the alleles are dispersed between the two parents, the estimate of k_1 becomes

$$k_1 = \frac{kr^2}{1+V\alpha}$$

where r is the coefficient of degree of dispersion. Thus only under the condition of $r = \pm 1$ and $V_\alpha = 0$, $K_1 = k$, otherwise $k_1 < k$ and thus the estimate of effective factors is underestimated.

We have already seen the effect of linkage on the estimate of D, the additive genetic variance and thus the estimate of number of effective factors will be underestimated.

Serebrovsky (1928) considered the effect of dominance on the estimate of effective factors as

$$k = (1.5 - 2h(1-h))k_1$$

where h is the degree of dominance and $h = 0$ or 1 indicates complete dominance. Thus it can be seen that the estimate of k is lowered.

Finally, in case of additive x additive interaction, the estimate of effective factors becomes

$$k = \frac{[d]}{D+1}$$

and is thus underestimated. Most types of genetic interactions except when the number of effective factors is less in the middle of the range than near the extremes (Wright 1968) are likely to produce a downward bias in the estimate of number of effective factors as would F_2 or backcross 'break down' entailing increased variance due to segregation of aneuploid genotypes or from non-Mendelian or non-genetic causes.

16.4 Estimate of k Using Dominance Genetic Effect and Dominance Genetic Variance

We can also estimate k using the estimates of dominance effect and dominance genetic variance. Using the two parental lines and their F_1, the differences between F_1 and the mean of the two parents will be equal to kh. Given the estimate of dominance genetic variance $H(=kh^2)$, the ratio of square of the difference between the F_1 and the mean of the two parents to the dominance genetic variance will yield the estimate of the number of effective factors.

$$\frac{[\overline{F}_1 - 12/(\overline{P}_1 + \overline{P}_2)]^2}{H} = \frac{(kh)^2}{kh^2} = k$$

The estimate of k is again underestimated in the presence of inequality of dominance effects and by the direction of dominance at different loci.

16.5 Use of F_2 Derived Generation Variances

Panse (1940 a and b) proposed a method for estimation of the number of genes as the ratio of the square of within F_3 families variances to the variance of the between F_1 families variances

$$\frac{(V_2F_3)^2}{V(V_1F_3)} = \frac{1/2(kd^2)^2}{1/4(kd^4)} = k$$

This estimate of number of genes (k) is equivalent to k_2 of Mather and Jinks (1971).

16.6 Estimation of Number of Effective Factors in Random Mating Population

Park (1977 a and b) proposed procedure to estimate the number of genes from measures of response and additive genetic variance from recurrent selection experiments in which gene frequencies of alleles at segregating loci are 0.5 but not long enough to measure maximum progress. Assuming no epistasis and linkage disequilibrium, the average contribution of genotypes from all segregating loci ($\overline{Y}$) and the additive genetic variance $D(\sigma_g^2)$ following Comstock and Robinson (1948) become

$$\overline{Y} \text{ (genetic mean)} = \sum_{i=1}^{n} (2q_i - 1)\mu_i +$$

$$\sum_{i=1}^{n} 2q_i(1+q_i)a_i\mu_i$$

Additive genetic variance

$$(\sigma_g^2) = \sum_{i=1}^{n} 2q_i(1-q_i)[1+(1-2q_i)a_i]^2 \mu_i^2$$

where a_i is a measure of dominance at the ith locus, n is the number of segregating loci, q_i is the frequency of a favourable allele at ith locus, u_i is the $1/2\,(\overline{P}_1 - \overline{P}_2)$, the difference between two homozygotes.

When $a_i = 0$, i.e. no dominance and all μ's are equal and all q's are equal, the above two quantities reduce to

$$\overline{Y} = n(2q-1)\mu$$

and

$$\sigma_g^2 = n\,2q(1-q)\mu^2$$

Taking derivatives and reducing q

$$n = -\overline{Y}\,\frac{\delta \overline{Y}}{\delta \sigma_g^2}$$

which upon integrating the differential equation results in estimate of the number of effective factors as

$$n = Y_c^{-2} - Y_b^{-2}/2(\sigma_{gb}^2 - \sigma_{gc}^2)$$

where b and c are two generations in a recurrent selection experiment and $c > b$. When $qb = 0.5$ and $qc = 1.0$ or 0.0 the above formula reduces to

$$n = \frac{R^2}{2\sigma_{go}^2} \qquad \text{Comstock (1969)}$$

where R is the maximum response attainable by selection upward (or downward) and σ_{go}^2 the initial additive genetic variance D. When $q = 0.5$ at all segregating loci, response in a selection experiment is equivalent to half the difference in values between extremes of up and down lines, i.e.

$$R = 1/2(\overline{P}_1 - \overline{P}_2)$$

Hence

$$n = \left(\frac{(\overline{P}_1 - \overline{P}_2)^2}{8\,\sigma_{go}^2}\right) \qquad \text{(Falconer, 1960)}$$

where σ_{go}^2 can be estimated as

$$\sigma_{go}^2 = \sigma_{F_2}^2 - \sigma_{F_1}^2$$

or

$$= \sigma_{F_2}^2 - \sigma_P^2$$
$$= \sigma_{F_2}^2 - \sigma_E^2$$

Thus n, the number of genes is estimated as

$$n = \frac{(\overline{P}_1 - \overline{P}_2)^2}{8(\sigma_{F_2}^2 - \sigma_{F_1}^2)} \qquad \text{(Castle, 1921)}$$

or

$$n = \frac{(\overline{P}_1 - \overline{P}_2)^2}{8(\sigma_{F_2}^2 - \sigma_P^2)} \text{ or } \frac{(\overline{P}_1 - \overline{P}_2)^2}{8(\sigma_{F_2}^2 - \sigma_E^2)}$$

or

$$\frac{(\overline{P}_1 - \overline{P}_2)^2}{8\sigma_{F_2}^2 - (1/2\,\sigma_{F_1}^2 + 1/4\,\sigma_{P_1}^2 + \frac{1}{4}\sigma_{P_2}^2)}$$

(Wright, 1934, 1968)

where $\sigma_{F_2}^2$, $\sigma_{F_1}^2$ and σ_P^2 are the variances of F_2, F_1 and parental generation, respectively and σ_E^2 is the environmental variance. This later formula was preferred by Wright because if the environmental variance in a population is related to its heterozygosity, weighing the parental and F_1 populations according to the frequencies of their genotypes in the F_2 will partly account for the effect.

Park further studied the bias in the estimate of number of genes resulting from various factors such as random drift, variation in gene effects, linkage disequilibrium, variation of dominance, sampling variation and epistasis. Bias due to random drift decreases the value of number of genes and is reduced by increase in population size, heritability or number of generation of selection. Bias due to dominance, would be quite small given any degree of positive heterosis or a small amount of negative heterosis. Unequal gene effect is a source of downward bias and the amount of bias is related to the coefficient of variation among μ's (Comstock, 1969). Sampling variation results in upward estimate but it can be reduced by longer generation (c) in a one way selection which will decrease the variances of $\overline{Y}_C$ and $(\sigma_{gb}^2 - \sigma_{gc}^2)$. Bias from linkage disequilibrium can be decreased by long generation of selection and finally, bias in estimation of gene number from a population of finite size will be reduced when the number of generations and the selection differential in the one way selection experiment are raised. Gene number estimates are considerably more reliable when heritability is high. The effect of heritability on sampling variance is particularly important when the gene number is high. If the heritability is low and K is very large, many generations of selection are required to obtain a reasonably precise estimate of the number of genes. Generally, the most effective way of decreasing the variance of a gene number estimate will be (i) to increase the number of generations in a selection experiment aiming at the improving the mean phenotypic value, (ii) to increase the number of generations in a two way selection experiment and (iii) to increase the population size. Lande (1981) showed that the original method, 'method of moments' of Wright (1934) for estimation of number of genes can still be applied to crosses between genetically heterogeneous (or wild) populations. The requirement of inbred lines which is sometimes violated in practice may produce some unwanted complications of inbreeding depression in the mean and developmental stability of line. The use of heterogeneous populations minimises the extent of inbreeding depression and reduces the time necessary to perform the experimental crosses. However, this method is expected to be of little value when the range of genetic variation in F_2 population exceeds the mean differences between the parental populations. The effective number of genes estimated by this method cannot exceed the number of chromosomal segments segregating independently in one generation, i.e. the recombination index of Darlington (1937) which equals the haploid number of recombination events per gamete. In most higher plants and animals, the number of recombination is limited to one or a few chromosomes per generation so that the recombination index is usually on the order of times the haploid number of chromosomes. The number of effective factors is also related to an upper bound on the magnitude of factor of largest effects.

Standard error of estimate of effective factors:
Variance of estimate of number of genes can be obtained as

$$\text{Var}(K) \underline{\Omega} K^2 \left(\frac{4(\sigma_{P_1}^2/NP_1 + \sigma_{P_2}^2/NP_2)}{(\overline{P}_1 - \overline{P}_2)^2} + \frac{\text{Var } \sigma_s^2}{\sigma_s^4} \right)$$

Where N is the sample size; var (σ_s^2) the sampling variance and σ_s^2 is genetic variance of F_2 population. The sampling variance of the estimate of variance is twice the square of variance divided by the number of degree of freedom from the estimates. The estimate of K will be reasonably accurate if the sample sizes of the parental and F_1 populations are at least 20 or 30 and those of F_2 and backcross populations are around 100 or more.

16.7 Genotype Assay

In the genotype assay of F_n generation derived from continuous selfing of a cross between two pure lines, the proportion of individuals that are heterozygous at one or more loci is determined by a progeny test using their F_{n+2} grand progeny families. The observed proportion of heterozygous F_n individuals is then equated to a theoretical expectation which is a function of the number of genes involved if they are all unlinked or, the number of effective factors. Genotype assay is based on the assumptions that there is no differential viability between genotypes and no selection.

16.7.1 Theory

The frequency of heterozygotes at anyone locus in the nth generation of selfing (F_n) following a cross between two pure breeding lines is $1/2 n - 1$. For k loci the probability (P_{Het}) of individuals in this generation being heterozygous at at least one locus becomes

$$P_{\text{Het}} = \left[1 - \frac{(2^{n-1} - 1)^k}{2n - 1} \right]$$

Heterozygotes can be detected only by segregation within their progenies and this segregation can be detected only by their differences in mean and variance of families derived from the segregants that is by 'Genotype assay'. Not all pairs of progeny chosen at random from the progeny of a selfed heterozygote will have different genotypes. At best, therefore, we shall detect only a portion of the heterozygote present. The probability of detecting a heterozygote will depend on the number of loci at which it is heterozygous. The probability that a heterozygote in the nth generation will be heterozygous at r loci where r can take all values from 1 to k, has the expectation.

$$P_{\text{Het}r} = \frac{1}{(2^{n-1})^k} \frac{k!}{r!(k-r)!} (2^{n-1} - 1)^{k-r}$$

or

$$\frac{1}{(2^{n-1})^k} \, ^kC_r \, (2^{n-1} - 1)^{k-r}$$

$$\sum_{r=1}^{k} P_{\text{Het}r} = P_{\text{Het}}$$

The $P_{\text{Het}\,r}$ thus depends on n (the generation) and k_1, (the number of loci) only. The probability that a pair of individuals chosen randomly from the progeny of a selfed heterozygote will differ, is related to the number of loci r at which it is heterozygous as

$$1 - (3/8)^r$$

Thus the frequency of heterozygotes in the nth generation that is detectable by progeny testing two random progenies of each individual in that F_n generation is

$$P_{\max} = \frac{1}{(2^{n-1})^k} \sum_{r=0}^{k} {}^kC_r (2^{n-1}-1)^{k-r} \left(1 - \left(\frac{3}{8}\right)^r\right)$$

This $P_{\max}$ is a maximal estimate because it assumes that all genotypic differences are capable of detection as phenotypic differences. But then genotypic differences may fail to be expressed as phenotypic differences because of internal balance and relational balance (Mather, 1943, 1973). Internal balance will maximize the number of genotypes having identical phenotypes when the additive effects at all loci are equal i.e. $da = db = dc \ldots = dk$. And relational balance will have its maximum effects when there is complete dominance at every locus, i.e. $ha = hb = hc \ldots = hk$. Thus under these two conditions the frequency of heterozygote in the nth generation that will be detected by the above procedure becomes

$$P_{\min} = \left(\frac{1}{2^{n-1}}\right)^k \sum_{r=0}^{k} {}^kC_r (2^{n-1}-1)^{k-r} \left(1 - \left(\frac{1}{16}\right)^k\right)$$

$$\sum_{s=0}^{r} (3^s \, {}^rC_s)^2$$

$$= 1 - \left(1 - \frac{1}{2^{n-1}}\right)^k \sum_{r=0}^{k} {}^kC_r \frac{\sum_{s=0}^{r} 9^s ({}^rC_s)^2}{(2^{n+3} - 16)^r}$$

This is the minimal estimate and thus, in practice the true value of frequency of heterozygotes detectable in that generation must lie between these two limits. Solving these equations for $P_{\max}$ and $P_{\min}$ for different values of k will give the expected proportions of heterozygous individuals. The relationships between the number of genes, k and the proportion of pairs of families in the F_{n+2} generation that are expected to differ because of heterozygosity of their F_n ($= F_5$) grand-parent is given in Figure 16.1. The relationships are given for two limiting sets of assumptions which maximize ($P_{\max}$) and minimize ($P_{\min}$) the proportion, respectively.

In the actual experiment, a number of individuals are randomly selected in the F_n generation and selfed. Two randomly chosen progenies from each individual are raised which constitute F_{n+1} generation and selfed again. Thus two families from each individual selected in the F_n generation are raised which constitute the F_{n+2} generation. A number of quantitative traits are recorded and means and variances for pairs of families are compared. The frequency of heterozygote in F_n generation is calculated as the ratio of the number of pairs of families showing significant differences to the total number of pairs of families studied and thus from the graph the member of effective factor is known.

Theoretically the two families could be different because of either differences in means or differences in variances between them. In practice, however, this problem does not arise because we rarely, if ever, detect a difference between families variances without also detecting a difference in the family means because of the greater sensitivity of the later. Further, the number of effective factors increases steadily over successive generations of selfing following an initial cross. This is because of successive rounds of recombination that occur during production of F_2, F_3, F_4 and F_5 gametes. In other words, the linkage disequilibrium initially generated in a cross between a pair of lines is subsequently resolved during successive rounds of recombination. Finally, there

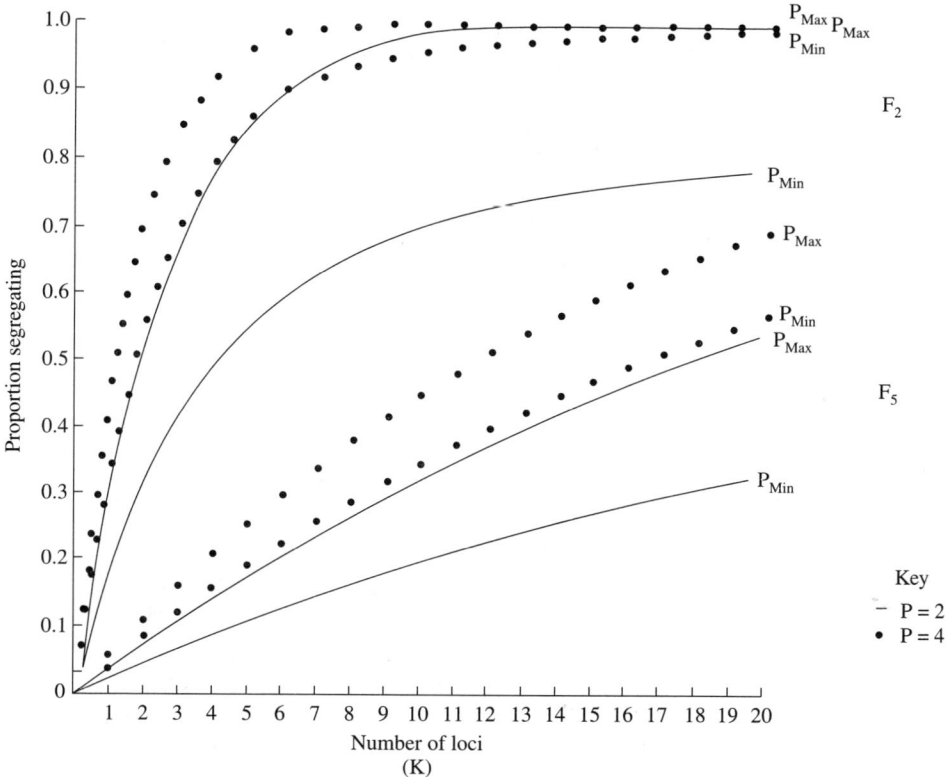

Fig. 16.1 The relationship between the proportion of detectable segregation and k, the number of loci for the maximum and minimum curves and for the two intermediate for $p = 2$ and $p = 4$, the of F_{n+1} individuals from a selfed individual in generation n using F_2 and F_5 generations.

should be an optimally designed experiment in order to obtain a precise estimate of number of effective factors. The reliability with which we can estimate the frequency of heterozygotes depends on m, the number of individuals to be selfed in any generation but the sensitivity with which we can detect a heterozygote will depend on the magnitude of error variances which in turn depends on (i) the family size, (ii) the unit of randomization (single plant or plot), i.e. the experimental design used for measuring error variation and (iii) variation within families arising from genetic segregation and uncontrolled residual variation in the environment. Hill and Avery (1978) showed how P values (P_{max} or P_{min}) change as a result of linkage and initial coupling or repulsion. They showed that unless all the genes are on different chromosomes or many generations of inbreeding are used, the estimate of gene number is biased downwards.

Although there is no definite information on the number of genes determining yield but the general consensus is that the number is large. Student (1934) reported that at least 20 to 40 loci and possibly 200 to 400 loci are conditioning oil percentage in corn.

16.8 Use of Dihaploids in the Estimation of Number of genes

The following methods can be used for estimation of number of genes in case of dihaploids.

1. Use of estimates of the range and genetical variance of an F_1, F_2-derived *DH* populations.

2. Use of the variances of an F_2-derived *DH* populations.

3. Genotype assay.

16.8.1 Use of range and genetical variance of an F_1, F_2 derived DH populations

The difference between the two extreme lines in the DH population is $2[d] = 2\,kd$ where k is the number of segregating loci and d is the additive genetic effect of an individual locus. The genetical variance of dihaploid population comprises only the additive genetic variance D in the absence of additive x additive epistasis and equals kd^2.

The number of genes, k is thus estimated as

$$\frac{(d^2)}{D} = \frac{(kd)^2}{kd^2} = k$$

The factors affecting the estimate of k are (i) epistasis (ii) linkage (iii) inequality of gene effects (iv) the generation from which DH lines are derived and (v) finite population size. Epistasis will not affect the numerator, 2 [d] but the additive genetic variance will contain additive × additive interaction and thus the estimate of D is inflated and so the estimate of k will be underestimated. Linkage will reduce the number of genes and the extent of bias will depend on the degree of linkage disequilibrium in the population. With linkage the number is underestimated if DH lines are produced from a cross of two preponderantly associated parents and is over-estimated if DH lines are produced from a cross of two preponderably dispersed parents. The bias due to the linkage disequilibrium can be reduced by developing the DH population from the F_2-rather than F_1. As the probability of obtaining the most extreme DH lines increases in an F_2 derived population, one can obtain a better estimate of the number of segregating loci using DH from F_2 rather than from F_1 plants. Mather and Jinks (1971) and Comstock (1969) showed that if d varies from locus to locus then the number of genes would be seriously underestimated due to an increase of additive genetic variance by a quantity V which is the squared coefficient of variation of d divided by 100. There seems to be no practical solution to eliminate or reduce the amount of bias resulting from inequality of gene effects. But Comstock (1969) indicated that if one believes that the distribution of d is likely to be fairly represented by a single tailed normal distribution, one might accept the biased estimate as a guide to the number of relatively important loci. A single tailed normal distribution would imply many loci with small effects and at the other extreme, a small number with effects many times as large.

The probability of obtaining the two most extreme lines in a finite sample decreases rapidly as the number of genes determining a character increases. The probability of having at least one line homozygous for all desirable or undesirable alleles in the sample following Mather (1963) is

$$1 - \left[\frac{2^n - 2}{2^n}\right]^{n'}$$

where n' is the sample size and n is the number of segregating loci. From the above equation it can be shown that at least the probability of obtaining at least one homozygous lines for all six ($n = 6$) desirable or undesirable alleles in a sample of 98 DH lines ($n' = 98$) is 0.96. But the results have shown that even with a population of 100 DH lines there is no surety that the best recombinant is included in the population where six or more genes are segregating (Choo and Reinbergs, 1982). Thus bias is introduced by the use of finite sample size which may not contain extreme genotypes, Therefore, the estimation method using the difference between the highest and the lowest lines is less accurate.

16.8.2 Use of variances of F_2-derived population

The estimate of the number of genes can also be obtained from the ratio of the square of the true variance between means of families derived from F_2 individuals $\left(= \frac{1}{2}D = \frac{1}{2}kd^2\right)$ and the variance of the variances between DHs within F_2-derived families $\left(\frac{1}{4}D^2 = \frac{1}{4}kd^4\right)$

$$\frac{\left(\frac{1}{2}kd^2\right)^2}{\frac{1}{4}kd^4} = k$$

Again, this estimate of k is equivalent to the k_2 estimate of Mather and Jinks (1982). The large error associated with the variances and variance of variances rather than means decreases the precision of the estimate of k.

16.8.3 Genotype assay

Snape et al. (1984) applied the method of Jinks and Towey (1976) for estimating the number of segregating genes using F_2-derived DH population and $p = 2$, i.e. two DH lines are developed from each F_2 individual. The $P_{max.}$ and $P_{min.}$ are obtained as follows:

$$P_{max} = \frac{1}{2} k \sum_{r=0}^{k} k_{C_r} \left[1 - \left(\frac{1}{2}\right)^r\right] = \left[1 - \left(\frac{3}{4}\right)^k\right]$$

where $1 - (1/2)^r$ is the probability that two DH lines developed from an F_2 individual which is heterozygous at r loci, are genetically different.

$$P_{min} = \frac{1}{2} k \sum_{r=0}^{k} k_{C_r} \left[1 - \sum_{r=0}^{k} \left(r_{C_m} \left(\frac{1}{2}\right)^r\right)^2\right]$$

They showed that the only cause of underestimation of the number of factors is the sensitivity of the experiment in detecting genetical differences between pairs of means. Greater sensitivity can be obtained by developing more than 2 DH lines from each F_2 individual. They concluded that genotype assay method is the best method for distinguishing between few and many genes.

16.9 Estimation of the Number of Effective Factors in Haploids

An estimate of the number of effective factors K segregating in haploid cross can be obtained as the ratio of the square of the difference between the extreme F_1 progeny individuals $(P_H - P_L)$ and the genetic variance σ_G^2 as

$$K = \frac{(P_H - P_L)^2}{4\sigma_G^2}$$

16.10 Method for Locating Genes

Various methods are available for locating genes: genetic marker (multiple genetic marker stocks), deficiencies, chromosomal interchanges, chromosomal inversions and trisomics. Isolation of individual factors is facilitated by chromosome assay techniques whereby the net effect of all the genes within a chromosome is assayed. The various banding methods such as Feulgen staining, Q banding, R banding, G banding and C banding of which the two most generally used are Acid-Saline-Giema (ASG) technique which reveals G (Giemsa stain) bands and Quinacrine mustard technique which produces fluorescent Q bands, have been developed for characterising individual segment of the plant's chromosomes. Nowadays molecular markers are being used for locating genes and counting the number of gene which are discussed in Chapter 21. Finally, once we have isolated the individual factors, our objective would be to know the relative magnitude of their effects.

References

Breese, E.L. and Mather, K. 1957. The organisation of polygenic activity within a chromosome in *Drosophila*, 1. Hair characters Heredity. **11**: 397–403.

Castle, W.E. 1921. An improved method of estimating the number of genetic factors concerned in case of blending inheritance. Science, **54**: 223.

Comstock, R.E. 1969. Number of genes affecting growth in mice. Genetic lectures (Oregon State University), **1**: 137–148.

Comstock, R.E. and Robinson, H.F. 1948. The components of genetic variance in populations of biparental progenies and their use in estimating the average degree of dominance. Biometrics, **4**: 254–266.

Darlington, C.D. 1937. The biology of crossing over. Nature, **140**: 759–761.

Dobzhansky, Th. 1936. Studies on hybrid sterility 3. Localization of sterility factors in *Drospophila pseudoobscura* hybrids. Genetics, **21**: 113–135.

Falconer, D.S. 1989. Introduction to Quantitative Genetics. Longman, Burnt Hill.

Gate, C.E. 1963. Discussion: Some considerations in variance components and partitioning methods of genetic analysis. In: Statistical Genetics and Plant Breeding, NAS-NRC, pp. 982.

Hill, W.G. and Avery, PJ. 1978. On estimating number of genes by genotype assay. Heredity, **40**: 397–403.

Jinks, J.L. and Towey, P. 1976. Estimating the number of genes in a polygenic system by genotype assay. Heredity, **37**: 69–81.

Kearsey, M.J. and Pooni, H.S. 1996. The Genetical Analysis of Quantitative traits. Chapman and Hall, London.

Lande, R. 1981. The minimum number of genes contributing to quantitative variation between and within populations. Genetics, **99**: 541–553.

Law, C.N. 1967. The location of genetic factors controlling as a number of quantitative characters in wheat. Genetics, **56**: 445–461.

Mather, K. 1943. Polygenic inheritance and natural selection. Biol. Revs. **18**: 32–64.

Mather, K. 1949. Biometrical Genetics. 1st edition. Methuen, London.

Mather, K. 1973. Genetical structure of populations. Chapman and Hall, London.

Mather, K. 1979. Historial Review: Quantitative variation and polygenic system. In: Quantitative Genetic Variation. Academic Press, New York.

Mather, and Harrison, B.J. 1949. The manifold effects of selection. Heredity, **3**: I–52.

Mather, K. and Jinks, J.L. 1982. Biometrical Genetics. 3rd edn. Chapman and Hall, London.

Panse, V.G. 1940. The application of genetics to plant breeding. 2. The inheritance of quantitative characters and plant breeding. J. Genet., **40**: 283–302.

Park, Y.C. 1977. Theory for the number of genes affecting quantitative characters. I. Estimation of and variance of the estimation of gene number for quantitative traits controlling additive genes having equal effect. Theor. Appl. Genet., **50**: 153–161.

Park, Y.C. 1977. Theory for the number of genes affecting quantitative characters. 2. Biases from drift, dominance, inequality of gene effects, linkage disequilibrium and epistasis. Theor. Appl, Genet., **50**: 163–172.

Powers, LeRoy. 1934. The nature and interaction of genes differentiating habit of growth in a cross between varieties of *Triticum vulgare*. J. Agr. Res., **49**: 573–605.

Powers, LeRoy 1963. The partitioning method of genetic analysis and some aspects of its application to plant breeding. In: Statistical Genetics and Plant Breeding, NAS–NRC.

Serebrovsky, A.S. 1928. An analysis of the inheritance of quantitative transgressive characters. Z.I.A.V., **48**: 229–243.

Smith, H.H. 1937. The relation between genes affecting size and color in certain species of Nicotiana. Genetics, **22**: 361–375.

Snape, J.W. and Simpson, E. 1986. The utilization of doubled haploid lines in quantitative genetics. Bull. Soc. bot. Fr., 133., Actualites bot., 4: 59–66.

Snape, J.W., Wright,, A.J. and Simpson, E. 1984. Methods for estimating gene numbers for quantitative chracters using doubled haploid lines. Theor. Appl. Genet., **67**: 143–148.

Student. 1934. A Calculation of the minimum number of genes in Winters, selection experiment. Ann. Eugenics. **6**: 77–82.

Thoday, J.M. 1961. Location of polygenes. Nature, **191**: 368–370.

Thompson, J.N., Jr. 1975. Quantitative variation and gene number. Nature, **258**: 665–668.

Towey, P. and links, J.L. 1977. Alternate ways of estimating the number of genes in a polygenic system by genotype assay. Heredity, **39**: 399–410.

Wehrhahn, C. and Allard, R.W. 1965. The detection and measurement of the effects of individual genes involved in the inheritance of a quantitative character in wheat. Genetics, **51**: 109–111.

Wright, S. 1934. The result of the crosses between inbred strains of guinea pigs differing in the number of digits. Genetics, **19**: 537–561.

Wright, S. 1968. Evolution and the Genetics of Populations. Vol. 1. Genetic and Biometric foundations. The University of Chicago Press, Chicago and London.

17

Analysis of Skewness and Kurtosis

There are four cumulants which assist in the description of a distribution. K_1 (the mean) specifies where the distribution is centered; K_2 (the variance) describes the degree of concentration of a distribution about mean; K_3 (the skewness) describes the degree of departure of a distribution from symmetry and K_4 (the Kurtosis) characterizes the peakedness of a distribution. By definition, the first cumulants K_1 is the expected value (or the mean) of a random variable and the three other cumulants are the second, third and the fourth moments about the mean (Yule and Kendall, 1950). For a normal distribution $K_1 = \mu$, $K_2 = \sigma^2$ and $K_3 = 0$ and $K_4 = 0$. The formulas for estimating mean, variance, skewness and kurtosis are given in chapter 3. As we have seen nearly all of our current quantitative genetic theories are based on first and second degree statistics. Fisher et al. (1932) called to our attention the usefulness of third degree statistics in the study of quantitative traits.

17.1 Genetical Causes of Skewness and Kurtosis

Skewness could result when (i) certain combinations of genes are lethal and hence one end of the curve is nonexistent or reduced expression of gene action at one end of the range of character due to physiological restriction or physiological barrier to gene expression on the upper side of the range, (ii) there is incomplete linkage of certain genes controlling the trait, (iii) there is epistasis, (iv) there is nonadditive genetic effects (dominance or over-dominance), (v) there is genotype × environment interaction, (vi) genes are showing percentage or multiplicative effects, (vii) one gene has a very much larger effect than others, (viii) competition and (ix) environmental factors such as stress and fertility may induce skewness.

Kurtosis will occur if either a few genes are contributing to the phenotypic distribution or there are inequalities in the additive genetic effects at different loci. Traits for which data show leptokurtic distribution are usually those under control of relatively few segregating genes whereas data showing a platykurtic distribution usually represent characters that are controlled by many genes. Strong leptokurtosis is theoretically possible where one alternative of each component is rare but extreme platykurtosis may arise when the effects of binomial components are in geometric progression but it requires that the leading one determining something like one half or more of the total variance.

17.2 Information on Gene Action and Interaction

Fisher, Immer and Tedin (1932) showed that, for one locus 2 alleles system, the skewness of the frequency distribution of F_2 plants is equal to $-\frac{3}{4}hd^2$, where d is the additive genetic effect and h, the dominance effect. Assuming no epistasis and linkage and extending the model to a polygenic system, the skewness of F_2 population is simply the sum of $-\frac{3}{4}hd^2$ terms of each locus, $-\frac{3}{4}\Sigma hd^2$. Thus by studying the skewness of the frequency distribution of F_2 generation plants, one can determine whether the average h is positive or negative. When the average h is positive it will make the frequency distribution skewed to the left indicating that the increasing alleles are dominant whereas negative average h produces rightward skewness with the decreasing alleles being dominant. Some statistics of the third degree alongwith their expectations in terms of d and h are given in Table 17.1.

Table 17.1 Statistics of the third degree

Statistics	Expectation
K_3 of F_2	$-(3/4)\,hd^2$
Mean K_3 of F_3 progenies	$-(3/8)\,hd^2$
Covariance of mean and variance of F_3	$h/32\,(2d^2 + h^2)$
Covariance of F_2 parental value and variance of F_3	$(3/8)\,hd^2$
K_3 of means of F_3 families	$-(3/16)\,hd^2$
Mean K_3 of biparental progenies	$-(1/8)hd^2$
Covariance of parental values and variances of the biparental progenies	$-h/32\,(2d^2 - h^2)$
Covariance of biparental progenies and biparental product	$-(1/8)\,d^2$
K_3 of means of biparental progenies	$-(3/16)hd^2$
Mean K_3 of maternal progenies	$-(3/8)hd^2$

From the comparison of the mean value of K_3 of F_3 progenies with the covariance of K_1 and K_2 of F_3, it is possible to distinguish between the theories that whether the heterosis is due to complete or incomplete dominance or due to true over-dominance. Four times this covariance will be greater than equal to or less than mean value of K_3 (with sign reversed) according as h (supposed positive) is greater, equal to or less than $+d$ but with true overdominance the four times covariance should have a positive value exceeding the negative average value of K_3 within F_3 progenies and in case of strong positive bias of dominance it should at most be equal to this value. Thus if the covariance is less than this critical value it shows dominance either incomplete or complete. This test is applicable particularly where it is difficult to produce sufficient quantity of F_1 seeds and thus the same information cannot be obtained from the comparison of F_1 and F_2 means.

The K_4 statistic can be applied to genetic variance component analysis, (I) in the unbiased estimation of variance of estimates of variance components and (II) in the estimation of the number of genes or factors controlling the inheritance of a quantitative character (Robson, 1956).

17.3 Estimation of Coefficients of Skewness and Kurtosis in Population of Pure Breeding Lines

Pooni et al. (1977) worked out the expected value of coefficient of skewness, g_1, and coefficient of kurtosis, g_2 in the presence of digenic interaction for the population of pure breeding lines derived by single seed descent from the F_2 of a cross between two pure breeding lines as

$$g_1 = [6\,\Sigma\,d_j dk_{ijk} + 6\,\Sigma\,i_{js}iksi_{jk}]/[\Sigma\,d_j^2 + \Sigma\,i_{jk}^2]$$

and

$$g_2 = \Sigma\,d_j^4 + \Sigma\,i_{jk}^4 + 6\,\Sigma\,d_j^2\,d_k^2 + 6\,\Sigma\,d_j^2\,\Sigma\,i_{jk}^2$$
$$+ 6\,\Sigma\,i_{jk}^2\,i_{st}^2 + 24\,\Sigma\,d_j d_k\,(\Sigma\,i_{js}i_{ks} + \Sigma\,i_{js}i_{sk}$$
$$+ \Sigma\,i_{sj}isk + \Sigma\,is_j i_{ks}) + 24\,\Sigma\,i_{js}k i_{js}\,i_{js}i_{kt}\,i_{st}$$
$$+ 24\,\Sigma\,i_{jk}i_{ks}i_{jt}l_{st} + 24\,\Sigma\,i_{js}i_{jt}i_{ks}i_{st}$$
$$+ 24\,\Sigma\,i_{js}i_{jt}\,i_{ks}\,i_{kt}\;\text{to}\;\{\Sigma\,d_j^2 + i_{jk}^2\},\;\text{minus 3}$$

where d_j and d_k represent the additive effects of alleles at ith and kth locus and i_{jk}, i_{js}, i_{jt}, i_{ks}, i_{kt} and i_{st} are respectively the additive × additive interaction effects of jth and kth, jth and sth, jth and tth, kth and sth and tth loci.

Assuming $d_j = d$ over all the k loci and $i_{jk} = i$ for everyone of $1/2k\,(k-1)$ pairs of loci, the expectations of g_1 and g_2 reduce to

$$g_1 = \{3k(k-1)\,d_i^2 + k\,(k-1)\,(k-2)i^3\}/$$
$$\{kd^2 + 1/2\,k(k-1)\,i^2\}3/2$$

and

$$g_2 = [(\{k(1+3)\,(k-1)\}d^4 + \{1/2k(k-1)$$
$$(1 + 3(1/2k(k-1) - 1 + 6\,(k-2)\,(k-3))]i^4$$
$$+ \{3k\,(k-1)(5\,k-8)\}d^2i^2)/\{kd^2$$
$$+ 1/2k(k-1)\}i^2\}^2] - 3$$

Thus it is shown clearly that the magnitude and the direction of g_1 and g_2 directly reflects the magnitude and the direction of the epistatic component (i). Figures 17.1 (a–c) show the simulated distribution each for a random sample size of 500 pure breeding lines differing at $k = 10$ loci and $h_n^2 = 100$. Skewness at is always positive for complementary epistasis and negative for duplicate epistasis (Snape and Riggs, 1975).

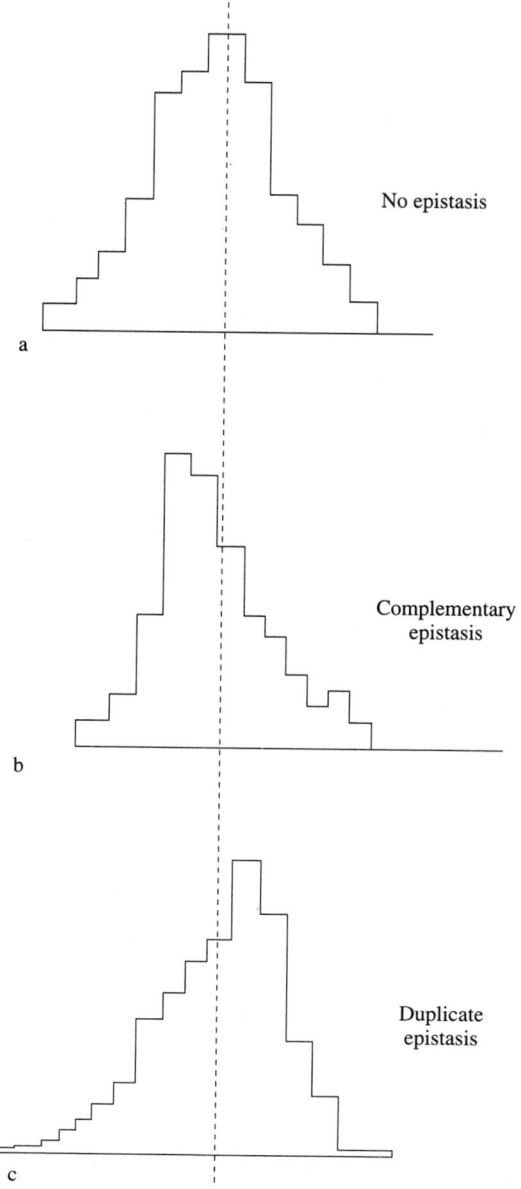

Fig. 17.1 (a-c). Distributions of sample of Inbreds lines.

17.4 Effects of Random Environmental Variation and Genotype × Environment Interaction

Theoretical estimates of g_1 and g_2 for various levels of epistasis and heritability (h_n^2) in the simulated populations show that there is a direct and predictable relationship between the occurrence, magnitude and direction of skewness and the presence, magnitude and sign of epistasis. Kurtosis is also related to epistasis although in a complicated way. Both skewness and kurtosis decrease with the decrease of h_n^2, i.e. with the increase of random environmental variation. Kurtosis, however, is more sensitive to increase in random environmental variation irrespective of the level and direction of epistasis and for moderate level of epistasis (0.25) it reaches its smallest value at intermediate level of heritability. Below heritability value less than 0.5 ($h^2 < 0.5$), the magnitude of epistasis is so low that it is unlikely that, in practice, it will lead to significant epistasis.

In case of genotype × micro environmental interaction, the environmental variation instead of reducing the non-normality due to epistasis may produce itself non-normality. These two causes of non-normality may then either reinforce or cancel out each other's contribution to non-normality depending upon whether they have the same or opposing signs. The direction of $G × E$ interaction will be determined by the sign of the covariance between the mean and the variance of the pure breeding lines and has the expectation $2 \Sigma\ e_j d_i g_{ij}$ where e_j is the additive environmental effect and g_{ij} is the $g × e$ interaction. The sign of the covariance is determined by the sign of e_j and g_{ij} as d_i is always positive. If both have same sign whether positive or negative, the covariance will be positive and hence the skewness will be positive. If, on the other hand, they have different sign, the skewness will be negative. When the average variance is associated with average mean and below average variance with high and low extreme means equally or vice-versa, the covariance will be zero and will produce no skewness even though $G × E$ interaction is present. The significance of this $g × e$ interaction covariance is tested as the correlation coefficient. In the early stage of inbreeding, information on the same can be obtained from the covariance of means and covariance among the P_1, P_2 and F_1 families (Perkins and Jinks, 1970) or amongst the families of an F_2 TTC (Perkins and Jinks, 1971). Also, the difference between the variances between the two parents will provide information on the sign and the magnitude of the $g \times e$ interaction covariance as

$$1/2(VP_1 - VP_2) = 2 \Sigma\, e_j d_i g_{ij}$$

17.5 Estimation of Skewness and Kurtosis in Dihaploid Population

K_3 and K_4 statistics can be used for detecting gene interaction and identifying the nature of gene interaction in dihaploid population. Skewness and Kurtosis of a dihaploid population are useful in detecting complementary and duplicate type of gene action. They are even move powerful than the ANOVA technique in revealing multiplicative type of gene action (Choo and Reinberg, 1982). Most of the genetic variance is additive. The amount of additive × additive genetic variance is so small that the ANOVA technique seemingly cannot detect its existence. $K_3 = 0$ in the absence of epistasis ($i^2 = 0$) and $K_3 \gtreqless 0$ in the presence of complementary and duplicate interactions, respectively. The 4th degree statistic is always negative or near zero in the absence of epistasis and is positive only in the presence of epistasis.

The expected values of K_3 and K_4 in DH population are

$$K_3 = 3k(k-1)id^2 + k(k-1)(k-2)i^3$$

and
$$K_4 = [k + 3k(k-1)]\, d^4$$
$$+ 3k(k-1)(5k-8)\, i^2 d^2$$
$$+ \frac{k(k-1)}{2}\left\{1 + 3\left[\frac{k(k-1)}{2} - 1\right]\right.$$
$$\left. + 6(k-2)(k-3)\right\} i^4 - 3k_2^2$$

The expected value of K_4 is slightly different from that given by Pooni et al. (1977). The later contained three redundant summation and a summation term that equals zero.

17.6 Effect of Skewness on Selection

Selection for high values of a variable when applied to a symmetrical population generally shifts the value of K_3 in the negative direction whereas selection for low values should shift it in the positive direction and the amount of these changes will depend on the number of factors present. The rate at which the skewness is modified for a given change in the mean is evidently greater other things being equal, the smaller the number of factors to the segregation of which the variance of F_2 is to be ascribed. The formula for calculating the response due to selection is based on the assumption that value of a character (P) under consideration is the sum of the two independent normal variates namely genetic value (G) and environmental value (E). If their distributions are not normal, they will affect the prediction of response and thus both have to be taken into account while predicting gain from selection. Cochran (1951) and Finny (1962) pointed out that the genotypic distribution was positively skewed after one generation of upward selection although initially the genotypic distribution was normal and phenotypic distribution can be predicted using environmental estimates obtained from genetically uniform population. The distortion of normality will be greatest if heritabilities are higher and the proportion to be selected is small but the recombination and the changed environmental variation will perhaps tend towards the partial restoration of the normality. Ronningen (1969) used Monte-Carlo technique to show in animal breeding studies that the amount of skewness of genotypic distribution in the next generation increases with heritability and decreases with the fraction selected. Nishida and Abe (1974) observed that the amount of gain from upward selection is positively correlated with the quantity (skewness (G)-skewness (E)). Now if the expected gain is correlated assuming G and E being normally distributed, but in reality G is positively skewed and E is negatively skewed then the realized gain will be higher than the expected gain while the negative skewness of G and positive skewness of E will result in low gain than expected. The observed gain from selection is often less than the expected gain because of several possible genetic causes (Falconer, 1960). Skewness of G and E can cause asymmetry of response to selection (Kelker and Kelker, 1986). They noted that in the absence of intergenotypic competition the distribution of environmental value is not affected by selection. However, the distribution of the genotypic values becomes more positively skewed with the subsequent cycles of upward selection. If the distribution of the genotypic value is normal but the distribution of environmental value is positively skewed then upward selection will give a lower gain than that expected if both distributions are normal. On the other hand, if the distribution of genotypic values is negatively skewed then gain from selection will be larger than expected gain. The result will be reversed with downwards selection. Thus there is asymmetry of response if both, upwards and downwards, selection are considered. But after one round of selection the distribution of genotypic value becomes positively skewed. The phenotypic distribution will be more positively skewed and thus the gain from positively skewed genotypic value will cancel the loss from positively skewed environmental values and hence the gain from selection will be close to the expected gain. Thus a constant gain over several generations has been frequently observed. Where expected responses to selection are obtained, the skewness of the data should be checked.

Skewness is not likely to be important for a characteristic like net merit or index as in case of index selection. When the population shows negative skewness, the gain is faster with mild selection and less rapid with very intense selection. The reverse is true if the population shows positive skewness. The type and extent of skewness found in actual populations seems unlikely to affect seriously the relative efficiency of the three methods of selection.

References

Choo. T.M. and Reinberg, E. 1982. Analyses of skewness and kurtosis for detecting gene interactions in doubled haploid populations. Crop Sci., **22**: 231–235.

Fisher, R.A., Immer, F. R. and Tedin, O. 1932. The genetical interpretation of the statistics of the third degree in the study of quantitative inheritance. Genetics, **17**: 107–124.

Kelker, D. and Kelker, H. 1986. The effect of skewness on selection in a plant breeding program. Euphytica, **35**: 303–309.

Nishida, A. and Abe, T. 1974. The distributions of genetic and environmental effects and the linearity of heritability. Can. J. Genet. Cytol., **16**: 3–10.

Perkins, J.M. and Jinks, J.L. 1970. Detection and estimation of genotype-environmental, linkage and epistatic components of variation for a metrical trait. Heredity, **25**: 157–177.

Perkins. J.M. and Jinks, J.L. 1971. Analysis of genotype × environment interaction in triple test cross data Heredity, **26**: 203–209.

Pooni, H.S., Jinks, J.L. and Cornish, M.A. 1977. The causes and consequences of non-normality in predicting the properties of recombinant inbred lines. Heredity, **38**: 329–338.

Snape, J.W. and Riggs, T.J. 1975. Genetical consequences of single seed descent in the breeding of self-pollinated crops. Heredity, **35**: 211–219.

Snedecor, G.W. and Cochran, W.G. 1967. Statistical Methods. Oxford and IBH Publishing Co., New Delhi.

Sokal, R. and Rohlf, F, 1969. Biometry. W.H. Freeman and Company, San Francisco.

Robson, D.S. 1956. Applications of the K_4 statistic to genetic variance component analysis. Biometrics, **12**: 433–444.

Ronningen, K. 1976. A method for estimation of appropriate selection intensity from skewed distributions. Acta Agricultural Scandinavica, **26**: 82–86.

Yule, G.U. and Kendall, 1950. An introduction to the theory of statistics. Charles Griffin and Co. Ltd., London.

18

Transformation of Scale

Quantitative characters, as we can see, can be measured on different scales, for example, plant height in centimetre, maturity in number of days, number of tillers or branches in numbers, oil or protein in per cent, yield in grams or harvest index and leaf length to leaf width in ratios. We can also have data of counts of insects in field plot or counts of bacterial colonies in a plate which may follow poisson distribution. Further, resistance traits can be scored on a scale of 0–5 or 0–9. A plant growth factor tends to contribute constant percentage increments rather than constant absolute ones and if such a character is measured in arithmetic scale then it will show a positively skewed distribution. While assuming that a quantitative trait shows normal distribution we assume that there is a scale on which both genetic and environmental factors are approximately acting additively. Thus deviation from anyone or both will result in non-normal distribution. The deviations from the additive effects of genes could be multiplicative gene action, dominance, epistasis genotype × environment interaction and unequal gene effects. For analysis of data the essentiality of normality is a must. That is why we see that the basic assumptions under ANOVA are that (i) the treatments and environmental effects are additive and there is no interaction between genotype and environment and (ii) the experimental errors are random, independently and normally distributed about mean zero and variance σ^2. And if the above conditions are not met, the conventional tests of hypothesis and methods of defining confidence intervals may not be strictly valid. In this situation either we can use non-parameteric procedures of analysing data, ignore the condition or transform the data.

18.1 Tests of Non-normality

Deviation from normality can be tested by obtaining third and fourth degrees statistics whose calculation is a tedious process. A much simpler method will be to plot the running sum of the proportional frequencies upto each class limit against the later. In case of normal distribution

$$pr^{-1}pi = (V_i - M)/\sigma$$

where pri^{-1} pi is the inverse probability integral and Vi is the class limit; M and σ are the mean and standard deviation of the distribution and are given by intercepts with the axis. When pis, the proportion of the total number is plotted against Vis, the points will fall in a straight line. This relationship applies to both continuous and meristic distributions. Any systematic deviation from straight line will be apparent and will indicate the presence of systematic non-additive effects and thus deviation from linearity due to gross heterogeneity can be examined. The term probit (Bliss, 1935) is used for $pri + p + 5$ to avoid the negative values taken by the inverse probability integral of percentage less than 50. This method can be used for testing transformations of scales for their capability to normalize distributions.

$$pri^{-1} p = (f(x) - M)/\sigma$$

The $pri^{-1}p$ can be plotted against the different transformation (function) and the most suitable transformation can be obtained. In this inverse probability transformation, the original scale, $f(x)$ is changed to logarithmic or reciprocal or any other effective scale to give a normal distribution and the proportion of individual $f(y)$ into normal deviates to remove the sigmoid shape of the normal curve. Now if $f(y)$ is plotted againt $f(x)$, a straight line will be obtained. This type of transformation has use in relating mortality to dosage of toxic substance where the transformation of a variate has value in giving a simpler relation with a second variate than dose, the untransformed form.

18.2 Relation of Variability to Mean

When the genes are acting in a multiplicative manner it forms a geometric series such as 2, 4, 8, 16 representing the effects of 1, 2, 3, 4 genes, respectively where an additional gene doubles the phenotypic value of an individual which differs from genes showing additive effects which produce effects in arithmetic series such as 2, 4, 6, 8 in which the additional gene increases the phenotypic value by an account which is constant. In case of multiplicative type of gene action, the variability in high line population containing all increasing alleles will be much higher than population of low line containing all decreasing alleles and thus when the variability differs between high and low line genes with multiplicative effects can be said to be operative.

On the other hand under additive action of genes variability does not differ greatly between the two populations. When high and low lines are crossed, F_1 and F_2 means are nearer to one of the parental means and the F_2 distribution tends to be skewed because the geometric mean of the two numbers is the square root of their product which is unlike when additive gene action is there in which F_1, will fall midway between the two parents and F_2 will show a normal distribution. Deviation from normality due to multiplicative action of genes can be removed by log transformation (Mather, 1949). As a result of log transformation the gene effects become additive and dominance changes to partial dominance. Butler (1941) and Khambanonda (1949) reported genes with multiplicative effects determining fruit size in tomato and red pepper, respectively. In case of normal distribution means and variances are independent but in case of multiplicative gene action, the means, variances or standard deviations are dependent or correlated and increase in mean is proportional to increase in variance or standard deviation. Thus although theoretically the environmental variances among different populations should be the same there results heterogeneity of variances among populations. The homogeneity of such variances can be tested by Bartlett test. The heterogeneity among the variances and among the means of individual factors of a quantitative trait is due to the presence of some genes having large effects resulting in deviation from normality. The relation between mean and variance or standard deviation is determined by a quantity, coefficient of variation ($CV = 100 \ \sigma/M$). The magnitude of CV thus determines the degree of departure from normality. The distribution can be normalized using logarithmic transformation which would stabilize the variance. Now if the CV is less than 20% there is no need for transformation of data but if it is greater than it, then it would be worth while to go for transformation (Falconer, 1960). Plant growth which shows deviation from normality on arithmetic scale can be normalized by using log transformation. Similarly, logarithmic transformation can be used in case of characters such as harvest index, leaf length/ leaf width, etc which are recorded as ratio.

Wright (1968) observed that the near normal distributions can be satisfactorily normalized by some sort of logarithmic transformation, frequently one which corrects for damping by a lower limit, but occasionally by an upper limit or by both. Fisher et al. (1932) and Mather (1949) suggested that gene effects at different loci can be made independent through a change of scale but there are two problems. First, there may be types of interaction that cannot be removed by transformation of scale. These include complementary and duplicate types of epistasis and the type of interaction postulated by Wright (1935) wherein there are two scales, primary and secondary for value of an individual genotype. Out of two characters say seed size and number of seeds per plant, the later constitutes the selective value of an individual. There is additivity of gene effects on the primary scale (of first character) but the maximum value on the second scale of second character constituting the selective (natural) value or net advantage to organism) is associated with an intermediate optimum value on the primary scale, and that resulting from the threshold. For the later situation transformations that shorten the lower end of the scale relative to the upper would be effective. However, Comstock and Robinson (1955) showed that the types of interaction most amenable to the transformation device represent the least extreme deviations from the no interaction model and for many purposes deviations that are not a source of serious disturbances. The second problem is that of finding an optimum scale. Mather (1949) suggested the criterion that regression of performance on percent homozygosity be linear but then materials at different levels of homozygosity differ so much in rate of development that the regression can never be linear, a result that Sentz (1954) obtained in maize. Thus we see that transformation may eliminate or reduce epistasis, genotype × environment interaction and thus simplifies the interpretation. But then no transformation of scale can be expected to make all factors (genetic and environmental) effects strictly additive. As deviation from normality can be due to genetic or environmental reason or both, a scale which can correct non-additive genetic effect may be different than one which can correct non-additive environmental variation. Different populations differ in their genetic constitution and the same character may differ in their genetic control in different populations or within a population different characters are under different genetic control so the scale appropriate for one population may not be appropriate for other population for the same character. Also, within a population different characters may require different scales as the development of different characters differ in paths of development.

Which scale is the best? The answer for this can be provided by the relationship between the mean and the variance. The transformation for which the relation is minimum is likely to be the most appropriate (Steel and Torrie, 1988). Once the data is analysed using a changed scale, all comparisons should be made on the transformed scale. Means of the populations should be transformed back to the original scale as means on transformed scale may not convey any clear meanings but it is not worth-while to transform the variances or standard deviations back to the original scale. In case of linearly related scales such as kilogram, grams, pounds and centimetre, inch, metre, as one scale can be converted into another scale simply by multiplying or dividing by the correct conversion factor, so it makes no difference which scale is used in the analysis of data. The mean and variance (or standard deviation) which are expressed in the units of scale can be converted into another scale as described above but other parameter such as heritability which is expressed as ratio and has got no units will remain the same irrespective of the scale used.

18.3 Scale Effect

If a character shows epistasis on original scale but no epistasis on a transformed scale then it is better to recognise epistasis a part of genetic determination rather than ignoring it and labelling the epistasis as a scale effect.

18.4 Variance Stabilizing Transformation

Assuming that the observed variable Y has mean μ_Y and variance σ_Y^2, we take the transformation $Z = g(Y)$ such that

$$g(\mu_Y) \alpha \int \frac{d\mu_Y}{\sigma_Y}$$

18.5 Poisson Distribution

When Y is the count of number of occurrences of a certain phenomenon following a poisson distribution, the data is transformed by the square root, $\sqrt{Y} = Z$. The variance of Z, $V(Z) = 1/4$. Also, when we have percentage data based on counts and common denominator and where the percentage ranges from either 0 to 20 or from 80 to 100 percent, the above square root transformation can be used. In case of percentage ranging from 80 to 100, each observation is subtracted from 100 before the transformation. When the observations take very small or zero values the appropriate transformation will be $Z = \sqrt{Y + 0.5}$.

18.6 Binomial Distribution

In case data showing binomial distribution the

transformation used is angular or inverse sine, $Z = \sin^{-1} \sqrt{Y}$ with Var $(Z) = \frac{1}{4} n$ where $Y = \frac{r}{n}$ (observed r out of n with a particular attribute). Angular transformation is also used for percentage data when it covers a wide range of values. But if the range of percentage is 30 to 70, no transformation is required.

References

Bartlett, M.S. 1946. The use of transformations. Biometrics, **1**: 39–53.

Bliss, C.I. 1935. The calculation of the dosage-mortality curve. Annals of Applied Biology, **22**: 136.

Bliss, C.I. 1935. The comparison of dosage-mortality data. Annals of Applied Biology, **22**: 307.

Butler, L. 1941. The inheritance of fruit size in the tomato. Can. J. Res. C., **19**: 216–224.

Falconer, D.S. 1989. Introduction to Quantitative Genetics. Longman, Burnt Mill.

Fisher, R.A., Immer, F.R. and Tedin, O. 1932. The genetical interpretation of statistics of the third degree in the study of quantitative inheritance. Genetics, **17**: 107–124.

Homer, T.W., Comstock, R.E. and Robinson, H.F. 1955. Non-allelic gene interactions and interpretation of quantitative genetic data. Tech. Bull., N.C. Agric. Exp. Sta., No. **118**: 1-117 pp

Khambanonda, I. 1950. Quantitative inheritance of fruit size in red pepper (*Capsicum frutense*). Genetics, **35**: 322–343.

Mather, K. and Jinks, J.L. 1982. Biometrical Genetics, 3rd edn, Chapman & Hall, London

Powers, L. 1950. Partitioning method of genetic analysis applied to quantitative characters of tomato crosses. U.S. Deptt. of Agric. Tech., Bull. No. 998.

Sentz, J.C., Robinson, H.F and Comstock, R.E. 1954. Relations between heterozygosis and performance in maize. Agron. J., **46**: 514–520.

Steel, R.G.D. and Torrie, J.H. 1980. Principles and Procedures of Statistics. McGraw Hill & Koqakusha, Ltd.

Wright, S. 1935. The analyses of variance and the correlation between relatives with respect to deviations from an optimum. J. Genetics, **30**: 243–256.

Pederson, D.G. 1986. Effects of logarithmic and site mean transformation on the relative fields from a variety trials. Euphytica, **35**: 169–174.

Kendall, M. and Stuart, A. 1976. The Advanced Theory of Statistics. Vol. **3**. Design and analysis and time series. Charles Griffin & Co., Ltd. London.

Fox, P.N. and Rosielle, A.A. 1982. Reducing the influence of environments main effects on pattern analysis of plant breeding environment. Euphytica, **31**: 645–656.

19

Genetic Structure of Population

Population refers to the community of sexually interbreeding or potentially interbreeding individuals. It is characterized by (i) gene pool and (ii) gene frequency. The gene pool refers to the sum total of all the genes in the reproductive gametes of a population and it is the gene frequency which determines the genetic structure of a population.

19.1 The Hardy-Weinberg Equilibrium

Considering single locus with two alleles A and a there will be three genotypes in the population namely, AA, Aa and aa. Let the frequency of allele A, $f(A)$ be u and the frequency of allele A, $f(A)$ be v and $f(A) + f(a) = 1$. Further assuming that the individuals are mating at random, the genotypic frequencies of these three classes (Aa, Aa and aa) of individuals will be AA (u^2), $Aa(2u)$ and aa (v^2) as shown in Table 19.1.

Table 19.1 Genotypic Frequencies in Random Mating population.

♂ ♀	$A(u)$	$a(v)$
$A(u)$	AA u^2	Aa uv
$a(v)$	Aa uv	aa v^2

Since the genotypic frequencies total equals unity, i.e. $u^2(AA) + 2uv(Aa) + v^2(aa) = 1$ (this equation can be regarded as the binominal expansion of $(u + v)^2$) the gene frequencies can then be calculated from the genotypic frequencies as:

$$f(A) = (u^2 + 1/2\ 2uv)/(u^2 + 2uv + v^2)$$

.
.

$$= u$$

and $f(a) = (v^2 + 1/2\ 2uv)/(u^2 + 2uv + v^2)$

.
.

$$= v$$

Again, in the next generation assuming random mating the genotypic frequencies of the three classes of individuals will remain constant as u^2, $2uv$ and v^2 and the gene frequencies will also remain constant as u and v. The population can thus be said to be in the Hardy-Weinberg equilibrium. Hardy in England and Weinberg in Germany independently proposed this equilibrium in 1908. Genotypic and gene frequencies in the equilibrium population are called equilibrium genotypic and gene frequencies, respectively. The population will thus continue with this constitution generation after generation till the following assumptions hold:

1. There is random mating in the population
2. No selection
3. No mutation
4. No migration
5. No drift and the population size is infinite.

19.1.1 Testing goodness of fit

Using the observed frequencies of three genotypes, the frequency of gene can be estimated and thus the expected equilibrium genotype frequencies can be found and the test of goodness of fit can be made using χ^2. If required, Yates correction can be used. The failure will result if all the assumptions are not fulfilled. On the other hand, a reasonably close fit does not prove *panmixia* and other assumptions as unusual patterns of evolutionary change can give a close fit. Also, under various degrees of inbreeding, genetic changes can occur which are significant in evolutionary time but too small to be detected in the chosen samples. The individuals belonging to each of the three genotypic classes can be identified using immunological/isozymatic study or simple tests. The inability to taste PTC (phenyl theo carbamide) is under control of single recessive gene. Having known the frequency of nontaster (homozygotes) the frequency of homozygotes and heterozygotes taster can be estimated in the population. In the distribution of M-N blood groups, the two antisera; anti-M and anti–N can distinguish MM, MN and NN individuals. Anti-M will react with MM red cells and anti–N will react with red cells of NN individual whereas red cells from MN individuals will react with both.

There is a relationship between genotypic and gene frequencies. With $u = v = 0.5$, the frequency of heterozygotes Aa is maximum (i.e. $2uv$ is maximum) whereas the frequency of homozygotes AA or aa increases or decreases with the increase or decrease of gene frequency of A, $f(A)$ or a, $f(a)$, respectively. The total gene frequency $(f(A) + f(a))$ being 1.0. Further it can be shown that no matter what is the constitution of the starting population, i.e. be it made of only homozygotes, 1/2AA and 1/2aa or any deviation from the equilibrium genotypic frequencies of 1/4AA: 1/2Aa: 1/4aa such as 1/3: 1/3: 1/3 or 2/5:1/3: 1/3 or 3/7:1/7:3/7, the Hardy-Weinberg equilibrium is achieved after a single generation of random mating.

19.1.2 Extension of H. W. equilibrium to cases of multiple alleles, sex-linked genes and polygenic inheritance

Hardy- Weinberg equilibrium can also be applied to cases of multiple alleles, sex-linked genes and polygenic inheritance. With two alleles system as we have seen the zygotic frequencies $u^2(AA) + 2uv(Aa)$ + $v^2(aa)$ are obtained by expanding the binomial $(u + v)^2$. With multiple alleles, say alleles having the gene frequencies $u_1, u_2, ..., u_n$ and $\sum_{i=1}^{n} u_i = 1$, the frequencies of the different genotypes in equilibrium can be described by the multinomial expansion, $(u_1 + u_2 + ... + u_n)^2$. Thus for any two alleles of a multiple allelic series the equilibrium is attained after a generation of random mating. With three alleles A_1, A_2 and A_3 with their respective frequencies u_1, u_2 and u_3, the six genotypes at H.W. equilibrium will be in the proportion of $u^2 A_1A_1: 2u_1u_2 A_1A_2: u_2^2 A_2A_2: 2u_1u_3A_1A_3: 2u_2u_3A_2A_3$ and $u_3^2 A_3A_3$. Thus when all the three alleles are of equal frequency 2/3 of the population is heterozygous. With four alleles this proportion will be 3/4 and with n alleles it will be $(n-1)/n$. That is why in case of single locus multiple allelic sporophytic incompatibility system, the frequency of heterozygous is so high that virtually sib mating is compatible. For three alleles case, we have example of A, B and O blood groups in human where the six genotypes AA, AB, AO, BB, BO and OO can be grouped into four phenotypes A, AB, B and O, as A is dominant over O, B is dominant over O whereas AB shows codominance. Assuming p, q and r being the frequencies of A, B and O allele, respectively the equilibrium genotypic frequencies of the six genotypes will be $p^2(AA): 2pr(AO): 2pq(AB): q^2(BB): 2qr(BO): r^2(OO)$. The gene frequencies can be obtained from the observed proportions of the different genotypes in human population as:

$$r = \sqrt{r^2} = \sqrt{f(OO)}$$
$$p = 1 - (q+r) = 1 - \sqrt{(q+r)^2}$$
$$= 1 - \sqrt{q^2 + 2qr + r^2}$$
$$= 1 - \sqrt{f(BB) + f(BO) + f(OO)}$$

and

$$q = 1 - (p+r) = 1 \sqrt{(p+r)^2}$$
$$= 1 - \sqrt{p^2 + 2pr + r^2}$$

$$= 1 - \sqrt{f(AA) + f(AO) + f(OO)}$$

where $f(AA)$, $f(BB)$, $f(AO)$, $f(BO)$ and $f(OO)$ are the observed frequencies of AA, BB, AO, BO and OO individuals, respectively. Since it is assumed that $p + q + r = 1.0$, if the calculated frequencies do not sum to 1.0, the Bernstein formula can be used to correct the estimated gene frequencies as:

Estimated gene frequency	Corrected gene frequency
$\hat{p}$	$p(1 + 1/2d)$
$\hat{q}$	$q(1 + 1/2d)$
$\hat{r}$	$(r + 1/2d)(1 + 1/2d)$
Sum = $\hat{p} + \hat{q} = \hat{r}$	

where $d = 1.0 - (\hat{p} + \hat{q} + \hat{r})$. Having calculated the correct gene frequency the expected equilibrium genotypic frequencies can be calculated and test of goodness of fit can be done using χ^2 and if required Yates correction can be applied.

19.1.3 Sex-linked loci

In case of sex-linked gene, for example color-blindness in man a recessive trait under control of one gene with 2 alleles, there will be five genotypic classes of individuals assuming the homogametic sex to be female and the heterogametic sex to be male. The genotypes along with their respective frequencies are given below:

Female			Male	
$X_A X_A$	$X_A X_a$	$X_a X_a$	$X_A Y$	$X_a Y$
u^2	uv	v^2	u	v

$X_A X_A$, $X_A X_a$ and $X_A Y$ individuals will be normal while $X_a X_a$ and $X_a Y$ individuals will be color blind. The frequency $v = f(a)$ in female population will be

$$(2R + Q)/2(P + Q + R)$$

where P, Q and R are the observed numbers of individuals of $X_A X_A$, $X_A X_a$ and X_{aa} genotypes, respectively. The frequency of $v = f(a)$ in male population will be

$$T/(S + T)$$

where S and T are the observed number of individuals of $X_A Y$ and $X_a Y$ genotypes, respectively. The overall frequency of v will be

$$(2R + Q + T)/(2(P + Q + R) + (S + T))$$

When male and female populations are considered separately, the H.W. equilibrium is achieved after one generation of random mating. But assuming that the sexes are equally frequent, the equilibrium frequencies come to be $1/2\ u^2$: $1/2\ (2uv)$: $1/2 v^2$ for females and $1/2\ u$ and $1/2\ v$ for males, respectively in order that the sum of the frequencies of five genotypes equals unity (Mather, 1973). The equilibrium is not attained after one generation of random mating as we saw in case of autosomal genes. However, the population will approach very close to H.W. equilibrium after 5 to 6 generations of random mating.

19.1.4 Polygenic traits

The H.W. equilibrium obtained in case of one locus with 2 alleles system and discussed above can be extended to cases of two or more loci (polygenes) determining a trait (Li, 1955). Assuming two loci A and B with two alleles at each locus, the two parents would be $AABB$ and $aabb$ which upon random mating will produce $AABB$, $aabb$ and $AaBb$ genotypes, respectively as shown below:

♂ ♀	AB	ab
AB	AABB	AaBb
ab	AaBb	aabb

In the next generation of random mating the gametes produced by these individuals would be AB, Ab, aB, and ab. If C_1, C_2, C_3 and C_4 are the proportions of gametes, the population generated will be in equilibrium if $C_1 \times C_4 = C_3 \times C_2$. Initially $C_1 C_4 \neq C_2 C_3$ but the difference between the two $C_1 C_4 - C_3 C_2$ will gradually decrease in successive rounds of random mating but then how fast or slow it is will depend on the recombination frequency and the tightness of linkage (see chapter on linkage). Thus H.W. equilibrium is not attained after one generation of random mating.

19.2 Changes in Gene Frequency

Out of the five forces (random mating, selection, mutation, migration and drift) mating system does not impel changes in gene frequency but four other forces can change gene frequency and thus these are the evolutionary forces which determine the genetic structure of a population. In population genetics any change in gene frequency which results in a change in the genetic structures of a population is defined as the evolution. The mathematical treatment of these forces were developed mostly by R.A. Fisher, J.B.S. Haldane and S. Wright in the 1920s. The processes of selection, mutation and migration bring about changes in allelic frequency which are predictable in magnitude and direction, i.e. if the initial gene frequency and the selection differential, the mutation rate or the migration rate are known, the magnitude and direction of the change in allele frequency can be predicted. All these processes will eventually lead to the fixation or loss of an allele unless they are opposed and a stable equilibrium is reached which we shall see later in the section. These processes have been described by various workers as deterministic, directed or systematic in their mode of action.

19.2.1 Effects of non-random mating

Inbreeding and assortative mating are the deviations from random mating. While self-fertilization is the most extreme form of inbreeding which occurs in hermaphrodite species, the other mild forms of inbreeding are mating of close relatives such as full-sib mating, half-sib mating, parent-offspring mating, double first cousins, single first cousins, etc.

19.2.2 Self-fertilization

Considering one locus with 2 alleles, the three genotypes (*AA*, *Aa* and *aa*) in the F_2 population will be in the proportion of 1/4: 1/2: 1/4. The frequency of heterozygote is 1/2. Now if selfing is started in this F_2 population, then upon selfing the individuals *AA* will breed true whereas the heterozygote *Aa* will produce *AA*, *Aa* and *aa* individuals.

Thus the frequency of heterozygote will reduce by half in every generation i.e. to 1/4 in F_3, to 1/8 in F_4 and so on whereas the frequency of homozygote will increase to 3/4 in F_3, to 7/8 in F_4 and so on. In terms of gene frequency u and v, the frequency of heterozygote in the nth generation of selfing will be equal to $(1/2^n 2uv$ and the frequency of homozygotes *AA* and *aa* will be $u-uv(1/2)^n$ and $v-uv(1/2)^n$, respectively. Thus after a large number of generations of selfing, the heterozygote will be lost and the population will compose of individuals of homozygotes *AA* and *aa*.

19.2.3 Sib mating

Unlike self-fertilization which is possible only in plants, sib mating is possible in plants as well as animals. Like selfing, the regular mating of sib leads to homozygosis though distinctly more slowly. The frequency of heterozygote under full-sib/parent-offspring mating decreases according to the series 1/2, 1/2, 3/8, 5/16, 8/32 and thus the frequency of homozygote increases in the series 1/2, 1/2, 5/8, 11/16, 24/32, etc. in the population starting as an F_2. The homozygosis is slower than full-sib/parent-offspring mating in half-sib/double first cousins and is followed by single first cousins. These mating systems have been discussed in detail by Fisher (1949). Formulas for calculating inbreeding coefficient have been given by Wright (1921).

19.2.4 Assortative mating

When like phenotypes preferentially mate, it is called positive assortative mating. The matings *AA* × *AA*, *Aa* × *AA*, *Aa* × *Aa* and *aa* × *aa* are positive assortative mating. Because of dominance of *A* over *a*, *AA* and *Aa* will be phenotypically similar. The two genotypes will look similar because of environmental effects. Now if the phenotypic assortative mating is between like genotypes then mating will lead to inbreeding and when it is carried out for long enough, the heterozygote will be eliminated and the population will comprise of two groups (*AA* and *aa*) of individuals and thus the variance increases. The original variance in an F_2 population was $1/2D + 1/4H$ which is now only *D*. In this system of mating inbreeding affects only those loci which control the character and for which the mating is assortative. This is unlike inbreeding which leads to homozygosis for all the genes in the genotype. In case of no dominance *AA*, *Aa* and *aa* individuals are clearly distinct phenotypically and the consequence of assortative mating will be similar to that of inbreeding. In case of complete dominance it can be shown that two generations of phenotypic assortative mating will raise homozygosis to the level achieved by one generation of selfing or one generation of assortative mating where dominance is incomplete. The inbreeding thus has a greater effect than assortative mating on the homozygosity.

If two or more than two genes having similar effects as in polygenic system, are determining the phenotypic assortative mating, the progress to homozygosity for these loci will be slower than for one locus as the same phenotype will be produced by different combinations of genes at the various loci. For example, considering two loci *A* and *B*, *AABB*, *AABb* and *AaBb* will be phenotypically similar. The efficacy of assortative mating is reduced further because of environmental effects. Thus in case of polygenic traits, the effect of assortative mating is not to increase the homozygosity but rather to increase population variability. The additive genetic variance is only or mainly increased. The dominance genetic variance hardly changes at all and epistatic component changes very little (Crow, 1952). Thus assortative mating coupled with selection can increase the rate of progress in the population. The effect of assortative mating depends on the degree of resemblance (can be measured by correlation coefficient). If the correlation is high, $r = 1.0$, the population can become homozygous. This is in contrast with most forms of inbreeding which eventually lead to complete homozygosity.

19.2.5 Negative assortative mating

Negative assortative mating is the mating of unlike phenotypes. Phenotypic differences could be due to the difference in sex, incompatibility system, physiological or morphological. Unlike positive assortative mating, negative assortative mating tends to maintain the differences. The mating $AA \times aa$ and $Aa \times aa$ will be called negative assortative mating. Thus at equilibrium the genotypic frequencies will be 1/2 *Aa* and 1/2 *aa* and the gene frequencies come to 1/4 for *A* allele and 3/4 for *a* allele. Although with these gene frequencies, the frequency of heterozygote in H.W. equilibrium will be 3/8. Thus negative assortative mating is similar to random mating in effect in that it also produces a half heterozygosis. We have example of sporophytic heteromorphic incompatibility wherein pin (*ss*) × thrum (*Ss*) combination is compatible. If the negative assortative mating is restricted to between two different homozygotes such as $AA \times aa$, the maximum frequency of heterozygote that can be maintained by adjusting the mating system is 2/3 the attainment of which in case of polygenic system would require an unrealistic restriction on mating. In case of sex difference where *XX* is female and *XY* is male and in a population of equal males and females the frequency of $X, f(X)$ will be 3/4 whereas the frequency of $Y, f(Y)$ will be 3/4. But here it must be recognised that it is group of genes (linked) rather than the single gene which decides whether or not to mate whereas in the determination of sex mechanism, chromosome can be involved. The negative assortative mating leads to dimorphism in that the two alternative phenotypes are maintained in the population at reasonable frequency which cannot be maintained by mutation.

The different mating systems thus do not affect the gene frequencies but they do affect the distribution of genes between homozygotes and heterozygotes.

19.3 Mutation

Assuming one locus with two alleles *A* and *a* there are two possibilities: either *A* can mutate to *a* (a forward mutation i.e. mutation from wild or normal type to mutant allele) or *a* can mutate back to *A* (a reverse mutation i.e. mutation from mutant type to wild or normal type). Assuming μ being the mutation rate per generation, μ_A is the rate with which *A* mutates to *a* and μ_a is the rate of reverse mutation. At equilibrium, rate of forward mutation will equal to the rate of back mutation. If *u* and *v* are the initial gene frequencies of *A* and *a*, respectively, the frequency of A allele $f(A)$ will be reduced by an amount $u\mu_A$ but because of back mutation the frequency will be increased by vu_a after one generation of mutation. Likewise, the frequency of a allele, $f(a)$ will be reduced by $v\mu_A$; but increased by $u\mu_A$ and at equilibrium

$$u\mu_A = v\mu_a$$

and

$$\frac{u}{v} = \frac{\mu_a}{\mu_A}$$

This shows that when $\mu_A = \mu_a$, i.e. when mutation rates are equal, the gene frequencies u and v are identical and when the mutation rates differ, the equilibrium gene frequencies will also differ. Thus whether u or v is higher at equilibrium depends on whether dominant mutation is higher or lower to recessive mutation. Mutation thus unlike mating system affects gene frequencies rather than affecting the distribution between homozygotes and heterozygotes. As mutation rates are generally of the order of 10^{-5} or 10^{-6} or even lower, the attainment of equilibrium would be a slow process. Further, between the forward and the reverse mutation, the later would appear to be much rarer and thus at equilibrium u will then be lower than v. As a consequence of mutation then the population would come to be dominated by mutant phenotypes—a result which is not found in nature. The very low frequency of mutant phenotype that is observed indicates that the frequency of mutant allele is rare. This points to some other force such as selection which is opposing mutational force. As mutant genes are deleterious and in extreme case can be lethal, mutant phenotypes are less fit and do not survive.

19.3.1 Balance between mutation and selection

The best examples of balance between mutation and selection are that of achondroplasia, a form of dwarfing and a dominant mutation and of phenylketonuria, a recessive mutation in man. The model of fitness in case of dominant and recessive mutation will take the following form:

I. Dominant mutation

Genotype	AA	Aa	aa
Frequency	u^2	$2uv$	v^2
Fitness	1.0	$(1-s)$	$(1-t)$

II. Recessive mutation

Genotype	AA	Aa	aa
Frequency	u^2	$2uv$	v^2
Fitness	1.0	1.0	$1-t$

In case of dominant mutation the homozygous genotype aa is usually lethal. Heterozygous genotype carrying the dominant allele a is defective and is less fit in comparison to aa genotype while AA genotype is of normal phenotype. In case of recessive mutation both AA and Aa genotypes are normal whereas aa genotype has lower fitness. Assuming the net mutation to be in the direction of a from A the gene frequencies of allele A, $f(a)$ and allele a, $f(a)$ in the next generation will become

$$f(A) = (u^2 + uv(1-s) - v\mu)/1 - 2uvs - v^2t$$
$$= (u - uvs - u\mu)/(1 - 2uvs - v^2t)$$

and $f(a) = (v^2(1-t) + uv(1-s) + u\mu)/(1 - 2uvs\ u^2t)$
$$= (v - uvs - v^2t + u\mu)/(1 - 2uvs - v^2t)$$

Since it is a rare defective mutant, v is small; v^2 will be very small and the aa phenotype will be virtually non-existent in the population and thus we can safely ignore the term v^2t in the denominator.

Thus, $f(A) = \dfrac{u - uvs - u\mu}{1 - 2\,uvs}$

and at equilibrium

$$= \frac{u - uvs - u\mu}{1 - 2\,uvs} = u$$

Since v is small, u will be approximately equal to 1.0 and then the above equilibrium equation reduces to

$$v = \mu/s$$

The expected frequency of mutant phenotype (Aa) in the population will be

$$2v = 2\mu/s$$

The frequency of mutant phenotype can thus be increased by either increasing μ, the mutation rate or decreasing s, the unfitness of heterozygote. μ can be increased by exposing the individuals to radiation whereas s can be decreased by treating medically the affected individuals.

In case of recessive mutant, the gene frequencies $f(A)$ and $f(a)$ in the next generation will become

$$f(a) = (uv + v^2)(1-t)/(1-tv^2)$$
$$= v(1-tv)/(1-tv^2)$$

and $f(A) = (u^2 + uv)/(1 - v^2 t) = u/(1 - tv^2)$

The change in gene frequency Δa, due to selection will be

$$v - \frac{v(1-tv)}{1-tv^2} = \frac{tv^2(1-v)}{1-tv^2}$$

The change in gene frequency Δa due to mutation will be $u\mu$. Then at equilibrium

$$u\mu = \frac{tv^2(1-v)}{1-tv^2}$$

Since v is small; v^2 will be very small and so tv^2 will also be very small and can be neglected and thus

$$u\mu = tv^2(1-v)$$

Again, since v is small $(1-v)$ approaches 1.0 and then the above equation reduces to

$$v^2 = \mu/t$$

and $v = \sqrt{\mu/t}$

Thus the frequency of mutant phenotype in case of recessive mutation can be increased by either increasing μ; the mutation rate or decreasing t, the unfitness of homozygotes aa but unlike dominant mutation the increase in the frequency of mutant phenotype will be less rapid with the decrease in s. In general, the increase will be somewhat slower when it depends on the reduction in s than when it depends on increase in μ.

We also have a case of sickel cell anemia. It is caused by a single recessive mutant gene (h). Homozygous (hh) recessive individuals suffer from sickel cell anemia which causes abnormal development commonly leading to early death but shows resistance to malaria and have S-type haemoglobin. Homozygous dominant (HH) individuals for this gene have normal haemoglobin and do not suffer from anemia but are susceptible to malaria. Heterozygous individuals (Hh) have both types of haemoglobin—normal and S—type haemoglobin (the normal round shape of RBC changes to sickel shaped, S-type haemoglobin differs from normal haemoglobin in that the glutamic acid is replaced by valine at position 6 in it).

The heterozygous individuals show heterozygous advantage in that they do not suffer from anemia and simultaneously show resistance against malaria. This is an example of polymorphism where both homozygotes are less fit than the heterozygote but only in the presence of malaria. Let us now see how the very delerious gene can be kept at higher frequency in the population. The model of fitness takes the following form:

Genotype	AA	Aa	aa
Frequency	u^2	$2uv$	v^2
Fitness	$1 - s_A$	1.0	$1 - s_a$

The frequency of allele A, $f(A)$ and allele a, $f(a)$ after one generation will become

$$u' = f(A) = u^2(1-s_A) + uv$$
$$v' = f(a) = v^2(1-s_a) + uv$$

Assuming mutation rate to be very small $u\mu$ can be safely ignored in the calculation. At equilibrium,

$$\frac{u'}{v'} = \frac{u^2(1-s_A) + uv}{v^2(1-s_a) + uv} = \frac{u}{v}$$

which upon simplification becomes

$$\mu s_A = v s_a$$

or $$\frac{u}{v} = \frac{s_a}{s_A}$$

From the above equation, the equilibrium gene frequencies can be calculated as:

$$\frac{u}{u+v} = \frac{s_a}{s_A + s_a} = u(f(A))$$

and
$$\frac{v}{u+v} = \frac{s_A}{s_a + s_A} = v(f(a))$$

Further it can be shown that very deleterious gene can be kept in population at higher frequency. In case of balance between selection, mutation and heterozygous advantage, the selection maintains the status quo of population by removing recurrent mutants and segregants from favoured heterozygote. Rare variants that arise from artificial mutation or recombination or deleterious gene or gene combinations are thus maintained in the population by a balance between mutation, migration or recombination and selection.

19.4 Selection

Selection is the most important force of evolution as bulk of the changes in gene frequency is caused by it. While discussing the H.W. equilibrium, it was assumed that an individual's contribution to the next generation is independent of its genotype. In other words, the individuals of these three genotypes AA, Aa and aa do not differ in viability and fertility or more precisely in fitness. Fitness refers to the ability to survive and reproduce. Fitness or selective value of an individual can be defined as the proportionate contribution of offspring to the next generation (Falconer, 1960). In other words, it refers to the relative reproductive success of an individual. For example, if A genotype produces 100 offspring and a genotype produces 90 offspring then the selective value of A would be 1.0 whereas that of a will be 0.9 and thus the selection coefficient will be 0.0 for A and 10/100 = 0.1 for a. Now if these three genotypes differ in fitness this will bring about a change in the gene frequency which we will see here. Here, the selection practised is natural selection but in practice, we are doing similar thing when, in a population, we are allowing only a particular phenotype (or genotype) and not others to make the next generation population. In other words, the genotype selected for, makes greater contribution to the next generation population whereas the genotypes selected against make a smaller contribution. The contribution of any individual to the next generation population will vary according to whether the population size is increasing or decreasing or stable and according to the contribution made by individuals of other genotypes.

Let us now consider a population consisting of three genotypes AA, Aa and aa. Assuming complete dominance AA and Aa will be phenotypically identical and can be clearly distinquished from aa individuals. Further assuming that selection is against aa genotype in the population and t being the selection coefficient the model takes the following form

Genotype	AA	Aa	aa
Frequency	u^2	$2uv$	v^2
Fitness	1	1	$1-t$

t can take values from 0 to 1.0. When t is zero, all the three genotypes in the population are contributing equally to the next generation population but when $t = 1.0$, the recessive homozygotes are selected against and thus not allowed to make any contribution to the next generation population. In artificial selection t is called the selection pressure. The gene frequency $f(a)$ changes when the selection is against aa genotype and becomes

$$v' = f(a) = uv + v^2(1-t)/(1-v^2 t)$$

where $1 - v^2 t$ is the total gene frequency. The change in gene frequency (a), the Δa becomes

$$\Delta a = v' - u$$

$$= \frac{v^2(1-t) + uv}{1 - v^2 t} - v$$

$$= \frac{-tv^2(1-v)}{1 - tv^2}$$

For small value of t, $1 - tv^2$ can be replaced by 1 and then Δa closely approximates to

$$\Delta a = -tv^2(1-v)$$

This can be rewritten as a differential equation by replacing Δa with dv/dt

$$\frac{dv}{dt} = tv^2(1-v)$$

Setting this equation to zero and solving for v, the $\frac{dv}{dt}$ (or Δa) is at a maximum when v is 2/3 (Li,1955). This shows among other things that when a favourable gene appears in the population, it spreads very slowly at the beginning i.e. the rate of change is slow but becomes faster when the gene frequency is at intermediate level. The v decreases at a rate approximately equal to the product of selection coefficient and the term $v^2(1 - v)$ from which it can be said that with constant selection pressure t, the change in gene frequency is a function of gene frequency in the previous generation. Selection is most effective at intermediate gene frequency i.e. when $u = v = 0.5$. Finally, Δv becomes very small when v approaches 0 or 1.0. In other words, selection is least efective when either v is very large or small.

19.4.1 Complete elimination of recessive homozygotes

In this situation the selection coefficient $t = 1.0$ and the model of fitness takes the form as:

Genotype	AA	Aa	aa
Frequency	u_0^2	$2u_0v_0$	v_0^2
Fitness	1	1	0.0

The frequency of allele a, $f(a)$ after one generation of selection becomes

$$v_1 = \frac{u_0 v_0}{u_0^2 + 2u_0 v_0} = \frac{v_0}{1 + v_0}$$

Starting with the v_1, the frequency of a, v_2 after the second generation of selection will become

$$v_2 = \frac{v_1}{1 + v_1}$$

Similarly, the frequency of allele a after third and fourth generation of selection will be

$$v_3 = \frac{v_2}{1 + v_2}$$

and

$$v_4 = \frac{v_3}{1 + v_3}$$

respectively. Thus in general, the frequency of allele a after nth generation of selection will become

$$v_n = \frac{v_{n-1}}{1 + v_{n-1}}$$

v_2, v_3, v_n can be expressed in terms of starting gene frequency v_0 as

$$v_2 = \frac{v_0}{1 + 2v_0},$$

$$v_3 = \frac{v_0}{1 + 3v_0},$$

and

$$v_n = \frac{v_0}{1 + nv_0}$$

This is obtained by substituting equation $v_1 = \frac{v_0}{1 + v_0}$ into the recurrent series one term at a time as we go to generation 2, then generation 3 and so on. Now, by rearranging the equation $v_n = \frac{v_0}{1 + nv_0}$ the number of generations required to achieve a given change in the frequency becomes

$$n = \frac{v_0 - v_n}{v_0 v_n} = \frac{1}{v_n} - \frac{1}{v_0}$$

The gene frequencies V_o or V_n can be calculated as

$$v_0 \text{ or } v_n = \sqrt{\frac{\text{Number of } aa \text{ individual observed}}{2 \times \text{population size}}}$$

The time required to change the allele frequency at different s is

$$t = \frac{2}{s} \ln \frac{v_t(1 - v_0)}{v_0(1 - v_t)}$$

For small s (i.e. when selection is weak) the time required for a specified change in allele frequency is inversely proportional to the intensity of selection. With natural selection, in case of most evolutionary changes, the value of s is 0.001 or less and so the rates of changes are constant but in long periods of time even this very small selection pressure can produce large changes in allelic frequencies. But in case of diseases and insects resistance, the evolutionary changes in pathogen/pest have been rapid because of higher value of selection coefficient because of intensive use of fungicide/insecticide and resulted in rapid evolution of pests having pesticide resistance. Resistance is acquired by natural selection.

19.4.2 Balance between selection and inbreeding

When two forces are acting simultaneously the change in gene frequency of an allele in a population can be predicted as the sum of the changes due to the two forces separately. If the forces are antagonistic, v will approach an equilibrium value $\hat{v}$ where Δv is zero.

In case of inbreeders, the heterozygotes upon selfing produce homozygous pure breeding lines which are as vigorous as the heterozygous individuals but in case of outbreeders, the inbred individuals produced by inbreeding are commonly less vigorous and less fertile in comparison to the heterozygous (non-inbreds) individuals and in many cases one cannot continue with the process of inbreeding indefinitely as after a certain generation of selfing, depending upon the crop species (the genetic organization of species) the inbreds produced would be too weak to continue further inbreeding. Thus it is worthwhile to examine the effect of fitness of homozygote on the process of inbreeding.

Considering the heterozygous advantage that the heterozygotes have the model of fitness takes the following from:

Genotype	AA	Aa	aa
Frequency	u^2	$2uv$	v^2
Fitness	$1-s$	1	$1-s$

In the above model, the fitness of homozygous genotypes (*AA*, *aa*) is the same as $1-s$ in comparison to 1 for heterozygous genotype *Aa*. The value of s can range from 0 to 1.0. Now if $s = 0$ i.e. when homozygotes do not differ from heterozygote in fitness, selfing will ultimately lead to population of homozygotes (*AA* and *aa*). When $s = 1.0$ i.e. when the fitness of homozygote is zero, half the population will continue generation after generation and will consist of heterozygotes. Thus under condition of heterozygous advantage selection in favour of heterozygote can prevent full homozygosis being attained. Under the condition $s > 1/2$ i.e. when the homozygotes are at least 'half as fit as the heterozygotes, selfing will ultimately lead to the population consisting solely of homozygotes (*AA* and *aa*) although the rate of progress towards homozygosis will be slower than when s is zero. Thus when $(1-s)$ ranges from 0 to > 0.5, complete homozygosis is attained sooner or later but when $(1-s)$ is less than 0.5 i.e. when s lies in between 0.5 to 1.0 which in turn means that homozygotes are less than half as fit as heterozygotes, full homozygosis will never be attained no matter how long the selfing continues. The proportion of homozygotes obtained falls from 1.0 at $s = 0.5$ to 0.5 when $s = 1.0$. Thus under selfing a value of $s > 0.5$ prevents loss of heterozygotes while in the other milder forms of inbreeding such as full-sib mating or parent offspring mating and half-sib mating $s > 0.257$ and $s > 0.19$ prevent loss of heterozygotes (Hayman and Mather, 1953).

19.5 Competitive Selection

So far we have assumed fitness of an individual to be independent of other individuals in the population but then selection can come about as a result of competition among individuals of the population. Competition can be between individuals of the same genotype (intra-genotypic) i.e. between individuals of *AA* or *Aa* or *aa* genotype or between individuals of different genotypes i.e. between *AA* and *aa*, *AA* and *Aa* or *aa* and *Aa* (inter-genotypic competition). The impact of competition thus will vary with the genetic structure of the population. With limited resources, the intensity of competition increases with increase of population size and the competition is thus density dependent. Thus fitness of an individual which obviously depends on the intensity of competition in turn depends on the relative density of population and is thus density dependent. An equilibrium is established under competitive selection and the alleles will be maintained in the population (Mather, 1973). Competition will also depend on the frequency of these genotypes *AA*, *Aa* and *aa* in the population which in turn depends on the gene frequencies $u = f(A)$ and $u = f(a)$. The competitive selection is therefore frequency dependent as well. In frequency dependent selection, the fitness of the alleles is not constant but changes with their frequencies. For example, in case of *S*-locus with multiple allelic system of incompatibility in *Oenothera* organism an allele is at selective advantage (i.e. has higher fitness) when its frequency is low but at a disadvantage when its frequency is high but an equilibrium is established which is stable. But not all frequency dependent selections lead to equilibrium for if the selective advantage of an allele increases with the increase of its frequency, it will inevitably lead to extinction of the other alleles. Thus the selection as we have seen cannot only lead to the fixation of one allele and elimination of another but can also maintain the two alleles at some intermediate equilibrium point between 0 and 1 by striking a balance with countervailing forces like mutation or migration. An equilibrium will also be established between competitive selection and inbreeding.

19.6 Migration

Moving out and moving in of individuals from one population to another is called migration. When the two populations differ in gene frequency, migration brings about a change in gene frequencies in the same way as mutation. Suppose there are two populations. One having gene frequencies $f(A) = u$ and $f(a) = v$ and another having gene frequencies $f(A) = \bar{u}$ and $f(a) = \bar{v}$. Thus for gene frequency of allele *a*, $f(a)$ the difference between the two populations will be $v - \bar{v}$. Let *m* be the proportion of individuals being exchanged per generation. The frequency of *a* allele, $f(a)$, after one generation of exchange will become

$$(1 - m) v + m\bar{v}$$

The change in gene frequency Δa will then become

$$(1 - m) v + m\bar{v} - v = - m (v - \bar{v})$$

Thus the change in gene frequency due to migration depends on the number of immigrants and the difference in gene frequency between the populations. The difference between the two populations in gene frequency after one generation would become

$$(1 - m) (v - \bar{v})$$

After many generations, the gene frequencies in the two populations would be the same and this would be fast if *m* is large.

19.6.1 Balance between selection and migration

If there is selection operating on the boundary, a balance can struck between migration and selection, i.e.

$$- m(v - \bar{u}) = tv^2(1 - v)$$

where the change in the gene frequency Δa due to migration is $-m(v - \bar{v})$ and the change in the gene frequency due to the selection is $- tv^2(1 - v)$. At equilibrium, $- m (v - \bar{v}) - tv^2(1 - v) = 0$.

From the selection model we know that when the selection is against homozygous recessive, the change in gene frequency is maximum at $v = 2/3$. Using this value of $v = 2/3$ in the equation $- m(v - \bar{v}) = tv^2 (1 - v)$, the value of $\Delta v = 0.15t$ which shows that migration is at least equally important a force in evolution and usually more so than selection pressures of equal magnitude.

19.6.2 Models for studying the population structure

Considering a population distributed over a large area, the whole population can be assumed to be made up of local subpopulations which could be formed as a result of presence of different ecological niches. These subpopulations thus constitute a case of isolation by distance. Isolation leads to differences in genetic structure between subpopulations and migration reduces such difference. There are two different models to study the population structure. Wright's island model assumes that all subpopulations exchange individuals/gene at the same rate regardless of the relative distances, i.e. the immigrants come randomly from the rest of the population. 'Stepping stone' model of Kimura is based on the assumption that only neighbouring subpopulations exchange individual or gene, i.e. immigrants are more likely to come from subpopulation close by. In this model subpopulations are assumed to be spread in a rectangular grid and immigrants into a subpopulation come from one of the four adjacent subpopulations. These models thus represent long and short range gene flow. While the assumption behind island model is unrealistic, the finite stepping stone model does not lead to simple solution. The estimates of GST which measures the degree of population subdivision in these models, do not differ greatly. A small increase in migration rate in island model has as expected proportionally larger effect than changes in migration rate in 'Stepping stone model' in reducing genetic differentiation. A single migrant in an 'island model' is about as effective as 12 to 20 migrants in 'Stepping stone model'. The difference in the effectiveness of migration between these two models is thus not as great as expected (Crow, 1992). These two models represent the two extremes and most natural populations probably fall between these two extremes and some use can be made of GST. However, the applications of these models have been so far limited.

Migration differs from mutation in that first, the rate of immigration can be high unlike mutation rate and thus it can maintain the immigrant allele even though it is aberrant at high frequencies and secondly, unlike mutation it can change the gene frequency at many loci simultaneously and the whole genotype (individuals) rather than a particular gene as in mutation is involved in the process.

19.7 Drift

While discussing the systematic processes we assumed an ideal population (an ideal population is a random mating population with no mutation, migration and selection and with no difference in the viability and fertility among individuals) of infinite size although in practice populations are not infinite and in small populations quite large changes in allelic frequency may occur by chance. The random changes in gene frequency through sampling error were called genetic drift by Wright and it arises because each generation of surviving offspring is produced from a sample of all possible gametes, the gene pool. Wright (1931, 40 and many other papers) showed the importance of random genetic drift (random fluctuation of gene frequencies) in determining the genetic structure of the population. Changes in gene frequencies are not linked to the starting gene frequencies. The magnitudes of the changes produced by drift are predictable but their direction is not. Such processes have been described as stochastic, non-directional or dispersive processes. We will first examine the effect of small population sampling using probability theory on the structure of the derived population.

Considering a large random mating population of a self-fertilizing species in H.W. equilibrium with genotypic frequencies as $u^2(AA)$, $2uv(Aa)$ and $v^2(aa)$ and with equal gene frequencies i.e. $u = f(A) = v = f(a) = 0.5$, the genotypic frequencies will be 1/4: 1/2: 1/4. Now if only one individual survives by chance or is randomly selected to breed then there is a 50% chance that the individual would be Aa which upon random mating of uniting gametes A and a, produced in equal numbers, will produce population with the same gene and genotypic frequencies as in the parental population. There is also a 50% chance that the surviving or the selected individual will be either AA or aa and so the offspring produced would be either AA or aa and thus there is a loss of an allele either a or A. Table 19.2 shows the effect of sampling on allele and genotype frequencies.

Table 19.2 Sampling effects on allele and genotype frequencies

Probability	Genotype of parents	Effects on allele and genotype frequencies
1/4	$Aa \times Aa$	No change
1/8	$AA \times aa$	No change
1/8	$AA \times AA$ or $aa \times aa$	One allele is lost completely
1/2	$Aa \times AA$ or $Aa \times aa$	Allele and genotype frequencies change

It can be seen from Table that in 1/8 cases, one allele either A or a is lost and in 3/8 cases there will be no change in allele frequencies and in 5/8 cases there will be change in gene and genotype frequencies. Likewise, the probability of fixation or loss of an allele can be calculated for increasing number of survivors and it can be shown that in a single generation, the probability of an allele being eliminated decreases rapidly as the number of surviving or randomly selected parents increases and at N (the number of parents) = 50, the chance of loosing an allele is very rare and the changes in gene frequency will be very little. The magnitude of the effect is thus proportional to the size of the population. Thus we see that in the variable sampling of gene pool, each generation will result in random genetic drift. The change in gene and genotype frequencies is due to sampling error and the genetic drift is thus caused by decrease in population size.

Now we will study the statistical procedure through which the effect of genetic drift can be roughly estimated. Consider that a large intermating population with equal gene frequencies ($u = f(A) = v = f(a) = 0.5$) is separated into a number of sub-populations of size N with equal gene frequencies ($u_0 = v_0 = 0.5$). After one generation, the frequency of allele A in the sub-population is expected to follow the binomial distribution with a mean, $u_1 = u_0 = 0.5$ and a variance of $v(u) = u_0 v_0/2N$ and the standard deviation, $SD(u_1) = \sqrt{u_0 v_0/2N}$. Since our aim is to estimate the magnitude of change in the gene frequency of some allele a, Δv after one generation, due to chance alone, it is essential to see the distribution Δv in a number of sibs of size N and the dispersion of Δv among all the sub-populations is measured by the variance. If the distribution is random then the mean of Δv among the sub-populations will be zero and the populations have the same Δv value. The expected variation in gene frequency due to sampling errors between sub-populations is inversely proportional to N, the sub-population size. Assuming that the binomial distribution can be approximated by the normal curve, it can be found from the normal probability integral that 95% of the sub-populations have an allele frequency in the range of $u_1 \pm 2SD(u_1)$ and the rest have allelic frequencies falling outside this range. Figures 19.1 and 19.2 show distribution of allele frequencies in populations of size 5 and 50, respectively.

Now if the population remains small in size and the sampling error is effective in each generation, then the drift is continuous and it can be shown that the average or expected gene frequency overall

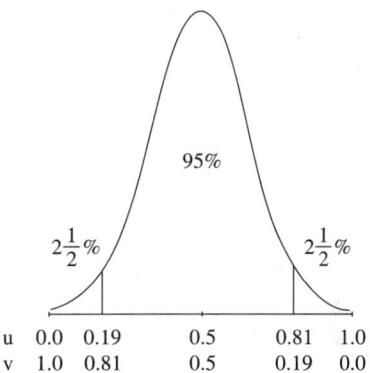

Fig. 19.1 Distribution of allele frequencies in populations of size 5 after one generation of drift, with an initial allele frequency of 0.5

In 5% of populations, ($f(A) = u < 0.19$ or > 0.81

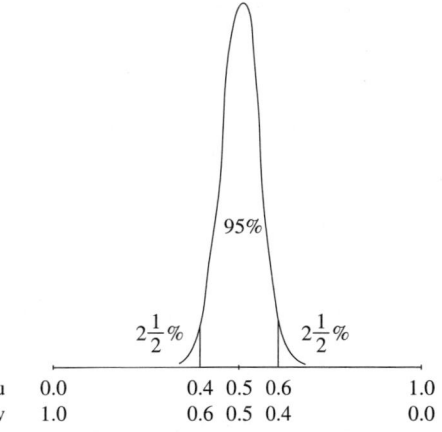

Fig. 19.2 Distribution of allele frequencies in populations of size 50 after one generation of drift, with an initial allele frequency of 0.5

In 5% of populations, $f(A) = u < 0.?$ or > 0.06

sub-populations and the variance in gene frequency among sub-populations after t generations are

$$u_t = u_0$$
$$v(u_t) = u_0 v_0 (1 - (1 - 1/2N)^t)$$

This shows that as t increases, the variance becomes larger with each successive generation. After many generations, the mean gene frequency over all sub-populations remains same as the starting gene frequency. The variance in allelic frequencies among sub-populations increases and tends to $u_0 v_0$. This implies that eventually A will be fixed in a proportion u_0 of populations and that a will be fixed in the remaining v_o populations. This further leads to an important conclusion that if a selectively neutral allele arises by mutation, at least once in a population of size N, the probability that this allele is fixed is $1/2N$. This probability is higher in smaller population but lower in larger population and tends to zero as the population size tends to infinity. Further, as the drift proceeds, the probability that both alleles coexist within a population is reduced. That is, the drift is accompanied by a reduction in the frequency of heterozygote in each population as the allelic frequencies become more extreme. The mean time required for the absorption of an allele (until an allele is either fixed or lost i.e. the frequency of heterozygote is reduced to zero) can be shown to be

$$\bar{t} = -4Nu_0 \log_e u_0 + v_0 \log_e v_0 \text{ generations}$$

when $u_0 = 0.5$, the value of t is maximum, $2.8N$ generations. The mean absorption time is shorter when u_0 is near 0.0 or 1.0.

The mathematical treatment of drift thus shows the following three mean effects of random genetic drift on the frequency of neutral allele in small populations assuming no mutation or migration:

(i) The different populations differ in gene frequencies.
(ii) There is less genetic variation within a population than between populations.
(iii) There is loss of heterozygotes.

19.7.1 Effective population size

We have just seen above that if the population size is not infinite, there will be random genetic drift and its magnitude of effects is proportional to the size of the population. The population size that is relevant when describing the drift is not the total number of individuals, the census population size but the number of mature, mating individuals who contribute the offspring to the next generation and which is called the effective population size. Wright (1973) showed that if the census population size changes with time, the effective population size can be calculated from the population size in each successive generation.

$$\frac{1}{Ne} = \frac{1}{t} \left[\frac{1}{N_1} + \frac{1}{N_2} + \ldots \frac{1}{N_t} \right]$$

where $N_1, N_2, \ldots N_t$ are the effective sizes at successive rounds of mating. Considering 10, 10^2, 10^3, 10^4, 10^5 and 10^6 being the population size in six successive generations, the value of effective population size comes to 54 individuals. Thus even though the population size increases from 10 to a million in six generations, the effective population size behaves as one with a constant size of 54 individuals. This shows that when a population size is small, it will have a disproportionate effect. A bottleneck in population size for example, caused by harsh biotic or abiotic stress environment, can cause the drift to occur.

Under the condition when all individuals (males and females) contribute equal number of gametes to the next generation, Li (1955) pointed out that the effective population size (Ne) becomes

$$Ne = 2N$$

where N is the number of individuals. In case of controlled mating when the males and females are different individuals and contribute unequally to the next generation, the effective population size is calculated as:

$$Ne = 4Nf \cdot Nm/(Nf + Nm)$$

where Nf is the number of female parents and Nm is the number of male parents. The effective population size is proportional to the harmonic mean of Nm

and Nf and the harmonic mean is more influenced by the smaller values. Therefore, if the number of females (Nf) is not equal to the number of males (Nm) then whichever is smaller will determine the effective size of the population. When the lines to be recombined are inbred lines such as S_2 then the effective population size (Ne) is reduced. In case of inbred parents, the effective population size takes the following form:

$$Ne = 2\left(\frac{N}{1+Fp}\right)$$

where Fp is the inbreeding coefficient of parents.

When $F = 1.0$ i.e. when parents are completely homozygous lines, $Ne = N$, Falconer (1960) expressed the inbreeding coefficient as a function of Ne, the effective population size as:

$$F_t = \frac{1}{2N} + \left(1 - \frac{1}{2N}\right)F_{t-1} = 1 - (1 - \Delta F)^t$$

where t corresponds to the cycle of selection and $\Delta F = \frac{1}{2}N$. In the absence of self fertilization

$$F = \frac{1}{2N+1}$$

is a preferable approximation. Further the difference in fitness of parents will reduce the effective population size.

19.7.2 Random genetic drift in natural populations

Random genetic drift undoubtedly influences allele frequency in small populations. In large populations, chance events occur but changes in opposite directions tend to cancel out so that both alleles persist. However, even in relatively large populations there is a small but finite chance that fixation may occur. Random genetic drift is more important when an allele is rare and alleles which have frequencies near zero or 1:0 may become lost or fixed as a result of chance fluctuations even in the presence of strong opposing selection pressure but then mutation may soon regenerate the lost allele in large populations. Although random genetic drift is often advocated as an important evolutionary force, there are contradictory views on its role in evolution. The question is whether or not the drift can be maintained in the face of operation of other evolutionary forces, especially selection. The models of drift described above are based on the assumption that no other evolutionary forces are operating. This implies that alleles are selectively neutral with respect to each other i.e. alleles have neither positive nor negative selective values. There are different views about the importance of neutral mutations i.e. mutations having no effect on fitness, in evolution and similarly, there are different views on the role of random genetic drift in evolution. If the mutations are effectively neutral then the drift must be responsible for their frequencies and fixation in neutral populations. Some believe that neutral alleles are common whilst others doubt that truely neutral alleles can exist in nature.

The model of genetic drift considered above gives the amount of change in gene frequency which is the upper limit in the absence of other evolutionary forces. As these evolutionary forces are operating in real population, they will diminish the effect of the genetic drift and the extent will depend on the magnitude and direction of selection, mutation and migration. This shows that even if the population size is small, drift will have less influence on gene frequency if the alleles are subjected to higher selective pressure and conversely drift will have a greater effect on relatively large population if the selection pressure is less. Further, an allele selected against can be fixed by drift so long as it is not disadvantageous as to lead to the extinction of its carriers. Conversely an allele selected for, may be lost by drift before it could show its full effect. However over many populations the selected alleles will be fixed more often than the other. Thus in small population with intense selection, a population nearly always becomes the homozygous for the advantageous alleles.

Considering mutation, migration and selection separately in a small population, as a rule of thumb, the random genetic drift will predominate if $4N\mu$, $4Nm$ or $4Ns$ are less than 1.0 where N is the population size and μ, m and s are the mutation rate, migration rate and the selection coefficient, respectively. If any of these expression is greater than 1.0, the population will behave as if it were large and random genetic drift will be swamped by the effects of mutation, migration and selection.

A change in selection, a change in environment, an unusual favourable mutation, rare hybridization, and an unusual swamping by mass immigration can result in change of gene frequency in an unpredictable way.

Role of random genetic drift

In theory the genetic drift might play an effective role in the evolution of small natural populations in three situations. When the populations remain small in size and sampling error is effective in each generation, continuous drift is said to be operating. In the two other situations there is reduction of population size. When the population is occasionally reduced to a size small enough to allow drift to occur and drift here is of intermittent type. If mortality is at random at the time of reduction of population size, the sample of survivors can have a different genetic composition due to chance alone (the bottleneck effect). Further, if the population remains small for at least 2 generations i.e. if the drift continues in the two successive generations, the process of continuous drift is then initiated.

19.8 The Founder Principle

Random drift is perhaps most typically exemplified by Mayr's (1954) 'founder principle' which designates the establishment of a new population of a species or form by a small number of individuals, the founding immigrants who carry only a small fraction of the genetic variability of the parental population and hence the allelic frequency of the founding immigrants may deviate from it if chance events occur. Thus new populations started from different founders will be genetically distinct from one another and also from the parental population. The consequences of founder effect are genetic divergence, reduced genetic variation and increased homozygosity in a populations founded by a small number of individuals. The resulting genetic revolution will enable new forms or species to be formed. The founder principle is of potential importance in the origin of species. Founder effect in crop plant evolution indicates the value and the breeding potential of the genetic variability in its wild relatives. Founder effect has been shown to have operated in the evolution of allopolyploids cultivated crops. Allopolyploid crops. plants derived from a limited number of interspecific hybridization followed by chromosome doublings show narrow genetic variability (consequence of founder effect) in the crop population compared to its wild progenitors. When alloploids evolve from a single diploid hybrid, its descendents will possess a very small fraction of the genetic variability of the diploid species. Founder effects can be suspected to be operating when the chromosomal variation in the wild progenitors is greater than the cultivated counterpart. Also, indication of isozyme polymorphism in the wild species as compared to cultivated points, the founder affect. Lewontin (1965), however, criticizes that only rare alleles will be lost during a founder effect and these contribute only marginally to the overall genetic variability. Further, remarks that although a genetic revolution is possible within an extremely small and isolated population, this type of population will probably go extinct soon after being established. These points have cast doubt on the real significance of founder effect in the origin of species (Barton and Charlesworth, 1984). However, the high incidence of the genetically determined disability, porphyria, in certain South African people can be explained by the founding principle. For further information on founding principle see Plant Breeding (Roy, 2012).

19.9 Gametic Selection

So far we have considered selection at the diploid or zygote level. There selection changes the gene and genotypic frequencies of diploid genotypes but selection for or against a particular allele can occur in the gemete i.e. at haploid level. Suppose u be the frequency of allele A and v be the frequency of allele a in the gametic pool. Assuming $1 - s$ be the fitness of a allele relative to 1.0 of A allele the model of fitness takes the following form:

Genotype	A	a
Frequency	u	v
Fitness	1	1 − s

The frequency of allele a after a generation of selection will become

$$v(1-s)/((u+v)(1-s))$$

and the change in gene frequency Δv will be

$$v(1-s)/(u+v)(1-s) - v$$

which upon simplification becomes

$$\frac{-vs(1-v)}{1-vs}$$

If s is small, vs will be smaller and the denominator approximates 1.0 and thus the change in gene frequency, Δv becomes $-vs(1-v)$. This shows that the change in gene frequency depends on the selection coefficient and the starting gene frequency.

19.10 Meiotic Drive

The unequal gene frequency can result because of inability of certain class of gametes to conjugate according to the usual pattern, the phenomenon termed meiotic drive. The phenomenon of meiotic drive is an example of gametic selection in which the heterozygote Aa instead of transmitting an equal number of two types of gene, A and a transmits an excess of one and it deviates from Mendel's laws of heredity and thus it causes changes in gene frequencies. In Drosophila female, the shorter chromosomes, a homologous pair is preferentially included in the egg. In maize there is often non-disjunction of *B-chromosomes* during the final cell division in the pollen tube and the nucleus with the extra *B-chromosomes* somehow regularly fertilizes with the eggs. The number of *B-chromosomes* increases until it starts affecting the viability and thus an equilibrium is achieved which can be described as mutation selection balance (Crow, 1992). Segregation Distorter allele (SD) in *Drosophila melanogaster* is a well studied example of meiotic drive. Although the *SD* allele in homozygous condition causes self destruction of sperm it is transmitted to 95% or more of the progeny. The *SD* acts through another locus, R^S, the sensitive responder on the homologous chromosomes and the chromosome having *SD* in cis-phase with R^S locus, acts as a killer chromosome. As *SD* chromosome occurs in less than 10% of population, it can be said that it is subject to some opposing selective forces and their low frequency in population further suggests the operation of some kind of equilibrium determined by selection.

These phenomena can have the evolutionary significance. On one hand if a chromosome causing significant distortion has no harmful effects then it will sweep through the population to fixation and it will carry along the effect of the linked hitch hiking genes but on the other hand if the chromosome shows harmful effects then the whole population will be wiped out.

19.11 Genetic Load

We have seen that the genotypes AA, Aa and aa in a population differ in fitness considering the rare deleterious mutants or polymorphism. Some individuals show optimum fitness whereas some others have lower fitness and thus the average fitness of the individual in the population reduces and this reduction in fitness is known as the genetic load. The genetic load thus refers to the reduction in population fitness resulting from the presence of genetic variation. The genetic load can be defined as the proportionate decrease in average fitness relative to that of the fittest population. Mutation thus contributes to the genetic load. With 10,000 loci in an organism the total genetic load which will be equal to the sum of genetic load for each locus will be of the order of 10^{-1} or 10^{-2}. Any organism keeping a greater number of alleles in the class $0.01 < P < 0.05$ represents deleterious recessive and constitutes the mutational load. As the mutation rate is of the order of 10^{-5} or 10^{-6} the mutational load it produces is of the order of 10^{-5} or 10^{-6}.

19.11.1 Segregational load

In case of the heterozygous advantage, the gene frequencies of recessive and dominant genes at equilibrium are

$$u = \frac{s_a}{s_A + s_a} \text{ and } v = \frac{s_A}{s_A + s_a}$$

The segregational load thus becomes

$$u^2 s_A + v^2 s_a$$

Putting the values of u and v in the above equation the genetic load becomes

$$\frac{s_A s_a}{s_A + s_a}$$

As the values of selection coefficients $s_1(s_A)$ and $s_2(sa)$ are much greater than the mutation rate, it can be shown that the segregation load produced by a single polymorphism would be as great as that from a very large number of loci undergoing mutation. Further, considering 10,000 loci, the segregational load will be so high that population would have to go extinct many times over to achieve such a level of polymorphism. In the above case alleles at different loci are treated as isolated, independent units and selection is acting on individual locus. Considering the individual as unit of selection and not the locus and that the environment acts upon the total finished phenotype rather than upon the loci separately and further assuming that the alleles at different loci are acting in a cumulative fashion or interacting, the cost of maintaining many loci in a polymorphic state could be lower. Study of biochemical traits like protein/isozymes in many populations has revealed that a typical species having about one third or more of all loci, in a single population, have two or more alleles in a polymorphic state and individuals in the population are heterozygous for about 10% of its loci. The finding is in contrast to the classical evolutionary theory that predicted that the most efficient form of an isozyme should over time become predominant in isolated populations with an occasional rare allele produced through mutation. As stabilizing selection is required for maintaining a locus in polymorphic state how can enough selection occur to keep this percentage of loci polymorphic. How the population could carry such as enormous load of unfitness? The cost of maintaining such a high percentage of loci in the polymorphic state would be very high. Further, as with n loci $(1/2)^n$ of the individuals will be simultaneously heterozygous for all the loci and knowing that an organism contains a very large number of genes, the number of individuals heterozygous at all loci will be very small. With $n = 40$ only one in million will be heterozygous at all the 40 loci, so the individuals having optimum fitness will have heterozygosity at less number of loci and the frequency of such individual will be high and so the load calculated will be misleadingly high. The high level of genetic load gives the impression that only a few loci could have substantial impact on fitness and that the number of loci whose variation is maintained by selection must therefore be linked-to a few dozen loci. Consideration of genetic load and the number of loci that might be maintained by selection led some biologists to propose that the majority of the genetic variation was adaptively neutral.

Though the genetic load is measured as deviation from the optimum genotype, the optimum genotype may itself vary in time and space or may even differ in the same place. A population with little or no genetic load may become extinct within a short period with a change of environment but a population with a relatively large genetic load when subjected to new environment may survive because the deleterious genes are now at an advantage. In case of genetic load which lowers fitness of genotype and is calculated as

$$L = \frac{W_{max} - W}{W_{max}}$$

where W_{max} is the fitness of the fittest genotype and W is the average fitness of the entire population, the individuals having lower fitness have reduced reproductive ability in comparison to the optimum genotype and thus contributing enormously to the, genetic load of the population. In extreme case individual because of sterility or inability to find a mate may not be contributing at all to the population and these are thus lost. This condition refers to as the genetic death of individual.

Although the genetic load can arise from a variety of reasons such as mutation, migration, incompatibility of mother and foetus as in the case of Rh factor, non-disjunction of chromosomes, hybridization of otherwise discrete population, polymorphism and so on. The two types of loads—mutational load arising from deleterious mutations and the segregational load due to segregation in polymorphic population when the different alleles are maintained by the heterozygous advantage are most important and are discussed as follow.

19.11.2 Mutational load

We saw in case of balance between rare deleterious mutation and selection that at equilibrium the frequency of recessive gene is μ/s whereas the frequency of dominant gene is $2\mu/s$. As the effect of mutation is to lower the fitness of an individual and if their reduction is by an amount, s, then the frequencies of affected individual multiplied by their loss of fitness (s) provides the estimate of the genetic load. From this the load for recessive gene in the population can be calculated as $\mu/s \times s = \mu$ and for dominant gene as $2\mu/s \times s = 2\mu$. This shows that the genetic load is dependent on the mutation rate μ, but is independent of the fitness s. It is no surprising as when such mutation occurs. It almost always confers a lower fitness at least in the homozygous condition. Homozygotes are, therefore eliminated or held at low levels within the individuals resulting in death of the individual and thus contribute to genetic load of the population but as the component of fitness includes longevity, time to age at first reproduction, fertility besides survivorship, the lower fitness of an individual can be due to production of a comparatively less number of offspring per generation. It will also contribute enormously to the genetic load or if an individual is comparatively more competitive in a new habitat or niche, the less competitive genotype(s) will contribute to the genetic load.

As selection involves complex interactions between an organism and its biological and physical environment, while calculating the fitness in the estimation of load, it is assumed that the selection is of Wallace's hard selection type (hard selection results from interaction between individual and its environment) i.e. the relative fitnesses of different genotypes are constant but, infact, it is not and thus the estimate of load is biased. Further considering Wallace's soft selection (which is the result of competitive interactions between individuals of the population and that the carrying capacity of a particular environment is fairly fixed for a species) the relative values of fitness depends on the number and the frequency of other genotypes. The genotypic fitnesses are likely to be frequency and density dependent. Frequency dependent selection does not produce any *load* as all the genotypes are equally fit when they are present in equilibrium proportion. Through soft selection unlimited number of polymorphic loci could be selectively maintained. If the change in environment makes the predominant genotype lethal or semi-lethal the population will not be able to maintain itself as the only rare genotype which can survive will not be able to fill the carrying capacity of the environment and thus the load i.e. reduction in population size is produced by hard selection. Thus the population will have to go extinct before evolving to the present day level. On the other hand, if the change in environment makes the predominant genotype comparatively less fit than the rare genotype, then the rare genotype will be able to maintain its number and thus soft selection will incur no load and the substitution of rare genotype could take place and thus the evolution takes place without incurring load. Extending the case of heterozygous advantage in sickel cell anemia to a polygenic case, the load will be extremely high i.e. the population size will be reduced greatly due to hard selection and thus the cost of maintaining polymorphism will be very high.

From the above consideration it can be said that the term fitness is relative and conditional. It is relative in that the fitness of an individual depends on the genotype of the individual with which it is competing and it is conditional because it is density dependent. Further, fitness is a comparative quality relating the environment (physical or biotic) in which it is grown. The environment as we know is heterogeneous in space and time and thus the selection coefficient is not constant as assumed throughout the discussion. Further information on these aspects can be found in Li (1967), Crow and Kumura (1970) and Cook (1971).

References

Allard, R.W. 1988. Genetic changes associated with the evolution of adaptedness in cultivated plants and their wild progenies. J. Hered., **79**: 225–238.

Allard, R.W. and Hansche, P.E. 1965. Population and biometrical genetics in plant breeding. In: Genetics

Today, Vol. 3, Proc. XIth Int. Congo Genetics, The Hague, The Netherland, 1963. Greets, S.1. (ed). Pergamon Press, Oxford, pp, 665–668.

Barton, N.N. and Charlesworth, 8. 1984. Genetic revolution, Founder effects and Speciation. Ann. Rev. Ecol. Syst., **15:** 133–164.

Bradshaw, A.D. 1984. The importance of evolution — any ideas in ecology and vice versa. In: Evolutionary ecology, Sharrock, B. (ed.), Blackwell Sci., Publ., Oxford.

Breese, E.L. 1960. The genetic assessment of breeding material. In: Proc. VIIIth Int. Grassland Cong., pp. 45–49.

Breese, E.L. 1983. Exploitation of genetic resources through plant breeding. Lolium species. In: Proc. Symp. on Genetic Resources of Forage Plants. Bray, R.A. and Mclvor, J.G. (eds.), CISRO, Melbourne, pp. 275–288.

Crow, J.F. 1986. Basic concepts in population, quantitative and evolutionary genetics. W.H. Freeman and Company, New York.

Crow, J.F. and Kimura, M. 1970. An Introduction to Population Genetics Theory. Harper and Row, New York.

Falconer, D.S. 1989. Introduction to Quantitative Genetics. Longman, Burnt Mill.

Fisher, R.A. 1930. The Genetical Theory of Natural Selection, Clarendon Press, Oxford.

Fisher, R.A. 1949. The Theory of Inbreeding. Oliver and Boyd, Edinburgh

Hardy, G.H. 1908. Mendelian proportions in mixed population. Science. **28:** 49–50 (Reprinted in the collections of Gabriel and Fogel and Peters).

Hayward, M.D. 1985. Adaptation, Differentiation and Population Structure in *Lolium perenne*. In: Genetic Differentiation and Dispersal in Plants. NATO ASI Series, G. Vol. 5. Jacquard, P. and Heims, G. (eds.) Springer- Verlag, Berlin, pp. 83–93.

Hayward, M.D. 1990. Genetic strategy and future prospects for breeding cross pollinated species. Norwegian Agricultural Research, Supplement. **9:** 77–84.

Hayward, M.D. and Breese, E.L. 19 68. Genetic organisation of natural populations of *Lolium perenne*. III Productivity. Heredity, **23:** 357–368.

Hayward, M.D. and Breese, E.L. 1993. Population structure and variability. In: Plant Breeding. Hayward, M.D.; Bosemark, N.O. and Romagosa, I (eds.). Chapman and Hall, London. pp. 16–29.

Kimura, M. 1970. The length of time required for a selectively neutral mutant to reach fixation through random frequency drift in a finite population. Genet. Res., Cambridge, **15:** 131–133.

Lewontin, R.C. 1965. Discussion of paper by Dr. Howard. In: The Genetics of colonizing species. ed. H.G. Baker; G.L. Stebbins. p. 481, New York Academic.

Li, C.C. 1955. Population Genetics. University of Chicago Press. Chicago. Chaps. 1–10.

Mather, K. 1953. The genetical structure of populations. Symp. Soc. Exp. Bio., 1. **7:** 66–95.

Mather, K. 1973. Genetical structure of Populations, Chapman and Hall, London.

Mayr, E. 1954. Change of genetic environment and evolution. In: Evolution as a Process. ed. J.S. Huxley, A.C. Hardy, E.B. ford. pp. 156–180. Allen and Unwin, London.

Wilson, E.O. and Bossert, W.H. 1971. A Primer of Population Biology. Sinauer Associates. Inc. Publishers, Sunderland, M.A.

Wright, S. 1921. Systems of mating. Genetics, **6:** 1I1–178.

Wright S. 1931. Evolution in Mendelian Populations, Genetics, **16:** 97–159.

Wright, S. 1940. Breeding structure of populations in relation to speciation. Amer. Nat., **74:** 232–248.

Weinberg, W. 1908. Uben Den Nachweis der Vererbung bein Memschen.

20

Selection Theory

When we are practising selection in a population we are selecting a group (or percentage) of individuals or families having phenotypic values greater than a particular level (the truncation level) and allowing it to form the next generation population. The difference between the mean of the selected group of individuals and the mean of the original population is called the selection differential and the mean of the progenies of the selected group of individuals minus the mean phenotypic value of the original population is called the response due to selection.

20.1 Response to Selection

What we have seen above is that there are two populations, selected parental population and its progeny population. As phenotypic value of an individual can be partitioned into genotypic value and environmental value, we have

$P = G + E$ for parental population

$p = g + e$ for progeny population

Now the response due to selection will depend on the relationship of P with g provided the parental and the progeny environments are uncorrelated. This relationship can be measured by the regression coefficient bg/P, the regression of g, the genotypic value of progeny on the parental phenotypic value and estimated

$$bg/P = \frac{\text{Cov}(P, g)}{\sigma_P^2} = \frac{\text{Cov}(G, g)}{\sigma_P^2}$$

In case of parent-offspring and mid parent-offspring the cov. $(G.g)$ is $\frac{1}{2}\sigma_A^2$ (or $\frac{1}{4}DR$) and σ_A^2 (or $\frac{1}{2}DR$), respectively. Thus the regression coefficient in case of mid parent-offspring is a measure of heritability in narrow sense (h_n^2) which is shown in Figure 20.1 where X represents the mid-parental values and the Y represents the offspring values. $\overline{X}$ represents the mean of mid-parental values and $\overline{Y}$ represents the offspring mean. Now if individuals with higher means ($\overline{X}_S$) are selected then the selection differential S becomes $\overline{X}_S - \overline{X}$ and by extrapolating, the regression line, the response due to selection R becomes

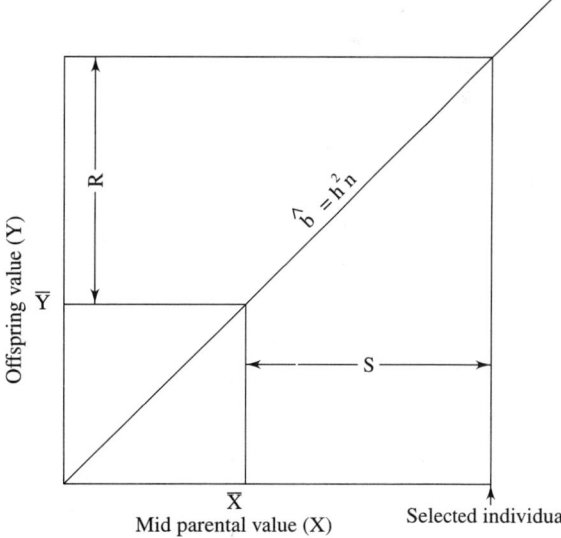

Fig. 20.1 Figure showing response due to selection, S is the selection differential and $\overline{X}$ is the mean of the mid parental values and $\overline{Y}$ is the mean of offspring values.

$$R = h_n^2 \times S$$

In other words, the response due to selection is the portion of selection differential which is transmitted to the progeny and if heritability is 100% then all S will be transmitted to the progeny. The estimate of heritability obtained as

$$h_n^2 = R/S$$

is called the realized heritability. Larger population tends to give greater response to selection due mainly to larger realized heritability. Frankham et al. (1968) observed no consistent effect of selection intensity on realized heritability. For populations with the number of individual scored, less intense selection gave greater realized heritability.

20.1.1 Selection differential

The selection differential S measured in standard deviation unit is called the intensity of selection i.

$$i = \frac{S}{\sigma_P}$$

where $\sigma_P = \sqrt{V_P}$ and V_P is the phenotypic variance of the population.

The selection differential is a function of the selection intensity. For various S, the selection differential, the selection intensity is given below:

Selection differential	Selection intensity
2.56	1%
2.42	2%
2.05	5%
1.75	10%
1.40	20%
1.16	30%

In case of small size population, the intensity of selection can be known from the Table 20 of ordinal or ranked data of Fisher and Yates, 1943, given the selected top q' of the population. For example, as shown in Table 20.1, if one individual out of 20 is selected then the selected individual's mean will deviate from the population mean $\overline{X}$ by 1.87 σ_P i.e. S, the selection differential equals 1.87 σ_P.

Assuming the population to be normally distributed, the response due to selection can be shown as in Figure 20.2 given q, the per cent selected population. The population mean being $\overline{X}$ and the mean of selected population is $\overline{X}_S$ and S is the selection differential $(\overline{X}_S - \overline{X})$. X', is the point of truncation. The distance from $\overline{X}$, the mean of distribution to X', the point of truncation equals d. The $\overline{X}_S$, the mean value of the proportion q of the population falling above X' is

$$\overline{X}_S = \frac{1}{q}\frac{1}{\sqrt{2\pi}} \int_{X'}^{\infty} xe^{-x^2} dx = \frac{1}{q} = \frac{1}{\sqrt{2\pi}} e^{\frac{X'^2}{2}}$$

$$= \frac{1}{q} f(X') = \frac{Z}{q}$$

where Z is the height at d which is the ordinate of the standard normal frequency distribution (Z-distribution) at the point X'. Thus the selection differential in standard deviation unit i becomes

$$i = \frac{Z}{q}$$

and the selection differential S becomes

$$S = \frac{Z}{q}\sigma_P = i\sigma_P$$

Now, the response due to selection R becomes

$$R = i\sigma_P h^2$$

Thus response due to selection depends on the intensity of selection, phenotypic standard deviation (or variance) and the heritability. Here the response R is expressed in some unit and to make it free of unit in order to compare the response of different traits and different populations, R should be divided by the phenotypic standard deviation σ_P which then takes the form:

$$R = ih\sigma_A$$

$h = \sigma_A/\sigma_P$ and $\sigma_A = \sqrt{\sigma_A^2}$, where σ_A^2 is the additive genetic variance.

Considering that heritability is a function of a number of parameters (see Chapter 17), the rate of speed of selection response will finally depend on: (i) selection intensity, (ii) number of genes or effective factors, (iii) variation in the magnitudes of additive

Table 20.1 Intensity of selection in small size population

Number of selected individual	Sample size										
	20	21	22	23	24	25	26	27	28	29	30
1	1.87	1.89	1.91	1.93	1.95	1.97	1.98	2.00	2.01	2.03	2.04
2	1.41	1.43	1.46	1.48	1.50	1.52	1.54	1.56	1.58	1.60	1.62
3	1.13	1.16	1.19	1.21	1.24	1.26	1.29	1.31	1.33	1.35	1.32
4	0.92	0.95	0.98	1.01	10.4	1.07	1.09	1.11	1.14	1.16	1.18
5	0.75	0.78	0.82	0.85	0.88	0.91	0.93	0.96	0.98	1.00	1.03

genetic effects, (iv) variation in the magnitudes of dominance genetic effects, (vi) linkage relation, (vii) magnitude of E_1 and E_2, (viii) sampling error, (viii) initial amount of free and potential variability. Thus response to selection is determined by genetic and nongenetic variance, selection intensity and the size of the experiment. As heritabilities estimated from parent-offspring regressions and full-sib components of variance are biased upwards by epistatic and non-genetic maternal variance, so predicted response will be greater than realized response.

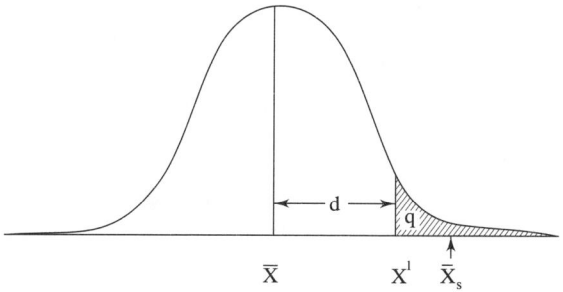

Fig. 20.2 The distribution curve showing the population mean $\overline{X}$, mean of selected population, $\overline{X}s$, the point of truncation, X' and q, the percent population selected.

As the selection results in reduction of genetic variances of all traits under selection and also modifies the covariances between traits, the phenotypic variance and covariance will change in subsequent generations. If the proportion of selected individual is large, change in genetic variance will be obvious only after a number of generations and correlations being more important but if the proportion of selected individual is less than 0.1, both this effect and the non-normality which also develops as a result of selection could complicate the formula of response to selection.

20.1.2 Response due to selection in tetraploid

For any ploidy level the response to selection measured at equilibrium depends only on A, additive genetic variance but the immediate response is proportional to the covariance of offspring and parent which differs between diploid and tetraploid. In case of diploid whether parents or self progenies are used to reconstitute the new generation population the response is the same in both cases in diploid crops. But in case of tetraploid the response will be different as there is effect of inbreeding on heterozygosity. Recombining selfed progeny for n generations will lead to an inbreed coefficient of $F_n = (1 - (5/8)^n/13 = 1/13$ and which although looks smal but the effect of inbreeding on heterozygosity could be much greater. Thus recombine only parents and not their selfs (Gallais, 1981).

Selection induced a gametic-phase disequilibrium that reduces the observed additive variance. The decrease in additive genetic variance can be offset by increase in the variance from random mating.

20.2 Correlated Response

When we are practising selection for a particular trait we should not expect any improvement in other trait(s) if the traits are uncorrelated. But, if the trait under selection is genetically correlated with other trait(s) we should expect improvement in the correlated trait(s) as well. Thus improvement in a trait can be made by direct selection as well as indirect selection. Supposing two correlated traits, X and Y, we must workout what would the amount of response in Y if the selection is practised in character X. When the selection is applied independently in X and Y then the response due to direct selection in $X(R_x)$ and in $-Y(R_y)$ will be follows:

$$R_x = i_x h_x \sigma_{A_x}$$
$$R_y = i_y h_y \sigma_{A_y}$$

When X and Y are correlated, the change in Y due to change in X can be measured by the regression coefficient b_{yx} of additive genetic value (or breeding value) of $Y(A_y)$ on the additive genetic value of $X(A_x)$ which becomes

$$b_{yx} = \frac{\text{Cov}(A_x \cdot A_y)}{\sigma^2_{A_x}}$$

$$= r_A \frac{\sigma_{Ay}}{\sigma_{Ax}}$$

where r_A is the additive genetic correlation coefficient between X and Y. Now, the change in Y due to change in X as a result of selection which is the correlated response in Y (CR_y) becomes

$$CR_y = b_{xy} \cdot R_x$$

$$= r_A \frac{\sigma_{Ay}}{\sigma^2_{A_x}} i_x h_x \sigma_{Ay}$$

As $\sigma_{Ay} = h_y \sigma_{Py}$ so substituting σ_{Ay} by $h_y \sigma_{Py}$ in the above equation the CR_y takes the form as

$$CR_y = i_x h_x \sigma_{Ax} r_A h_y \sigma_{py}$$

The advantage of indirect selection over direct selection for Y can be measured by the ratio CR_y/R_y which becomes

$$\frac{CR_Y}{R_y} = r_A \frac{i_x h_x h_y \sigma_{Ax} \sigma P_y}{i_y h_y \sigma P_y}$$

$$= r_A \frac{i_x h_x}{i_y h_y}$$

If this ratio is greater than one, indirect selection will be superior to direct selection and if the ratio is less than one, the reverse will be true. The above ratio would be greater than one when either $i_x > i_y$ i.e. when higher selection intensity can be applied in character X than in Y or $h_x > h_y$ i.e. when the heritability of character X, h_x is greater than the heritability of character Y, h_y. Thus the circumstances under which indirect selection of trait would be superior to its direct selection are (i) when the desired trait for direct selection cannot be precisely measured and, (ii) when the desired trait appears only in one sex but the other trait linked to the desired trait appears in both sexes and thus higher selection intensity can be applied in indirect selection. Study of this type will provide information on whether or not the direct selection for yield would be effective in practical plant breeding programme. Falconer (1960) touched the problem of direct and indirect selection in a different way. As the growing condition differs in the selection field than in commercial field, the character under selection should be a character that is correlated with the character of ultimate interest. Thus it is the correlated response which is decisive for the success of selection under field condition. The efficiency of such indirect selection is then given by the ratio h_s/h_c, where $h = \sqrt{h^2}$ and pg is the genetic correlation between the characters in the selection field s and the same character in commercial field c. The indirect selection can be said to be more effective only when this ratio pgh_s/h_c is greater than one which has not been proved so far.

Response due to selection will be greater if the trait under selection is not correlated (either genetically or environmentally) with other traits other wise it will give a correlated response which consists of reduction in fertility, variability and other fitness characters. Selection intensity applied to a desired trait is divided between the selection for the two traits. Relationships among traits that slow the breeding progress are (i) symmetry (or harmony) in size of plant parts, (ii) compensation among plant parts (component compensation) and (iii) linkage. Realized response is less than expected response because of the correlated response. Both the rate of progress and the limit are less than expected.

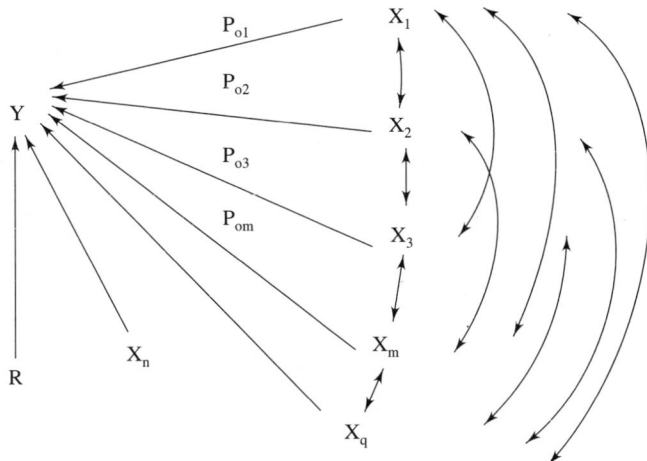

Fig. 20.3 Path diagrams showing direct effects of independent variables (X_s) on dependent variable Y.

20.3 Path Analysis

Correlation coefficients measure the absolute value of correlation between variables in a given body of data. Correlation does not say anything about the cause and effect relationship. Wright (1921) designed the method of path analysis for the purpose of interpretation of a system of correlation coefficients in terms of paths of causation. The theory underlying path analysis is that a variable Y is represented as completely determined by a number of immediate factors $X_1, X_2, X_3, \ldots X_m, R$ all of which except the residual R are represented as inter-correlated as shown in Figure 20.3. The path diagram shows Y as either completely determined by certain other factors or as the ultimate factor. In plant breeding context Y, the dependent variable (effect) could be the yield and $X_1, X_2, \ldots X_m$, the independent variables, could be the yield components and are inter-correlated among themselves and also correlated to Y and R the residual factor is not included in the experiment when all the relations are linear i.e. when Y is treated as a linear function of a number of others, X_1 to X_m and the residual factors R we have

$$Y = b_0 + b_{01} X_1 + b_{02} X_2 + b_{03} X_3$$
$$+ \ldots + b_{0m} X_m + b_{0R} R$$

where b_{0i}'s are the partial regression coefficients and they are defined as path regression coefficients and which measure the concrete contribution that $X_1, X_2, \ldots X_m$ make directly to Y from the point of view represented in the figure. The coefficient b_{OR} refers to the coefficient of residual factors not included in the study. When these variables are standardized i.e. when they are measured from their means in units of standard deviation ($Y^1 = (Y - \bar{Y})/\sigma y_1$ $X_l^1 = (X_1 - \bar{X})/\sigma x_1$ etc.) the above equation takes the form as:

$$Y' = P_{01} X'_1 + P_{02} X'_2 \ldots + P_{om} X'_m + P_{OR} R'$$

where P_{oi}'s are the standarized path regression coefficients.

The correlations between Y and the known variables ($X_1, X_2, X_3, \ldots X_m$) may be written as a series of simultaneous linear equations equal in number to the unknown path coefficients:

$$r_{YX_1} = P_{01} + r_{X_1 X_2} P_{02} + r_{X_1 X_3} P_{03} + \ldots + r X_1 X_m P_{0m}$$
$$r_{YX_2} = r_{X_1 X_2} + P_{01} + P_{02} + r_{X_2 X_3} P_{03} + \ldots + r X_2 X_m P_{0m}$$
$$\vdots$$
$$r_{YX_m} = r X_1 X_m P_{01} + r X_2 X_m P_{02} + r_{X_2 X_m} P_{03} + \ldots + P_{0m}$$

Thus the correlation between Y and any variable X_q takes the form as

$$r_{YX_m} = \sum_{i=1}^{n} P_{0i} r_{iq}$$

This is the basic equation of path analysis.

Let us now see how these equations have been derived. Considering $Y = X_1 + X_2 + X_3$ the correlation between Y and X_1, r_{YX_q} would become

$$r_{YX_1} = \frac{\text{Cov} \cdot (X_1, Y)}{\sqrt{\text{Var}(X_1) \cdot \text{Var}(Y)}}$$

$$= \frac{\text{Cov}(X_1, X_1 + X_2 + X_3)}{\sigma_{X_1} \sigma_Y}$$

$$= \frac{\text{Cov } X_1 X_1}{\sigma_{X_1} \sigma_Y} + \frac{\text{Cov } X_1 X_2}{\sigma_{X_1} \sigma_Y} + \frac{\text{Cov } X_1 X_3}{\sigma_{X_1} \sigma_Y}$$

$$= \frac{\sigma X_1^2}{\sigma X_1 \sigma Y} + \frac{\sigma X_1 X_2}{\sigma X_1 \sigma Y} + \frac{\sigma X_1 X_3}{\sigma X_1 \sigma Y}$$

$$= \frac{\sigma X_1}{\sigma Y} + \frac{\sigma X_1 X_2}{\sigma X_1 \sigma Y} \cdot \frac{\sigma X_2}{\sigma X_2} + \frac{\sigma X_2 X_3}{\sigma X_1 \sigma Y} \cdot \frac{\sigma X_3}{\sigma X_2}$$

$$= P_{01} + r_{X_1 X_2} P_{02} + r_{X_1 X_3} P_{03}$$

where

$$P_{01} = \frac{\sigma X_1}{\sigma Y}, P_{02} = \frac{\sigma X_2}{\sigma_Y} \text{ and } P_{03} = \frac{\sigma X_3}{\sigma Y}$$

Thus considering $Y = X_1 + X_2 + X_3 + \ldots X_m$ it can be shown that

$$r_{YX_1} = P_{01} + r_{X_1 X_2} P_{02} + r_{x_1 x_3} + \ldots + r_{X_1 X_m} P_{0m}$$

Here we see that the correlation coefficient between variables is partitioned into (i) direct effect measured by path coefficients and (ii) indirect effect. The path coefficient P_{01} measures the direct effect of X_1 on Y where as P_{0m} measures the direct effect of X_m on Y. The path coefficient is concrete in the sense that it measures the influence of one variable on the other from a particular view point indicated in Figure 20.3. $r_{X_1 X_2} P_{02}, r_{X_1 X_1} P_{03}, \ldots r_{X_1 X_m} P_{0m}$ are the indirect effects of X_1 on Y through $X_2, X_3 \ldots X_m$, respectively. Thus the method of path coefficient which is a form of multiple regression, can be applied to find the best linear expression of one variable in terms of a number of others from the knowledge of correlation coefficients.

The above simultaneous equations can be written in the form of matrices as follows:

$$\begin{bmatrix} ry\,X_1 \\ ry\,X_2 \\ \vdots \\ ry\,X_m \end{bmatrix} = \begin{bmatrix} r_{X_1 X_1} & r_{X_1 X_2} & \cdots & r_{X_1 X_m} \\ r_{X_2 X_1} & r_{X_2 X_2} & \cdots & r_{X_2 X_m} \\ & & & \\ r_{X_m X_1} & r_{X_m X_2} & \cdots & r_{X_m X_m} \end{bmatrix} \times \begin{bmatrix} P_{01} \\ P_{02} \\ \\ P_{0m} \end{bmatrix}$$

$$\underset{\sim}{A} \qquad \underset{\sim}{B} \qquad \underset{\sim}{C}$$

The path coefficient values are estimated as

$$\underset{\sim}{C} = \underset{\sim}{B}^{-1} \underset{\sim}{A}$$

20.3.1 Determination of residual variability

Considering the equation $Y = X_1 + X_2 + X_3$ it can be shown that

$$\sigma_Y^2 = \sigma_{X_1}^2 + \sigma_{X_2}^2 + \sigma_{X_3}^2 + 2\sigma_{X_1 X_2} + 2\sigma_{X_1 X_3} + 2\sigma_{X_2 X_3}$$

Now, dividing both sides by σ_Y^2

$$\frac{\sigma_Y^2}{\sigma_Y^2} = \frac{\sigma_{X_1}^2}{\sigma_Y^2} + \frac{\sigma_{X_2}^2}{\sigma_Y^2} + \frac{\sigma_{X_3}^2}{\sigma_Y^2} + \frac{2\sigma X_1 X_2}{\sigma_Y^2}$$

$$+ \frac{2\sigma_{X_1 X_3}}{\sigma_Y^2} + \frac{2\sigma_{X_2 X_3}}{\sigma_Y^2}$$

or $1 = (P_{01})^2 + (P_{02})^2 + (P_{03})^2 + 2r_{X_1 X_2} P_{01} P_{02}$

$$+ 2_{rX_1 X_3} P_{01} P_{03} + + 2r_{X_2 X_3} P_{02} P_{03}$$

Now extending the model to include residual factors $R (= X_1 + X_2 + X_3 + \ldots + X_m + R)$, the degree of determination of Y by the residual factors R, can be estimated as

$$R^2 = 1 - \sum_{i=1}^{m} P_{01}^2 - 2 \sum_{j,k=1}^{m} P_{0i} P_{0k} r_{jk}, k > j$$

20.3.2 Path coefficient vs correlation coefficient

Path coefficients differ from correlation coefficients in that they may exceed by + 1 or − 1 in absolute value as there is no restriction on the relative magnitudes of the variances of an effect and a cause. Further, we can have situation when the correlation coefficient between the two variable is zero yet the path coefficient from one variable to the other is not zero.

20.3.3 Application of path analysis

If the correlation coefficient say r_{YX_1} equals path coefficient P_{01}, the selection based on the trait X_1 will be effective. If the correlation coefficient is positive, but direct effect is negative or negligible then indirect causal factors are to be considered simultaneously. In case correlation coeficient is negative but direct effect is positive and high then restricted selected index should be applied which will nullify the undesirable indirect effect in order to make use of direct effect. As path analysis is based on the assumption that all the variation is due to additive gene action, this analysis has all the limitations of any linear model method in a non-linear world.

20.4 Yield and Yield Components

Yield, a complex character. is the end point of a process of which the successive stages are represented by the observed primary characters. A group of two or more characters which completely determine the complex character are termed component traits. Thomas and Graffius (1976) defined the components as strictly those characters which when multiplied together give yield exactly. $X_1, X_2 ... X_n$ can be said to be component traits yield 'Y' if

$$Y = X_1 \cdot X_2 \cdot X_3 ... X_n$$

The components have the strongest influence on the genetic variation of the complex character. There is no way in which yield can be changed without changing one or more of the components. Further, all changes in components need not be expressed in changes in yield but all changes in yield must be accompanied by changes in one or more of the components.

Correlation coefficients provide incomplete information about the nature of the relationship between yield and its components because the components are mutually correlated. The coefficients of correlation, r or determination r^2 do not quantify the contribution of individual components to the variance of the complex trait because the complex trait does not have a linear relationship with its components. Thus the fact that a trait is correlated with yield does not imply that it is a component of yield. Some times there is component compensation i.e. there is influence of the variation of one component on that of another. Willams (1959) suggested that the maximum level at which a single component can function separately is far in excess of what can be achieved by their product. Infact there are physiological limitations, at least partly imposed by the environment that are expressed as negative correlation between components and the preceeding primary traits. The negative correlations between yield components are often interpreted as indicating that yield is limited by the supply of assimilates and the other explanation is that yield component compensation occurs. For example, in maize there is negative correlation among the principal yield components such as ears per plant, number of kernal rows, number of kernels per row and kernel weight, in beans there is negative correlation among number of pods per plant, average number of seeds per pod and average seed size (weight) and in tomato there is negative association between number of fruits per plant and weight per fruit. A genetic correlation can be due to (i) genetic linkage, (ii) pleiotrophy or from developmentally induced relationship between components that are only indirectly the consequence of gene action. Under non-competitional or non-stress environment component correlations are near zero and the absence of the correlation thus supports the hypothesis of genetic independence among the components. Negative correlation in competitional environment might have arisen in response to competitional forces operating on developmentally flexible components. The attainment of characteristic form and function in a crop plant depends upon a chain of interrelated events which are sequential in time, gene regulated at critical sites and times and subject to the modifying influences of non-genetic forces and these events follow an integrated pattern.

Negative correlation arises primarily from developmentally induced relationship. Such a relationship occurs when two or more developing structures compete for a common limited nutrients supply. If one structure is favoured for, any reason, over the other in amount of nutrients received a negative correlation develops which Stebbins (1950) called developmental correlation. Developmental plasticity enables plant to take alternate pathways in the attainment of the final adult form or metric values. Developmental plasticity of yield components could facilitate the maintenance of a more stable yield level if in development variation on one component tended to compensate for variation in another. Thus component compensation (leads to negative association) might be expected almost as a regular feature of the development. On the other hand, positive association will occur if component and the preceeding primary characters react favourably to the variation in the environment or genetic background.

Selection criteria must be chosen among the components on the basis of mutually independent effects in yield i.e. on the bases of their complementary determination cd values (Bos and Sparnaaij, 1993). The primary characters represent successive ontogenical or chronological stages in the process leading to Y, the complex character where

$$Y = X_1 \cdot X_2 \cdot X_3 \ldots X_n$$

An increase in the r^2 value from one stage, say X_1, to the next stage (say X_2) represents the influence of the intervening component (X_2). The difference between the r^2 values of the two consecutive primary traits can be taken to be the complementary determination (cd) variation in Y by the variation in the intervening component. Eaton and Kyte (1978) used sequential multiple regression analysis of the log-transformed data for yield and its components assuming

$$\ln Y = \ln X_1 + \ln X_2 + \ln X_3 + \ln X_4$$

The assumption under this analysis is that the components should be introduced in the correct order. Choosing the component is a problem sometimes. Yield components in some crops are as follows:

Crops	Yield components
Maize	Number of ears/plant, number of kernel rows, number of kernel per row, kernel weight.
Wheat	Number of spikes per unit area, number of kernels/spike and average kernel weight.
Rice	Number of panicles/unit area, number of spikelets/panicle.
Legumes	Number of pods/plant, number of seeds/pod. 100-seed weight.
Tomato	Number of fruits/plant, weight per fruit.
Sugarcane	Number of milleable canes/unit area, cane weight, sugar percent.
Sunflower	Head diameter, number of kernels/head, average seed weight.
Rapeseed	Number of primary branches, number of secondary branches, seed weight, siliqua length, number of seeds siliqua.

20.5 Types of Selection

When an individual is selected in a population, the individual's performance will be equal to the sum of deviation of its family mean P_f from the population mean and the deviation of the individual from the family mean P_w.

$$P = P_f + P_w$$

These values can be shown by Figure 20.4(a-b) in which $\bar{X}$ represents the population mean. There are thus three different ways of carrying out selection:

1. Individual selection
2. Family selection
3. Within family selection

In the individual selection both P_f and P_w components are given equal weight and only individual with high phenotypic value is selected. Family selection is based on the component P_f alone and families with high mean phenotypic values are selected whereas in the within family selection, the selection is based on P_w alone i.e. individual showing highest deviation from its family mean is selected. The families may be of full-sibs or half-sibs or selfed progenies. In the combined selection P_f and P_w are given different weight.

20.5.1 Efficiency of different methods

The efficiency of a method can be judged by the response to selection. Assuming intensity of

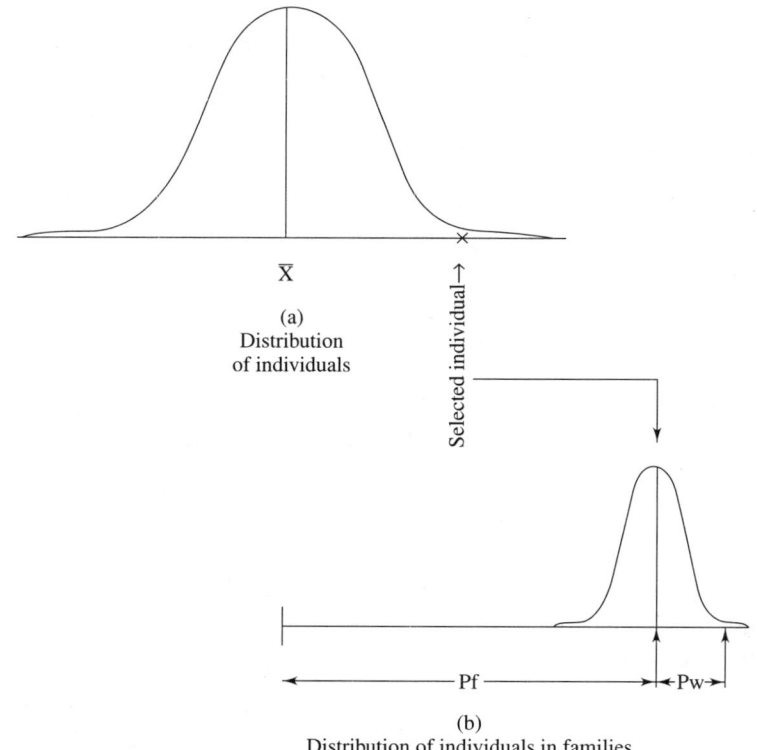

Fig. 20.4 (a-b) Figures show the possible three types of selection. Individual selection ($P_f + P_w$), family selection (P_f alone) and within family selection (P_w alone).

selection and phenotypic standard deviation constant, the response due to selection will depend on the heritability of trait under consideration. Individual selection appears to perform better at intermediate (0.25 to 0.50) heritability than at low (0.1) or high (0.75) heritability. Family selection is most useful when heritability h^2n is low, E_2 (the common environmental effect) is low and the family size is large. Even if E, the environmental variation, is large, it will cancel out and the mean phenotypic value of a family will be close to its mean genotypic value. Further, larger the family size, the closer will be correspondance between phenotypic mean and the genotypic mean. However, if P_f is largely due to E_2, the common environmental effect or maternal effect and thus the heritability is low because of this reason then within family selection can be useful.

The estimates of heritability will differ in different methods of selection and are estimated as described in Chapter 17. The heritability estimate in case of within family selection can be obtained as

$$h_n^2 = \frac{(1/4)DR}{(1/4)DR + (3/16)HR + E}$$

For comparison of the various methods one will have to keep the total experimental size constant and keeping the intensity of selection constant although not possible as the proportion of individual selected in different methods cannot be the same, the relative response then depends on heritability of a character and the family size. The family size, in turn, depends on the number of families that one wants to accommodate in the experiment. The greater the family size, the lower will be the number of families to keep the total experimental size constant. For the same total experimental size say 1000 plants we can have 100 families with 10 progenies each or 50 families with 20 progenies each or 200 families with

5 progenies each. Thus we can see that the selection intensity in individual selection in population of 1000 plants will be higher than in family selection and thus the family selection will give a lower selection differential. The optimum number of families and family size can be determined that will maximise the selection response by combined selection for a fixed total size of experiments (Robertson, 1957b). The composition of the genetic variance has a limited influence on the optimal size of the progenies. If the epistatic variance is important then the number of families has to be reduced and the number of F_3 and F_4 families must be increased. The same is true if non-genetic variance is increased. The gain from selection in segregating population depends on the relation of additive to epistatic variance. With additive type of gene action a combination of mass, family and progeny test selection would be most effective. In the presence of epistasis mass or family, selection can make use of additive type of epistasis but will not be effective with others.

20.5.2 Combined selection

In the combined selection an individual is selected on the basis of its family performance as well as its own performance in the family. The extra gain from application of combined selection is not worth preferring it over other methods. In practice, we are applying the combined selection when we are advancing the best individual from the best family. Theoretically the combination of family and within family selection should bring more rapid genetic progress and should be the best method to use (Empig et al. 1972).

20.6 Selection Criterion

In any selection method, one can practice selection for one character or two or more characters at a time and thus the selection can be of the following types:

(i) Single trait selection

(ii) Multi-trait selection.

20.6.1 Single trait selection

Single trait selection is similar to tendem selection in which the selection is practised for one trait at a time. When the desired level of improvement is obtained in that trait then selection is practised for improving another trait and thus characters are improved one after another. The gain due to selection can be calculated using the formula described in Section 20.1. The problem with this type of selection is that it will take very many number of years to improve a number of traits which we need to improve. Tendem selection is not useful when there is negative correlation between traits.

20.6.2 Multitrait selection

In this type of selection we can select a number of traits simultaneously. The multi-trait selection can be of two types: (i) Independent culling level, and (ii) Selection index.

20.6.2.1 Independent culling level

Here the individual or family with phenotypic values falling below the limit set for each trait is rejected. Infact, this is the type of selection which most plant breeders are practising. Plant breeders select plants on the basis of visual observation on a number of traits as they have rough idea about the economic importance, heritability, correlation between traits and phenotypic variation of the different traits. This method is easy in the sense that whenever an individual shows undesirable trait(s) it is rejected.

20.6.2.2 Selection index

In this method of selection a phenotypic score or index I is given to individual or family and those with I's equal or greater than the set level of index value, say I', are selected. In this method of selection individuals with high merit in one characteristic are selected even though they are a bit inferior in other characteristics. Smith (1937) used Fisher's (1936) concept of discriminate function to develop an index for selection of plants whereas Hazel (1943) developed index for selection for animals application to describe what linear function of the observable characters may best indicate the genetic value of an individual or family. Following Henderson (1963) the phenotypic index of an individual or family is made up of two components as follows:

$$I = T + E$$

Phenotype | Additive genetic merit | All other effects
or | or | (E_1 + non-additive
Phenotypic index | True breeding value | genetic effect)

In case of n traits the phenotypic index takes the form as:

$$I_\alpha = b_1 X_1 + b_2 X_2 + \ldots + b_n X_n$$

where I is the aggregate phenotype, $X_1, X_2 \ldots X_n$ are the phenotypic score of characters on an individual and b_1, b_2, b_n are their respective coefficients and are called weight. In case of multiple traits selection, the true breeding value T is redefined as the linear function of the additive genetic merits for the traits under selection. Thus the aggregate genotype now takes the form as:

$$T = V_1 Y_1 + V_2 Y_2 + \ldots + V_m Y_m$$

where Y_i's are the additive genetic values for the m traits and V_i's are the economic weight of the trait. The objective in index selection is to maximize the average genetic worth of a population. The economic weight refers to relative importance of one character to another. The product of relative economic value, heritability and phenotypic standard deviation is a measure of the importance of a trait in the selection programme. The basic index equation takes the following form:

$$b_1 \sigma_{X_1} + b_2 \sigma_{X_1 X_2} + \ldots + b_n \sigma_{X_n X_n}$$
$$= V_1 \sigma_{Y_1 Y_1'} + V_2 \sigma_{Y_2 Y_1'} + \ldots + V_m \sigma_{Y_m Y_1'}$$

$$b_2 \sigma_{X_2 X_1} + b_2 \sigma_{X_2}^2 + \ldots + bn \sigma_{X_2 X_n}$$
$$= V_1 \sigma_{Y_1 Y_2'} + V_2 \sigma_{Y_2 Y_2'} + \ldots + V_m \sigma_{Y_m Y_2'}$$

$$b_1 \sigma_{X_n X_1} + b_2 \sigma_{X_n X_2} + \ldots + b_n \sigma_{X_n}^2$$
$$= V_1 \sigma_{Y_1 Y_n'} + V_2 \sigma_{Y_2 Y_m'} + \ldots + V_m \sigma_{Y_m Y_n'}$$

The above equations can be written in the form of matrices as follows:

$$\mathbf{Pb = GV}$$

From which the estimates of b's are obtained as

$$\hat{b} = \mathbf{P^{-1} GV}$$

where Y is a vector of additive genetic values for the m traits, V the vector of relative economic weight of these traits, X is a vector of phenotypic measure for the n traits included in the index, b is a vector of weighting factor used in the index, $\mathbf{P}$ is $n \times n$ matrix of phenotypic covariances between the X_i, $\mathbf{C}$ is an $m \times m$ matrix of genotypic covariances between the. m traits in $\mathbf{Y}$ and $\mathbf{G}$ is $n \times m$ matrix of genotypic covariances between the n variables in $\mathbf{Y}$ and the m traits in $\mathbf{Y}$.

20.6.2.2.1 Other information on selection index
(1) The variance of the index σ_I^2 is estimated as

$$\sigma_I^2 = \mathbf{P^1 \, PB}$$

where the variance of true breeding value, σ_T^2 is

$$\sigma_T^2 = \mathbf{V^1 \, CV}$$

(2) The correlation between phenotype and additive genetic merit, r_{TI} is obtained as

$$r_{TI} = \sqrt{\mathbf{b^1 \, Pb / V' \, CV}}$$

(3) The genetic gain from selection takes the form as

$$\Delta_g = r_{TI} D \sigma_T$$

where $\Delta_g = R$, $r_{TI} = \sqrt{\text{heritability}}$, $D = i$, intensity of selection and $\sigma_T = \sqrt{\text{additive genetic variance}} = \sigma_p$ of Falconer (1960). Thus after one round of selection the genetic gain would be

$$\Delta_g = \sqrt{\mathbf{b' \, Pb}} \cdot D$$

(4) The change produced in individual trait by one round of selection becomes

$$b_{yiT} = \sigma_{YiT} = \frac{\sigma_{Yil}}{\sigma_I^2} = \mathbf{b'} \text{ (ith colum of } \mathbf{G}) / \sigma_I^2$$

If b_{yiT} is zero it shows no change in the trait.

(5) The relative importance of the different variables in the index can be known by measuring the reduction in r_{TI} which would result from droping

that particular variable from the index. Thus r_{TI} original minus r_{TI} new (which is obtained after excluding a variable) will provide an idea about the importance of different variables. This procedure becomes complicated if large number of variables have been included in the index and under this situation, σ_I^2 original $- \sigma_I^2$ new will provide the same information which becomes

$$\sigma_I^2 \text{ (original)} - \sigma_I^2 \text{ (new)} = b_i^2/W_{ii}$$

where b_i is the ith weighting factor in the original index and W_{ii} is the corresponding diagonal element in P^{-1}. For further details the reader is referred to Cunningham (1969).

The assumptions under selection index are (i) the relative economic weights are known without error and are consistant over a range of variation encountered and are constant in time, (ii) the genotypic and phenotypic variances and covariances are estimated without errors and (iii) both index and aggregate genotype are normally distributed. The greatest source of error in the selection index is the poor estimates of phenotypic and genotypic variances and covariances. But even if those assumptions are violated selection index is the best method. Selection index is most useful in population improvement program aiming at improving the aggregate phenotype. The superiority of index selection increases with decreasing differences in relative importance of traits. The relative importance of trait can be estimated as a function of economic weight, heritability ($h^2 n$) and phenotypic variance(σ_p^2). Low heritability will not affect the efficiency of index. If n traits are uncorrelated, the weight of each trait is proportional to product of its economic weight and heritability but if the correlation is high, it will alter the correct weights in an unpredictable manner.

20.6.2.2.2 Characters to be included in selection index

One should carry out the multiple regression ($Y = b_1 x_1 + b_2 X_2 + b_3 X_3 + b_4 X_4 + b_5 X_5$) analysis and if for a character $b = 0.0$, then there is no use of taking that character in the formation of index. See the per cent of variation explained by one, two, three, four or five variables and suppose only three variables, explain 90% of the variation then include only three variables, in the selection index. Further, one should expect significant deviation from the theoretical expectation if discontinuous traits are included in the selection index. Disease resistance traits cannot be included because it is a simply inherited character. As with selection index, gain from selection for any given trait normally decreases as additional traits are added in the index so that traits to be included must be chosen objectively.

20.6.2.2.3 Restricted selection index

Kempthorne (1957) described a situation in which yield, the primary trait, the only character which is of direct economic importance is given a weight 1.0 and weights of all other are set to zero. This will lead to an optimum index if all other traits are correlated with primary trait. This situation is similar to optimising selection for a given aggregate genotype subject to the condition that no genetic change occurs in one or more of the traits and this can be achieved by applying the standard selection index procedure with the restriction that Cov $(I, Y_i) = 0$ for every trait Y_i specified and the selection index is now called the restricted selection index (Kempthorne and Nordskog, 1959). Restricted selection index can be applied when the population approaches the optimum level for a particular trait. Zero economic weight attached to any trait held constant is required for the proper evaluation of the selection index. The somewhat complicated matrix equation for computing restricted index can be replaced by a similar procedure suggested by Moen (1968). In this procedure a dummy variable is added to the index for each restriction and a corresponding row and column are added to the **P**. The row consists of the genotypic covariances of the other variables (the X_i) with the trait being restricted to zero change and the column is the transpose of the row and the diagonal element is zero. A row of 0 is added to the **G** and the economic values for those trait (Y_i) held constant, are set to zero in **V**. The index values are then obtained by the solution of these equations.

Tallis (1962), modified the restricted selection index for cases where it is desirable to improve

n_1 traits without any limit and n_2 the traits only to a predetermined limit. Thus Tallis modification required Cov $(I, Y_i) = K_1$, where K_1 is a pre-selected constant which specifies the amount of change desired in Y_i. Williams (1962) suggested the use of a base index that differs from the optimum index in that the traits are weighted directly by their economic value. When heritabilities are high (as with family selection) and genetic correlations are low, a base index (with the relative economic weight as the phenotypic weight) is expected to be nearly as efficient as the standard index in improving the aggregate genetic worth of the population. Pesek and Baker's (1970) index in which the desired gains are used to obtain phenotypic weights has value when the relative economic weight cannot be obtained but it may be considerably less efficient than base or standard index when the relative economic weights are known.

Changes in the structure of the genetic and environmental variation and covariation in the population over cycles of selection and of economic balance between traits through time will require periodic changes in the index and thus the requirement of estimation of phenotypic and genotypic variances and covariances among traits and/or specification of relative economic weight for each trait has limited the use of selection index in applied plant breeding.

The predicted response due to different criterions of selection is valid only for one generation though in practice it may hold for several generations. The reason is that the heritability is a population parameter and it can differ from individual to individual. Also, the value of the heritability declines as selection accumulates homozygotes and variation becomes fixed. In other words, the value of heritability will be higher in F_2 and much lower in F_8 or 9 generation. In addition, new linkage disequilibrium will be developed due to selection between genes of different characters which will affect the predicted response.

20.7 Comparison of Efficiency of Different Methods

Single stage index selection is superior to tandem selection which in turn is superior to independent culling. Selection based on n equally important, independent traits is $\sqrt{n}$ times as efficient as tandem selection for the same trait at one time (Hazel and Lush, 1942). The genetic gain in any one trait by index selection is only $\frac{1}{\sqrt{n}}$ times as much as if selection were directed at that trait alone. Selection for multi-traits by using independent culling levels for each is more efficient than tandem selection for each trait at a time, the relative efficiency increasing with the number of traits and intensity of culling. The aim of the selection is to maximise a single linear combination of X_i and this need not always be appropriate if one wishes to maximise 2 or more of the X_i (or two or more functions of X_i) simultaneously. Recognising the impossibility of this breeder may compromise by seeking to maximise a particular X_i subjected to the conditions that there shall be only small probabilities that a certain others of X_i fall below specified limits. In such circumstances independent culling level and tandem selection may compare more favourably with index selection. This is due to the insensitivity of the index coefficient due to manifold inter-relationship among the characters (Finny, 1962). Young (1961) showed the superiority of index over tandem and independent culling level. This superiority increases with the increase in the number of traits under selection and with decreasing the difference in the relative economic weight of the different traits. Swantaraden et al. (1975) compared the efficiency of the selection index with other modified forms and observed that conventional indices were efficient in bringing about satisfactory improvement in all traits. Base indices were less efficient than conventional index and selection indices based on desired gains were the least efficient among all the indices.

20.8 Selection Limit

With selection in different segregating generation populations derived from cross of two pure breeding lines or in different cycles of open pollinated population the progress continues for many generations but after sometime it starts tapering off and finally reaches a limit as shown in Figure 20.5. Selection limits may result from depletion of additive genetic variance through the fixation (Robertson,

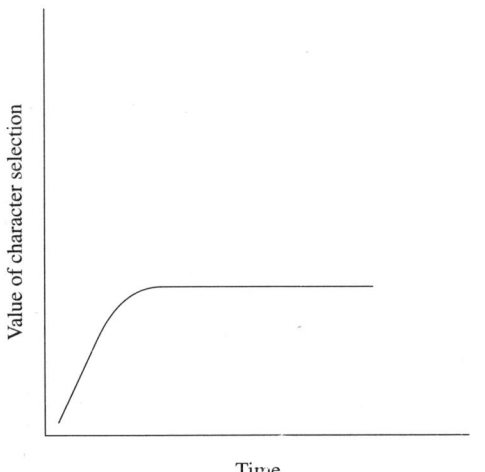

Fig. 20.5 Selection plateau.

1960). However, plateaus in selection response also may result from pleiotropic associations among components of fitness as defined by the intended plus natural selection from the interactions of genotypes with fluctuating environments between generations and from recurrent losses of selected favourable epistatic or non-genetic maternal effects (Dickerson, 1955, 1956). Plateaus can arise because of physiological and genetical limits, Favourable epistatic combinations and non-genetic maternal effects can be temporary responses from selection because they can occur without changes in gene frequency and are expected to be lost quickly when the selection is relaxed. A net response to selection continues as long as the total superiority of the progeny from selected over unselected parents exceeds the loss in performance from relaxing selection for a generation. The reverse selection reveals the persistance of genetical variation in the high selection lines long after they had stopped responding. The limits arise when there is additive genetic variation present in the population but there is negative genetic correlation between a desired trait and the other components of net reproductive capacity (the fitness traits) i.e. there is undesirable and deleterious correlated response. This represents the biological limit to selection progress. As we have seen, the correlated response can be due to genetic linkage or pleiotropy and if pleiotropy (i.e. same loci controlling the two traits) is the cause then the direction and the amount of change are determined by the type of gene action with respect to these loci and the unselected trait. But if two independent sets of loci are affecting the two traits and simultaneously there is a group of loci affecting both these two traits then in this situation the correlated response is not perfectly correlated. The genetic correlation may be positive initially and there is selection response but after some value of a desired trait there develops a negative correlation between the desired trait and the fitness of the organism and thus rate of progress will be lower and the ultimate limit to selection may also be less. Lerner and Dempester (1951) and Reeve and Robertson (1953) advanced two possibilities for reconciling presence of additive genetic variance with ineffectiveness of intravariety selection. The first rests on negative genetic correlations between a desired trait and the fitness character and the second envisages that gene frequencies at loci providing most of the additive genetic variance are at equilibrium between the forces of mutation and selection.

Limits of selection will be different in the two populations with same additive genetic variance but with different number of segregating loci. Between the two populations with the same mean, the population with larger additive genetic variance is likely to give the greater limit. Selection limits also depend on the intensity of selection. The weaker the selection the greater is the limit and for a given selection intensity, the larger the selected population the better is the ultimate response.

The selection plateau in case of yield can be broken by increasing the genetic variation in the population by way of introducting new and superior alleles at loci controlling the trait. Finding new alleles at loci that control such characteristics as disease and pest resistance can break yield plateau. Introducing new individuals i.e. wild relatives of the domesticated species from the gene centre will increase the genetic variability of the population. Finally, mutation can create variability (Gregory, 1968).

20.9 Natural Selection

Considering one locus with 2 alleles (A, a) system we have AA, Aa and aa genotypes in a population. Further, assuming additive gene action these three genotypes will represent three phenotypically different classes. Individuals from *AA* and *aa* classes represent the extreme types whereas individuals belonging to *Aa* class is of intermediate type. When we are considering k loci together as in case of quantitative trait, we see that a quantitative trait follows a normal distribution as shown in Figure 20.6. Although normal distributions are common but not universal among continuously varying characters. Loosely, we can classify the curve into 3 parts (or class), the two tails represent the two extremes expression of a trait and the third represent the expression of average population. Now depending upon which one or more than one classes of individuals is favoured by selection (natural or artificial) we can have three types of selection: (i) Stabilizing selection, (ii) Directional selection, (iii) Disruptive selection.

20.9.1 Stabilizing selection

In the stabilizing selection, individuals with means equal to or very close to population mean are selected and thus become the parents of the next generation population as shown in Figure 20.6. The next generation population will have the same mean as the previous one but with a narrow range (or reduced variance). Stabilizing selection is generally considered as the most prevalent type of natural selection. The evidence for this type of selection to have occurred is seen when the majority of the population closely approximates the optimum phenotype. Phenotypes close to the mean have highest fitness and decline in fitness increases with the departure of phenotypes from the mean. As a result of selection, individuals which have low fitness are eliminated and loss of viability or fertility (which are fitness character) is due to: (i) ecological control of population size and (ii) to differential effects between genotypes (competitive ability) i.e. to selection. Those which survive have better fitness and better competitive ability. The selection affects the genetic components of variation. There is reduction in additive genetic variance. There is reduction in non-heritable variance but then it depends on the G × E interaction. The developmental component of variation is also reduced. The reduction in additive genetic variance is due to the development of repulsion phase linkage of

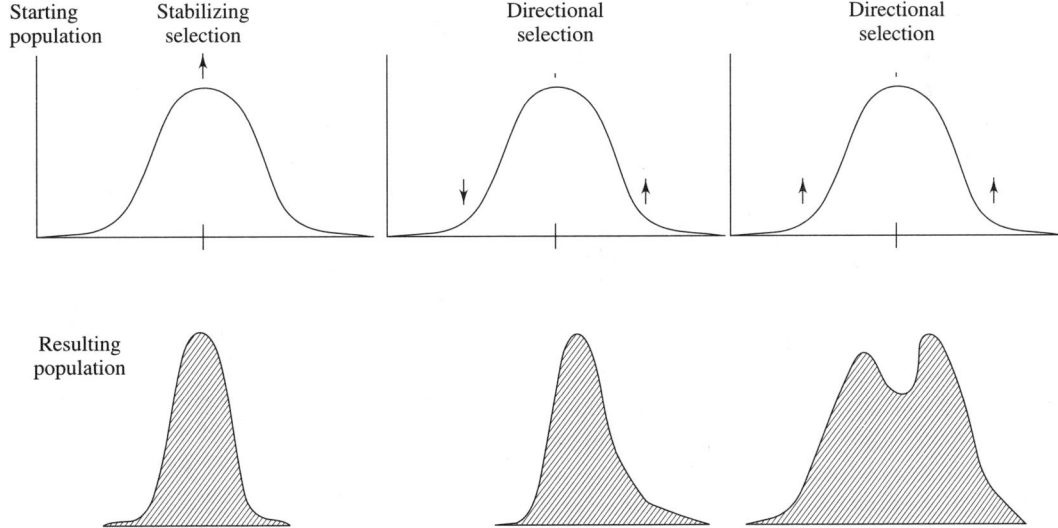

Fig. 20.6 Effects of different types of selection on the population frequency distribution of a phenotypic character. The arrow (↑ or ↓) shows selection pressures on different parts of the population frequency distribution. In these figures, the ordinates represent the frequencies of individuals in the population and the abscissas, the phenotypic variation.

favourable relevant genes. Thus stabilizing selection results in the decrease of phenotypic variation. Non-fitness characters are under stabilizing selection. Breese and Mather (1960) showed that stabilizing selection for character will tend to produce ambidirectional dominance and interaction if it is prevalent at all. Further, Mather (1974) showed that [h], [i], [j] and [l] tend towards O. Also, as neither + allele nor-allele has, under stabilizing selection, any advantage over the other, the average value of $u - v$ taken over all loci tends to 0 also. This will result in the reduction of DR and HR. Mather (1987) discussed the consequences of the stabilizing selection on the genetic structure of the random mating population. Assuming random mating and absence of epistasis, the stabilizing selection acting on a pair of alleles (A, a) can have any of the three possible outcomes depending upon the relative values of m and h which characterize the effects of the gene differences on the primary character viz. (i) a stable equilibrium in the population where in respect of the primary character, Aa is near to the optimum than both homozygotes (AA, aa), (ii) fixation of the fitter alleles where Aa is intermediate between the homozygotes in its departure from the optimum and (iii) a theoretical unstable equilibrium leading to fixation of the commoner allele where Aa departs further from the optimum than both homozygotes but only the first outcome can lead to the conservation of variation in the population. We have seen in case of balanced polymorphism that Aa individuals are favoured over the homozygotes AA and aa and this shows that stabilizing selection is operative. Every quantitative trait in any species having an intermediate optimum and natural selection keeps the population near this optimum. With intermediate optimum, both alleles are maintained in the population and hence the population variability is increased. Mutations (and possibly overdominance, epistatic pleiotropic effects) maintain variability. The phenotypes contain a great deal of additive genetic variance which allows it to move quickly to a new optimum whenever the environment changes. Thus evolution is through the continual stabilizing selection toward a slowly changing optimum rather than the result of directional selection.

20.9.2 Directional selection

In the directional selection, selection is for an extreme expression of a trait. However, as we can see in Figure 20.6 the two tails represent the two extremes (one in positive and the other in the negative direction) expression of a trait shown in directional selection, selection can be either in the direction of higher mean or lower mean. For a trait like yield, directional selection is for higher yield but for a trait like maturity, directional selection can be for either early maturity or late maturity. Thus in directional selection only individuals representing one extreme are selected over generations. The mean of the next generation population is shifted in the direction of the selection pressure applied but the range may be somewhat contracted though the reduction in variance due to selection is very small. In the natural population, the traits such as the viability and fertility are under directional selection through the changes in environmental conditions or heterogeneous environments. Directional selection tends to produce unidirectional dominance and duplicate type of interaction and favour coupling type of linkage. With duplicate interaction there is reduction in the apparent additive (DR) and dominace (HR) components and thus assuming other things being equal, the genetical contribution of the variance of the character in population would be expected to be lower and consequently the heritability h^2 to be lower. Thus the fitness character should have lower heritabilities than non-fitness characters which experience a predominance of stabilizing selection.

20.9.3 Disruptive selection

In this type of selection, the next generation population is formed by individuals, selected from more than one class as shown in Figure 20.6. In case of natural population when selection favours more than one phenotypic optimum (and different genotypes) and discriminates against intermediate, it is called disruptive selection. The consequences of disruptive selection (Mather, 1955) are (i) increase of genetic and/or developmental variability followed by an increase of phenotypic variability and the appearance of bimodal or multimodal distribution of the selected trait while the mean value of the selected character remains unchanged, (ii) breakage of linkage or maintenance of linkage disequilibrium (favouring coupling type of linkage), (iii) establishment of polymorphism, some of which are very similar in genetic principle to sex dimorphism and mimicry polymorphism in possessing switching super genes and modifing (enhancing) genetic backgrounds, (iv) divergence and reproductive isolation. Disruptive selection can produce and maintain divergence between two populations between which there is a very high ratio of gene flow and it can also split a population into two between which there is considerable reproductive isolation, (v) correlated response to selection and (vi) breakdown of self incompatibility.

In practical plant breeding programme, disruptive selection can be used for breaking of tight linkage, for correlated response to selection, for release of latent genetic variability and for transfer of polyploidy from wild type to cultivated plants. The development of wild, intermediate and cultivated populations with gene flow between them ensures the production and increase of the very large amount of variability needed for the development of a cultivated crops.

Different species adjust to the variation in environments in different ways: (i) Some become highly specialized, restricted in range to a uniform and stabilized area and (ii) some have incorporated genotypes with wide reaction norms—genotypic and phenotypic flexibility. Individuals adjust developmentally and physiologically to changes in their environments that may occur at various times. When a random mating population is raised in constrasting environments because of the spatial variation among the individuals of the population, the different individuals will be subjected to different selection pressures and thus a population may show two or more optima (i.e. genotypes having different optimal phenotypes in different environments). When two optima coexist their relative frequencies may be critical to the function of each. Genetics but some times environment can also play a role in few systems of this nature. In other case, we can have a functional dependencies between the two optima. When the optima alternate in time, a common genotype may characterize a population over several generations but produce distinct phenotypes under different environmental conditions-a case of seasonal polymorphism (Mather, 1973).

Disruptive selection can be of two types: (i) D^+ selection (divergent directional selection) and (ii) D^- selection. In D^+ selection, individuals with either increasing alleles for a trait mate (High × High) or decreasing alleles mate (Low × Low), and there is no gene flow between these two groups of populations. This can be due to distance (geographical isolation) or the barrier. This type of disruptive selection will lead to phenotypic and genotypic divergence and this represents a model of allopatric speciation. Within these two groups of population, there will be increase in additive genetic variance but decrease in environmental and G × E interaction. In case of D^- selection, individuals having increasing and decreasing alleles mate which then leads to genetical polymorphism and this populations will show not only increase in additive genetic variance but also in environmental and genotype × environmental interaction variance. When the populations without geographical isolation are in two different habitat or niche, making different selection pressures, the genetical divergence will be produced provided that (i) the density dependent factor regulating the population size operates separately in the two niche and (ii) the selection differentials are large (at least 30%). This is a model of sympatric speciation. Geneticists differ in views on whether divergence among subpopulations can lead to reproductive isolation in the face of gene exchange among groups, and thus evolution of a new species. Only very strong disruptive selection forces can lead to such situation. Whether that intensity of selection occurs in nature in not known. Thoday and Co-workers, working with Drosophila, showed that (i) polymorphism could be established in a population under disruptive selection even with high gene flow between diverging groups, (ii) the divergent groups could be maintained distinct under high gene flow with negative associative mating and (iii) the reproductive isolation occurred between the divergent groups in spite of full opportunity for random mating.

References

Baker, R.J. 1986. Selection Indices in Plant Breeding, CRC Press, Boca Raton.

Breese, E.L. and Mather, K. 1960. The organisation of polygenic activity within a chromosome in Drosophila. 1. Hair Character. Heredity, **11**: 373–393.

Breese, E.L. and Mather, K. 1960. The organisation of polygenic activity within a chromosome in Drosophila. 2. Viability. Heredity, **14**: 375–400.

Cochran, W.G. 1951. Improvement by means of selection. In: Proc. Second Berkeley Symp. Math. Stat. Prob., 449–470.

Cunnigham E.P. 1969. Animal Breeding Theory. Institute of Animal Genetics. and Breeding, Landbruksbokhandelen, Universite sforlaget 1969 Vollebekk, Oslo.

Dewey, D.R. and Lu, K.H. 1959. A correlation and path coefficient analysis of components of crested wheat grass seed production. Agron. J., **42**: 515–517.

Empig, L.T., Gardner, C.O. and Compton, W.A. 1972. Theoretical gains for different population improvement procedures. Nebraska Agric. Exp. Stn. Bull, M.P. 26.

Falconer, D.S. 1989. Introduction to Quantitative Genetics. 3rd edn. Longman, Burnt Mill.

Fisher, R.A. 1936. The use of multiple measurements in taxomic problems Annals of eugenics, **7**: 179–188.

Hallauer, A.R. and Miranda, F.O., J.B. 1988. Quantitative Genetics in Maize Breeding. Iowa State Univ. Press, Ames, Iowa, USA.

Hazel, L.N. 1943. The genetic basis for constructing selection index. Genetics, **28**: 476–490.

Hazel, L.N. and Lush, J.L. 1943. The efficiency of three methods of selection J. Hered., **33**: 393–399.

Henderson, C.R. 1963. Selection index and expected genetic advance. In: Statistical genetics and Plant Breeding, NAS-NRC, 982.

Kearsey, MJ. and Pooni, H.S. 1996. The Genetical Analysis of Quantitative traits. Chapman and Hall, London.

Kempthorne, O. 1957. An introduction to Genetic Statistics. Wiley, New York.

Kempthorne, O. and Nordskog, A.W. 1958. Restricted selection index, Biometrics, **15**: 10–19.

Lerner, I.M. and Dampester, E.M. 1951. Alternation of genetic progress under continued selection in poultry. Heredity, **5**: 75–94.

Mather, K. 1987. Consequences of stabilizing selection for polygenic variation, Heredity, **58**: 267–277.

Mayo, O. 1980. The Theory of Plant Breeding. Claredon Press, Oxford.

Moen, R. 1968. Selection index for performance tested boars, Unpublished report.

Pesek, J. and Baker, R.J. 1969. Desired improvement in relation to selection indices. Canadian 1. Plant Sci., **49**: 803–804.

Reeve, E.C.R. and Robertson, F.W. 1953. Studies in quantitative inheritance. 2. Analyses of strain of *Drosophila melanogaster* selected for long wings. J. Genet., **51**: 267–316.

Robertson, F.W 1957b. Studies in quantitative inheritance. XI. Genetic and environmental correlation between body size and egg production in *Drosophila melanogaster*. J. Genet., **55**: 428–443.

Smith, H.F. 1936. A discriminate function for plant selection. Ann. Eugen. London, **7**: 240–250.

Suwantaradon, K., Eberhart, S.A., Mack, J.J.; Wens, J.C.; and Guthrie, WD. 1975. Index Selection for several agronomic traits in the BSSS2 Maize population. Crop Sci., **15**: 826–833.

Tallis, G.M. 1962. A selection index for optimum genotype. Biometrics, **18**: 120–122.

Thoday, J.M. 1972. Disruptive selection. Proc. R. Soc. London, B, **182**: 109–143.

Williams, J.S. 1962. The evaluation of a selection index. Biometrics, **18**: 375–393.

Wricke, G. and Weber, WE. 1986. Quantitative genetics in selection in plant breeding. Water de Gruyper, Berlin.

Wright, S. 1921. Systems of mating. 3. Assortative mating based on somatic resemblance. Genetics, **6**: 144–161.

Wright, S. 1968. Evolution and the Genetics of Populations. Vol I. The University of Chicago Press, Chicago

Young, S.S.Y. 1961. A further examination of the relative efficiency of three methods of selection for genetic gains under restricted conditions. Genet. Res., **2**:106–121.

21

QTL Analysis

21.1 QTL Analysis

Cloning and transformation techniques are feasible only for qualitative traits under control of single genes but most of the agronomically important traits are quantitative traits under control of polygenes or quantitative trait loci (QTL) (Gelderman, 1975) and there is thus a need to locate, characterize and eventually clone QTL.

All the classical quantitative genetics models (Chapters 5 and 6) for analysing quantiative variation assume that

 (i) unknown number of genes are determining a quantitative trait,
 (ii) each gene is having small effect,
 (iii) gene effects at loci are additive
 (iv) these genes are showing $g \times e$ interaction.

Further, estimates of genetical component of variation were obtained assuming equal effects, complete dominance, no epistasis and no linkage and also in the presence of various levels of dominance and assuming epistasis and linkage. Proponents of biometrical analysis of quantitative traits assumed this because at that time there were no way in which one could determine the location and effect of each gene determining quantitative variation although it was known that the polygenes are inherited in fashion as Mendelian factors do. Because of the above problems there were no unanimity about the explanation behind overdominance, pleiotropy and other genetical phenomena. Now with the coming of the molecular marker technology, the scenario has changed and one can now precisely and accurately locate and measure the effect of the individual QTL. We can not only count the number of QTL determining a quantitative trait but also we can have estimate of the gene effects without the restrictive assumptions of equal effect, complete dominance, no epistasis and no linkage. QTL mapping will throw light on genetics of various phenomena such as heterosis, epistasis, pleiotopy, etc. Besides these genes for resistance to disease and pest, tolerance to drought, heat, cold and other adverse conditions such as alkalinity, salinity, flooding, etc. and nutritional value could be mapped and introgressed into domestic lines from exotic relatives. With known number of QTL determining yield and other agronomic traits and their gene effects, the method of exploitation of quantitative variation will change. Further one will have information regarding the QTL which is not interacting with environment or which is responsive to better management and thus the breeding strategy for developing high yielding and more stable or with wider adaptibility will change. Finally, QTL mapping will throw light on the role of polygenes in the evolutionary process. Speciation could be explained by determining the number and nature of genes involved in the reproductive barriers. Methods of detection of linkage between QTL and marker, estimation of genetic effects and applications of QTL mapping are discussed in various sections.

21.1.1 Linkage between QTL and molecular marker

For effective marker-assisted selection (MAS) and determining QTL effect, it is desirable to tag an agronomically important gene with closely linked molecular markers. There are different approaches to associate QTL with molecular markers. One approach would be to map the trait relative to a set of molecular markers in a segregating population. As the desirable trait has transcended to this segregating population from a different population, that different populations must be scored for segregation of some of the molecular markers where also the gene determining the desirable trait is segregating. Thereafter one can start analysing the derived segregating generation population for different molecular markers for finding the association of gene determining the desirable trait with the molecular marker. The segregating generation population could be F_2 or back-cross population. Here it is possible to use a modified approach to detect QTL. Instead of analysing each individual in the segregating generation population, marker analysis is performed only on individuals in the two extreme traits of the distribution (i.e. those with the lowest and highest values of the character, Tanksley, 1993). If the allele frequency at any molecular marker locus differs significantly between the two extreme sub-populations, it can be inferred that a QTL is linked to a character of interest.

In the second method of obtaining information on linked molecular marker, near isogenic lines (NIL) are produced for a number of traits and if there is sufficient polymorphism for molecular markers between the parental lines (donor and recurrent parent), molecular markers can be rapidly screened to locate those which are in the introgressed segment and thus polymorphic between recurrent parent and the desired near isogenic line.

The third method called bulked segregant analysis (BSA) involves pooling of DNA samples and is a rapid method of identification of linkage (Michel more et al. 1985, 1991). Considering one locus with two alleles, the F_2 population derived from two parents differing at one locus will consist of homozygous dominant, heterozygous dominant and homozygous recessive individuals. In this bulked segregant analysis, the F_2 population is separated into two groups—one group comprising homozygous dominant genotype and the other consisting of homozygous recessive genotype. Heterozygous dominant individuals are discarded on the basis of progeny test. DNA samples obtained from individuals within a group are pooled and thus there are two DNA pools. These samples are then subjected to marker analysis (RAPD). As the two pooled DNA samples differ from one another in the chromosomal region around that locus, any polymorphism seen in the RAPD analysis should represent markers in that region near that gene in question. This gene is question can determine resistance against pathogen or any other quality trait. BSA has advantages over NILS, in that the later requires many generation of back-crossing and selection to develop. Further BSA does not suffer from the linkage drag problem (i.e. genes incorporated into lines by backcrossing that are flanked by DNA segment introduced from donor parent) associated with NILS (Young and Tanksley, 1989).

21.1.2 Mapping methods

All mapping methods which we will see are based on least squares, maximum likelihood and method of moments. The principles involved in these methods are described in Chapters 3 and 4, respectively. Here we would see how these methods can be applied in detecting and estimating the effects of quantitative trait loci. Like in any biometrical genetics studies, progenies in a population (e.g. F_2, backcross or any other population) are scored for the quantitative trait of interest (denoted by Y) and the markers available (denoted by X_i, with $i = 1$ to m for a set of m markers). A linear model is constructed to establish a relationship between the trait value (Y) and each of the markers ($X_i's$). The model takes the form as

$$Y_{jk} = \beta_0 + \beta_1 X_{ij} + \beta_2 X_{ij}^2 + \varepsilon_{jk}$$

where $j = 0, 1, 2$ denotes the three possible X_i values for each marker in case of F_2 derived from a cross between two inbred parents, $k = 1$ to n_{ij} denotes the number of progeny in the marker j, β_1 and β_2 are the additive effects respectively of the locus near the marker defining X_i and ε is the residual.

This model can be analysed using least squares or maximum likelihood for each of the m markers to produce m analyses.

(a) Least squares

The principle behind least squares estimation is to find the parameter estimates that minimize the residual sum of squares. The residual sum of the squares as a function of other terms can be written as follows:

$$\varepsilon_{jk}^2 = (Y_{jk} - \beta_0 + \beta_i X_{ij} + \beta_2 X_{ij}^2))$$

The least squares parameter estimates can then be derived by differentiating the above equation with respect to β_0, β_1 and β_2 and equating these partial differentials to zero. The values of β_0, β_1 and β_2 that satisfy these equations are the least squares estimates of these parameters. If a significant relationship is found between Y and X_i for an individual model then we usually conclude that a trait locus is near the marker that defines the X_i in that model. The main objective of least squares estimation is that the model should explain as much of the variation in the dependent variable as possible and thus a model with zero mean squared residuals would perfectly fit data. Least squares estimates have the desirable properties of minimum residual variance and unbiasedness. The significant X_i's can be used in prediction equation as in indirect selection criterion (e.g., marker-assisted selection).

The choice between least squares and maximum likelihood depends on the nature of the residuals. If the residuals are normally distributed and have equal variance in all levels of X_i' then the least squares and maximum likelihood analyses are equivalent. But if the residuals have heterogeneous variance then the maximum likelihood analysis can be advantageous. Further, in case of complicated models such as non-linear models, maximum likelihood can be used (Bridges and Sobral, 1996; Weller, 1992). Finally, both single and flanking markers method can use the least squares method.

A quantitative trait loci must be evaluated simultaneously to determine the relative main effects of the QTL and the effects of interactions among the QTL on the quantitative trait value, the use of multiple regression is suggested. The multiple regression model takes the form

$$Y = \beta_0 + X_i + X_2 \ldots + X_m + \varepsilon$$

where X_i represents the term $\beta_i X_i$ and $\beta_2 X_i^2$ for marker (i). The model again can be analysed using least squares or maximum likelihood and if X_i's are found to be significantly related to Y, then the markers defining those X_i's are said to be near QTL. Further, X_i's can be used in a prediction equation for marker assisted. selection.

b. Maximum likelihood method

We can have maximum likelihood estimate for a single parameter or multiple parameters. In this method of estimation, the principle is to find parameter estimates that are best match of the sample of data. It requires that the type of distribution from which the data were sampled, are known. It is a parametric method of estimation and unlike least squares maximum likelihood parameters estimates may be biased and do not necessarily yield least squares residuals. It can be applied to both linear and non-linear models and yields parameter estimates within the parameter space. It does not yield negative estimates of variance components (Weller, 1993). The basic involved in maximum likelihood estimation for single parameter is discussed in Chapter 4.

In case of maximum likelihood estimation of multiple parameters simultaneously, it is necessary to maximize the likelihood with respect to all the parameters. This is done by taking the partial derivatives of the log likelihood with respect to each parameter and setting each derivative equal to zero. Thus it is necessary to solve a system of equations equal to the number of parameters to be estimated. The likelihood function for estimation of m parameters ($\phi_1, \phi_2 \ldots \phi_m$) from a sample of n observations ($X_1, X_2, X_3 \ldots X_n$) takes the forms as

$$\begin{aligned} L &= p(X_1, X_2, \ldots Xn/\phi_1, \phi_2 \ldots \phi_m) \\ &= p(X_1/\phi_1, \phi_2 \ldots \phi_m)\, p(X_2/\phi_1\, \phi_2 \ldots \phi_m) \\ &\quad p(X_n/\phi_1\, \phi_2 \ldots \phi_m) \\ &= \Pi p(X, \phi_1, \phi_2 \ldots \phi_m) \end{aligned}$$

where $p(X, \phi)$ represents the probability of obtaining X, conditional on ϕ and Π denotes the product $p(X_i/ \phi_1 \phi_2, ... \phi_m)$ from X_1 through X_n. In case of continuous distribution, $p(X, 0)$ will be replaced with $f(X, \phi)$ i.e. the density of X_1 conditional on ϕ, Maximum likelihood can be used to test the hypothesis by a 'likelihood ratio test.' In a likelihood ratio test, the maximum likelihoods obtained under the two alternative hypotheses (e.g. in case of QTL analysis, QTL vs. no QTL) are compared. Under the assumption of no difference, the natural log of the likelihood ratio will be asymptotically distributed as $(½) \chi^2$ where χ^2 is the chi-square statistic with one degree of freedom (Simpson, 1989).

Often it is not possible to derive an analytical solution to the resultant system of equations. Several methods have been developed but three, namely the Fisher's method of scoring (Dentine and Cowan, 1990), expectation maximization (EM) (Dempester et al., 1977) and the Newton-Raphson's method (Dahlquist, 1974) have been applied to QTL mapping. Out of these three, EM is generally considered the method of choice and has been described by Luo and Keassey, 1992 in case of interval mapping of the QTL.

21.1.2.1 Single marker case

Weller (1986) used the method of maximum likelihood for estimating QTL effects along with frequency of recombination in the F_2 npopulation in case of QTL associated with a single marker. This maximum likelihood technique is based on the assumption that the genotypic values of AA, Aa and aa at the QTL are normally distributed with means μ_{11}, μ_{21}, and μ_{22} and variances σ^2_{11}, σ^2_{12} and σ^2_{22}, respectively. Thus means are assigned values such as $\bar{X}_{11}$, $\bar{X}_{12}$ and $\bar{X}_{22}$ and the variances are assigned the values such as S_{ll}, S_{12} and S_{22} respectively. Further, a value is assigned to the recombination frequency. With these assigned values, the likelihood of occurrence of the observed performance of F_2 individual are obtained and through iteration maximum likelihood estimates of the parameters are obtained. The likelihood function for the entire F_2 population takes the form as

$$L = \prod_{i=1}^{i} f(MM_i)] [\prod_{j=1}^{j} f(M_{mj})] [\prod_{k=1}^{k} f(mm_k)]$$

where $f(MM_i)$, $f(Mm_j)$ and $f(mm_k)$ are the probability density function of individuals i, j and k belonging to the marker classes MM, Mm and mm, respectively. I, J and K are the members of individuals in the marker classes. The density of a mixture of several distributions can be calculated as the density of each distribution times, the probability that the individual has been sampled from the distribution. Thus the probability density of marker classes are:

$$f(MM) = (1-r^2)f(AA) + 2r(1-r)f(Aa) + r^2 f(aa)$$

$$f(Mm) = r(1-r)f(AA) + (1-2r(1-r)f(Aa)) + r(1-r)f(aa)$$

$$f(mm) = r^2 f(AA) + 2r(1-r)f(aa) + (1-r^2)f(aa)$$

where $f(AA)$, $f(Aa)$ and $f(aa)$ are the probability density functions of genotypes AA, Aa and aa at the QTL (Table 21.6) and r is the probability of recombination. Maximum likelihood estimates of r, the three means and the three variances can be obtained by taking partial differentials of equation (1) (or as is usually done, the partial differentials of the log of the likelihood) for all seven parameters, setting these equal to zero and solving the resultant set of equation.

There are three problems associated with one marker approach (Lander and Botstein, 1987; Kearsey and Farquhar, 1998). First, false positives will occur if the significance level is set too low. Second, because all genes on a chromosome will show some linkage among themselves, any one QTL will be associated with several markers. Third, because the QTL will not necessarily be allelic with any given marker, its exact position and its exact effect cannot be known although the strongest association will be with the closest marker. This approach of analysing single markers one at a time does not define the likely position of the QTL.

21.1.2.2 Flanking markers

Lander and Botstein (1989) developed the maximum likelihood method of interval mapping to calculate the likelihood of a putative gene underlying a quantitative trait being located in a given chromosome interval flanked by genetic markers. They suggested the exploitation of power of high-density (saturated) linkage map for searching the entire genome for genes controlling a quantitative trait and hence enabling to locate QTL systematically. They briefly described the main idea and algorithm of this technique for analysing experimental data from back-cross design. In the BC_1 generation, the linear equation relating phenotype ϕ_1 and genotype g_1 takes the form as

$$\phi = \mu + bg_i + E\varepsilon_i,$$

where ϕ is the trait value (phenotype) of the individual, g_i is the indicator variable and it takes values 1 and 0 for genotypes AA and Aa, respectively, ε is a normal random residual variable with mean 0 and variance σ^2, μ is the genotypic value of Aa and the remaining loci contributing to the quantitative trait (the population mean) and b is the estimated phenotypic effect of a single allele substitution at a putative QTL. Here μ, b and σ^2 are unknown parameters. The likelihood function for a genotype (L_{gi}) is the probability density function of a normal distribution.

$$L_{gi}(\mu, b, \sigma^2) = \frac{1}{\sqrt{2\pi}\,\sigma^2} e^{-\frac{(\phi - bg_i)^2}{2\sigma^2}}$$

The likelihood function for all individuals falling in the marker class $k (L_k)$ is

$$L_k(\mu_1\, b_1\, \sigma^2) = \Pi(P_i(1)\, L_i(1) + P_i(0)\, L_i(0))$$

where k (1 to 4) refers to the marker class $M_1M_1M_2M_2$, $M_1M_1M_2m_2$, $M_1m_1M_2m_2$ and $M_1m_1M_2m_2$ in BC_1 generation and $P_i(1)$ and $P_i(0)$ refer to the probability of AA and Aa occurring in the marker class k, respectively e.g., $P_i(1) = r_2$ and $P_i(0) = r_1$ for marker class 2, $M_1M_1M_2m_2$ (Table 21.2) and $L_i(1)$ and $L_i(0)$ are likelihood functions for genotypes AA and Aa, respectively.

The likelihood function for all scored individual is

$$L(\mu, b, \sigma^2) = \Pi k L_k(\mu, b, \sigma^2)$$

Maximum likelihood estimates with missing data are obtained using *EM* algorithm (Dempester et al. 1977). Since flanking markers are used in the population, intervals between adjacent pairs of markers along a chromosome are scanned and the likelihood profile of a QTL being at a particular point in each interval is determined, or to be more precise, the log of the ratio of likelihood (LOD) of there being one vs. no QTL at a particular point. The application of log-odds proposed by Lander and Botstein has been used in the study of linkage in human pedigree analysis. The log 10 of the odds ratio is calculated with the following expression.

$$\text{LOD} = \log_{10}[L(\mu, b, \sigma^2)/L(\mu_0, 0, \sigma_0^2)]$$

In the case of testing 60 flanking markers in 1200 cM, the LOD threshold value for avoiding a false positive was estimated to be about 2.4. When the LOD scores exceed the predetermined threshold value, the null hypothesis is rejected and the existence of a QTL is suggested. A simple explanation of the principle of interval mapping for locating and mapping QTL using F_2 population derived from either selfing or sibbing of the F_1 derived from two inbred lines P_1 and P_2 and the EM algorithm used was given by Luo and Kearsey (1992). If the marker allele in P_1 is denoted by M and the marker alleles in P_2 by m, there will be three genotypes for each quantitative trait locus and each marker gene, respectively. Assuming that the genome is separated into a series of chromosomal segments on the basis of marker linkage map, each chromosome segment will be flanked by two marker loci. Now if a QTL is present in the ith interval as shown in Figure 21.1, its location can be summerised as follows:

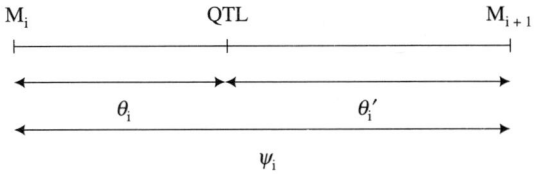

Fig. 21.1 Show QTL lying between M_i and M_{i+1} markers.

In Figure 21.1, ψ_i is the map distance between the two markers loci, M_i and M_{i+1} and θ_i and θ_i' are the mapping distances from QTL to its left and right markers, respectively. Following the mapping function of Haldane (1919),

$$\psi = \theta_i(1 - \theta'_i) + \theta'_i(1 - \theta_i)$$

Where θ'_i can be solved for a known ψ_i and a given θ_i as

$$\theta_i^1 = \frac{\psi_i - \theta_i}{1 - 2\theta_i}$$

Now the first step would be to detect the presence of the QTL and once its presence is confirmed, the second step would be to locate the QTL. For detecting the presence of the QTL, one will have to test the likelihood of the hypothesis that there is a QTL at a given site in the interval against the likelihood that the hypothesis is not true. The logarithm of the ratio of these two likelihoods, i.e. LOD score is used as a measure of the reliability of the hypothesis. If the LOD score is higher than a given threshold, the hypothesis is accepted and thus a QTL is declared to be present, otherwise it is rejected. LOD scores corresponding to all possible sites throughout the interval are calculated and the site at which the LOD score has its maximum value will be the most likely location of the QTL.

In the F_2 generation population, a sample of n plants is scored for their quantitative phenotype denoted by $Y = \{y_1, y_2 \ldots y_n\}$ and their marker genotype. The effect of three genotypes QQ, Qq and qq of a particular QTL are assumed to be $N(a_1, \sigma^2)$, $N(a_2, \sigma^2)$ and (a_3, σ^2) respectively but then these distributions parameters may vary for different QTL. The distribution of the effect of the QTL on the quantitative trait will be a mixture of three normal distributions with frequencies denoted by p_1, p_2 and $1-p_1-p_2$, respectively. Its probability density function can be written as

$$fY(y/\phi) = p_1 u(y - a_1) + p_2 u(y - a_2) + (1 - p_1 - p_2) u(y - a_3)$$

where $\phi = (p_1, p_2\ a_{1,2}, a_3, \sigma^2)^T$ represents the vector of unknown parameters and

$$u(z) = \frac{1}{\sqrt{2\pi}\ \sigma^2} e\left(-\frac{z^2}{2\sigma^2}\right)$$

Thus the estimate of the unknown parameter vector ϕ is needed to separate out the individual genotypic effects of the putative QTL. The maximum likelihood estimate of ϕ can be obtained by using EM algorithm (Dempester, et al. 1977). In case of a QTL flanked by two markers, there will be nine different gametes and 27 F_2 genotypes with the expected frequencies $\overline{\omega}_{jk}^{(0)}$ ($j = 1, 2, \ldots n$, $k = 1, 2, 3$ for F_2 population) in the ith interval as shown in Table 21.1.

In initial estimate of the unknown parameter, ϕ can be derived using the following formulae (Hosmer, 1973; Louis, 1982; Leytham, 1984; Little and Rubin, 1987) using the frequencies and observations of the quantiative trait.

$$p_k^{(0)} = \frac{1}{n} \sum_{j=1}^{n} \overline{w}_{jk}^{(0)}$$

$$a_k^{(0)} = \frac{1}{np_k^{(0)}} \sum_{j=1}^{n} \overline{\omega}_{jk\ yi}^{(0)}$$

and $$\sigma^{(0)^2} = 1/n [\sum_{k=1}^{3} \sum_{j=1}^{n} \overline{\omega}_{jk}^{(0)} (y_i - a_k^{(0)})^2]$$

The EM logarithm cogarithm consists of two steps: 1. Expectation step 11. Maximization step. In the expectation step the EM algorithm uses these initial estimates, $p_k^{(0)}$, $a_k^{(0)}$ and $\sigma^{(0)^2}$ and computes the new

$$\overline{\omega}_{jk}^{(1)} = \frac{p_k^{(0)} u\ (y_i - a_k^{(0)})}{\sum_{j=i}^{3} p_k^{(0)} u\ (y_i - a_k^{(0)})}$$

In the maximization step the new estimate $\overline{\omega}_{jk}^{(1)}$ yields the following new estimates

$$p_k^{(1)} = 1/n \sum_{j=1}^{n} \overline{\omega}_{jk}^{(1)}$$

$$a_k^{(1)} = \frac{1}{np_k^{(1)}} \sum_{j=1}^{n} \overline{\omega}_{jk}^{(1)} y_i$$

and $$\sigma^{(1)^2} = 1/n [\sum_{k=1}^{3} \sum_{j=1}^{n} \overline{\omega}_{jk}^{(1)} (y_i - a_k^{(1)})^2]$$

which will increase the value of the following likelihood function (Hosmer, 1973; Louis, 1982; Titterington et al., 1985)

$L(Y/p_k, a_k, \sigma^2)$

$$= \prod_{j=1}^{n} \frac{1}{\sqrt{2\pi\sigma^2}} \left[\sum_{k=1}^{3} p_k \exp\left[-\frac{(y_i - a_k)^2}{2\sigma^2} \right] \right]$$

These above two steps are repeated iteratively and thus a sequence of estimates $\{p_k^{(i)}, a_k^{(i)}, \sigma^{(i)^2}\}$, $i = 1, 2 ...$ will be generated. These sequence will converge to its limit at which the derivatives of the likelihood function mentioned above take a value of zero and hence this point is called a global maximum of the function which provides the maximum likelihood estimates required. Now these estimates will be represented by $(P_k^*, a_k^*, \sigma^{*2})$ then the likelihood ratio of locating the QTL at the site of θ_i cM away from its left marker locus M_i will be

$$\text{LOD}(\theta_i) = \log_{10}\left[\frac{L(Y\{p_k^*, a_k^*, \sigma^{*2}\})}{L(Y\{1, a^{**}, \sigma^{**2}\})} \right]$$

where $A^{**} = \frac{1}{n}\sum_{j=1}^{n} y_i$

and $A^{**2} = \frac{1}{n}\sum_{j=1}^{n} (y_i - a^{**2})$

A curve of LOD scores against the searched chromosomal location will be obtained for every chromosome if every interval of each chromosome of the genome is scanned for the presence of the QTL at all possible locations.

(a) Appropriate threshold for LOD scores

A threshold is required for statistical inference i.e. those regions of the chromosome for which the LOD score exceeds this threshold, may be regraded as the potential sites of a QTL while its most likely position is where the LOD score is maximum. Following the principle of maximizing the probability of type II errors Lander and Botstein (1989) found that the appropriate threshold depends not only on the size of the genome but also on the density of the genetic markers mapped. Infact, as the number of intervals (*m*) tends to infinity, the required nominal significance level for each individual test approaches a non-zero limit independent of the number of markers genes scored. Typically, a LOD threshold of between 2 and 3 is required to ensure a false-positive ratio of 5 per cent.

There are two major problems in applying traditional hypotheses of testing to the detection of QTL. The first in the distribution of the test statistic is unknown because the conditions that ensure an asymptotic chi-square distribution for the test statistic are not satisfied (Churchill and Doerge, 1994). This effect is further compounded by small sample size, unknown distributional properties of the quantitative trait, the composition of the genome and the genetic map density (Darvasi et al., 1993; Churchill and Doerge, 1994). The second is the problem of multiple comparison implicit in the genome searchers for QTL (Churchill and Doerge, 1994; Haley et al., 1994; Jansen and Stam. 1994; Zeng, 1994).

Reliability of threshold values can be improved by reducing type I error (rejection of the null hypothesis when it is correct i.e. QTL are identified where none exists, false positive) and type II error (accepting the null hypothesis when it is wrong, false negative). The parametric statistic, chi-square based statistic (Gerber and Rodolphe, 1994; Carbonell et al., 1992), use of additional markers (Jensen, 1994; Rebai et al., 1994; Zeng, 1994) and non-parametric permutation based test (Churchill and Doerge, 1994) have been proposed but there is no clear picture of which method will provide the best threshold for the diverse experimental situations.

(b) Confidence Interval (CI)

The estimate of confidence interval for the QTL location is important because it may determine the strategies for further experimentation to get closer to the QTL or for using the QTL in breeding program. For example, while introgressing a QTL allele from a donor parent into a recepient parent using the molecular marker, one strategy would be to transfer a donor marker haplophyte that covers the 95% CI for the QTL. The methods used to calculate confidence interval are as follows:

Table 21.1 The probabilities of three QTL genotype for a given flanking marker genotype

Marker genotype	QTL genotype		
	QQ	Qq	qq
$\dfrac{M_i M_{i+1}}{M_i M_{i+1}}$	$\dfrac{(1-\theta)^2 (1-\theta')^2}{[A]^2}$	$\dfrac{2\theta\theta'(1-\theta)(1-\theta')}{[A]^2}$	$\dfrac{\theta^2 \theta'^2}{[A]^2}$
$\dfrac{M_i M_{i+1}}{M_i m_{i+1}}$	$\dfrac{\theta'(1-\theta)^2(1-\theta')}{[A][B]}$	$\dfrac{\theta(1-\theta)(1-\theta')^2 + \theta\theta'^2(1-\theta)}{[A][B]}$	$\dfrac{\theta'\theta^2(1-\theta')}{[A][B]}$
$\dfrac{M_i m_{i+1}}{M_i m_{i+1}}$	$\dfrac{(1-\theta)^2 \theta'^2}{[B]^2}$	$\dfrac{2\theta\theta'(1-\theta)(1-\theta')}{[B]^2}$	$\dfrac{\theta^2 (1-\theta')^2}{[B]^2}$
$\dfrac{M_i M_{i+1}}{m_i M_{i+1}}$	$\dfrac{\theta(1-\theta)(1-\theta')^2}{[A][B]}$	$\dfrac{\theta^2 \theta'(1-\theta') + \theta'(1-\theta^2)(1-\theta')}{[A][B]}$	$\dfrac{\theta\theta'^2(1-\theta)}{[A][B]}$
$\dfrac{M_i M_{i+1}}{m_i m_{i+1}}$	$\dfrac{2\theta\theta'(1-\theta)(1-\theta')}{[A]^2 + [B]^2}$	$\dfrac{[\theta^2 + (1-\theta)^2][\theta'^2 + (1-\theta')^2]}{[A]^2 + [B]^2}$	$\dfrac{\theta\theta'(1-\theta)(1-\theta')^2}{[A]^2 + [B]^2}$
$\dfrac{M_i m_{i+1}}{m_i m_{i+1}}$	$\dfrac{\theta\theta'^2(1-\theta)}{[A][B]}$	$\dfrac{\theta^2\theta'(1-\theta') + \theta'(1-\theta')^2(1-\theta)}{[A][B]}$	$\dfrac{\theta'(1-\theta)(1-\theta')^2}{[A][B]}$
$\dfrac{m_i M_{i+1}}{m_i M_{i+1}}$	$\dfrac{\theta^2 (1-\theta')^2}{[B]^2}$	$\dfrac{2\theta\theta'(1-\theta)(1-\theta')}{[B]^2}$	$\dfrac{(1-\theta)^2 \theta'^2}{[B]^2}$
$\dfrac{m_i M_{i+1}}{m_i m_{i+1}}$	$\dfrac{\theta^2 \theta'(1-\theta')}{[A][B]}$	$\dfrac{\theta\theta'^2(1-\theta) + \theta(1-\theta)(1-\theta')^2}{[A][B]}$	$\dfrac{\theta'(1-\theta)^2(1-\theta')}{[A][B]}$
$\dfrac{m_i m_{i+1}}{m_i m_{i+1}}$	$\dfrac{\theta^2 \theta'^2}{[A]^2}$	$\dfrac{2\theta\theta'(1-\theta)(1-\theta')}{[A]^2}$	$\dfrac{(1-\theta)^2(1-\theta')^2}{[A]^2}$

$A = \theta\theta' + (1-\theta')$ and $B = \theta(1-\theta') + \theta'(1-\theta)$.

1. LOD drop-off method (Lander and Botstein, 1989).
2. Boot strapping (E-fron, 1979, 1982; Visscher, Thompson and Haley, 1996).

1. LOD drop-off method

In the LOD drop-off method, the CI is calculated by finding the locations at either side of the estimated QTL location that corresponds to a decrease in the LOD score of 1 or 2 units. The total width corresponding to 1 or 2 LOD drop-off is then taken as the confidence interval and asymptotically, these should approximately equivalent to 96.8 and 99.8% CI, respectively (Mangin et al., 1994). However, using differences in likelihood between the estimated location of the QTL and the locations elsewhere on the chromosome to determine approximately 90 or 95% CI, may be biased for small and medium sized populations (Van Ooijen, 1992; Mangin et al., 1994) because then the distribution of the test statistic does not truly follow a chi-square distribution. For example, Van Ooijen (1992) in simulation study found that the proportion of CIs based on a support interval of 1 LOD that contained the QTL varied from 0.73 to 0.84 depending upon the size of the QTL and the type of population and experimental size. CIs are calculated only for samples that give significant evidence of a segregating QTL. Similarly, Mangin et al. (1994) observed in simulation study that the 90% CI was biased downward, i.e. the proportion of 90% CI which contained the simulated QTL was < 0.90 in particular for QTL that gave small amount of variation and for dense marker map. For example, in a back-cross population of 200 individuals, the empirical probabilities that 90% CI based on the LOD drop-off method contained the actual location of the QTL was ~0.84 and 0.74 for a map density of 20 and 5 cM, respectively, Hence, the 'One LOD drop-off concept' is not recommended for use as the actual drop-off needed, varies with each study and each QTL. Other methods (Van Ooijen, 1992; Darvasi et al., 1993, Mangin et al., 1994) based on simulation have been proposed for calculating CI work on a number of assumptions such as normality of residuals, large experimental size and evenly spaced markers which are not normally fulfilled in practice.

2. Bootstrapping method

For estimation of CI in QTL mapping by the bootstrapping resampling method, boot strap samples are created by sampling with replacement N individual observations. An observation consists of a marker genotype and a phenotype. At each boot strap sample, N observations out of the pool $y(N)$ original observation are drawn with replacement. Some records can appear more than once in a boot strap sample while others are not included at all. After n boot strap samples, the empirical central 90 and 95% CI of the QTL position is determined by ordering the n estimates and taking the bottom and top fifth and 2.5th percentile, respectively. Thus an empirical boot strap confidence interval is obtained for a particular population. If averaged over replicate populations, the proportion of empirical boot strap CI that contains the QTL is calculated. The method is considered working perfectly if the proportion would be 0.90 (0.95) when 90% (95%) empirical boot strap CI is determined. For comparing this method with LOD drop-off method, CIs were calculated using this LOD drop-off method. A putative QTL was fitted at 1 cM interval and a support interval corresponding to a drop in the test statistic of 2.71 and 3.84 for a 90% and 95% CI, respectively. Asymptotically these values correspond to a 90 and 95% support interval when the test statistic is distributed at a χ^2 with one degree of freedom. The values from the chi-square correspond to a drop in the LOD of 0.58 and 0.83, respectively. Visscher et al. (1996) using simulation for a backcross population found that, in general, the CIs were slightly conservatively biased. Correlations between the test statistics and the width of the CI were strongly negative and thus stronger the evidence for a QTL segregating, the smaller the empirical CI for its location. The size of the average CI depends heavily on the population size and the effect of the QTL whereas the marker spacing had only a small effect on the average empirical confidence interval. The empirical support interval calculated from LOD drop-off method gave confidence intervals which were generally too small, in particular, when confidence intervals were calculated only for samples above a certain significance threshold.

The various methods of detecting and locating a QTL have located QTL with poor precision (10–30 cM). This shows the uncertainty of a QTL's genetic location relative to DNA markers. Confidence intervals (CIs) for QTL with large effects remain near 10 cM (Darvasi et al. 1993). Simulation studies have indicated that even with dense maps (1–2 cM) and large population size (1000 gametes), the degree of genetic resolution has not improved. Infact, more markers beyond a density of one every 15 cM, do not help much. The factors limiting the resolution are (i) lack of recombinant gametes, (ii) genetic heterogeneity for regional restrictions to recombination, (iii) missing data for markers and traits, (iv) error in data collection, (v) linked QTL and (iv) QTL of truly minor effects (Lee, 1995). Thus even with high resolution mapping, it is difficult to prove that a QTL represents only one gene. A typical plant genome includes perhaps 10, 000 to 100, 000 genes which are scattered through a total of 10^8–10^{10} bp of DNA. With DNA markers using NIL location of gene might be determined to a fraction of chromosome perhaps 1–2% of the genome (Hanson, 1959). Further, with additional generation of crossing and high resolution, mapping location can be narrowed down to perhaps 0.1 % of the genome (Paterson et al., 1990a). A single gene responsible for an increamental improvement is ~10^3 bp and thus 0.1% of the genome would include an average of 10–100 genes. In other words, in trying to identify a particular 10^3 bp from DNA from 10^9 bp, DNA markers can narrow the field of candidates by

99.9% but then it still leaves a stretch of 10^6 bp to be dissected. Considering that CIs have still not been reduced to much less than 10 cM, this 10 cM equals to 300 kbp and 6000 kbp of DNA in Arabidopsis and wheat, respectively.

Several approaches have been explored to reduce the size of confidence intervals. The first is to increase the exprimental size and this can be easily achieved. But then increasing population size particularly in case of F_2 or back-cross generation population will not help as these populations are mixtures of genotypes and because the yield of individuals is scored in spaced planting which is not the commercial practice. Thus one must use plot trials with recombinant inbreds lines (RIL) or doubled haploid (DH) lines. Further, heritability of individual QTL can be increased by either minimizing the environmental variation by way of increasing the number of replications or identifying the residual variation due to other QTL and removing from error (Jansen, 1993; Jensen and Stamp, 1994).

21.2 Models for Estimating Genetic Effects of QTL

The different approaches for estimating gene action are as follows:

(i) Single marker approach
(ii) Interval maping (two markers) approach
(iii) Marker-regression approach

1. Single marker approach

In the traditional single marker approach (Sax 1923, Soller and Broody 1976; Tanksley, Medina Filho and Rick, 1982; Edward, Stuber and Wendel, 1987) for detecting a QTL near a genetic marker, the phenotypic means of the two classes of progeny, those with *MM* and those with marker genotype *Mm* are compared. The difference between the means provides an estimate of the phenotypic effect of substituting a *m* allele for *M* allele at the QTL. The idea of using single markers to systematically characterize and map individual polygenes controlling quantitative traits was put forward by Thoday in 1961. If the segregation of a single gene marker could be used to detect and estimate the effect of linked polygenes then it is possible to map and characterize all of the polygenes affecting a trait if single gene markers are scattered throughout the genome of an organism. Methodology for locating chromosome segments or loci affecting quantitative trait based on linkage to marker loci in backcross populations has been applied in Drosophila (Thoday 1961; Spickett and Thoday, 1966; Mather and Jinks, 1981) and in wheat (Law, 1966).

Considering a QTL locus A with two alleles *A* and *a*, the three genotypes and their genotypic values following Mather and Jinks (1971) will be as follows:

Genotype	Value
AA	+ da
Aa	ha
aa	– da

Now considering a marker locus *M*, there will be again three genotypes *MM*, *Mm* and *mm*, respectively. The two pure breeding lines P_1 and P_2 selected for obtaining F_1, F_2 and backcross generations will be of genotypes *MMAA* and *mmaa*, respectively for marker locus and QTL. Assuming *r* being the recombination frequency between QTL and marker locus, the F_1 (*Mm*, *Aa*) (Figure 21.2) will produce the following types of gametes:

Gametes	Frequencies
MA	1/2 (1 – r)
ma	1/2 (1 – r)
Ma	(1/2) r
mA	(1/2) r

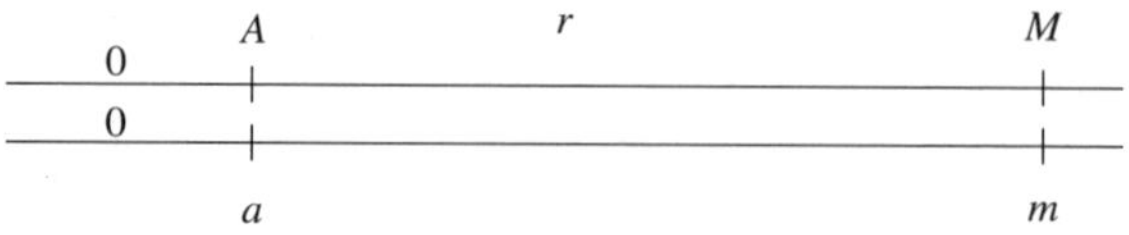

```
   A              r              M
0  +-----------------------------+
0  +-----------------------------+
   a                              m
```

Fig. 21.2 Recombination frequency *r* between marker locus (*m*) and QTL locus (*A*)

The genotypes along with gene frequencies and the genotypic values of individual of F_2 generation population are shown in Table 21.2.

Table 21.2 can be reorganised and written in the form given in Table 21.3

With respect to marker locus M, the F_2 population will consist of individuals of three types, namely, MM, Mm and mm and the expected means of three marker classes will be as follows:

Generation	Maker class	Expected class mean
F_2	MM	$(1 - 2r)\, da + 2r\,(1 - 2r)\, ha$
	Mm	$(1 - 2r + 2r^2)\, ha$
	mm	$(2r^2 - 1)\, da + 2r\,(1 - r)\, ha$

Table 21.2 Genotypes their frequencies and genotypic values in F_2 generation population

Genotypes	Frequency	Value
$MMAA$	$1/4\,(1 - r^2)$	$+ da$
$MmAA$	$1/2\,r\,(1 - r)$	$+ da$
$mmAA$	$(1/4)\, r^2$	$+ da$
$MMAa$	$(1/2)\, r\,(1 - r)$	$+ ha$
$MmAa$	$1/2\,(1 - 2r + 2r^2)$	$+ ha$
$mmAa$	$1/2\, r(1 - r)$	$+ ha$
$MMaa$	$(1/4)\, r^2$	$- da$
$Mmaa$	$(1/2)\, r\,(1 - r)$	$- da$
$mmaa$	$1/4\,(1 - r^2)$	$- da$

Table 21.3 Conditional probabilities for QTL genotypes in F_2 population

Genotype at QTL	MM (0.25)	Mm (0.50)	mm (0.25)
AA	$(1 - r)^2$	$2r(1 - r)$	r^2
Aa	$r(1 - r)$	$1 - 2r(1 - r)$	$r(1 - r)$
aa	r^2	$2r(1 - r)$	$(1 - r^2)$

Genotype at genetic marker locus M with probabilities in the bracket

Estimates of additive and dominance effects are obtained from the contrasts among marker classes in the following way:

$$\text{Additive effect} = 1/2\,(\overline{MM} - \overline{mm}) = (1 - 2r)\, da$$

$$\text{Dominance effect} = \overline{Mm} - 1/2\,(\overline{MM} - \overline{mm})$$

$$= (1 - 2r)^2\, ha$$

Dominance/Additive ratio = $\dfrac{\overline{Mm} - (\overline{MM} - \overline{mm})/2}{(\overline{MM} - \overline{mm})/2}$
(degree of dominance)

$$= (1 - 2r)\, ha/da$$

Now it can be seen that these estimates depend on r, the estimate of recombination frequency. Further, it can be shown that if $r = 0$ i.e. the marker locus is tightly linked with QTL, then the mean values of marker genotypes, MM, Mm and mm, reduce to the assigned value of $+ da$, ha and $-da$ of the QTL genotype AA, Aa and aa, respectively. If $r = 0.50$ i.e. the marker locus segregates independently of the QTL, then the mean of each of the marker class reduces to $ha/2$ which is similar to the expected mean of the F_2 population for a quantitative trait. Thus we see that estimates of additive, dominance and the degree of dominance are biased downward by the coefficients $(1 - 2r)$, $(1 - 2r)^2$ and $(1 - 2r)$, respectively. Higher the frequency of recombination between the marker and the QTL, the greater the bias. In other words, if the QTL does not lie exactly at the marker locus, the phenotypic effects may be seriously underestimated as a result of recombination or it might be missed out altogether. In particular, the variance accounted for by the marker locus decreases by a factor of $(1 - 2r)^2$ and consequently the number of progeny required to be increased increases by a factor of $1/(1 - 2r)^2$.

The basic problem with this approach is that the trait scrore of a particular genotype is a single value resulting from the combined allelic effects of many genes and the environment and the two individuals could have the same genotype but a different phenotype or vice-versa. Assuming the confounding effect from distant and close QTL associated with the marker, the estimate of additive and dominance effects will represent the following:

$$\text{Additive effect} = \sum_{i=1}^{n} d_i\,(1 - 2r_i)$$

and

$$\text{dominance effect} = \sum h_i\,(1 - 2r_i)^2$$

where $i = 1\ldots n$ and n is the number of QTL associated with the marker. Finally, if the trait under investigation is having lower heritability then effect of environment on the phenotypic value will be large and therefore it will require a large number of progeny (experimental size) to detect the presence of QTL. Further one cannot distinguish between tight linkage to a QTL with small effects and loose linkage to a QTL with large effects. The accurate estimates of genetic parameters can be obtained for a QTL at about 20 cM from the marker locus. This approach can be extended to S_1 generation population, backcross population, selfed backcross progenies and the estimate of additive and dominance effects can be obtained.

2. Interval mapping (two markers) approach

In this approach marker brackets or flanking markers are used. In other words, a QTL has marker on both sides. Interval mapping suggested by Lander and Botstein (1989) and Knapp et al. (1990) was introduced as a means of overcoming many of the problems associated with individual marker model.

Assuming two pure breeding parents P_1 and P_2 with flanking marker genotypes $M_1\ M_1\ M_2\ M_2$, and $m_1\ m_1\ m_2\ m_2$, respectively and the QTL genotypes AA, Aa and aa, respectively, the F_1 between the P_1 ($M_1\ M_1\ AA\ M_2\ M_2$) and P_2 ($m_1\ m\ aa\ m_2m_2$) will be $M_1\ m_1\ Aa\ M_2m_2$. Let us assume that r_1 and r_2 are the recombination frequencies between loci M_1 and a and between A and M_2, respectively

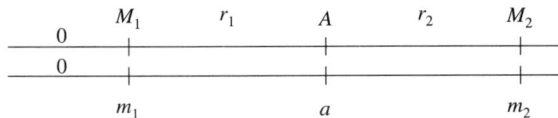

Fig. 21.3 Shows QTL locus (A) in markers ($M_1\ M_2$) bracket.

As the chance of double cross over is almost 1% in case of a 20 cM RFLP map (Lander and Botstein, 1989) double cross-over is not considered. Also, the error caused by D.C.O. is smaller than the misclassification error (Knapp et al. 1990). Knapp et al. (1990) assumed complete interference in applying their model. Thus assuming no double cross-over, the BC_1 and BC_2 populations genotypes and their frequencies are shown in Table 21.4.

Now assuming $r = r_1 + r_2$, where r is the recombination frequency between M_1 and M_2 and $p = r_1/r$, the estimates of additive and dominance effects, p and the genotypic values of the remaining genes contributing to the quantitative trait in generation BC_1 and BC_2 can be obtained by contrasting marker classes from backcross generation. Table 21.5 shows the different marker classes in BC_1 and BC_2 generations along with their expected values.

Table 21.4 Genotypes and their frequencies in case of flanking markers

Generation	Genotype	Frequency
BC_1	$M_1M_1\ AA\ M_2M_2$	$1/2\ (1 - r_1 - r_2)$
	$M_1M_1\ AA\ M_2m_2$	$(1/2)\ r_2$
	$M_1M_1\ Aa\ M_2m_2$	$(1/2)\ r_2$
	$M_1m_1\ AA\ M_2M_2$	$(1/2)\ r_2$
	$M_1m_1\ Aa\ M_2M_2$	$(1/2)\ r_2$
	$M_1m_1\ Aa\ M_2m_2$	$1/2\ (1 - r_1 - r_2)$
BC_2	$M_1m_1\ Aa\ M_2m_2$	$1/2\ (1 - r_1 - r_2)$
	$M_1m_1\ Aa\ m_2m_2$	$(1/2)\ r_2$
	$M_1m_1\ aa\ m_2m_2$	$(1/2)\ r_1$
	$m_1m_1\ Aa\ M_2m_2$	$(1/2)\ r_1$
	$m_1m_1\ aa\ M_2m_2$	$(1/2)\ r_2$
	$m_1m_1\ aa\ m_2m_2$	$1/2\ (1 - r_1 - r_2)$

Table 21.5 Marker classes in BC_1 and BC_2 generation with their values

Generation	Marker class	Expected value
BC_1	$M_1M_1M_2M_2$ (1)	$\theta = \mu_4 + da$
	$M_1M_1M_2m_2$ (2)	$\theta_2 = \mu_4 + (1-p)\ da + pha$
	$M_1m_1M_2M_2$ (3)	$\theta_3 = \mu_4 + pda + (1-p)\ ha$
	$M_1m_1M_2m_2$ (4)	$\theta_4 = \mu_4 + ha$
BC_2	$M_1m_1M_2m_2$ (5)	$\theta_5 = \mu_5 + ha$
	$M_1m_1m_2m_2$ (6)	$\theta_6 = \mu_5 + (1-p) \times ha - pda$
	$m_1m_1M_2m_2$ (7)	$\theta_7 = \mu_5 + pha - (1-p)\ da$
	$m_1m_1m_2m_2$ (8)	$\theta_8 = \mu_5 - da$

Now

$$da = (\theta_1 - \theta_4 + \theta_5 - \theta_8)/2$$

$$ha = (\theta_5 - \theta_8 - \theta_1 + \theta_4)/2$$

$$p = (\theta_1 - \theta_2 + \theta_3 - \theta_4)/4(\theta_1 - \theta_4)$$
$$+ (\theta_7 - \theta_8 + \theta_5 - \theta_6)/4(\theta_5 - \theta_8)$$

$$\mu_4 = (\theta_2 + \theta_3 - \theta_5 - \theta_5)/2$$

and

$$\mu_5 = (\theta_6 + \theta_7 - \theta_1 + \theta_4)/2$$

where θ_1, θ_2, θ_3 and θ_4 are the means of four marker classes of BC_1 generation and θ_5, θ_6, θ_7 and θ_8 are the means of marker classes of BC_2 generations. μ_4 and μ_5 are the genotypic values of the remaining loci contributing to the quantitative trait in BC_1 and BC_2 generations, respectively. The use of pairs of flanking markers for interval mapping with maximum likelihood analysis of data (Lander and Botstein, 1989) provides little extra power for detection of QTL close to a marker but gives much more accurate parameters estimates than analysis using only a single marker (Knott and Haley, 1992). Interval mapping requires fewer progeny than the single marker for detection of QTL of a given magnitude.

The estimates of effects of QTL and its map position to flanking markers can also be obtained in the following way considering only BC1 generation. The F_1 (M_1m_1 AaM_2m_2) when crossed to P_1 ($M_1M_1AAM_2M_2$) to generate BC_1 generation would produce gametes assuming no double crossingover as shown in table given below.

or quantitative trait loci, and QTL must be evaluated simultaneously in order to determine the relative main effects of the QTL and the effects of interaction among QTL on the quantitative trait value. Further, in case of interval mapping, the likelihood of a single putative QTL is assessed at each position on the genome. However, QTL located elsewhere on the genome can have interfering effect. As a consequence, the estimates of locations and effects of QTL may be baised. Therefore, it is obvious that multiple QTL could be mapped more efficiently and more accurately by using multiple QTL model. Thus here an alternative approach using multiple regression was developed by Haley and Knott (1992) for mapping QTL using flanking markers. Assuming a QTL (A) lying between two co-dominant flanking markers M_1 and M_2. they developed model for mapping in F_2 population derived from the cross of two inbred parents which carried different alleles for all three loci. Assuming r_1 being the recombination frequency between M_1

Table showing F1 gametes and the frequency of marker classes in BC1 generation and their contribution

F1 gametes	Frequency	Genotypic value	Marker class	Frequency of marker class	Contribution	Actual
$M_1 A M_2$	$1/2 (1-r_1-r_2)$	d	$M_1M_1M_2M_2$	$1/2 (1-r)$	$d(1-r)/2$	d
$M_1 A m_2$	$1/2 r_2$	d	$M_1M_1M_2m_2$	$r/2$	$(dr_2 + hr_1)$	$(dr_2 + hr_1)$
$M_1 a m_2$	$1/2 r_1$	h				
$m_1 A M_2$	$1/2 r_1$	d	$M_1m_1M_2M_2$	$r/2$	$(dr_1 + hr_2)$	$(dr_1 + hr_2)$
$m_1 a M_2$	$1/2 r_2$	h				
$m_1 a m_2$	$1/2 (1-r_1-r_2)$	h	$M_1m_1M_2m_2$	$1/2(1-r)$	$h(1-r)/2$	h

From this table given above one can estimate r_1, r_2, r and the effects of QTL as follows.
$$M_1M_1M_2M_2 - M_1m_1M_2m_2 = (d - h) \qquad (i)$$
$$M_1M_1M_2m_2 - M_1m_1M_2M_2 = (d - h)(r_2 - r_1)/r \qquad (ii)$$
The limitation with use of only one backcross generation is that the estimate of additive effect is biased when h is not zero and recessive or partial recessive QTL may not be detected.

3. Multiple regression approach

The single marker analysis can be effectively used if the trait is controlled by only a few loci (i.e, a qualititative trait) and there are not too many markers available. Both single markers and flanking markers analysis using least squares or maximum likelihood will not work if the trait is determined by polygenes

and A_1 and r_2 being the recombination frequency between a_1 and M_2 and further assuming no interference and $r = r_1 + r_2 - 2r_1 r_2$, the expected means in terms of the putative QTL for each F_2 marker genotype can be derived. For example, the gamete $M_1 AM_2$ has expected frequency $(1 - r_1)(1 - r_1)/2$ and the gamete $M_1 aM_2$ has expected frequency $r_1 r_2/2$.

21.14 Biometrical Genetics–Analysis of Quantitative Variation

The homozygous marker genotype $M_1M_1M_2M_2$ has an expected frequency of $(1-r)^2/4$ in the F_2 and the expected frequencies of the three possible QTL genotypes AA, Aa and aa with this marker genotype will be $(1-r_1)^2(1-r_2)^2/4$, $2(1-r_1)(1-r_2)r_1r_2/4$ and $r_1^2r_2^2/4$, respectively. Summing over QTL genotypes and scaling for the expected frequency of the marker genotype, the expected mean performance of an F_2 individual of homozygous marker genotype $M_1M_1M_2M_2$ will become

$$da\,[(1-r_1)^2(1-r_2)^2 - r_1^2r_2^2]/(1-r)^2$$
$$+ ha[2(1-r_1)(1-r_2)r_1r_2]/(1-r)^2$$

The expectation of mean genotypic value of a QTL for the nine flanking marker genotypes in the F_2 population are given in Table 21.6. In the similar fashion, the expectation for any other segregating generation population or populations of inbred lines can be worked out.

The estimates of da, the additive effect and ha, the dominance effect are obtained by multiple regression. For a given interval between two markers, numerical values for the coefficients of da and ha for each marker genotype are calculated for a putative QTL at several positions (e.g. 1 cM interval, 2 cM interval and so on) between two markers. Multiple regression is used to fit m, da and ha for each position separately using the numerical values as coefficients for da and ha. Thus we obtain estimates of da and ha and also we obtain regression and residual sum of squares and mean squares and thus a test for da and ha can be made using F test, the regression variance (F) ratio. Thus we can see that the estimates of da and ha are obtained by iteration. All these operations can be done by computer. Computer programmes are available. The position which gives the best fitting model i.e. which produces the smallest residual mean squares gives the most likely position of a QTL and the best estimate of the effect. The multiple regression analysis produces very similar

Table 21.6 The expection of mean genotype value of a QTL for the nine flanking marker genotypes in the F_2 population

Marker genotype	Expectation in terms of	
	da	ha
$M_1M_1M_2M_2$	$[(1-r_1)^2(1-r_2)^2 - r_1^2r_2^2]/(1-r)^2$	$[2r_1(1-r_1)r_2 - (1-r_2)]/(1-r)^2$
$M_1M_1M_2m_2$	$[(1-r_1)^2r_2(1-r_2) - r_1^2r_2(1-r_2)]/r(1-r)$	$[r_1(1-r_1)(1-r_2)^2 + r_1(1-r_1)r_2^2]/r(1-r)$
$M_1M_1m_2m_2$	$[(1-r_1)2r_2^2 - r_1^2(1-r_2)^2]/r^2$	$[2r_1(1-r_1)r_2(1-r_2)]/r^2$
$M_1m_1M_2M_2$	$[r_1(r-r_1)(1-r_2)^2 - r_1(1-r_1)r_2^2]/r(1-r)$	$[(1-r_1)^2r_2(1-r_2) + r_1^2r_2)(1-r_2)]/r(1-r)$
$M_1m_1\,M_2m_2$	0	$[r_1^2r_2^2 + r_1^2(1-r_2)^2 + (1-r_1)^2r_2^2$ $+ (1-r_1)^2(1-r_2)^2]/[r^2 + r(1-r)^2]$
$M_1m_1\,m_2m_2$	$[r_1(1-r_1)r_2^2 - r_1(1-r_1)(1-r_2)^2]/r(1-r)$	$[(1-r_1)^2r_2(1-r_2) + r_1^2r_2(1-r_2)]$ $/r(1-r)$
$m_1m_1\,M_2M_2$	$[r_1^2(1-r_2)^2 - (1-r_1)^2r_2^2]/r^2$	$[2r_1(1-r_1)r_2(1-r_2)]/r^2$
$m_1m_1\,M_2m_2$	$[r_1^2r_2(1-r_2) - (1-r_1)^2r_2(1-r_2)]/r(1-r)$	$[r_1(1-r_1)(1-r_2)^2 + r_1(1-r_1)r_2^2]/r(1-r)$
$m_1m_1\,m_2m_2$	$[r_1^2r_2^2 - (1-r_1)^2(1-r_2)^2]/(1-r)^2$	$[2r_1(1-r_1)r_2(1-r_2)/(1-r)^2$

results to LOD mapping both in terms of precision and accuracy but has the advantage of speed and simplicity of programming. It can be adopted to complex pedigrees and the model can be extended to include a wide range of effects such as sex difference and environments.

Jensen (1993) combined the interval mapping with the multiple regression method for detection and mapping QTL. This is achieved by fitting one QTL at a time in a given interval and simultaneously using (part of) the markers as cofactors to eliminate the effects of additional QTL. Thus the residual variation due to other QTL can be identified and removed from the error. In the regression analysis of Haley and Knott, the model fits a putative QTL at different places along the chromosome (e.g. at 1–cM intervals) and calculate the test statistic at each point. The position giving the largest test statistic is the most likely position for a QTL. Whittaker et al. 1996 obtained similar answers by performing a multiple regression analysis of phenotypes on pairs of flanking markers and transforming the estimated effects of the two markers in each regression to estimate the QTL effect and its location. So the search is over $(n - 1)$ pair of markers instead of search at 1 cM intervals. This method is preferred because of speed of calculation.

Multiple regression model can be used by both the single and flanking marker methods. Although multiple regression approximations are certainly more useful than one-at-a time analysis for determining the markers near QTL but the results from the multiple regression suffer due to multicollinearity caused by missing marker (and QTL) genotypes. Suppose that we are evaluating 60 markers and each has three possible genotypes in F_2 sample. Thus we will have 3^{60} genotypes to evaluate to determine the nature of relationship between all possible markers and the quantitative trait value. Similarly in case of *RI* or *DH* lines, there will be 2^{60} genotypes. A sample of 100–200 plants constituting the experimental population will not contain all the genotypes and thus there will be missing genotype problem which causes multicollinearity. The effect of the multicollinearily is that markers that define X_1 and X_2 appear to be correlated even if the markers are not actually linked and in this situation, it will be difficult to estimate the relative effect of possible QTL near X_1 and a possible QTL near X_2 on Y, the quantitative trait value. Further, if there is only one QTL near X_1 both X_1 and X_2 would appear to be significant and we can't say to which marker the QTL is actually near.

(a) Cross validation strategies

We will have to check whether or not multicollinearity exists. To test this we will have to conduct another study and if the same markers are found to be related to the quantitative trait value in both studies then multicollinearity is not a problem and we can generalize the result. But if different markers are found to be related to the quantitative trait value then we have the evidence of multicollinearity. The purpose of cross validation is to assess the effect of multicollinearity in a single study. There are different methods of cross validation. In one method, the data is randomly split into two halves and this method is called split sample validation. Thus we have two sub sets, say *A* and *B*. Multiple regression analysis is carried out to determine the markers near the QTL in set *A* and also in set *B*. Results from the two sets are compared and if the analyses include the same set of markers then we have some evidence that the chosen markers are near the QTL and multicolinearity is not there. But then in most cases markers are found near the QTL in the two subsets. Thus what is required is to conduct several studies and find out the markers which are consistently shown to be associated with the QTL and only those markers should be used to map the QTL.

As one does not have time and money to conduct different studies one can still build a valid prediction equation from single study and use the markers as indirect selection criteria. Let us see how we can go about it. Once we have found the markers associated with a QTL in one subset, say *A*, we use these markers in the prediction equation for *Y and* we can know how well it works for the other set B. In practice, what is done is to calculate the R^2 value for a subset *A* and then *Y* values for subset *B* is predicted using the prediction equation from subset *A*. Now the correlation coefficient between the observed *Y* values and the predicted Y values (based on information from sub set A) for sub set B is calculated. The square of the correlation is called cross validation, R^2 and is denoted as R^{2*}. The difference $R^2 - R^{2*}$ is called the shrinkage in R^2 (SR) on cross validation. This process can be repeated and thus we will have a second value of shrinkage. The shrinkage value is almost always positive. Small values of shrinkage indicate that the QTL model is correct and it can be used in marker aided selection whereas the larger shrinkage values indicate that the QTL model is unreliable.

The problem with split sample validation strategy is that the sample size becomes small and this creates another problem. To overcome this problem another resampling strategy called jackknifing or bootstrapping can be used. In the jackknifing strategy (Tukey, 1958) out of n individuals, $n - 1$ individuals are used to build a model and Y value for the left one individual is predicated and this process is repeated and thus we have different models to produce the n predicted values.

(b) Simultaneous mapping of multiple traits

Jiang and Zen (1995) extended the model to incorporate the simultaneous mapping of multiple traits. n individuals from an F_2 population derived from the cross of two inbred parents (P_1, P_2) are sampled and observations on m quantitative traits are recorded. Marker genotypes are assigned values of 2, 1 and 0 for homozygous P_1, heterozygous F_1 and homozygous P_2, respectively. Now if P_{jk} denotes the value of jth individual for kth trait then the model to test for a QTL on a marker interval $(i, i + 1)$ takes the following form:

$$P_{j1} = b_{01} + b_1^* X_j + d_1^* Z_j + \sum_i^t (b_{i1} X_{j1} + d_{i1} Z_{j1}) + e_{j1}$$

.
.
.

$$P_{jm} = b_{0m} + b_m^* X_j^* + d_m^* Z_j^* + \sum_i^t (b_{im} X_{j1} + d_{im} Z_{j1}) + e_{jm}$$

where $j = 1 \ldots n$, b_{0k} represents the mean effect of the model for trait k, b_k^* is the additive effect of the putative QTL on trait k, x_j^* is the number of alleles at the putative QTL from one of the two parents, d_k^* is the dominance effect of the putative QTL on the trait k, Z_j^* is the indicator variable of the heterozygosity at the QTL, X_{jl} and z_{jl} are corresponding variables for marker l other than i and $i + 1$ in individual j with regression coefficients b_{lk} and d_{lk} on trait k and e_{jk} is the residual effect on trait k for individual j.

The joint mapping of multiple traits increases the statistical power of detecting QTL, improves the precision of parameter estimates and develops formal statistical test for pleiotropy, QTL by environment interaction and pleiotropy vs. linkage.

4. *Marker regression approach*

Marker regression approach involves regressing the additive difference between marker genotype means at a locus against a function of the recombination frequency between that locus and a putative QTL (Kearsey and Hyne, 1995). A QTL is located, as by other regression methods, at that point where the residual mean squares is minimized. The estimates of location and gene effects are consistent and as reliable as conventional flanking marker methods.

Let us consider a pair of homologous chromosomes in an F_1 derived from cross between two pure breeding parents P_1 and P_2 and assuming this pair of chromosomes be heterozygous for alleles at k marker loci, M_{i1} and M_{i2} depending upon whether the allele came from P_1 or P_2, respectively and $i = 1, k$, situated at $C_i cM$ on the linkage map of that chromosome. Finally, let there be a single QTL (Aa) on this chromosome at X cM. An F_2 of N individuals is scored for a quantitative trait, Y_j (where $j = 1 \ldots N$) on its marker genotypes determined at each of the k marker loci. Following Mather and Jinks (1982), the mean trait score of the three possible QTL genotypes in the F_2 are as follows:

$$AA = m + d$$

$$Aa = m + h$$

$$aa = m - d$$

Because it is not possible to genotype the QTL, one has to rely on marker phenotype and the present procedure uses the mean scores of each of the three genotypes at every marker locus. Following Cowen (1988)

$$\overline{M_{i1}M_{i1}} = m + (1 - 2R_i)d + 2R_i(1 - R_i)h$$
$$\overline{M_{i2}M_{i2}} = m + (1 - 2R_i)d + 2R_i(1 - R_i)h$$
$$\overline{M_{i1}M_{i2}} = m + [1 - 2R_i(1 - R_i)]h$$

where $\overline{M_{i1}M_{i1}}$, etc. is the expected mean trait value of all these individuals having marker genotype M_{i1},M_{i1} where $i = 1$ to k and R_i, is the recombination frequency between the QTL and the ith marker. Therefore,

$$(\overline{M_i M_{i2}} - \overline{M_{i2} M_{i2}})/2 = (1 - 2R_i)d = \delta_i \quad (i)$$

and

$$\overline{M_{i1} M_{i2}} - 1/2 (\overline{M_{i1} M_{i1}} - \overline{M_{i2} M_{i2}})$$
$$= (1 - R_i)^2 h = \lambda_i \quad (ii)$$

where $(1 - 2R_i)^2 = e^m$ (Haldane, 1919) where $m = 1 (XC_i)/50!$ the mean chiasma frequency in that interval.

The present approach of detecting, locating and estimating the effect of QTL is based on the finding that values of X, d and h which best fit the observed values of δ_i and λ_i at map position C_i. The relation between δ_i λ_i and marker poistion in cM is shown in Figure 21.14a for a QTL at 50 cM with gene effect $d = h = 1$. In practice, however, there will be markers present and thus the observed outcome for a possible set of markers can be shown with the help of Figure 21.4b.

Now in the equation (i) that $\delta_i = (1 - 2R_i)d$ when $R_i = 0.5$ i.e. there is no linkage between the QTL and the ith marker then $\delta_i = 0$ and if $R_i = 0.0$, i.e. there is complete linkage then $\delta_i = d$. The equation (1) is thus a linear equation of form $Y = 0 + bX$. Therefore, if δ_i, i.e. Y is regressed on $(1 - 2R)$, i.e. X, one should obtain a straight line of slope $b = d$ passing through the origin. Similarly, the equation (2) $\lambda_i = (1 - R_i)^2 h$ is also a form of linear equation $Y = 0 + bX$ where $Y = \lambda_i$ and $X = (1 - 2R_i)^2$ therefore, regression yields $b = h$, δ_i and λ_i are orthogonal and are hence independent variable. Thus in the regression analysis

$$\hat{b} = \Sigma XY/\Sigma X^2 = \Sigma \delta_i (1 - 2R_i)/\Sigma (1 - 2R_i)^2$$

Regression SS $= \hat{b} \Sigma XY$

$$= [\Sigma \delta_i (1 - 2R_i)]^2/\Sigma (1 - 2R_i)^2$$

and residual $SS = \Sigma Y^2$-regression SS

$$= \Sigma \delta_i^2 - [\Sigma \delta_i (1 - 2R_i)]^2/\Sigma (1 - 2R_i)^2$$

with the degrees of freedom for the regression and remainder 1 and K, respectively.

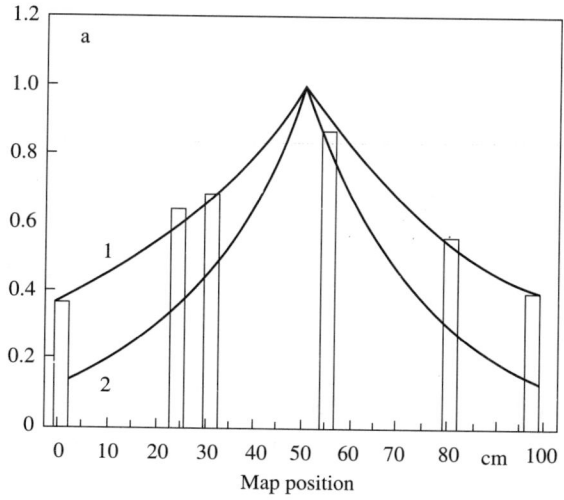
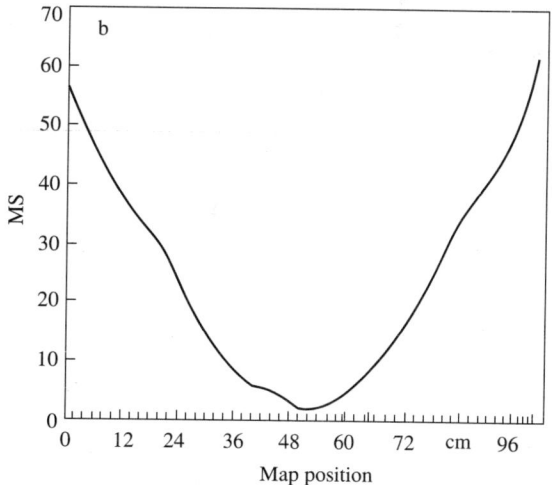

Fig. 21.4 (a-b) (a) Relation between δ_i, λ_i, and marker position, respectively; (b) Change in residual mean square for various putative QTL positions. Bars indicate the possible effect associated with marker loci (Kearsey and Hyne 1994).

21.18 Biometrical Genetics–Analysis of Quantitative Variation

Now as the true position of X, of the QTL is unknown, true values of R_i cannot be calculated but then the regression analysis of variance for δ_i may be carried out at a range of possible positions of the QTL along the chromosome from which R_i can be calculated. The estimated QTL location is then the position at which the residual sum of squares is at a minimum.

The above discussed approach can be extended to a case of more than one QTL effect in a linkage group. Suppose there are two QTL having a recombination frequency of R_{1i} and R_{2i} with marker i then

$$\delta_i = (1 - R_{1i}) d_1 + (1 - R_{2i}) d_2$$

Now if the two QTL are tightly linked then

$$R_{1i} = R_{2i} = R_i \text{ and hence}$$

$$\delta_i = (1 - 2R_i)(d_1 + d_2)$$

Now if δ_i is plotted against marker position in cM then it will yield the same curve as a single QTL with additive effect $(d_1 + d_2)$, as shown in Figure. 21.5a. If d_1 and d_2 are opposite in sign (i.e. QTL in dispersion) then δ_i would approach zero at all marker positions. If the two QTL are in association and further apart then the curve of δ_i will develop two peaks connected by a hanging valley as shown in Figure 21.5b. The problem then is how to distinguish between the two curves. If the QTL effects at the two loci are very unequal then it is possible to replace the curves outside the two peaks by a curve produced from a single QTL as shown in Figure 21.5c. Thus the discrimination between a model with one QTL

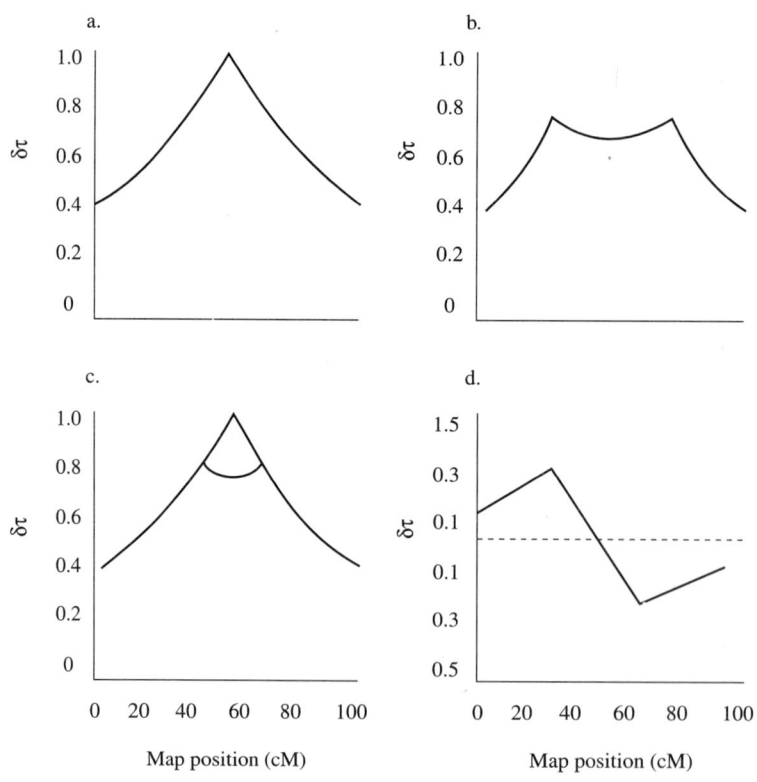

Fig. 21.5 (a-c) Relation between d_i and marker position for two QTL: (a) when two QTL are completely linked in association at 50 cM, (b) two QTL at 30 and 70 cM in association, (c) a comparison of one QTL at 50 cM with two QTL at 30 cM and 70 cM, (d) two QTL at 30 cM and 70 cM in dispersion.

and two QTL in association depends on information from markers between the two QTL. If the two QTL are in dispresion, the curve looks like as shown in Figure 21.5d. Marker regression approach has the advantages of speed and of integrating all the marker information in a single test. It can be shown that if there is just one QTL on a chromosome all the necessary information to locate and measure the effects of that QTL are available from the markers that flank that QTL. But then as we do not know which markers flank the QTL nor that there is just one QTL per chromosome, multiple marker approach does not provide an overall test of the model irrespective of the organization of the QTL on the chromosome.

Thus we see that there are a variety of statistical approaches to estimate QTL location and effects using multiple markers. Although these techniques yield comparable estimates of QTL position and effects, the reliability is extremely poor unless very large populations are used. The appropriate threshold value to use in the tests of significant is uncertain because numerous often non-independent tests are performed. Without a suitable criterion for determining the presence of one QTL, it is impossible to test whether the effect detected is truly one QTL or two QTL or more linked QTL. In addition, a QTL detected in one cross cannot be tested for correspondence to a QTL found in a different cross (Hyne and Kearsey, 1995).

(b) Comparison of estimation procedure

Hyne et al. (1995) compared the accuracy and reliability of the different methods using F_2 and DH populations. The parameters considered in the simulation study were, the number and position of the marker available, the accuracy of the marker position and the size of the experimental population besides QTL with differing heritabilities and locations along the chromosome. They found that the estimates of QTL position and effect obtained from 300 F_2 individuals and 150 DH lines were comparable albeit extremely unreliable. Even for the QTL of high heritability (10%) the confidence interval was 35 cM. QTL estimates were not significantly improved either by using the expected rather than the observed marker position or by using a dense map of markers rather than a sparse map. A QTL located asymmetrically in the linkage group resulted in accurate estimates of QTL position which were seriously biased at low heritability of the QTL. Keeping the experimental population constant this bias increases with the decrease in heritability.

21.13 Mapping Populations

The following types of population are used for mapping:

(i) Two inbreds or pure breeding lines and their F_1 and F_2 (F_2 population can be produced through either selfing or sibmating F_1 individuals), F_3 and backcross (BC_1 and BC_2) populations.

(ii) Recombinant pure breeding or inbred lines (RIL) derived from single F_2 or backcross individual using either SSD or brother - sister mating.

(iii) Dihaploid lines (DH) produced by self fertilizing doubled haploid derived from F_1 from the two pure breeding parents.

(iv) Test cross progeny derived from three-way cross, for example, F_1 of the two inbred parents (A, B) and a third parent (C).

Various variants of the above populations such as half-sib analysis, progeny testing, etc. can be used for mapping.

Selection of a particular population depends on biological, economic, statistical, and genetical considerations and molecular marker system availability and suitability. Selection of a population depends on the mating system of the plants. With inbreeders it is much more easier to produce F_2, F_3 and recombinant lines. With cross-fertileizing species, it is easy to produce back-cross and testcross. Further, production of DH varies with different crop species. Economic consideration refers to economic value of the possible cross. If the goal is to transfer a specific gene from a wild or weedy species into a cultivated plant, the backcross to the cultivated will be the best population to go for. Statistical consideration refers to selection of a particular type of population which can increase the power of test (in case of QTL analysis-maximum power to detect QTL effects). If the objective of the study is to measure dominance effects of QTL then homozygous lines or RIL or DH lines will not provide information although the advantage with homozygous or RIL or DH lines is that only one individual from single genotype is to be genotyped and phenotypic mean can be based on all the measurements of the individuals within a line. DH lines offer the opportunity to control environmental variation through replication and to repeat the experiment in different environments. Finally, the choice of an appropriate mapping population depends on the type of marker system to be applied. With codominant marker system, maximum information can be obtained from F_2 population (Mather, 1938). An F_2 population is advantageous for detecting QTLs even if few marker loci are available as the linkage disequilibrium is maximized in the F_2 population but then the probability of genotypic classes at a marker locus may reflect the effects of multiple QTLs as quite high. After a number of rounds of recombination, the size of the genomic region associated with each marker allele would be reduced considerably from that found in an F_2 population. Thus recombinant inbreds produced by sib-mating experiment, double the number of informative cross over that occurs in selfing and this offers a great scope for fine mapping. Dominant markers will supply similar information but then one will have to go for some kind of progeny test such as production of F_3 or F_2 BC population through which heterozygous individual can be identified. Both types of markers provide similar information in inbreds, RIL or DH lines or backcross population in coupling phase (Burr et al. 1988). Under condition of tight linkage ($r < 10\%$), dominant and co-dominant markers in RIL provide more information than either marker type in backcross population (Reiter et al., 1992). If recurrent and donor parents in backcross are homozygous and have contrasting polymorphic marker alleles then backcross populations can be useful for mapping dominant markers. Information obtained from backcross population using either type of marker is less than F_2 population as only one rather than two recombinant gametes are sampled per plant. Backcross populations are more informative (at low marker saturation) when compared to RIL populations as the distance between linked loci increases in RIL population (i.e. about $\geq 15\%$ recombination) (Staub et al., 1996). Increased recombination can be beneficial for resolution of the tight linkage but may be undesirable in the construction of maps with low marker saturation.

Since economic significance of a map depends on marker trait association as many simply inherited traits as possible should be included in the genetic stock for map construction. Further, one must consider the source of parents (adapted vs. exotic) used in the mapping population so that the map has uses in marker assisted selection. The precision of QTL estimates without loss of power can be improved by relaxing linkage disequilibrium through random sib-mating (Beavis et al., 1992). But then it takes more time to develop such population and also it requires a greater saturation of genetic markers. If the molecular marker techniques are to be useful for predictive or selective purposes, the genome should be fully saturated with markers and or strong disequilibrium is required. An ideal population for studying QTL using the molecular markers would be one with closely spaced, uniformly distributed marker loci about every 10 cM in a population which is in linkage equilibrium.

21.4 Population size

Detection of QTL with molecular markers normally requires the analysis of fairly large segregating populations (i.e. > 100 individuals) (Tanksley, 1993). There was little increase in reliability to be obtained from using 300 rather than 200 F_2 individuals and 100 doubled haploid lines gave similar results to 150 (Hyne et al. 1995).

21.5 Experimental Design

Like in biometrical analysis of quantitative trait, our aim in QTL analysis is to obtain estimates of the components of variation, dominance ratio, $g \times e$ interaction variance, heritability, etc. To obtain the estimates of these parameters, one will have to develop progeny using mating designs and put these progeny in the field using a suitable experimental design for recording observations. Estimation of parameters in QTL analysis differs from biometrical analysis in that the QTL genotypes cannot be observed and so the frequencies of hypothesized QTL genotypes are inferred from the marker phenotypes and further there is the requirement of recombination frequency between the QTL and the markers. So one will have to consider three models in QTL analysis rather than two as in case of biometrical genetical analysis. A linear model for experimental design effects and another linear model for the QTL genotypes are constructed and finally, a third linear model is constructed by substituting the effect of QTL genotypes with the effects of marker genotypes as the QTL genotypes are not observed. In QTL analysis, data can be balanced or unbalanced. Data are called unbalanced when the number of observations or replications of QTL genotype are unequal but are called balanced when the number of progeny of all QTL genotypes is equal. The another problem besides QTL genotypes being unbalanced, is that many QTL genotypes may be missing and these cause problems in estimations of various parameters.

The experimental design generally used is the randomized complete block design (RBD). The parameters are estimated by analysing the least squares means or maximum likelihood estimates of means of lines instead of the original observations. If the experimental design is balanced then the least squares and arithmetic means of lines are equal (Searle, 1971). The experiment then is first analysed ignoring marker genotype but not other factors (effect of all other QTLs which are not part of the model). The least squares means of the lines obtained are used for subsequent QTL analysis. The least squares methods combine linear least squares (Haley and Knott, 1992) and linear model theory for unbalanaced linear model (Knapp et al., 1993). The advantage with using least squares means of lines for the QTL analysis is that the observation from different experimental designs can be pooled and one QTL analysis can be performed for any experimental design. When analysis is done by using the original observation and ignoring the experimental design then error and line nested in the marker or QTL genotype variances are confounded and can be less powerful.

There are four methods of (Type I, II, III, IV) of estimating sum of squares (Freund et al. 1986). Type I, II, III and IV sum of squares are equal when the data are balanced but they are not necessarily equal for unbalanced data and missing cells (Knapp et al., 1992). In case of two QTL(Q_1, Q_2) model with interaction, Type I SS in case of unbalanced data is calculated by fitting Q_1 before Q_2 or fitting Q_2 before Q_1 and the interaction is normally fit after fitting Q_1 and Q_2. There is another way of estimating SS for unbalanced data which is to fit each effect after fitting the other effect and this leads to the type III SS for the two locus model with interaction.

Without a genetic map and marker phenotype, the analysis of variance in case of RBD using doubled haploid lines takes the form given in Table 21.7.

In the presence of information on recombination frequency and the molecular marker phenotypes, the genetic parameters can be defined as functions of marker or QTL parameters. Now the effects of lines are the sum of the effects of QTL genotypes and lines tested in QTL genotypes, the later are the effects between lines which are left over after estimating the effects between QTL genotypes. These are the effects of all the genes which are not part of the model. Thus sum of squares between lines is the sum of squares between QTL genotypes and between lines tested in the QTL genotypes. When estimate of recombination frequency is known and the marker phenotypes are known, the QTL effects can be estimated by using linear differences among the marker means. The ANOVA takes the form given in Table 21.8 and the hypothesis of no difference between QTL genoytpe means is tested by using the F-statistic, $F = M_Q/M_{G:M}$ and the hypothesis of no difference between marker genotype means is tested using F-statistic, $F = M_M/M_{G:M}$

When the experiment is conducted in RBD over many environments the ANOVA takes the form given in Table 21.9 (Knapp, 1994).

In case of multienvironment experiment, the sum of squares for lines is the sum of squares for QTL genotypes and lines tested in QTL genotypes and the sum of squares for lines by location is the sum of the squares for QTL genotypes × locations and for the lines nested in QTL genotypes by locations. Now considering marker genotypes theANOVA takes the form given in Table 21.10 (Kanpp, 1994). The hypotheses to be tested here are as follows:

(i) Whether or not there are significant differences between lines and between line × locations

Table 21.7 Type III sum of squares and expected mean squares for the lines (doubled haploid) using QTL genotypes

Source	D.F.	S.S.	M.S.	E.M.S.
Block	$B-1$	$R[b|\mu, g]$	M_B	$\sigma_E^2 + N\sigma_B^2$
Line (G)	$N-1$	$R[g|\mu, b]$	M_G	$\sigma_E^2 + r\sigma_G^2$
QTL (Q)	$q-1$	$R[q|\mu, b]$	M_Q	$\sigma_E^2 + r\sigma_{G:Q}^2 + \phi_Q^2$
G:Q	$N-q$	$R[g(q)|\mu, g, b]$	$M_{G:Q}$	$\sigma_E^2 + r\sigma_{G:Q}^2$
Residual	$(N-1)(b-1)$	$R[e|\mu, b, g]$	M_E	σ_E^2

where q is the number of QTL genotype, $N = \sum_{i=1}^{q} n_i$ is the number of lines where n_i is the number of lines of ith QTL genotype and b is the number of replications or blocks. The genetic variance between doubled haploid lines is calculated as:

$$\sigma_G^2 = 2\sigma_A^2 = \frac{M_G - M_E}{r}$$

and the heritability as

$$H = 1 - \frac{M_E}{M_G}$$

Table 21.8 Type III sum of squares and expected mean squares using marker genotypes

Source	D.F.	S.S.	M.S.	E.M.S.
Block	$b-1$	$R[b|\mu, g]$	M_B	$\sigma_E^2 + N\sigma_B^2$
Line (G)	$N-1$	$R[g|\mu, b]$	M_G	$\sigma_E^2 + r\sigma_G^2$
Marker (m)	$m-1$	$R[m|\mu, b]$	M_m	$\sigma_E^2 + r\sigma_{G:M}^2 + \phi_M^2$
G:M	$N-m$	$R[g(m)|\mu, m, b]$	$M_{G:M}$	$\sigma_E^2 + r\sigma_{G:M}^2$
Residual	$(N-1)(b-1)$	$R[e|\mu, b, g]$	M_E	σ_E^2

m is the number of marker genotypes
$G \times E$ interaction experiment

(ii) Whether or not there are significant differences between QTL genotypes and between QTL genotypes × locations

The hypothesis of no difference between marker genotype means across locations is tested using F statistic = $(M_M + M_{G:ML})/(M_{G:M} + M_{ML})$. The hypotheses of no difference between hypothesized QTL genotypes and no difference between QTL genotypes by locations are tested using F-statistics $F = (M_Q + M_{G:ML})/(M_{G:M} + M_{ML})$ and $F = M_{QL}/M_{G:ML}$, respectively.

The various sum of squares are obtained by sequential model fitting.

A complete QTL model explains a maximum of 100 H per cent of the line-mean phenotypic variance or 100 per cent of the between line variance where H is the family mean heritability. There could be two types of QTL, QTL showing significant effect across environments and QTL showing significant g × e interaction effect. While constructing model both types of QTL should be retained in the model.

21.6 Mapping in Polyploids

Polyploids differ from diploids in that it contains more than two copies of each chromosome. Autopolyploids differ from allo-polyploids in that

Table 21.9 Degrees of freedom, Type II SS and expected means squares in $G \times E$ interaction experiment

Source	D.F.	S.S.	M.S.	E.M.S.
Block	$b - 1$		M_B	$\sigma_E^2 + lN\sigma_b^2$
Location (L)	$l - 1$		M_L	$\sigma_E^2 + r\sigma_{GL}^2 + rN\sigma_L^2$
Line (G)	$N - 1$	$R[g|\mu, 1, b, g1]$	M_G	$\sigma_E^2 + r\sigma_{GL}^2 + rl\sigma_G^2$
QTL	$q - 1$	$R[q|\mu, 1, b, g1]$	M_Q	$\sigma_E^2 + r\sigma_{G:OL}^2 + rl\sigma_{G:Q}^2 + r\bar{m}\sigma_{QL}^2 + \phi_Q^2$
G:QTL	$N - q$	$R[g(q)|\mu, 1, b, g1]$	$M_{G:Q}$	$\sigma_E^2 = r\sigma_{G:QL}^2 + rl\sigma_{GQ}^2$
G × L	$(N - 1)(l - 1)$	$R[gl|1\ \mu, 1, b, g]$	M_{GL}	$\sigma_E^2 + r\sigma_{GL}^2$
QTL × L	$(q - 1)(l - 1)$	$R[ql|u, 1, b, g]$	M_{QL}	$\sigma_E^2 + r\sigma_{G:Q}^2 + \sigma_E^2 + \sigma_{G:QL}^2 + m\sigma_{QL}^2$
G:QTL × L	$(N - q)(l - 1)$	$R[g(q1)|\mu, 1, b, g, q1]$	$M_{G:QL}$	$\sigma_E^2 + r\sigma_{G:QL}^2$
Residual	$(Nl - 1)(b - 1)$	$R[e|u, l, b, g, g1]$	M_E	σ_E^2

where l is the number of locations and

$$\bar{n} = \frac{N - \frac{\Sigma n_i^2}{N}}{q - 1}$$

Table 21.10 Expected mean squares in $G \times E$ interaction experiment using marker genotype

Source	D.F.	S.S.	M.S.	E.M.S.
Block	$b - 1$		M_B	$\sigma_E^2 + lN\sigma_b^2$
Location (L)	$l - 1$		M_L	$\sigma_E^2 + r\sigma_{GL}^2 + rN\sigma_L^2$
Line (G)	$N - 1$	$R[g|u, 1, b, g1]$	M_G	$\sigma_E^2 + r\sigma_{GL}^2 + rl\sigma_G^2$
Marker (M)	$m - 1$	$R[m|u, 1, g, g1]$	M_M	$\sigma_E^2 + r\sigma_{G:ML}^2 + rl\sigma_{GM}^2 + r\bar{m}\sigma_{ML}^2 + \phi_M^2$
G:M	$N - m$	$R[g(m)|u, 1, b, g1, m]$	$M_{G:M}$	$\sigma_E^2 + r\sigma_{G:ML}^2 + rl\sigma_{GM}^2$
G × L	$(N - 1)(l - 1)$	$R[gl|u, b, g1]$	M_{GL}	$\sigma_E^2 + r\sigma_{GL}^2$
M × L	$(m - 1)(l - 1)$	$R[ml|u, 1\ b, g1]$	M_{ML}	$\sigma_E^2 + r\sigma_{G:ML}^2 + r\bar{m}\sigma_{ML}^2$
G:M × L	$(N - m)(l - 1)$	$R[g(ml)|u, 1, b, g, m1]$	$M_{G:ML}$	$\sigma_E^2 + r\sigma_{G:ML}^2$
Residual	$(Nl - 1)(b - 1)$	$R[e|u, 1, b, g, g1]$	M_E	σ_E^2

several sets of chromosomes are readily distinguishable (see Chapter 12) and a particular chromosome may pair with different partners in different regions. Further autopolyploids are different than diploids in segregation and recombination whereas allopolyploids are similar to diploids in segregation and recombination. Thus what we find is that autopolyploids can have from 0 to x copies of an ancestral gene (x being the number of copies of each chromosome). The study of dosage effects of gene and epistasis is much more complicated than in diploids. Because of multiple opportunity for chromosome pairing, the study of recombination frequency is quite complicated in autopolyploids and there is just not enough molecular markers to distinguish each of the either four chromosomes (e.g., in potato and alfalfa) or three chromosomes (e.g. banana) and so only a fraction of total number of recombination events can be studied which suggests the use of higher experimental size in the study.

21.6.1 Methods of linkage analysis

The first step in linkage analysis in autopolyploids is to assess the dosage or ploidy level (e.g. simplex, duplex, triplex, etc.) of each polymorphism. This is achieved by examining the genetic segregation ratio. Table 21.11 shows the expected gametic segregation. ratio for different dosages in disomic and polysomic octoploids assuming no double reduction (Jorge et al. 1996).

The agreement between the observed and the expected gametic ratio is tested using χ^2. Once the dosage of each marker is identified, the next step would be to run linkage test to determine linkage relationship and finally the homology map can be fused into a chromosome map by using repulsion-phase linkage.

There are two types of linkage analysis depending upon the dosage of the markers involved in the test:

(i) Use of single dose markers
(ii) Use of multi-dose markers

1. Single dose marker

This method of linkage analysis in polyploids (Wu et al., 19991) involves the identification of band that represents single dose restriction fragment (SDRF). In this approach, each restriction fragment is analysed for its presence or absence in the progeny. If the fragment is represented by a single dose, then this single dose marker will segregate 1: 1 (presence : absence) in the gametes or 3: 1 in the selfed progeny in both auto-and allopolyploids. The dose restriction fragment markers in autopolyploids represent the genotype of one parent or the other and thus do not discriminate between the individuals which are heterozygous from those which are homozygous for parental fragment. In other words, the markers are dominant and in this respect SDRFs and other SD markers are no different than PCR based markers

Table 21.11 Genetic segregation ratio in disomic and polysomic actoploids assuming no double reduction

| Parent | Gametes | | | | | Ratio |
	aaaa	Aaaa	AAaa	AAAa	AAAA	A:aaaa
Disomic octoploid						
Aa aa aa aa	1	1	–	–	–	1:1
AA aa aa aa	–	1	–	–	–	–
Aa Aa aa aa	1	2	1	–	–	3:1
AA Aa Aa aa	–	1	1	–	–	–
Aa Aa Aa aa	1	3	3	1	–	7:1
AA AA aa aa	–	–	1	–	–	–
AA Aa Aa aa	–	1	2	1	–	–
Aa Aa Aa Aa	1	4	6	4	1	15:1
Polysomic octoploids						
A aaaaaaa	1	1	–	–	–	1:1
AA aaaaaa	3	8	3	–	–	11:3
AAA aaaaa	1	6	6	1	–	13:1
AAAA aaaa	1	16	36	16	1	69:1
AAAAA aaa	–	1	6	6	1	–
AAAAAA aa	–	–	3	8	3	–
AAAAAAAa	–	–	–	1	–	–

(except microsatellites) in diploids. Estimation of recombination distance between two SDRF markers is similar to that for diploids. The only limitation is that the SD method explores only in coupling phase. For detection of linkage in repulsion phase, an extremely large population size is required (Wu et al., 1992). The SD markers have been used to study the type of inheritance in *S. spontaneum* SES 208 (Al Janabi et al., 1993; Da Silva et al., 1993, 1995). The principle of SDRF markers was used to construct a linkage map of potato (Bonierbale et al., 1988) from a cross between the heterozygous diploids. Estimation of the linkage distance depends on the population size. Thus a large difference between allo-and autopolyploids in detection of repulsion phase linkage and a comparison of the number of coupling and repulsion phase linkages will provide an indication of whether the species is allo or autopolyploids. Detection of only coupling phase linkages results in a linkage map with 2n number of linkage groups.

2. *Multiple dose markers*

Markers showing segregation ratio 1:1 are likely to be in two or more doses. In case of disomic octoploid, a double dose marker will segregate 3:1 for presence: absence (if two alleles are on homologs that do not pair) as compared to 11:3 in a polysomic octoploid. A triple dose marker will segregate 7:1 in a disomic octoploid as compared to 13:1 in a polysomic octoploid. After having identified DD (double dose) and TD (triple dose) markers, two point linkage tests are conducted using the expected frequencies which reflect the possibility of each chromosome pairing configuration at metaphase 1, for each of the four genotypic classes. Once the type of chromosome pairing (random or preferential) and ploidy level have been characterized for a given species, the agreement between the expected gametic segregation and the observed gametic segregation for DD and TD markers is tested.

In diploids information from both recombination classes (coupling and repulsion) are used to calculate the recombination frequency. In polysomic polyploids as many more gametic phases are possible, the amount of linkage information that can be obtained from each gametic phase depends on our ability to distinguish the ratio from phenotypic classes under linkage from that same ratio under independent segregation. Hence coupling phases provide more linkage information than repulsion phases. The estimate of r, the recombination frequency is associated with large standard error.

Triple dose markers are more useful to identify true homologous chromosomes since it is easier to distinguish disomy from polysomy. In a polysomic octoploid, a TD marker located on non-homologous chromosomes will show a disomic segregation 7:1 as compared to a polysomic ratio 13:1 which is expected when the doses are located on homologous chromosomes. Further, double dose markers are more efficient than single dose markers in identifying homologs. However, with double dose markers, distinction between the polysomic and disomic segregation requires a much larger population. But then for detecting pairing partner the ability to discern the true linkages is much more important then accurately estimating r, the recombination frequency. The multidose markers complement the single dose markers in that it identifies pairing partners, indirectly detects single dose marker and uses the data which is otherwise not available for mapping.

21.7 Genetic Mapping

The genetic mapping involves (i) the determination of linear order with which the genes are arranged with respect to one another (gene order) and (ii) the determination of the relative distances between genes (gene distance). The unit of distance expresses the probability that crossing-over will occur between the two genes under investigation. One unit of map distance (1 cM) is therefore equivalent to 1% crossing over. On an average, a chromosome is about 100 cM long. As crossing-over equals half the chiasma per cent i.e., each chiasma produces 50% cross-over products. 50% cross-over is equivalent to 50 map units. If the mean number of chiasmata is known for a chromosome pair, the total length of the map for that linkage group can be predicted by the formula: Total length = Mean number of chiasmata × 50. The additivity of the map distance is accepted which allows us to place genes in proper order. If double crossovers do not occur then map distances may be treated as completely additive units. Double crossovers usually do not occur between genes less than 5 map units apart. In a two point linkage experiment, the greater the unmarked distance between two genes, the greater the probability of double crossing-over occurring without detection and there is thus underestimation of the map distance. Therefore the most reliable estimates of the amount of crossing over will be obtained from closely linked genes. In Drosophila, the DCOs do not occur within a distance of 10 to 12 cM. The minimum DOC distance varies between species. Within this minimum distance (1–10 map units) recombination percentage is equivalent to the map distance but then this relationship becomes non-linear outside the minimum distance. The true map distance will thus be underestimated by the recombination fraction and at large distances (40–50 map units) they virtually become independent of each other (Kosambi, 1944). As the frequency of C.O usually varies in different segments of the chromosome therefore the actual physical distances between linked genes bear no direct relationship to the map distance calculated on the basis of C.O percentage (Stansfield, 1969; Swanson et al., 1990). Genes that are loosely linked (> 20 map units) can be placed on a map but their location is much more tentative. However, the linear order of genes in the physical map and genetic map should theoretically be identical. Further as the maximum recombination between two linked genes is 50%, the genes very far apart on the same chromosome may behave as though they were on different chromosomes.

Relation between physical and genetic distance.
Physical distance refers to the number of base pairs between two loci and is not the same as genetic distance whereas genetic distance is based on the observed, θ, the referred to as the probability of observing odd number of crossoverss and thereby production of recombinant chromatids. When two loci are located on different chromosomes or on the same chromosome but very far apart, then

$$\theta = 1 - \theta = a\ 0.5 \text{ and } b = 0.0$$

where a is the probability of observing two or more crossovers and b is the probability of observing zero crossovers. Thus when θ = 0.5, we expect 50% recombinant chromatids and 50% non-recombinant chromatids. When two loci are located on the same chromosome and are very close (i.e., linked) then both θ and a are low but b is very high. Thus when θ < 0.5 we expect to observe less recombinant (<50%) than non-recombinant chromatids. As θ considers only recombinant events (the odd ones, they are not simply additive across regions).

Considering three consecutive loci A, B and C the recombination fraction, θ between A and C, (θ_{AC}) = $\theta_{AB} + \theta_{BC} - 2\ \theta_{AB}\ \theta_{BC}$ assuming no interference.

Unlike θ, genetic distance, m, is an additive measure but it can not be directly measured and mapping functions must be used to predict m based on observed θ. Genetic distance is expressed in units of Morgans (Ms) or centiMorgans(cMs) where 100cMs = 1M. One mapping function is Halden's functin where

$$M = -1/2\ [\ln(1 - 2\theta)]$$

Halden's genetic distance of 5cM(m = 0.05) corresponds to θ = 0.02. Another mapping function, Kosambi function allows the non-independence of crossingovers.

Kosambi Morgans (KM) = $\left(\frac{1}{4}\right) \ln[(1 - 2\theta)]/(1 - 2\theta)$
and
$$\theta = 1/2[e^{4KM-1}]/[E^{4KM+1}]$$

Halden Morgans (HM) = $-1/2[\ln(1-2\theta)]$ and
$$\theta = 1/2 \ln(1-e^{-2HM})$$

With the advent of molecular markers, the relationship between DNA marker map on one hand and cytological map and physical on the other hand is being worked out. The use of aneuploids such as monosomics, trisomics and substitution lines can be made in order to correlate DNA make up with the specific chromosomes. Nucleic acid hybridization with a mapped clone indicates its chromosome location by observing the loss of a band (in the case of nullisomics) or a change in the relative signal on a radiogram (McCouch et al., 1988). This type of analysis requires a second DNA clone of previously determined chromosome location so that even very small changes in the relative intensity of a band can be compared. In case of substitution lines where known chromosomes or chromosome arms are substituted with homologous segment from alien species probing a clone onto a blot containing restriction, digested DNA from a complete set of substitution lines easily identifies the chromosome location of that clone (Sharp et al., 1989) as substitution corresponding to the location of a clone shows a different restriction fragment pattern compared to other substitution lines.

As, in general, recombination is inhibited near centromeres and in heterochromatic regions and introgressed regions whereas it is relatively very high in other regions, this makes many DNA markers to cluster at site probably near the centromere or heterochromatic region while in other parts of maps, markers are separated by large gaps even after hundred of markers have been placed on the map (Young, 1994). Although there is a need to describe the distance between DNA markers on the linkage map in terms of recombination frequency and actual physical distance for effective map based cloning but so far no solution has come to describe the relation. With the increase in the density of DNA markers linkage map and availability of techniques for analysing the large DNA fragments, it is hoped that a correlation between genetic map and physical distance could be found.

The ratio, DNA base pairs/centimorgan may vary 10,000 fold among different regions of the genome (Meagher et al., 1988). The distribution of recombination may be very uneven and seemingly non-random along the cytological length of the chromosome with a tendency to be higher in distal regions and lower in centromeric regions (Lukaszewski and Curtis, 1993). The recombination may vary considerably with genetic background (Lukaszewski, 1992), between related taxa Rick, 1969) and with sexual origin of the gametes (Philips, 1969). The rate of recombination seems to be much higher in expressed genes (Thurieaux, 1977; Dooner, 1994).

Information from interspersion patterns of sequence classes, ratio of physical and genetic distance and fine mapping has a bearing on the prospects for chromosomal walks for direct isolation of genes from the crops, the nature and putative products of recombination and the source of genetic information. The information on the ratio, DNA base pairs/centimorgan is more important to those who are interested in positional cloning and introgression strategies. In case of former, the high ratio could mean the difference between successful chromosomal walks and a seemingly endless forced march, in case of latter, the information—a favourable chromosome position for the target gene (i.e. a relatively highly recombined gene region) or a prior knowledge of pattern of the distribution of recombination might provide clues regarding the most efficious approach of gene transfer.

In maize, molecular and genetic fine structure analysis of several loci (bz, wx, Adh, al and R) has indicated that the physical/genetic distance ratio, averaged over the entire genome is approximately 100 times higher than those values for regions within genes (Dooner, 1994). This shows that recombination occurs at much greater rate within genes and may even be restricted to genes in eukaryotes. The physical/genetic distance ratios in various crops are as follows:

(i) Tomato, Tm–2a (4Mb; Ganal et al., 1989).
(ii) Maize, al sh_2 interval (1460 kb; Civardi et al., 1994).
(iii) Maize, bzl (43 kb for the region, within the locus), al (217 kb within the locus).
(iv) Wheat, alpha-Amyl-1 (lMb; Cheung et al., 1991).

Interspersion patterns of DNA classes have predicted bleak prospects for chromosomal walking in crops with large, complex genome using methods currently available (Springer et al., 1994).

(a) Construction of map

For successful map construction, one should start with selection of the most appropriate mapping population. Then using appropriate molecular marker technology pairwise recombination frequencies are calculated from which linkage groups are established and map distances are estimated and finally gene order is determined. As different types of population and different molecular marker technology can be used for map construction, various computer packages such as Linkage I (Suiter et al., 1983), G Mendel (Echt et al., 1992), Map maker (Lander and Bostein, 1986; Lander et aI., 1987), Map Manager (Manly and Elliot, 1991) and Join Map (Stam, 1993) have been developed for the analysis of data for map construction.

'Join Map' software package considers the estimates of recombination between a given pair of markers of different origins (data sets /mapping population), calculates the weights and then generates a single recombination value. After averaging weights to all pairwise combinations, it comes with the best fitting linear arrangement of the marker loci. The programme further calculates goodness of fit criterion corresponding to the hypothesized levels of interference (positive and negative) and thus makes possible to test each synthetic map.

Traditionally gene orders have been determined from either two- or three-point test cross data. Three linked genes may be in any of three orders depending upon which gene is in the middle of the linkage group. Without the middle marker the true map-distance between the two genes is underestimated. When multiple cross-over occurs with much greater than random frequency (i.e., localized negative interference) gene order of closely linked sites can be obtained using three factor reciprocal crosses.

(b) A complete genetic map

When should a genetic map be considered complete? Wherever cytogenetic maps are not available, the DNA genetic map can be considered complete when either there are no remaining unlinked markers or number of DNA marker linkage group equals the number of chromosomes observed. Mapping is complete only and only when even the ends of chromosomes have been marked. As satellite sequences tend to be located near the ends of chromosomes (Ganal et al., 1988), these sequences may provide a method of cloning adjacent low copy sequences which could not be used to establish definitely the chromosome ends.

(c) Optimal genetic map

Multiple maps are now being constructed in different crops using various molecular marker systems, types of populations and generations. If two or more genetic maps possess a minimal number of common markers, they can be merged to construct a more informative map (Hauge et al., 1993). But, as the types of information and the estimates of recombination frequencies often differ greatly between populations and data set, these must be weighted while merging the different maps and constructing the most likely or optimal map. Chromosome pairing and recombination rate can be severely suppressed in wide crosses (adapted × exotic) and generally yield reduced linkage distances (Albini and Jones, 1987; Zamir and Taldmor, 1986). Segregating population from wide crosses will show a relatively large array of polymorphism incomparison to segregating population from adapted × adapted cross. Map constructed from wide cross must be colinear (i.e., order of loci similar) with map constructed from narrow cross then only it will have use in plant breeding.

21.8 Information from QTL Mapping

(a) Number of genes involved in determination of a quantiative trait

Methods for estimating the number of effecti ve factors (Chapter 16) are based on a number of assumptions which are generally violated and thus making the estimate suspect. Furthermore, there was no practical way to independently verify or falsify one's estimate. With the advent of molecular marker, this problem has been solved although the molecular marker approaches for estimation of number of polygenes are straight forward but not without limitations and the statistical biases. The number of segregating polygenes affecting a particular trait in a population is equal to the sum of the number of QTL detected in a particular study (Tanksley, 1993). The limitations with molecular approaches are: (i) only those QTLs with sufficiently large phenotypic effects to be detected significantly would be counted. Genes with small effects will fall below the threshold of detection depending upon the size of segregating population. The larger the population size, the more likely it is to statistically detect genes with smaller effect. So if there are a large number of QTL, by definition of quantitative trait, they cannot all have a large effect or individually constitute a large proportion of the genetical variation and thus true number of QTL will be underestimated, (ii) because only the significant estimates are used in the calculation that the poorer the power of test, the greater the bias i.e. the more stringent the significance level, the greater the bias, (iii) two or more polygenes close together than ≈ 20 cM will usually be detected as a single QTL in a typical population of 500 individuals and thus the bias is towards underestimating the number of genes, (iv) QTLs at the ends of chromosome will tend to be located further from the ends than they actually are because all locations beyond the end of chromosome will be excluded (Hyne et al. 1995). Similar effects may occur in the vicinity of markers using any interval mapping. Thus all these biases will result in underestimation of the true number of QTL.

Studies so far conducted with QTL have shown the involvement of 1 to 16 with a mean number of 4 QTLs for a particular trait. In 94% cases 8 or fewer QTLs have been found and in only about 2% cases more than 12 QTLs have been reported. Further, only few studies detected individual chromosome with more than one QTL but there are examples with at least 3 QTLs per chromosome (Van de Schaar et al., 1997).

(b) Per cent genetical variation explained by the QTL

The advantage with molecular marker approach for estimation of the number of polygenes is that the results are quantifiable and testable. They are quantifiable in that the % genetic and phenotypic variation attributable to the QTL detected can be estimated but the variation left unexplained is due to other QTLs or the environment. Survey of literature on this aspect has shown that 46% of variances is explained by the QTL identified although it varies from 10 to 95% in individual studies. Further, one might expect that those studies finding more QTLs would explain more variance but this is not the case. Although % variance explained does not increase significantly with increase in QTL number, the R^2 is only 6% while those cases in which most variation is explained involve just 1–6 QTL (Kearsey and Farquhar, 1998). In 19 cases where 70% or more of the variance is explained, 5 are biochemical traits controlled by 2–5 genes, 3 concern disease resistance, two pollen traits and the rest are agronomic traits. Both erucic acid and linoleic acid content in *Brassica napus* are largely explained by 2QTL each (Ecke et al., 1995; Tanhuanpaa et al., 1995). Individual QTL explains from 1% to 60% of the phenotypic variation.

The detection of a QTL with small effect depends on a number of factors (Tanksley, 1993), viz. (i) the map distance between a marker and a QTL, the closer is a QTL to a marker, the smaller effect that the QTL can have and still be detected statistically because the effects of QTL closer to the marker is less confounded by recombination events between the markers and the QTL, (ii) size of the segregating population, the larger the population size, the more likely it is that the QTL with small effects will reach statistical significance, (iii) heritability of the trait, lower the heritability of a trait, the less likely a QTL will be detected, (iv) significance level for detecting a QTL, higher the significance level, less likely a supurious QTL being reported or less likely a QTL with small effect being detected.

(c) Estimates of gene effects

Additive and dominance genetic effects are estimated through different QTL analyses. Because only the significant estimates are used in calculation of gene effects it will cause a greater bias on the estimates of dominance than on additive effects because dominance effects are more difficult to detect. Considering all the statistical biases, the estimates of additive and dominance effects will be over-estimated.

As most of the studies conducted so far have involved homozygous lines, recombinant inbred lines (RIL) or doubled haploid (DH) lines, they do not provide information on dominance effects of QTL. Of those that do some degree of dominance has been found in 53%. Survey has revealed that a total of 55% of the dominant QTL showed apparent over dominance (i.e. $d > a$) while for 23%, the overdominance was significant. These overdominant QTLs should be analysed further through high resolution mapping to confirm their status and if real, to identify their genetic status. Such wide occurrence of overdominance, if real, is not expected from the previous quantitative genetic studies. It could be possibly due to closely linked dispersed genes but then as the size of significant dominance effects is likely to be overestimated, it could be spurious overdominance and not the genuine one. Thus molecular marker studies will throw light on whether or not overdominace is involved in heterosis or hybrid vigour. The different theories of heterosis are described in Chapter 11.

(d) Epistasis

Information on various kinds of epistasis using classical ratios is described in Chapter 5 and biometrical explanation and estimations of epistasis are described in Chapters 6 and 10, respectively. Using molecular markers, the study of QTL can now confirm its presence. In the studies conducted so far, the number of statistically significant interactions has normally been close to the number expected by chance (De Vincente and Tanksley, 1993; Edwards, et al. 1987; Stuber et al. 1992; Tanksley, et al., 1982). Most QTL ×QTL interactions being near the probability threshold can be said to be a statistical artifact whereas a few of the specific QTL interactions did have very high probability and thus indicating real epistasis and not spurious. These results suggest that strong epistatic interactions are the exception rather than the rule for the polygenic system. These conclusions have been further supported by few studies in which individual QTLs have been genetically isolated from other QTL in nearly isogenic lines (NTL) and have been shown to continue producing their same individual effects.

For detecting QTL × QTL interaction, what is required is to detect significant QTL affecting a trait first and once that is achieved any two significant unlinked QTL can be used as independent variables and the quantitative trait as dependent variable. Thus in two way ANOVA, the interaction factor calculated will provide an estimate of interaction between the two QTLs in determining the phenotype. A significant interaction factor suggests that the effect rendered by the two QTLs together is not simply a sum of their, independent effects. Theoretically, two factors, 3 factor, 4 factor and multi-factor (n factor) or higher order interactions are all possible among the QTLs and can be estimated using n-way Anova. There are several factors affecting the estimate of interactions.

(i) Population size

Small segregating population size will result in inaccurate estimate of multifactor interactions. For example, considering two locus (A, B) with two alleles (A, a: B, b) system, the double homozygous such as AABB, aabb, etc., would appear in the F_2 population with an expected 1/16 frequency. Now even considering a population size of 500 F_2 individuals, the number of individuals in each of these classes would be 31 which is a relatively small sample size to measure a phenotypic effect and make comparison with the phenotypic effect of the similar classes of individuals. Considering 3 factor interaction, the frequency of the rarest class would be 1/64 and thus it would require thousands of individuals to make comparison more valid.

(ii) Level of significance for detecting epistasis

As the number of multiple factor interaction is very large, some of which will by chance reach statistical significance. For example, there will be 45 two factor interactions with two factors (A, B) and with a significance level of 0.05, one would expect two spurious interactions to be reported. One way to overcome such problem is to raise the level of significance to avoid reporting spurious interactions but then this would result in real interaction undetected —a problem of small population size.

(iii) Recombination frequency between marker and QTL

Although it is assumed that the QTL is tightly linked with the marker, the recombination between marker and QTL will result in underestimation of the interaction effect. This problem can be mitigated by using high density molecular linkage map for detecting and analysing QTL which will ensure that most QTLs are linked with at least one molecular marker.

(e) Genotype × environment interaction

The understanding of G × E interaction and the more fundamental issue of adaptation is more important in practical plant breeding. Various statistical and genetical methods of analysis of G × E interaction are discussed in Chapter 8. It is a fact established from studies that polygenes are interacting with the environment and further genes determining yield performance are different than genes determining stability i.e. both are under independent gene control. The breeders thus would like to isolate and use the following kinds of genes for successfully breeding a cultivar for its wide spread use:

(i) Genes that are responsive to better management, and fertility environment.

(ii) Genes that perform better in harsh environment.

With the help of molecular marker, the QTL determining stability, adaptability or responses to better management or giving higher performance in harsh environment can be identified, characterized and mapped and further introgressed into an other wise good cultivar. QTLs which are expressed in all the environments could provide stability to the performance of a particular variety. Similarly, several QTLs with different environmental specificities can be combined and put in a variety which will make it more buffered against the vagaries of environments. Sometimes the low performance of a genotype in an environment can be due to lack of a simply inherited trait, resistance to pests and diseases, photoperiod response, vernalization requirement and this can be overcome by transgenic plant technology and the improvement in yield can be achieved.

Baker (1988) characterized group of genotypes as showing cross-over type G × E interaction where ranks of means of genotypes change across environments and non-cross over G × E interaction where the difference between the means of genotypes changes but their ranks remain unchanged. Where cross-over G × E interaction arises the difference between QTL genotype means across environments is usually less than within environments. Non-cross over G × E interactions do not lead to fixation of unfavourable alleles as most QTLs might be selected which do not affect the trait across target environments or QTLs which might be missed as they affect the trait across target environments. Means of QTL genotypes across test environments are needed to be estimated to select a QTL for marker-assisted selection (MAS).

The quantitative genetic studies have shown that the polygenes respond differently to different environmental changes and that polygenes important in one environment may not be as important in determining the phenotype in another environment. QTL analysis in maize has shown that QTLs detected in one environment were frequently detected in the other environment thereby suggesting no G × E interaction (Stuber et al. 1992). Peterson et al. (1991) working with tomato found that 48% of the QTLs were detected in two of three environments. These studies thus suggest that a substantial portion of QTL determining a quantitative trait in one environment will be active in the other environment and this is especially true for QTL with large effects. Thus we see that while some QTLs effect varied across environments, the majority were stable.

Adaptibility has been assumed to follow the polygenic model of inheritance that has been the basis of quantitative genetics. Adaptibility has been defined as the process of transgene-rational changes in which organisms become better suited to the environment through changes of features or functions that solve or improve solutions to problem posed by the environment of integrating metabolism or of enhancing reproduction (Burian, 1992). Now the question is whether adaptation involves major genes or polygenes. Recently conducted molecular studies suggest that the genetic architecture or a QTL loci is compatible with a role for major gene in adaptation (Tai et al., 1994) and major genes have been found to be involved in some cases of adaptation involving visual differences among individuals and species (Charlesworth, 1994). Examination of genetical basis of morphological differentiation of closely related species has shown the involvement of a small number of major gene (Paterson, et al. 1988; Doebley and Slec, 1993). Further, studies of adaptation of insect to insecticides have shown higher selection intensity favouring with major gene responsible while lower or moderate selection favouring polygenic response in insects. Adaptation in response to high selection intensity is more likely to depend on the genetic variation present in a population.

The $G \times E$ interaction to produce the phenotype is conceptualized as the norm of reaction of an individual (Shlichting and Pigliucci, 1995) which could be assessed if the individual genotype could be cloned and the ramets assessed in different environment or across an environment gradient. An important question then can be asked whether selection in two or more environments (either spatially or temporally) results in adaptive phenotypic plasticity which has been the basis behind shuttle breeding for higher adaptibility (See Roy, 2012). Two genetic mechanisms have been proposed for phenotypic plasticity. Allelic sensitivity occurs when the effect of an allele varies in different environment and gene regulation can cause suites or cascades of genes to be expressed in some environments but not in others. Thus difference in adaptibility may also be due to the environmental sensitive QTLs. These two mechanisms may not be mutually exclusive but they could integrate if variants at a regulatory locus cause different level of gene products to be expressed in different environments. The two types of phenotypic plasticity recognised are: graded response and discrete (or switched) response (Via et al., 1995). Several studies have shown the importance of environmental regulation. The genes responsible for QTL effects could be environmentally responsive, regulatory genes (Doebley, 1993). Single gene, the teosinte branched gene in maize is responsible for shrubby appearance when plants are grown in open but a single straight stem is produced when the plants are shaded or grown in crowded condition (Doebley et al., 1995). Differences in density or shading also affect the allomentry of the plant.

(f) Abiotic stress

Responses to unpredictable change associated with year to year differences in growing conditions (e.g., drought) could best be studied in mapping population established on contrasting sites (e.g., QTLs that explain growth at one site but not at others). Similarly, QTLs determining response to salinity, flooding condition, etc. can be identified and isolated.

21.9 Comparative Mapping and Orthologous Poly Genes

One of the uses of the molecular marker would be to attempt to discern evolutionary relationship between species, genera or larger taxonomic groups. For this one will have to study similarity and dissimilarity among the texa using genetic markers. Such study will also reveal consequences of evolutionary divergence on chromosome organization. By defining the sites of chromosomal rearrangement, one also defines intervening regions in which genes are arranged similarly in different organisms. Besides uses in systematics and evolutionary studies, comparative mapping can be used for prediction for gene location. Prediction for gene location mapped in one species can be made in the second. One species can be used for cloning a gene required in another species. Finally, chromosome regions that are sparsely mapped in one species, might be filled by markers from the molecular map of other species (Burr, 1994). Mapping of genes controlling important traits in the polyploids can be expedited by extrapolating the information obtained from diploid species. Thus we see that knowledge gained from comparative mapping will have impact on plant breeding strategies, germplasm characterization, conservation and use.

For comparative mapping, two classes of marker loci are required: (i) coding loci that are conserved across taxa being compared and (ii) loci that are highly polymorphic in the species. The first class is comprised of cDNA sequences that can be analysed via specific PCR, DNA sequencing, or Southern hybridization whereas second class comprised of minisatellites, microsatellites, single stranded conformational polymorphism, arbitrary primed PCR polymorphisms, etc. and can also be analysed using Southern hybridization or PCR. New genes created by assembly of pieces of other genes, do they arise from pseudogenes, or are they truly created *do novo* from relatively unconstrained sequences like introns or gene flanking sequences. We can have answers for these questions from the study of comparative mapping. The comparative mapping will not only help in map-based cloning but will also identify the significant steps in genome evolution. It will also identify, isolate gene that can be used in trans-species improvement. Information from genome collinearity will help in identification of novel genes-genes would be called novel by their unique map position relative to the position of similar genes in other species that condition a similar process. Identification of truly new genes that have arisen in the ancestor of a subset of grasses can be also made. Thus one can go for conducting comparative genetic mapping among sexually incompatible species which have so far been studied in isolation. By using a common set of markers comparative linkage maps can be constructed for different crops and finally compared for map positions of a loci or QTL for the same or similar characters. The characters can be qualitative and quantitative. Coincidence of map positions would support the hypothesis that loci underlying natural quantitative variation have been conserved during long periods of evolutionary divergence (i.e. they are orthologous lines). The best evidence for orthologous QTL determining seed weight comes from mungbean and cowpea (Fatokun et al., 1992). The single most significant QTL determining seed weight, maps to the same chromosome locus in both genomes. *B. campestris* and *B. oleracea* show a very high degree of conserva-tion of linkage arrangements. Extensive collinearity between maize and sorghum has been demonstrated (Hulbert et al., 1990; Whitkus et al., 1992).

Candidate loci have also been identified in the vicinity of QTL. Some of the same genetic factors influencing quantitative traits in two distantly related species have been found. For example, QTLs controlling variation in heading data in cereals have been located close to known loci controlling photoperiod and vernalization (Laurie et al., 1994; Bezant et al., 1995). QTLs have also been located in similar position in different population (Lin et al., 1995 a; Thomas et al., 1995) while QTLs in wheat have been found in similar position in homologous chromosome reinforcing the view that they are the same loci. Similarly QTL controlling flowering time in Brassicas, map to similar region in homologous chromosome both within and between species (Lager crantz et al., 1996; Oshorn et al., 1997). Similar syntenous relationships are found in maize and sorghum. They do not only share common linkage groups but also recombination rate is equivalent in these conserved groups (Whitkus et al., 1992). There is a large amount of conservation between these two genomes. One can make use of such study by identifying sorghum drought tolerant QTLs that map in the same place as maize drought tolerant QTL. Because sorghum has much superior tolerance to water and heat stress, sorghum QTL might provide better drought resistance than some maize QTL. Some of the duplicated Brassica regions show very close physical similarity to one end of chromosome 5 of *A. thaliana* which is known to contain several flowering related genes including constans (B-ohuon et al. 1998). Three linkage groups of *B. oleracea* (O_2, O_3 and O_9) and probably O_4 contain this same Arabidopsis region and all carry QTL for flowering time. Such syntenous regions suggest that the same few QTLs may be involved for at least some quantitative traits both within and between species. But then the large confidence interval on QTL positions and the fact that physical identity of chromosome tracts often involve very different map distances, cautions in nominating candidate loci.

21.10 Identification of Links between Genotype and Phenotype

It is important to know the links between the variation at the genetic level and the variation at the phenotypic level. A better understanding of these links will help in manipulating genes effectively in practical plant breeding.

There are four predominant interpretations to the relationship between genotype and phenotype (Wright 1980, 1982). The first relationship represents a one to one mapping of genotype to phenotype i.e. there is one-to-one correspondence between genes and traits. This relationship applies to Mendelian genes, the major genes of classical genetics (Haldane, 1932). Chapter 2 showed how a particular genotype reflected the phenotype. The other two relationships between genotype and phenotype were explained using quantitative genetic principles. One relationship represents polygenes with small additive effects on phenotypic variation, a view put forth by Fisher

(1930) and the other relationship represents genetic relationship with pleiotropic effect producing a variety of interactions influencing the phenotype (Wright 1931; 1932). The same genotype can result in different phenotypes and likewise different genotypes can give rise to the same phenotype. Here the traits are highly influenced by the environment. The relationship has also been explained on the population genetics principles of neutral theory of evolution (Kimura, 1983) where the variation at the genotype has no relationship to variation observed at the phenotypic level.

There are two approaches to the study of relationship between genotype and phenotype:

(i) Unmeasured genotype approach
(ii) Measured genotype approach

(i) Unmeasured genotype approach

In the unmeasured genotype approach, the loci affecting a given phenotype are unknown but the links between phenotype and genotype are studied through correlations between relatives, response to selection and hybridization and controlled crosses (Fisher, 1912; and Wright 1921; Falconer, 1989) and these are explained in Chapter 5. The use of quantitative genetic approach for studying the relationship between genotype and phenotype offers little insight into the genetic architecture of the trait which refers to the number of genes determining a quantitative trait, number of functional alleles at each gene and their relative frequencies, the arrangement of these alleles into genotype and the effect of alleles and genotypes on the trait of interest and other traits (Boerwinkle et al. 1986; Sing et al. 1988, Sing et al. 1992 a; Haviland et aI., 1995 a). These questions will be answered through the application of molecular marker technology by studying quantitative trait loci (QTL).

(ii) Measured genotype approach

There are two approaches to identify the links between genotype and phenotype:

(a) Marker locus approach
(b) Candidate gene approach

The candidate gene approach identifies gene region with presumed functional relationship to the phenotype of interest. In other words, a gene known to influence a quantitative trait by virtue of role in the biochemical or physiological pathway associated with the phenotype while the marker locus identifies candidate gene affecting a given phenotype. The marker locus approach identifies casual links between the QTL and phenotype trait of interest and is discussed in Section 21.1.2. The candidate gene approach can be used to determine the number of functional alleles at a gene and its effect on the trait of interest. It can provide insight into epistatic interactions and pleiotropic effects among loci, influence of environment which have so far not been considered by molecular marker system because of stage of development of these methods (Crandall, 1996). Three factors have contributed to the applicability of condidate gene approach. The first is the ability to survey specific regions for restriction site or nucleotide sequence variation. Secondly, molecular techniques offer high resolution information on variation underlying candidate gene. Thirdly, information on the biochemical and physiological pathways of phenotype of interest allows for identification of broad spectrum of candidate genes. Recently developed analytical tools have permitted identification and localization of genetic variation associated with phenotypic changes. For example, for studying genetic variability in serum cholesterol levels, more than 30 genes have been identified, as candidate genes worthy of genetic analysis (Templeton et al. 1987; 1988; 1992; Templeton and Sing, 1993). Once a candidate gene is identified, it is analysed for genetic variation using RFLP analysis or nucleotide sequencing. In other words, the genetic variation is being surveyed to identify functional alleles whose average effects can be calculated. Candidate gene approach provides a statistical framework for exploring the association of phenotypic variation and the underlying genetic variation in natural population. This variation is then used to establish evolutionary relationship among haplophytes. Templeton et al. (1992) developed a cladogram estimation procedure for estimating phylogenetic relationship which takes into account different intraspecific phenomena including recombination.

References

1. QTL analysis

Bezant, J., Laurie, D., Pratchett, N., Chojecki, J. and Kearsey. M.J. 1995. Marker regression mapping of QTL controlling flowering time and plant height in a spring barley (*Hordeum vulgure*) cross. Heredity, **77**: 64–73.

Beavis, W.D. 1994. The power and deceit of QTL experiments. In: Forty-Ninth Proc. of the Annual Corn and Sorghum Industry Research Conf. Willkinson, D. (ed.). American seed Trade Association, Washington, DC.

Beavis, W.D. and Keim, P. 1995. Identification of QTL that are affected by environment., In 'New Perspective on Genotype-by-Environment Interaction'. Kang M. (ed.) CRC Press, Boca Raton, FL.

Brzustowicz, L.M., Merette, C., XIE, X., Townsend, L., Gilliam, T.C. and Ott, J. 1993. Molecular and statistical approaches to the detection and correction of errors in genotype databases. Am. J. Human Genet., **53**: 1137–1145.

Carbonell, E.A., Asins, M.J., Baselga, M., Balansand, E. and Gerig, T.M. 1993. Power studies in the estimation of genetic parameters and the localization of quantitative trait loci for backcross and doubled haploid populations. Theor. ApI. Genet., **86**: 411–416.

Carbonell, E.A., Gerig, T.M., Balansand, E. and Asins, M.J. 1993. Interval mapping in the analysis of non-additive quantitative trait loci. Biometrics, **48**: 305–315.

Churchill, G.A. and Doerge, R. W. 1994. Empirical threshold values for quantitative trait mapping. Genetics, **138**: 963–971.

Darvasi, A., Weintreb, A., Minke, Y. Weller, J. and Soller, M. 1993. Detecting marker QTL linkage and estimating QTL gene effect and map location using a saturated genetic map. Genetics, **134**: 943–951.

Dempster, A.P., Laird, N.M. and Rubin, D.B. 1977. Maximum likelihood from incomplete data via the EM algorithm. J.R. Stat. Soc., **39**: 1–58.

Dasilva, J.A.G. and Sobral, W.S.B. 1996. Genetics of polyploids. In: The impact of Plant Molecular Genetics. Sobral, W.S.B. (ed.) Birkhatiser.

Echt, C., Knapp, S. and Liu, B.H. 1992. Genome mapping with non-inbred crosses using G. Mandel 2.0 Maize Genet. Coop. Nwsl. **66**: 27–29.

Ecke, W., Uzunova, M. and Weissleder, K. 1995. Mapping the genome of rapeseed (*Brassica napus* L.) II. Localization of genes controlling erucic acid synthesis and seed oil content. Theor. Appl. Genet. **91**: 972–977.

Edwards, M.D., Helentjaris, T., Wright S. and Stuber, C.W. (1992) Molecular marker facilitated investigations of quantitative trait loci in maize. 4. Analysis based on genome saturation with isozyme and restriction fragment length polymorphism markers. Theor. Appl. Genet., **83**: 765–774.

Edwards, M.D., Stuber, C.W. and Wendel, J.F. 1987. Molecular marker facilitated investigations of quantitative trait loci in maize. 1. Number, genomic distribution, and type of gene action. Genetics, **116**: 113–125.

Georges, M., Nielsen, D., Mackinnon, M., Mishra, A. Okimoto, R., Pasquino, A.T., Sargeant, I.S., Sorensen, A., Steele, M.R., Zao, X., Womack, I.F and Hoeschle. I. 1995. Mapping quantitative trait loci controlling milk production in dairy cattle by exploiting progeny testing. Genetics. **139**: 907–920.

Guo, S.W., and Thompson, E.A. 1992. A Monte Carlo method for combined segregation and linkage analysis. Am. 1. Hum. Genet., **51**: 1111–1126.

Mackett, C. 1997. Model diagnostics for fitting QTL models to traits and marker data by interval mapping. Heredity, **79**: 319–328.

Halden, J.B.S. 1919. The combination of linkage values and the calculation of distance between loci of linked factors. J. Genet., **8**: 299–309.

Haley, C.S. and Knott, S.A. 1992. A simple regression method for mapping quantitative trait loci in the crosses using flanking markers. Heredity, **69**: 315–324.

Haley, C.S., Knott, S.A. and Elsen, J.M. 1994. Mapping quantitative trait loci in crosses between outbred lines using least squares. Genetics, **136**: 1195–1207.

Hayes, P.M., Liu, B.H., Knapp, S.J., Chen, F, Jones, B., Blake, T., Franchowiak, 1., Rasmusson, D., Sorrells, M., Ullrich, S.E., Wesenberg, D. and Kleinhofs, A. 1993. Quantitative trait locus effect and environmental interactions in a sample of north American Barley germplasm. Theor. Appl. Genet., **87**: 392–40l.

Hyne, V. and Kerarsey, M.J. 1995. QTL analysis; Further uses of marker regression. Theor. Appl. Genet. **91**: 471–476.

Hyne, V., Kearsey, M.J., Pike, D.J. and Snape, J.W. 1995. QTL analysis: unreliability and bias in estimation procedures. Mol. Breed., **1**: 273–282.

Jansen, R. 1999. Estimations of recombination parameters between a quantitative trait loci (QTL) and two marker gene loci. Theor. Appl. Genet., **78**: 613–618.

Jansen, R. 1993 b. Maximum likelihood in a generalized linear finite mixture model by using the E.M. algorithm. Biometrics.

Jansen, R. 1992. A general mixture model for mapping quantitative trait loci by using molecular markers. Theor. Appl. Genet., **85**: 252–260.

Jansen, R.C. 1993. Interval mapping of multiple quantitative trait loci. Genetics, **135**: 205–211.

Jansen, R.C. 1994: Controlling type I and type II errors in mapping quantitative trait loci. Genetics, **138**: 871–881.

Jansen, R.C. and Stam, P. 1994. High resolution of quantitative trait into multiple loci via interval mapping. Genetics, **136**: 1447–1455.

Jiang C. and Zeng, Z.B. 1995. Multiple trait analysis of genetic mapping for quantitative trait loci. Genetics, **140**: 1111–1127.

Kearsey, M.J. and Farquhar, A.G.L. 1998. QTL analysis in Plants: where are we now? Heredity, **80**: 137–142.

Kearsey, M.J. and Hyne, V. 1994. QTL analysis: a simple marker regression approach. Theor. Appl. Genet. **89**: 698–702.

Kearsey, MJ. and Pooni, H.S. 1996. The Genetical Analysis of Quantitative Traits. Chapman and Hall, London.

Knapp, S.J. 1991. Using molecular markers to map multiple quantitative trait loci: Models for backcross, recombinant inbred and doubled haploid progeny. Theor. Appl. Genet., **81**: 333–338.

Knapp, S.J. and Bridges, W.C. 1990. Using molecular markers to estimate quantitative trait locus parameters: Powers and genetic variances for unreplicated and replicated progeny. Genetics, **126**: 769–777.

Knapp, S.J., Bridges, W.C. and Birken, D. 1990. Mapping quantitative trait loci using molecular marker linkage map. Theor. Appl. Genet. **79**: 583–592.

Knapp, S.J., Bridges, W.C. and Liu, B.H. 1992. Mapping quantitative trait loci using non-simultaneous estimates and hypothesis tests. In. J.S. Beckmann and T.S. Osborn (eds.) Plant Genomes: Methods for Genetic and Physical mapping, pp 209–237. Kluwer Academic Publishes Dordrecht, The Netherland.

Knott, S.A. and Haley, C.S. 1992b. Maximum likelihood mapping of quantitative trait loci using full-sib families. Genetics, **132**: 1211–1222.

Knott, S.A. and Haley, C.S. Aspects of maxim urn likelihood methods for the mapping of quantitative trait loci in line crosses. Genet. Res. **60**: 139–151.

Korol, A.B., Ronin, I.Y. and Kirzhner, V.M. 1995. Interval mapping of quantitative trait loci employing correlated trait complexes. Genetics. **140**: 1137–1147.

Lander, E.S. and Botstein, D. 1989. Mapping Mendelian factors underlying quantitative traits using RFLP linkage maps. Genetics, **121**: 185–199.

Lander, E.S., Green, P., Abrahamson, J., Barlow, M.J., Daly, S.E., Lincoln, S.E. and Newburg, L. 1987. Mapmaker: An interactive computer package for constructing primary genetic linkage maps of experimental and natural populations. Genomics, 1: 174–181.

Laurie, D.A., Pratchett, N., Bezant, J.H. and Snape, J.W. 1994. Genetic analysis of a photoperiod response gene on the short arm of chromosome 2 (2H) of *Hordeum vulgare* (barley). Heredity, **72**: 619–627.

Law, C.N., Snape, J.W. and Worland, A.J. 1983. Aneuploidy in wheat and its use in genetical analysis. In: Luption, F.G.H. (ed.) Wheat Breeding and its Scientific Basis. Chapman and Hall, London.

Lebreton, C.M. and Haley, C.S. 1998. A non parametric boot-strap method for testing close linkage vs. pleiotropy of coincident QTLs Genetics, in press.

Lebreton, C.M. and Visscher, P.M. 1998. Empirical non- parametric bootstrap strategies in QTL mapping conditioning on the genetic model. Genetics, in press.

Mangin, B., Goffinet, B. and Rebai, A. 1994. Constructing confidence intervals for QTL location. Genetics **138**: 1301–1308.

Manly, K.F. and Elliot, R.W. 1991. Map Manager, a micro computer programme for analysis of data from recombinant inbred strains. Mammalian Genome, **1**: 123–126.

Martinez, O. and Curnow, R.N. 1992. Estimating the locations and the sizes of the effects of quantitative trait loci using flanking markers. Theor Appl. Genet., **85**: 480–488.

Nilsson, N.Q., Sail, J. and Bengtsson, B.O. 1993. Chiasma data and recombination in plants-Are they compatible? Trends Genet., **9**: 344–348.

Paterson, A.H., Damon, S, Hewitt, J.D.; Zamir, D., Rabinowitch, H.D. Linclon, S.E., Lander, E.S. and Tanksely, S.D. 1991. Mendelian factors underlying quantitative traits in tomato: Comparison, across species, generations and environments. Genetics, **127**: 181–197.

Paterson, A.H., Lander, E.S. Hewitt, J.D., Peterson, S., Lincoln, S.E. and Tanksley, S.D. 1988. Resolution of quantitative traits into Mendelian factors by using a complete linkage map of restriction fragment length polymorphism. Nature, 335: 721–726.

Paterson, A.H., Tanksley, S.D. and Sorrells, M.E. 1991. DNA markers in plant improvement. Adv. Agron., **46**: 439–890.

Ramsay, L.D., Jennings, D.E., Bohuon, A.E. Lydiate, D.J. Kearsey, M.J. and Marshall, D. F. 1996. The construction of a substitution library of recombinant backcross lines in *Brassica oleracea* for the precision mapping of quantitative trait loci (QTL). Genome, **39**: 558–567.

Rebai, A., Goffinet, B. and Mangin, B. 1994. Approximate thresholds of interval mapping tests for QTL detection. Genetics, **138**: 235–240.

Rodolphe, F. and Lefort, M. 1993. A multi-marker model for detecting chromosomal segments displaying QTL activity. Genetics, **134**: 1277–1276.

Ronin, V.L., Kirzhner, Y.M. and Korol, A.B. 1995. Linkage between loci for quantitative traits and marker loci: multi-trait analysis with single marker. Theor. Appl. Genet., **90**: 776–786.

Routman, E. and Cheverud, J.M. 1994. Individual genes underlying quantitative traits: Molecular and analytical methods. In: Molecular Ecology and Evolution; Approaches and Applications. Schierwater B, Strait B, Wanger, G.P. and Desalla, R. (eds). Basil, Switzerland, Birkhauser Verlag.

Stam, P. 1993. Construction of integrated genetic linkage maps. by means of a computer package; Join map. The Plant. J. **5**: 739–744.

Tanhuanpaa, P.K., Vilkki, J.P. and Vilkki, H.J. 1995. Association of a RAPD marker with linolenic acid concentration in the seed oil of rapeseed (*Brassica napus L.*). Genome, **38**: 414–416.

Tanksley, S.D. 1993. Mapping polygenes. Ann. Rev. Genet., **27**: 205–233.

Thomas, W.T.B., Powell, W., Waugh, R., Chalmers, K.J., Barua, U.M., Jack, P., Lea, V., Forster, B.P., Swanston, J.S., Ellis, R.P., Hanson, R.P. and Lance, R.C.M. 1995. Detection of quantitative trait loci for agronomic yield, grain and disease characters in spring barley (*Hordeum vulgare L.*) Theor. Appl. Genet., **91**: 1037–1047.

Tuckey, J.W. 1958. Bias and confidence in not quite large samples (Abstract). Annals of Math. Stat., **29**: 614.

Van Ooijien, J.W. 1992. accuracy of mapping quantitatie trait loci in autogamous species. Theor. appl. genet. **84**: 803–811.

Van der Schaar, W., Alonso, Balanco, C,; Leoni, Kloosterzieil, K.M. Jansen, R.C., Van Qoijen, J.W. and Koomeef., M. 1997. QTL analysis of seed dormancy in *Anabildopsis* using recombinant inbred lines and MQM mapping Heredity, **79**: 190–200.

Visscher, P.M., Thompson, R. and Haley, C.S. 1996. Confidence Intervals in QTL mapping by bootstrapping. Genetics. **143**, 1013–1020.

Weller, J.1. 1986. Maximum likelihood techniques for the mapping and analysis of quantitative trait loci with the aid of genetic markers. Biometrics, **42**: 627–640.

White-head Institute 1993. Mapping Genes Controlling Quantitative traits Using Map marker QTL. Version 1.1: A tutorial and reference Mannual.

Wu, K.K., Burnquish, W., Sorrells, M.E., Twe, T.L., Moore, P.H. and Tanksley, S.D. 1992. The detection and estimation of linkage in polyploids using single dose restriction fragment. Theor. Appl. Genet. **83**: 294–300.

Zeng, Z.B. 1993. Theoretical basis of precision mapping of quantitative trait loci. Proc. Natl. Acad. Sci. **90**: 10972–1 0976.

Zeng, Z.B. 1994. Precision mapping of quantitative trait loci. Genetics, **136**: 1457–1468.

2. Comparative genome mapping

Ahn. S. and Tanksley, S.D. 1993: Comparative linkage maps of the rice and maize genomes. Proc. Natl. Acad. Sci. USA., **90**: 7980–7984.

Ahn, S.; Anderson, J.A., Sorrells, M.E. and Tanksley, S.D. 1993. Homoeologous relationships of rice, wheat and maize chromosomes. Mol. Gen. Genet., **241**: 483–490.

Bennetzen, J.L. 1996. The use of comparative genome mapping in the identification, cloning and manipulation of important plant genes. In: The Impact of Plant Molecular Genetics. Sobral, B.W.S. (ed.), pp. 71–85, Bir-Khauser, Boston.

Bennetzen, J.L. and Freeling, M. 1993. Grasses as a single genetic system: genome composition, collinearity and compatibility. Trends Genet., **9**: 259–261.

Binelli, G., Gianfranceschi, L., Pe, M.E., Taramino, G., Busso, C., Stenhouse, J. and Ottaviano, E. 1992. Similarity of maize and sorghum genomes as revealed by maize RFLP probes. Theor. Appl. Genet., **84**: 10–16.

Bonierbale, M.W., Plaisted, R.L. and Tanksley, S.D. 1988.: RFLP maps based on a common set of clones reveals modes of chromosomal evolution in potato and tomato. Genetics, **120**: 1095–1103.

Devos, K.M., Millan, T. and Gale, M.D. 1993. Comparative RFLP maps of homologous group 2 chromosomes of wheat, rye and barley. Theor. Appl. Genet., **85**: 784–792.

Gebhardt, G., Ritter, E., Barone, A., Debener, T., Walkemeier, B., Schachtschabel, U., Kaufmann, H., Thompson, R.D., Bonierbale, M.W., Ganal, M.W., Tanksley, S.D. and Salamini, F. 1991. RFLP maps of potato and their alignment with the homologous tomato genome. Theor. Appl. Genet., **83**: 49–57.

Lagercrantz, U., Putterill, J., Coupand, G. and Lydiate, D. 1996. Comparative mapping in Arabidopsis and Brassica, Fine scale collinearility and congruence of genes controlling flowering time. The Plant J., **9**: 13–20.

Lin, Y.R., Schertz, K.F. and Paterson, D.H. 1995. Comparative analysis of QTLs affecting plant height and maturity across the Poaceae, in reference to an interspecific sorghum population. Genetics, **41**: 391–411.

Melake-Berhan, A., Hulbert, H.S., Butler, L.G. and Bennetzen, J.L. 1993. Structure and evolution of the genomes of *Sorghum bicolor* and *Zea mays*. Theor. Appl. Genet., **86**: 598–604.

Pereira, M.G. and Lee, M. 1995. Identification of genomic regions affecting plant height in sorghum and maize. Theor. Appl. Genet., **90**: 380–388.

Pereira, M.G., Lee, M., Bramel-Cox, P., Woodman, W., Doebley, J. and Whitkus, R. 1994. Construction of an RFLP map in sorghum and comparative mapping in maize. Genome, **37**: 236–243.

Teutonico, R.A. and Osborn, T.C., 1994. Mapping of RFLP and qualitative trait loci in *Brassica rapa* and comparison to the linkage maps of *B. napus*, *B. oleracea*, and *Arabidopsis thaliana*. Theor. Appl. Genet., **89**: 885–894.

Whitkus, R., Doebley, J. and Lee, M. 1992. Comparative genome mapping of sorghum and maize. Genetics, **132**: 1119–1130.

Zamir D. and Tanksley, S.D. 1988. Tomato genome is comprised largely of fast evolving, low copy-number sequences, Mol. Gen. Genet. 213: 254–26l.

3. *Identification of links between genotype and phenotype*

Boerwinkle E., Chakraborty, R. and Sing, C.F. 1986. The use of measured genotype information in the analysis of quantitative phenotypes in man. Ann. Hum. Genet., **50**: 181–194.

Bohuon, E.J.R., Ramsay, L.D., Craft, J.A., Arthur, A.E., Lydiate, D.J., Kearsey, M.J. and Marshall, D.F. 1998. The association of flowering time QTL with duplicated regions and candidate loci in *Brassica oleracea*. Genetic.

Hallman, D.M., Visvikis, S., Steinmetz, J. and Boerwinkle, E. 1994. The effect of variation in the apolipoprotein B gene on plasma lipid and apolipoprotein B Levels. I. A Likelihood-based approach to cladistic analysis. Ann. Hum. Genet., **58**: 3564.

Haviland, M.B., Ferrell, R.E. and Sing, C.F. 1995a. A cladistic analysis of the relationship between variation in the low density lipoprotein receptor gene region and interindividual variation in plasma lipid, lipoprotein and apolipoprotein levels. Am. J. Hum. Genet., in Press.

Sing, C.F., Boerwinkle, E., Moll, P.P. and Templeton, A.R. 1988. Characterization of gene affecting quantitative traits in humans. In: Proceedings of the Second International Conference on Quantative Genetics, Weir B.S., Eisen, E.J., Goodman, M.M. & Namkoong, G., eds. Sanderland, MA: Sinauer Associates.

Sing, C.F., Haviland M.B., Templeton, A.R., Zerba, K.E. and Reilly, S.L. 1992a. Biological complexity and strategies for finding DNA variations responsible for inter-individual variation in risk of a common chronic disease, coronary artery disease. Ann. Med., **24**: 539–547.

Sing, C.F., Haviland, M.B., Zerba, K.E. and Templeton, A.R. 1992b. Application of cladistics to the analysis of genotype-phenotype relationships. Eur. J. Epidemiol., **8**: 3–9.

Templeton, A.R. 1995. A cladistic analysis of phenotypic associations with haplotypes inferred from restriction endonuclease mapping or DNA sequencing. V. Analysis of case/control sampling designs; Alzheimer's disease and the apoprotein *E. locus*. Genetics, **140**: 403–409.

Templeton, A.R. and Sing, C.F. 1987. A cladistic analysis of phenotypic associations with haplotypes inferred from restriction endonuclease mapping IV. Nested analyses with cladogram uncertainty and recombination. Genetics, **134**: 659–669.

Templeton, A.R., Boerwinkle, E. and Sing C.F. 1987. A cladistic analysis of phenotypic associations with haplotypes inferred from restriction endonuclease mapping 1. Basic theory and an analysis of alcohol dehydrogenase activity in Drosophila, Genetics, **117**: 343–351.

Templeton, A.R., Crandall, K.A. and Sing, C.F. 1992. A cladistic analysis of phenotypic associations with a haplotypes inferred from restriction endonuclease mapping and DNA sequence data. III cladogram estimation. Genetics, **132**: 619–633.

Templeton, A.R., Sing., C.F., Kessling, A. and Humphries, S. 1988. A cladistic analysis of phenotypic associations with haplotypes inferred from restriction endonuclease maping. II. The analysis of natural populations. Genetics, **120**: 1145–1154.

Wright, S. 1980. Genic and organismic selection. Evolution, **34**: 825–843.

Wright, S. 1982. Character change, speciation, and the higher taxa. Evolution, **36**: 427–443.

4. *Systematics and evolution*

Aldrich, P.R., Doebley, J., Schertz, K.F. and Stec, A. 1992. Patterns of allozyme variation in cultivated and wild Sorghum bicolor. Theor. Appl. Genet., **85**: 451–460.

Anderson, E. 1945. What is Zea mays? A report of progress. Chron. Bot., **9**: 88–92.

Arnold, M.L., Bennett, B.D. and Zimmer, E.A. 1990a. Natural hybridization between *Iris fulva* and *I. hexagona*; Pattern of ribosomal DNA variation. Evolution, **44**: 1512–1521.

Bonierbale, M.W., Plaisted, R.L. and Tanksley, S.D. 1988. RFLP maps based on a common set of clones reveal modes of chromosomal evaluation in potato and tomato. Genetics, **120**: 1095–1103.

Coyne, J.A., Rux, J. and David, J.R. 1991. Genetics of morphological differences and hybrid stertility between *Drosophila sechellia* and its relatives. Genet. Res., Camb. **57**: 113–122.

Crawford, D.J. 1983. Phylogenetic and systematic inferences from electrophoretic studies. In: S.D. Tanksley and T.J. Orton (eds.), Isozymes in Plant Genetics and Breeding. Part A, pp. 257–287. Elsevier, Amsterdam.

Doebley, J. and Stec, A. 1991. Genetic analysis of the morphological differences between maize and teosinte. Genetics, **129**: 285–295.

Doebley, J., Stec, A. Wendel, J. and Edwards, M. (1990) Genetic and morphological analysis of a maize-teosinte F_2 population: implications for the origin of maize. Proc. Natl. Acad. Sci. USA. **87**: 9888–9892.

Doyle, J.J. 1992. Gene trees and species trees: molecular systematics as one character taxonomy. Syst. Bot., **17**: 144–163.

Fatokun, C.A., Menancio-Hautea, D.I., Danesh, D. and Young, N.D. 1992. Evidence for orthologous seed weight genes in cowpea and mung bean based on RFLP mapping. Genetics, **132**: 841–846.

Figdore, S.S., Kennard, W.D., Song, K.M., Slocum, M.K. and Osborn, T.C. 1988. Assessment of the degree of restriction fragment length polymorphism in Brassica. Theor. Appl. Genet., **75**: 833–840.

Gottlieb, L.D. 1984. Genetics and morphological evaluation in plants. Am. Nat., **123**: 681–709.

Guo, M., Lightfoot, D.A., Mok, M.C. and Mok, D.W.S. 1991. Analyses of *Phaseolus vulgaris* L. and *P. coccineus Lam.* Hybrids by RFIP; preferential transmission of P. vulgaris alleles. Theor. Appl. Genet., **81**: 703–709.

Hillis, D.M. and Moritz, C. (eds.). 1990. Molecular Systematic. Sinauer Associates, Inc., Sunderland, MA.

Keim, P., Paige, K.N., Whitham, T.G. and Lark, K.G. 1989. Genetic analysis of an interspecific hybrid swarm of Populus: occurrence of unidirectional introgression. Genetics, **123**: 557–565.

McGrath, J.M. and Quiros, C.F. 1991. Inheritance of isozyme and RFLP markers in *Brassica campestris* and comparison with *B. Oleracea*, Theor. Appl. Genet., **82**: 668–673.

Miller, J.C. and Tanksley, S.D. 1990b. RFLP analysis of phylogenetic relationship and genetic variation in the genus Lycopersicon. Theor. Appl. Genet., **80**: 437–448.

Neuhausen, S.L. 1992. Evaluation of restriction fragment length polymorphism in *Cucumis melo*. Theor. Appl. Genet., **83**: 379–384.

Paterson, A.H., Damon, S., Hewitt, J.D., Zamir, D., Rabinowitch, H.D. Lincoln, S.E., Lander, E.S. and Tanksley, S.D. 1991. Mendelian factors underlying quantitative traits in tomato: comparison across species, generations and environments. Genetics, **127**: 181–197.

Rhoades, M.M. 1951. Duplicated genes in maize. Am. Nat., **85**: 105–110.

Rieseberg, L.H. and Wendel, J.F. 1993. Introgression and its consequences. In: R. Harrison (ed.), Hybrid Zones and the Evolutionary Process, pp. 70–109. Oxford University Press.

Soltis, D.E., Soltis, P.S. and Milligan, B.G. 1992a. Intraspecific chloroplast DNA variation: systematic and phylogenetic implications, In: P.S. Soltis, D.E. Soltis and J.J. Doyle (eds.). Molecular Systematics of Plants, pp. 117–150. Chapman and Hall, New York.

Song, K.M., Osborn, T'C. and Williams, P.H. 1990. Brassica taxonomy based on nuclear restriction fragment length polymorphisms (RFLPs). 3. Genome Relationship in Brassica and related genera and the origin of *B. rape* (syn. campestris). Theor. Appl. Genet. **79**: 497–509.

Song, K.M, Suzuki, J.Y., Slocum, M.K., Williams, P.H. and Osborn, T.C. 1991. A linkage map of *Brassica rapa* (syn. campestris) based on restriction fragment length polymorphism loci, Theor. Appl. Genet., **82**: 296–304.

Tanksley, S.D., Bernatzky, R., Lapitan, N.L. and Prince, J.P. 1988. Conservation of gene repertoire but not gene order in pepper and tomato. Proc, Natl. Acad. Sci. U.S.A. **85**: 8419–6423.

Van de Ven, M., Powell, W., Ramsay, G. and Waugh, R. 1990. Restriction fragment length polymorphisms as genetic markers in Vicia. Heredity **65**: 329–342.

Whitkus, R., Doebley, J. and Lee, M. 1992. Comparative genome mapping of sorghum and maize. Genetics, **132**: 1119–1130.

Whitkus, R., Doebley, J. and Wendel, J.F. 1994. Nuclear DNA markers in systematic and evolution: In: DNA based markers in Plants. Philips, R.L. and Vasil, I (eds.), Kluwer Academic Publishers.

13. Genetic resources management

Avise, J.C. 1994. Molecular markers, Natural History and Evolution. New York. Chapman and Hall.

Crothers, B.L. 1992. Genetic characters, species concept and conservation biology. Cons. BioI., **6**: 314.

Erwin, T. 1991. An evolutionary basis for conservation strategies. Science, **253**: 750–752.

Hamilton, M.B. 1994. *Ex situ* conservation of wild plant species: time to reassess the genetic assumptions and implication of seed bank. Cons. BioI., **8**(1): 39–49.

Mortiz, C. 1994. Application of mitochondrial DNA analysis in conservation: a critical review. Mol. Ecol., **3**: 401–411.

O'Brien, S.J. 1994. A note for molecular genetics in biological conservations. Proc. Natl. Acad. Sci., **91**: 5748–5755.

Olmstead, R.G. and Palmer, J.D. 1994. Chloroplast DNA Systematic: A review of method and data analysis. Amer. J. Bot., 81(9): 1205–1224.

Westman, A.L. and Kresovich, S. 1997. Use of molecular marker techniques for description of plant genetic variation: In: Callow, lA., Ford-Llyod, J.A. and New bury, H.J. (eds.), Biotechnology and Plant genetic resources: Conservation and use, pp. 9-48. CAB International, Oxford, U.K.

22

Matrix

22.1 Definition and Types

A matrix is a rectangular array of numbers (also called scalars). The individual numbers in the matrix are known as the elements of the matrix. The elements of a matrix A are denoted by a_{ij} or A_{ij} where i refers to the row and j to the column. A matrix with r rows and c columns is called a $r \times c$ matrix and this is called the dimension of the matrix A. A 2×2 matrix i.e. matrix with 2 rows and 2 columns and a 3×3 matrix i.e. matrix with 3 rows and 3 columns will look like

$$\begin{bmatrix} a_{11} & a_{12} \\ a_{21} & a_{22} \end{bmatrix} \qquad \begin{bmatrix} a_{11} & a_{12} & a_{13} \\ a_{21} & a_{22} & a_{23} \\ a_{31} & a_{32} & a_{33} \end{bmatrix}$$

2×2 matrix $\qquad\qquad$ 3×3 matrix

A $r \times c$ matrix will take form as follows.

$$\begin{bmatrix} a_{11} & a_{12} & a_{1j}\cdots & a_{1c} \\ a_{21} & a_{22} & a_{2j}\cdots & a_{2c} \\ a_{i1} & a_{i2} & a_{ij}\cdots & a_{ic} \\ a_{r1} & a_{r2} & a_{rj}\cdots & a_{rc} \end{bmatrix}$$

The individual element in the matrix say for example a_{11} is 1×1 matrix and is called unidimensional scalar. The linear arrangements of the numbers are known as vectors. Thus a vector is a matrix with a single row or column. In the above general case if $c = 1$, the matrix is called a $r \times 1$ column vector. Likewise if $r = 1$, then the matrix will be called a $1 \times c$ row vector. In other words, a matrix with one column is called a column vector whereas a matrix with a single row is called a row vector. When $r = c$ i.e. number of rows = number of columns, the matrix A is a square matrix. The diagonal elements of a square matrix are the elements a_{ii} or A_{ii} where $i = 1, 2, 3, \ldots, n$. If $a_{ij} = a_{ji}$, then the matrix is a symmetrical matrix. A square matrix whose off diagonal elements are zeros is called a diagonal matrix and when the elements a_{ii} on the leading diagonal are 1, the matrix is called a unit matrix or identity matrix, I, as follows:

$$\begin{pmatrix} 1 & 0 & 0 \\ 0 & 1 & 0 \\ 0 & 0 & 1 \end{pmatrix}$$

A matrix whose all elements are zero, is called a null matrix.

22.2 Matrix Operations

The different matrix operations are as follows.

1. Addition

Two or more matrices can be added only if they have the same dimension. Given two matrices A and B,

$$(A + B)_{ij} = C_{ij} = (A_{ij} + B_{ij})$$

For example,

$$\begin{pmatrix} a_{11} & a_{12} \\ a_{21} & a_{22} \end{pmatrix} + \begin{pmatrix} b_{11} & b_{12} \\ b_{21} & b_{22} \end{pmatrix}$$

$$= \begin{pmatrix} a_{11} + b_{11} & a_{12} + b_{12} \\ a_{21} + b_{21} & a_{22} + b_{22} \end{pmatrix}$$

2. Subtraction

Subtraction is possible only when the two matrices have the same dimension. Thus given two matrices A and B,

$$(A - B)_{ij} = C_{ij} = (A_{ij} - B_{ij})$$

For example,

$$\begin{pmatrix} a_{11} & a_{12} \\ a_{21} & a_{22} \end{pmatrix} - \begin{pmatrix} b_{11} & b_{12} \\ b_{21} & b_{22} \end{pmatrix}$$

$$= \begin{pmatrix} a_{11} - b_{11} & a_{12} - b_{12} \\ a_{21} - b_{21} & a_{22} - b_{22} \end{pmatrix}$$

3. Multiplication

I. Multiplication by a scalar K

It is same as adding K identical matrices. Thus,

$$K \times A = KA$$

$$K \times \begin{pmatrix} a_{11} & a_{12} \\ a_{21} & a_{22} \end{pmatrix} = \begin{pmatrix} Ka_{11} & Ka_{12} \\ Ka_{21} & Ka_{22} \end{pmatrix}$$

i.e. each element of matrix A is multiplied by K.

II. Multiplication of the two matrices

Two matrices, say A and B can be multiplied only if the number of columns in the matrix A is equal to the number of rows in matrix B. Thus,

$$A(= a_{ij}) \times B(= b_{ij}) = C_{ij} = (AB)_{ij} = \sum_k A_{ik} B_{kj}$$

For example,

$$\begin{pmatrix} a_{11} & a_{12} \\ a_{21} & a_{22} \end{pmatrix} \times \begin{pmatrix} b_{11} & b_{12} \\ b_{21} & b_{22} \end{pmatrix}$$

$$= \begin{pmatrix} a_{11}b_{11} + a_{12}b_{21} = C_{11} & a_{11}b_{12} + a_{12}b_{22} = C_{12} \\ a_{21}b_{11} + a_{22}b_{21} = C_{21} & a_{21}b_{12} + a_{22}b_{22} = C_{22} \end{pmatrix}$$

Here the elements of the first row of A are multiplied by the corresponding elements of first column of B and are added to give C_{11} and elements of the first row of A are multiplied by the corresponding elements of the second column of B and are added to give C_{12} of the product matrix AB. In other words, the jth element of the product matrix $(AB)_{ij}$ is the sum of the products of the corresponding elements of ith row of A and the jth column of B matrix. If the dimension of A is $m \times n$ and that of B is $r \times s$, then the dimension of the product matrix $(AB)_{ij}$ will be $m \times s$.

Multiplication is not cumulative i.e.

$$A \times B \neq B \times A$$

However all diagonal matrices of the same dimension are cumulative. If matrix r is the product of matrices A and X i.e. $r = AX$ and s is the product of matrices B and X i.e. $s = BX$ then,

$$r + s = (A + B) X$$

Also, if $r = AX$ and $X = CY$ then,

$$r = ACY$$

Multiplication is associative i.e.

$$A \times (B \times C) = (A \times B) \times C$$

Also, multiplication is distributive i.e.

$$A(B + C) = AB + AC$$

Further if the product matrix of matrices A and B is a null matrix i.e. $AB = 0$, it does not necessarily mean that either $A = 0$ or $B = 0$. If a matrix A is premultiplied or post multiplied by an $r \times r$ identity matrix I, the matrix A remains unchanged i.e.

$$AI = IA = A$$

4. Division

We shall see this operation while calculating the inverse of the matrix.

5. Transpose

The transpose of a matrix A is denoted by A'. It is obtained by interchanging rows and columns i.e. the first row in the transpose of matrix A' is the first column of the matrix A and the second row is the second column and so on. Thus the elements in the transpose of matrix can be denoted as a_{ji}.

$$\begin{pmatrix} a_{11} & a_{12} \\ a_{21} & a_{22} \end{pmatrix} \begin{pmatrix} a_{11} & a_{21} \\ a_{12} & a_{22} \end{pmatrix}$$

$$\qquad A \qquad\qquad\qquad A^1$$

It shows that for a symmetrical matrix $A = A'$. And if A is an $r \times c$ matrix then A' is $c \times r$ which can be seen as follows:

$$x = \begin{pmatrix} x_1 \\ x_2 \\ \vdots \\ x_n \end{pmatrix}$$

$$x' = (x_1 x_2 \ldots x_n)$$

We can further see that

$$\sum_i x_i^2 = xx'$$

but when x is $n \times n$ matrix then

$$xx' \neq x'x$$

Given two matrices A and B it can be shown that

$$(AB)' = B'A'$$

Similarly, with three matrices A, B and C

$$(ABC)' = C'B'A'$$

6. Determinant

The determinant of a matrix A is denoted by $|A|$ or Δ and is a number or scalar. In case of 2×2 matrix

$$\begin{pmatrix} a_{11} & a_{12} \\ a_{21} & a_{22} \end{pmatrix}$$

the determinant $|A|$ is calculated as

$$a_{11} \cdot a_{22} - a_{21} \cdot a_{12}$$

which is usually written as

$$\begin{vmatrix} a_{11} & a_{12} \\ a_{21} & a_{22} \end{vmatrix}$$

In case of 3×3 matrix,

$$\begin{pmatrix} a_{11} & a_{12} & a_{13} \\ a_{21} & a_{22} & a_{23} \\ a_{31} & a_{32} & a_{33} \end{pmatrix}$$

the determinant is denoted by

$$\begin{vmatrix} a_{11} & a_{12} & a_{13} \\ a_{21} & a_{22} & a_{23} \\ a_{31} & a_{32} & a_{33} \end{vmatrix}$$

and equals to

$$a_{11} \begin{vmatrix} a_{22} & a_{23} \\ a_{32} & a_{33} \end{vmatrix} \times (-1)^{1+1} + a_{12} \begin{vmatrix} a_{21} & a_{23} \\ a_{31} & a_{33} \end{vmatrix} \times (-1)^{1+2}$$
$$+ a_{13} \begin{vmatrix} a_{21} & a_{22} \\ a_{31} & a_{32} \end{vmatrix} \times (-1)^{1+3}$$

or,

$$a_{21} \begin{vmatrix} a_{12} & a_{13} \\ a_{32} & a_{33} \end{vmatrix} \times (-1)^{2+1} + a_{22} \begin{vmatrix} a_{11} & a_{13} \\ a_{31} & a_{33} \end{vmatrix} \times (-1)^{2+2}$$
$$+ a_{23} \begin{vmatrix} a_{11} & a_{12} \\ a_{31} & a_{32} \end{vmatrix} \times (-1)^{2+3}$$

or,

$$a_{31} \begin{vmatrix} a_{12} & a_{13} \\ a_{22} & a_{23} \end{vmatrix} \times (-1)^{3+1} + a_{32} \begin{vmatrix} a_{11} & a_{13} \\ a_{21} & a_{23} \end{vmatrix} \times (-1)^{3+3}$$
$$+ a_{33} \begin{vmatrix} a_{11} & a_{12} \\ a_{21} & a_{22} \end{vmatrix} \times (-1)^{3+3}$$

where

$$\begin{vmatrix} a_{22} & a_{23} \\ a_{32} & a_{33} \end{vmatrix}, \begin{vmatrix} a_{21} & a_{23} \\ a_{31} & a_{33} \end{vmatrix}, \ldots \begin{vmatrix} a_{11} & a_{12} \\ a_{21} & a_{22} \end{vmatrix}$$

are called minor of a_{ij} i.e. they are minor of a_{11}, a_{12}, and a_{33}, respectively. The minor of a_{ij} is obtained by striking out the ith row and jth column of the matrix A which results in $(r-1)(r-1)$ matrix if the original matrix A has the dimension $r \times r$ and then its determinant is taken.

The cofactor of a_{ij} = minor of a_{ij} X $(-1)^{i+j}$

$= C_{ij}$, the matrix of cofactor

where $(-1)^{i+j}$ refers to the coefficient of minor where i refers to row number and j refers to the column number.

Thus we have a matrix of cofactor,

$$\begin{pmatrix} C_{11} & C_{12} & C_{13} \\ C_{21} & C_{22} & C_{23} \\ C_{31} & C_{32} & C_{33} \end{pmatrix}$$

The determinant, Δ then is worked out as

$$a_{11}C_{11} + a_{12}C_{12} + a_{13}C_{13}$$

or

$$a_{21}C_{21} + a_{22}C_{22} + a_{23}C_{23}$$

or

$$a_{31}C_{31} + a_{32}C_{32} + a_{33}C_{33}$$

7. Inverse

The inverse of a matrix A is denoted by A^{-1}. Where inverse of a matrix exists,

$$A^{-1}A = I = AA^{-1}$$

Thus inverse is unequal. Here the multiplication is cumulative. The steps involved in the calculation of inverse of a matrix are as follows:

(i) Obtain matrix of cofactor

In case of the above considered 3×3 matrix, the matrix of cofactor was

$$\begin{pmatrix} C_{11} & C_{12} & C_{13} \\ C_{21} & C_{22} & C_{23} \\ C_{31} & C_{32} & C_{33} \end{pmatrix}$$

(ii) Calculate the determinant
(iii) Transpose the matrix of cofactor which in the above case will be as

$$\begin{pmatrix} C_{11} & C_{21} & C_{31} \\ C_{12} & C_{22} & C_{32} \\ C_{13} & C_{23} & C_{33} \end{pmatrix}$$

(iv) Divide each element in the transpose by the determinant and the resultant matrix will be inverse of the original matrix

$$\begin{pmatrix} \dfrac{C_{11}}{\Delta} & \dfrac{C_{21}}{\Delta} & \dfrac{C_{31}}{\Delta} \\ \dfrac{C_{12}}{\Delta} & \dfrac{C_{22}}{\Delta} & \dfrac{C_{32}}{\Delta} \\ \dfrac{C_{13}}{\Delta} & \dfrac{C_{23}}{\Delta} & \dfrac{C_{33}}{\Delta} \end{pmatrix}$$

A matrix must be squared to have an inverse but not all square matrices have inverses. The inverse of a matrix exists only when the matrix is of full rank. The rank of a matrix is denoted by $r(A)$. When matrix is of full rank, r equals c, such a matrix is non-singular i.e. its determinant is not zero and there exists an inverse.

If X and Y are vectors and if A is a non-singular matrix and if the equation $Y = AX$ holds then

$$X = A^{-1}Y$$

In the following equations,

$$y_1 = a_{11}x_1 + a_{12}x_2 + a_{13}x_3$$
$$y_2 = a_{21}x_1 + a_{22}x_2 + a_{23}x_3$$
$$y_3 = a_{31}x_1 + a_{32}x_2 + a_{33}x_3$$

$$\begin{pmatrix} y_1 \\ y_2 \\ y_3 \end{pmatrix} = \begin{pmatrix} a_{11} & a_{12} & a_{13} \\ a_{21} & a_{22} & a_{23} \\ a_{31} & a_{32} & a_{33} \end{pmatrix} = \begin{pmatrix} x_1 \\ x_2 \\ x_3 \end{pmatrix}$$

$$Y \qquad\qquad A \qquad\qquad X$$

Y and X are column vectors. If the inverse of the matrix A, A^{-1} exists i.e. if A is a non-singular matrix then

$$A^{-1}AX = A^{-1}y$$
$$IX = A^{-1}Y$$
$$X = A^{-1}Y$$

8. Rank of a matrix

Considering the following equations,

$$x + 2y + 3z = 2$$
$$2x + 3y + z = 4$$
$$3x + 5y + 4z = 6$$

the simultaneous equations take the following form:

$$\begin{pmatrix} 1 & 2 & 3 \\ 2 & 3 & 1 \\ 3 & 5 & 1 \end{pmatrix} \begin{pmatrix} x \\ y \\ z \end{pmatrix} = \begin{pmatrix} 2 \\ 4 \\ 6 \end{pmatrix}$$

$$A \qquad\quad X \quad\; Y$$

But here the determinant $\Delta = 0$ and so we cannot get unique solution for x, y and z. The reason is that all the three equations are not independent as the third equation is just the sum of the first two equations and thus it is not independent. There are only two linearly independent equations and thus the ranks of the matrix which is defined as the number of independent rows and column, in the present case is 2. Further it can be shown that if A is $n \times n$ matrix and if $|A| = 0$, then the rank of the matrix is less than n. Thus the rows and columns of A are not independent. And if the rank of the matrix is $m<n$ then the number of linearly independent row and column is m.

22.3 Dispersion Matrix

It is a matrix composed of the variances and covariances of p variables.

$$\begin{pmatrix} S_{11} & SP_{12} & SP_{1p} \\ SP_{21} & S_{22} & SP_{2p} \\ SP_{p1} & S_{p2} & S_{pp} \end{pmatrix}$$

where $S_{11}, S_{22}, \ldots S_{pp}$ are the variances of variables 1, 2, ... and p, respectively and the SP_s are the covariances.

In case of simple regression $y = a + bx$, the parameters to be estimated are a and b, along with their variances. In the case of multiple regression $y = a + b_1x_1 + b_2x_2$, the normal equation takes the following form:

$$b_1 SSx_1 + b_2 SPx_1x_2 = SP(yx_1)$$
$$b_1 SPx_1x_2 + b_2 SSx_2 = SP(y_1x_2)$$

which in matrix notation takes the form as

$$S\hat{b} = R$$

where $S = \begin{pmatrix} SSx_1 & SPx_1x_2 \\ SPx_1x_2 & SSx_2 \end{pmatrix}$

$$b = \begin{pmatrix} \hat{b}_1 \\ \hat{b}_2 \end{pmatrix}$$

and $R = \begin{pmatrix} SP(yx_1) \\ SP(yx_2) \end{pmatrix}$

Thus the normal equation for any least squares regression procedure is $S\hat{b} = R$ from which $\hat{b}$ is obtained as

$$\hat{b} = S^{-1} R$$

The variance-covariance matrix is obtained as $\sigma^2 S^{-1}$ where σ^2 is the remainder mean squares in the ANOVA of regression and S which is called an information matrix as

$$\begin{bmatrix} \dfrac{n}{\sigma^2} & 0 \\ 0 & \dfrac{SSx}{\sigma^2} \end{bmatrix}$$

The inverse of this information matrix is the variance-covariance matrix or dispersion matrix.

$$\begin{pmatrix} \text{Var}(\hat{a}) & \text{Cov}(\hat{a}\hat{b}) \\ \text{Cov}(\hat{a}\hat{b}) & \text{Var}(\hat{b}) \end{pmatrix} = \begin{pmatrix} \dfrac{\sigma^2}{n} & 0 \\ 0 & \dfrac{\sigma^2}{SSx} \end{pmatrix}$$

Here a and b are independent.

In case of the multiple linear regression, the information matrix takes the form as

$$\dfrac{1}{\sigma^2}\begin{pmatrix} N & 0 & 0 \\ 0 & SSx_1 & SPx_1x_2 \\ 0 & SPx_1x_2 & SSx_2 \end{pmatrix}$$

and the inverse takes the form

$$\dfrac{1}{\sigma^2}\begin{pmatrix} \dfrac{1}{n} & 0 & 0 \\ 0 & \dfrac{SSx_1}{\Delta^1} & \dfrac{SPx_1x_2}{\Delta^1} \\ 0 & \dfrac{SPx_1x_2}{\Delta^1} & \dfrac{SSx_2}{\Delta^1} \end{pmatrix}$$

where $\Delta^1 = SSx_1 SSx_2 - [SP(x_1x_2)]^2$.

The variance-coveriance matrix is

$$\begin{pmatrix} \text{Var}(\hat{a}) & 0 & 0 \\ 0 & \text{Var}(\hat{b}_1) & \text{Cov}(\hat{b}_1\hat{b}_2) \\ 0 & \text{Cov}(\hat{b}_1\hat{b}_2) & \text{Var}(\hat{b}_2) \end{pmatrix}$$

As in the above multiple linear regression equation, b_1 and b_2 are independent of a, so all the covariances involving a are zero.

22.4 Orthogonal Matrix

A matrix A is said to be orthogonal if it produces identity matrix when multipled by its transpose i.e.

$$A \cdot A' = I$$

and if A is a $r \times r$ matrix, then

$$A' = A^{-\prime}$$

22.5 Diagonalization of Matrix

If A is a matrix then

$$|A| > 0$$

i.e. positive which is also symmetric then there exists orthogonal C matrix such that

$$CAC^1 = \Lambda = \begin{pmatrix} \lambda_1 & 0 \\ 0 & \lambda p \end{pmatrix} - a$$

diagonal matrix so that $CA = \Lambda_C$. This is an eigen vector equation of the matrix A. The $\lambda_1, \lambda_2, \ldots \lambda p$ are given by the solution

$$|A - \lambda_I| = 0$$

and are called the eigen values (also called latent roots or characteristic roots) of A.

References

Crow, J.F. and Kimura, M. 1970. An Introduction to Population Genetic Theory. Harper and Row, New York.

Kempthorne, O. 1957. An Introduction to Genetic Statistics. Wiley, New York.

Searle, S.R. 1966. Matrix Algebra for the Biological Sciences. Wiley.

Steel, R.G.D. and Torrie, J.H. 1980. Principles and Procedures of Statistics. McGraw Hill Kogakusha Ltd.

Problems in Biometrical Genetics

1. From the means and variances of two pure breeding lines(P1 and P2) and their F1, F2 and first backcross (BC1 and BC2) families, the following perfect fit estimates of the components of the family means were obtained. All means and variances are based on families containing more than 50 individuals.

$$m = 30.0 \pm 1.5$$
$$[d] = 2.6 \pm 1.2$$
$$[h] = + 10.7 \pm 2.5$$
$$[i] = 3.2 \pm 1.5$$
$$[j] = -1.7 \pm 0.9$$
$$[l] = -15.2 \pm 4.3$$

What can you say about the genetic control?
When the same model was fitted to these data with the addition of F3 and second backcross (BC11, BC12, BC21 and BC22) families the χ^2 was 20. Explain the result and suggest alternative model.

2. From an experiment involving F1, F2 and backcross generations derived from two inbred lines of tobacco the following estimates were obtained:

$$m = 86.5 \pm 2.3$$
$$[d] = 12.1 \pm 0.8$$
$$[h] = 0.7 \pm 0.9$$
$$\chi^2 = 3.7$$
$$D = 24.1$$

   a. Estimate the number of loci segregating in this cross listing any assumptions you are making in this process.

   b. The F2 generation individuals were randomly mated among themselves and one progeny from each cross was raised the next year along with a repeat sample of the F2. The variance of the F2 was found significantly greater than that of their randomly mated progeny.

What can you say about your estimate of the number of loci?

3. The two haploid strains of *Aspergillus* (P1 and P2) were crossed to give F1 haploid progeny. Duplicate cultures obtained by clonal propagation of P1, P2 and of a random sample of 100 individuals from the F1 progeny were grown in a single block with single cultures as the unit of randomization and their rates of growth determined.

The mean rates of growth of P1, P2 and F1 and SEs of these means are as follows

Generation	Mean	SE(mean)
P1	95.00	±2.20
P2	53.00	±2.20
F1	74.25	±0.22

Analysis of variance partitioning the variation among duplicate cultures of the 100 F1 progeny is as follows.

Source of variation	d.f.	Mean square
Between individuals	99	85.60
Between duplicates		
Within individuals	100	5.7

23.2 Biometrical Genetics–Analysis of Quantitative Variation

(I) Is there any evidence of epistasis?
(II) Estimate the additive component of the means and variances
(III) Estimate the number of effective factors
(IV) The extreme individuals in the F1 had rates of growth of 125.00 and 30.00. What can you deduce from this?

4. 200 bi-parental families(B.I.Ps) were raised from crosses among parents sampled at random from an outbreeding population. An analysis of variance of the parental and offspring scores produced the following statistics:

$$V_P = 125$$
$$V_O = 175$$
$$W_O/p = 100$$
$$P = 100$$

(i) What is the expected mean of the top 5% of the base population?
(ii) If these top 5% had been randomly mated among themselves, what would be the expected mean of their offspring?

5. (A) What are the expectations, in terms of D_R, H_R, E_1 and E_2 of the obtained from the analysis of variance of the following sets of twins?
   (i) Monozygotic raised together(MZ_T)
   (ii) Monozygotic raised apart(MZ_A)
   (iii) Dizygotic raised apart(DZ_A)
   (iv) Dizygotic raised together(DZ_T)

(B) In an analysis of MZ_T and DZ_T for psychoticism, a D_R, E_1 model was fitted to the following data.

Types of twins	Nos. of pairs	Source of variation	d.f.	MS	Model
MZ_T	300	Between families	298	0.82	$D_R + E_1$
		Within families	300	0.27	E_1
DZ_T	180	Between families	178	0.59	$3/4 D_R + E_1$
		Within families	180	0.44	$1/4 D_R + E_1$

The following estimates were obtained from these data by weighted least squares:

$$D_R = 0.509$$
$$E_1 = 0.274$$

Test the adequacy of the D_R, E_1 model, using the observed varainaces for computing the weights.

The theoretical error appropriate to the above analyses of variance proved to be 0.25. What further conclusions may now be drawn?

6. 20 highly inbred lines were obtained by selfing from a wild population of an annual plant and were used as parents in a 10 × 10 North Carolina II design. Two progeny were raised from every family in a single randomised block.

An analysis of variance of their scores produced the following:

Source of variation	d.f.	MS
Between male groups(M)	9	580.0
Between female groups(F)	9	560.0
MXF	81	75.0
Within families	100	25.0

(i) Write down the expected mean squares for this analysis and test their significance.
(ii) Derive algebraically the genetical and environmental expectations for these, recalling that the parents were inbred.
(iii) Estimate genetical and environmental components and interprete their significance

7. In a triple test cross experiment involving *Nicotiana rustica*, n randomly selected F2 plants were used as male parents, each being crossed to the two parental lines and their reciprocal F1's. From these 4n families, r sibs were raised in a single randomized block.

(i) Give the skeleton analysis of the resulting data, together with the expectations of mean squares and a brief interpretation of the items in the analysis.
(ii) Assuming that n male parents used were homozygous lines obtained from many generations of selfing among the original F2 plants, derive expressions(involving additive, dominance and environmental components of variation) for the 's in the previous analysis.
(iii) If, in situation described in (ii) above, the n inbred lines had also been raised in the same experiment, what further useful analyses would then be possible.

8. The array variances (V_r) and array covariance (W_r) of a 9 × 9 diallel set of crosses for the character flowering time are listed in the order of the flowering time of the common parents of the arrays.

Array	Vr	Wr
1	34.10	44.60
2	75.23	84.53
3	36.42	46.02
4	14.64	25.14
5	35.15	44.10
6	18.08	27.58
7	6.80	17.30
8	6.73	16.25
9	7.77	16.70

Plot W_r against V_r and draw conclusions from the inspection of the graph.

9. An experiment designed to investigate the effect of the environment on the yields of two inbred lines of brussels sprouts, P1 and P2 and their F1 hybrid cross, was replicated over a number of farms which were known to differ considerable in respect of their suitability for this crop. An analysis of the data of this experiment produces the following estimates, all of which are significant.

$$m = 100.0$$
$$[d] = 2.5$$
$$[h] = 5.0$$
$$b_d = 0.25$$
$$b_h = 0.20$$

(i) What are the expected mean yields of the inbred lines and those of their F1 and F2 descendants when these families are raised in the jth environment?

(ii) Is the yield of P1 always better than that of P2? If not, in which environment is their performance similar; what is their yield in this environment; and in what environments is P1 worse than P2.

(iii) Is the yield of the F1 hybrid always better than that of its parents? If not, in what environment does the F1 cease to display heterosis and what is its yield in this environment?

(iv) What advice would you give to the farmer on the basis of these data?

10. 20 true breeding lines were grown in each of nine environments, seven individuals of each line being grown in each environment in a completely randomized design. A joint regression analysis of variance for one of the characters scored is shown below. Lines and environments are fixed effects(model 1).

Source of variation	d.f.	M.S.	χ^2
Lines	19	428.2	3113.9
Environments	8	896.6	2745.0
Lines x Environments	152		
Heterogeneity between Regression	19	15.6	113.7
Remainder	133	13.7	698.0
Error	1080	2.6	

Explain what each of the items is testing and interprete the analysis. Comment on the peculiarities of this analysis relative to a conventional Joint regression.

11. Two inbred lines of barley and their F1 regularly produced the following average yields(in arbitrary units):

$$P1 = 39$$
$$P2 = 31$$
$$F1 = 51$$

In an experiment including F2 and backcross progeny, an m, [d], [h] model was fitted to the data. All parameters were highly significant and a test of goodness of fit yielded a χ^2 of 4.25 for 3 d.f. Furthermore, D, H, E were estimated as 40, 25.6 and 17.6, respectively.

(i) What can you conclude about the genetic control of yield in this cross?

(ii) What is the expected mean of the top 5% of inbred line that could be produced from the F2?

(iii) Find the expected score of the best inbred line that might be produced, indicating the consequences of failure of any assumption in your approach.

(iv) Estimate the number of effective factors and the degree of dispersion, if any, in the parents.

23.4 Biometrical Genetics–Analysis of Quantitative Variation

12. Seventy five plants from an F2 population were crossed to the original parents (P1, P2) and their F1 in a Triple Test Cross design. Ten plants were raised from each of the 225 families in a completely randomised experiment and their flowering times scored. Analysis of variance of these data yielded the following mean squares:

Source of variation	Mean squares
Additive	86.5
Dominance	82.5
Epistasis	7.3

   (i) Derive the appropriate errors and test the significance of these main effects.
   (ii) Estimate the genetical and environmental components of variation
   (iii) What do you conclude about the genetic control of flowering time in this cross?

13. The following data are the flowering times of the families produced by an 8×8 diallel cross on inbred lines of poppies, each datum being the mean of 10 individuals. Also presented are the uncorrected W_r and V_r values for 7 of the arrays.

    The error M.S. appropriate to the analysis on means was 5.09

   (i) Calculate the W_r and V_r values for the missing array
   (ii) What conclusions can you draw from the W_r, V_r values in this experiment?

14. The following are the estimates of the genetical components of mean performance and environmental sensitivity for final height for two pure breeding lines of N. rustica (P1 and P2) and their F1 cross.

$$m = 119.50$$
$$[d] = 3.0$$
$$[h] = 5.50$$
$$b_d = 0.35$$
$$b_h = 0.24$$

   (i) Derive equations giving the expected means of P1, P2 and F1 in the jth environment
   (ii) Predict the expected means of the F2, F_α, BC1 and BC2 generations of this cross in the jth environment.
   (iii) What are the values of the environments in which:
      (i) P1 and P2 have the same final height and
      (ii) the F1 shows no heterosis?

15. A randomly-mating population was investigated by carrying out a breeding programme based on a North Carolina Design I. Twenty males were chosen at random and each was crossed to a different set of three females. Five progeny were raised from every family in a completely randomised design and scored for character X.

    Analysis of variance of the scores yielded the following sum of squares

Source of variation	Sum of squares
Between male groups	2014.0
Between females groups within males	1840.0
Within families	6240.0

	1	2	3	4	5	6	7	8	W_r	V_r
1	12.00	13.12	18.21	15.90	18.09	14.45	21.14	14.94	-	-
2	13.50	7.32	13.20	18.98	12.94	12.38	18.57	8.09	9.91	15.60
3	17.51	14.01	11.29	16.10	16.54	15.90	16.59	14.34	1.27	4.25
4	15.05	17.26	16.48	13.90	18.78	18.18	17.21	16.96	-3.18	2.67
5	18.02	13.82	16.34	18.12	10.23	14.63	18.26	12.83	5.47	8.22
6	14.33	11.52	16.89	18.06	14.79	7.56	15.69	9.66	8.57	12.16
7	19.87	16.90	15.68	19.23	15.95	16.80	10.90	14.31	2.49	8.16
8	14.56	7.57	15.42	17.79	12.93	11.54	14.34	4.95	11.84	17.07

(i) Write down the degrees of freedom and expected mean squares for this analysis.

(ii) What are the genetical and environmental expectations of the σ^2's?

(iii) Estimate the narrow and broad heritability and the total phenotypic variance of this population

(iv) If the mean of the population for character X is 150, what is the expected mean of the top 5% of this population?

(vi) If the top 5% were selected and randomly mated, what is the expected mean of their progeny?

16. Derive the expectations for the contribution of linkage disequilibrium, in respect of a pair of loci A, a; B, b to the additive genetic component of variation in a population produced by random mating, taking as the initial gamete frequencies:

Gametes	AB	aB	Ab	ab
Frequencies	C_1	C_2	C_3	C_4

Where $C_1 + C_2 + C_3 + C_4 = 1$

The observed total variances (V) in the F2 generation of a cross between a pair of pure-breeding lines and in the generations produced by crossing the F2 to its parent(L3) and by randomly mating the F2(F2BIP) were:

$$V_{F2} = 150.95$$
$$V_{L3} = 101.36$$
$$V_{F2BIP} = 62.47$$

What can you deduce from their relative magnitudes?

17. The means and variances of the means of two true-breeding lines(P1 and P2) of tobacco and their F1, F2 and first backcross(BC1 and BC2) families, for the character final height, are as follows:

Generation	mean	Variance of mean
P1	91	1.0
P2	55	0.3
F1	55	0.7
F2	66	1.4
B1	75	1.0
B2	56	0.4

Each mean and variance of mean is based on measurements of ten individuals.

What can you deduce about the genetical control of family means?

18. The following parameters have been estimated from an F2 population.

P1 = 46.8   D = 16.604
P2 = 40.8   H = 2.657
F1 = 39.6   E 4.483

(i) If we were to inbreed the F2 population, how many inbred lines would flower earlier than
   (a) P2
   (b) F1

(ii) What would be the flowering time at the point of truncation corresponding to the earliest flowering 0.5%?

(iii) (a) What is the dominance ratio?
   (b) What is the potence ratio?
   (c) Why do (a) and (b) differ?
   (d) What is the mean of the earliest flowering inbred line?

(iv) If you select parents with a mean flowering time of 38, what is the expected flowering time of their progeny?

19. Forty plants raised in the experimental field of poppies were scored and 20 were chosen at random as seed parents, the remaining 20 being pollen parents. Seeds from all 20 crosses were germinated the next year and five plants from every family were raised in the field in a completely randomised design.

(i) Carry out an analysis of variance to see if there are significant differences between families and interpret the result.

(ii) Plot out the relationship between o and p.

(iii) Estimate the best fitting regression line of the offspring mean(o) onto mid-parent(p), and test the significance of this regression by an analysis of variance

23.6 Biometrical Genetics–Analysis of Quantitative Variation

Cross			Flowering time of		
i	♀(P1)	♂(P2)	Mid-parent(P_i) (P1+P2)/2	Mean of 5 offspring(O_i) of 5 ofspring	S.S.(4 d.f.)
1	41	46		41.5	73.2
2	47	48		44.2	11.1
3	47	42		42.0	10.0
4	44	39		44.6	69.0
5	40	39		42.4	24.9
6	43	40		42.8	54.8
7	42	40		39.2	74.8
8	49	32		39.4	89.2
9	54	38		41.6	79.2
10	42	53		44.0	10.2
11	50	47		45.6	35.2
12	40	39		42.2	54.8
13	35	40		40.0	14.8
14	40	29		38.4	14.7
15	42	43		43.4	11.2
16	50	40		43.0	11.0
17	36	40		39.0	30.0
18	32	33		36.9	12.8
19	43	32		39.0	17.6
20	47	47		44.6	45.2

(iv) Can you combine this regression analysis with the earlier analysis of variance in Q. 1.

(v) Similar regression of offspring mean onto mother and father proved to be 0.24 and 0.30, respectively. Why might these be expected to be less than the regression onto the mid-parent? What further can you conclude from these values?

20. From a biometrical genetical analysis of final height in a cross between two pure breeding lines of *N. rustica*, the following estimates of the components of means and variances were obtained.

Components of means	Components of variances
m = 74.0	D = 35.0
[d] = 2.0	H = 10.0
[h] = 5.0	E = 20.0

(i) What are the expected mean and variance of the F2 generation of this cross?

(ii) What is the expected heterosis in this cross?

(iii) What can you deduce about the cause of this heterosis from the potence ration and the dominance ratio?

(iv) What are the expected mean and variance of the pure-breeding lines that can be extracted from this cross?

(v) Are any of these pure-breeding lines expected to have a greater final height than the F1 of this cross?

21. The following MN blood group frequencies were found in two populations of American Indians.

Genotype	MM	MN	NN
Navaho	305	52	4
Pueblo	83	46	11

(i) Suppose (a) the migration rate is low (b) mating is random (c) the mutation rate is low what other factors might lead to significant departures from H-W equilibrium?

(ii) Are the populations in H-W equilibrium?

(iii) What conclusions can you draw?

(iv) Are the population consistent in their genotype frequencies?
(v) How can you explain the result?
(vi) What problems are there in the standard procedure for testing agreement with the H-W law?

22. In man the frequency of albinism is about 1 in 20, 000. Assuming that (a) all cases of albinism are genetically aa i.e. all cases are homozygous recessive at the same locus (b) that the H-W law holds approximately at this locus, find the frequency of individuals who are Aa i.e. normal, but carriers of the harmful gene.

(I) Calculate the proportion of all abnormals coming from carrier x carrier marriages.
    (a) Complete the table given below:

Marriage	Frequency	Expected proportion of children who are abnormal
Abnormal × Abnormal	v^4	1
Carrier × Abnormal		
Carrier × Carrier		
Others		

(c) Now find the proportion of all abnormals coming from carrier x carrier marriages
(d) What is this proportion for
    Albinism (frequency, 1 in 20, 000)
    Cystic fibrosis (frequency, 1 in 3000)

(II) Assuming (a) No albinos reproduce (b) Population is now in H-W equilibrium (c) The generation time in man is 28 years

Find out how long it would take to reduce the frequency of albinism to half to its present value by the following methods:

(A) From the frequency of albinos in the population now calculate the frequency (q_0) of the allele 'a'.
(B) From the frequency of albinos in the final, derived population, calculate the frequency(q_n) of the 'a' allele.
(C) Calculate the number of generations to go from q_0 to q_n as follows:
$$N = 1/q_n - 1/q_0$$
(D) Calculate the number of years this would take.
(E) Compare the frequency of carriers now with that after N generations.

The above set of questions is taken from the final examination of M.Sc. course in Applied Genetics of University of Birmingham, England.

Index

A

Ac-Ds system 1.4
Adequacy of the model 5.6
Allele 1.1
Allopolyploid 12.1
Alpha (α) inheritance 9.9
AMMI model 8.20
Amounts of information 5.4
Amphidiploids 12.1
Analysis of co-variance 2.17
Aneuploid 12.1
Aneuploidy 1.6
Arithmetic mean 3.5
Assortative mating 19.4
Augmented biparental mating design 14.4
Average effect of gene 5.7

B

Basic generations 6.3
Bayesian strategy 6.19
B chromosome 1.5
Beta (β) inheritance 9.9
Beta distribution 3.10
Binomial distribution 2.7
Binomial probability 2.2
Biparental mating design 6.4
Biplot display 8.20
Biplot method 8.18
Bootstrapping 21.9
Bottleneck hypothesis 5.8
Breeding value 8.5
Bulked segregant analysis 21.2

C

Candidate gene 21.34
Canonical analysis 4.18
Carrying capacity 4.21
Cell organelleles 4.3
Chi-square test 1.3
Chloroplasts 1.3
Chromatin 16.2
Co-dominance 11.16
Codon 4.13
Coefficient of coancestry 5.6
Coefficient of determination 7.10
Coefficient of dispersion 20.10
Combined selection 13.1
Combining ability analysis 13.7
Comparative mapping 21.32
Competition 5.6
Competitive ability 19.10
Competitive selection 4.12
Complementary interaction 4.9
Complete genetic map 21.28
Constitutive heterochromatin 20.3
Correlated response 10.2
Correlation coefficient 6.9
Coupling linkage 6.9
Covariance full-sib 6.9
Covariance half-sib 8.12
Covariance offspring 9.2
Cross validation strategies 21.15

D

D^2-statistics 4.19

Deoxyribonucleic acid 1.2
Dependent variable 20.5
Determinant 22.3
Developmental canalization 2.21
Developmental stability 8.5
Diagonalization of matrix 22.6
Diagonal matrix 22.1
Diallel 6.4
Dihaploid lines 21.19
Dihaploids 12.9
Direct effect 20.6
Directional selection 20.15
Discriminate analysis 4.19
Disomic inheritance 12.1
Dispersion matrix 22.5
Disruptive selection 20.15
Dominance 2.1
Dominance ratio 6.3
Double reduction 21.24
Drift 19.12
D scaling test 5.2
Duplicate interaction 5.6

E
E_1 14.1
E_2 14.1
Eb 14.1
Eberhert and russell's model 8.8
Ecovalence 8.10
Effective factors 16.1
Effective population size 19.14
Eigen vector 4.17
Emerson's method 2.14
Environmental correlation 6.21
Environmental covariances 6.21, 14.5
Environmental effects 5.1
Environmental heterosis 11.6
Environmental index 8.8
Environmental sensitivity 8.6
Environmental variance 6.1
Estimator 4.1
Euchromatin 1.5

Euclidean distance 4.20
Euheterosis 11.13
Euploid 12.1
Experimental error 14.4
Exponential growth curve 4.20
Expressivity 2.21

F
F_2 diallel 7.8
Factor analysis 4.18
Factorial experiment 4.10
Facultative heterochromatin 1.5
Family selection 20.8
Finlay-wilkinson's regression method 8.19
Fitness 19.8
Fixed model 4.6
Flanking markers 21.4
Flexible genotype 8.5
Founder principle 19.16
F-test 7.1

G
Gametic selection 19.16
Gamma distribution 3.10
Gene 1.1
Gene conversion 2.17
Gene distance 21.25
Gene frequency 19.1
Gene pool 19.1
General combining ability (GCA) 7.10
General competitive ability 13.3
Generation matrices 11.19
Genetical correlation 6.22
Genetic distance 11.10
Genetic effects 5.1
Genetic load 19.17
Genetic mapping 21.25
Genome 12.1
Genotype 1.6
Genotype assay 16.5
Genotype × environment interaction 8.1
Genotypic correlation 6.21

Genotypic covariance 6.21
Genotypic stability 8.10
GGE bilot 8.20

H
Half diallel 7.1
Half diallel excluding self 7.1
Haploids 12.8
Hardy-weinberg equilibrium 19.1
Heritability in the broad-sense 15.1
Heritability in the narrow-sense 15.1
Heterochromatin 1.4
Heterogametic sex 11.1
Heterosis 11.9
Heterosis in population cross 2.16
Homeostasis 3.11
Homoalleles 7.2
Homogametic sex 12.1
Hyman's analysis 12.1

I
Identity matrix 22.1
Immunoglobin 2.3
Inbreeding coefficient 11.16
Inbreeding depression 11.13
Inbreeding in polyploids 11.18
Incomplete dominance 2.21
Independent culling level 20.10
Independent variables 20.5
Indirect effect 20.6
Individual plant randomization 14.2
Individual selection 20.8
Intergenotypic competition 13.1
Intraclass correlation 6.23
Intragenotypic competition 13.1
Inverse 22.4

J
Joint regression analysis 4.14
Joint scaling test 5.3

K
Kosambi function 21.26
Kurtosis 3.6

L
Lag phase 4.22
Least square methods 4.2
Linkage 10.1
Linkage disequilibrium 10.1
Linked epistasis 10.13
local control 14.3
LOD threshold 21.5
Logarithmic transformation 18.2
Logistic growth curve 4.21
Logistic growth curve 4.20
Log phase 4.22

M
Macro-environment 8.4
Malthusian parameter 4.21
Mapping populations 21.19
Maternal effects 9.1
Maternal effects in heterosis 11.7
Maternal influence 2.16
Matrix 22.1
Matrix of cofactor 22.3
Matromorphy 12.10
Maximum likelihood method 4.2
Mega-environments 8.21
Meiotic drive 19.17
Mendelian genes 2.1
Meristic variability 3.9
Mermal 1.1
Mesokurtic 3.7
Messenger RNA 1.6
Microenvironment 8.4
Migration 19.11
Mini max strategy 6.19
Minimum variance estimator 4.1
Mitochondria 2.17
Mixed model 4.7
Mixture diallels 13.3
Mode is 3.6
Model fitting 5.6
Multimodal distributions 3.11
Multinomial probability 2.7
Multiple alleles 1.1

Multiple correlation 4.15
Multiple diallels 7.17
Multiple factor hypothesis 1.1
Multiple regression 4.15
Multiple regression approach 21.13
Multitrait selection 20.10
Multivariate analysis 4.17
Mutation 19.5
Mutational load 19.19
Muton 1.4

N
Negative assortative mating 19.5
NIL 21.2
Non-allelic interaction 5.4
Non-central chi-square 4.5
Non-central F 4.5
Non-linear genotype × environment interaction 8.15
Non-parametric statistics 4.16
Normal distribution 3.4
North Carolina Mating designs 6.4
Nucleo protein 1.5
Nucleoside 1.3
Nucleotide 1.2

O
One gene-one enzyme hypothesis 1.3
One gene-one polypeptide chain hypothesis 1.4
Operator 1.4
Operons 1.4
Optimal genetic map 21.28
Orthogonal matrix 22.6
Orthogonal polynomials 13.8
Orthologous QTL 21.33
Outcrossing rate 2.2
Overdominance 2.2

P
Parameters 3.7
Paramutation 2.17
Parthenogenetic haploids 12.8
Partial correlation coefficient 4.15

Partial diallel 6.4
Path analysis 20.5
PCR 21.32
Penetrance 2.21
Perkins and Jinks Model 8.5
Phenotype 1.6
Phenotypic correlation 6.21
Phenotypic covariance 6.21
Phenotypic stability 8.24
Phenotypic value 5.1
Phenotypic variance 6.2
Plasmagenes 2.16
Plasmon 2.15
Platykurtic 3.7
Pleiotropy 10.7
Poisson distribution 3.9
Poly embryony 12.8
Polygenic concept 3.3
Polyploids 12.1
Population 19.1
Posterior or empiric probability 2.6
Principal component analysis 4.17
Principal co-ordinate analysis 4.19
Priori probability 2.14
Product ratio method 3.3
Progeny test 3.3
Pseudogenes 21.32

Q
Qualitative trait 2.1
Quantitative trait 3.1
Quantitative trait loci 21.1

R
Randomization 14.3
Randomized complete block design 4.7
Random model 4.6
Range 3.6
Rank of a matrix 22.4
Realized heritability 20.2
Reciprocal difference 9.1
Reciprocal effects model 9.6

Recombination frequency 2.13
Recon 1.4
Recurrence relation 11.15
Regression coefficient 4.12
Reliable genotypes 8.5
Repeatability 6.23
Replication 14.3
Response to selection 20.1
Restricted selection index 20.12
Ribonucleic acid 1.2
RIL 21.19

S
Satellite DNA 1.5
Scalar 22.1
Scale effect 18.3
Scaling tests 5.2
Segregational load 19.17
Segregation distorter 19.17
Selection criterion 20.10
Selection differential 20.2
Selection index 20.10
Selection intensity 20.2
Selection limit 20.13
Selfish DNA 1.5
Senescence phase 4.22
Sex-limitation 9.13
Sex linkage 9.1
Sib mating 19.4
Single l-df interaction contrast 8.3
Skewness 3.6
Somatic mutation 2.17
Southern hybridization 21.32
Specific combining ability (SCA) 7.10
Specific competitive ability 13.3
Square matrix 22.1
Stability variance 8.10
Stabilizing selection 20.15
Statistical probability 3.7
Statistics 4.1
Symmetrical matrix 22.1

T
Tassel seed 2.19
Tertiary trisomic 12.1
Test of goodness of fit 2.8
Tests of non-normality 18.1
Theories of heterosis 11.1
Theory of evolution 3.3
The t-test 4.4
Threshold variability 3.8
Transpose 22.2
Triallel analysis 7.20
Triple test cross 6.4
Triple test cross (selfed families) 6.13
Triploids 12.6
TRNA 1.7
TTC with population of inbred lines 6.13
Type I error 8.25
Type II stability 8.14

U
Unbiased estimator 4.1
Unimodal distributions 3.11

V
Variance 3.6
Variants of TTC 6.13
Varietal diallel 7.19
Varimax criterion 4.18
Vectors 22.1

W
Watson and crick 1.3
Weighted least square 5.3
Wilkinson's statistics 8.8
Within family selection 20.8

Y
Yates' correction 2.10

Z
Z-distribution 4.5

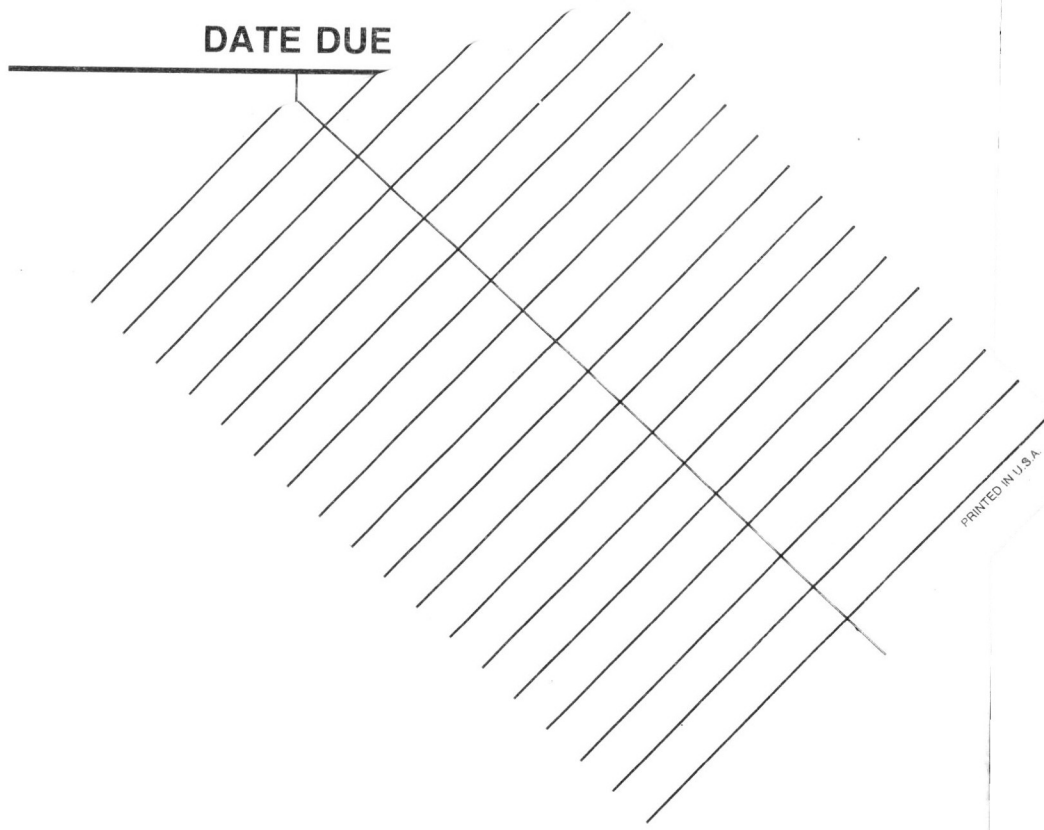